The Baculoviruses

THE VIRUSES

Series Editors

HEINZ FRAENKEL-CONRAT, *University of California*
Berkeley, California

ROBERT R. WAGNER, *University of Virginia School of Medicine*
Charlottesville, Virginia

THE VIRUSES: Catalogue, Characterization, and Classification
Heinz Fraenkel-Conrat

Other volumes in the series:

THE BACULOVIRUSES
Edited by Lois K. Miller

THE BUNYAVIRIDAE
Edited by Richard M. Elliott

THE CORONAVIRIDAE
Edited by Stuart G. Siddell

THE INFLUENZA VIRUSES
Edited by Robert M. Krug

THE PAPOVAVIRIDAE
Volume 1 • Edited by Norman P. Salzman
Volume 2 • Edited by Norman P. Salzman and Peter M. Howley

THE PARAMYXOVIRUSES
Edited by David W. Kingsbury

THE PARVOVIRUSES
Edited by Kenneth I. Berns

THE PLANT VIRUSES
Volume 1 • Edited by R. I. B. Francki
Volume 2 • Edited by M. H. V. Van Regenmortel and Heinz Fraenkel-Conrat
Volume 3 • Edited by Renate Koenig
Volume 4 • Edited by R. G. Milne
Volume 5 • Edited by B. D. Harrison and A. F. Murant

THE REOVIRIDAE
Edited by Wolfgang K. Joklik

THE RETROVIRIDAE
Volumes 1–4 • Edited by Jay A. Levy

THE RHABDOVIRUSES
Edited by Robert R. Wagner

THE TOGAVIRIDAE AND FLAVIVIRIDAE
Edited by Sondra Schlesinger and Milton J. Schlesinger

THE VIROIDS
Edited by T. O. Diener

A complete listing of volumes in this series appears at the back of this volume.

The Baculoviruses

Edited by

LOIS K. MILLER

The University of Georgia
Athens, Georgia

PLENUM PRESS • NEW YORK AND LONDON

Library of Congress Cataloging-in-Publication Data

The baculoviruses / edited by Lois K. Miller.
p. cm. -- (The viruses)
Includes bibliographical references and index.
ISBN 0-306-45641-9
1. Baculoviruses. I. Miller, Lois. II. Series.
QR398.5.B34 1997
579.2'436--dc21

97-17668
CIP

ISBN 0-306-45641-9

A Division of Plenum Publishing Corporation
233 Spring Street, New York, N.Y. 10013

http://www.plenum.com

10 9 8 7 6 5 4 3 2 1

Printed in the United States of America

Contributors

Bruce C. Black, American Cyanamid Company, Agricultural Products Research Division, Insecticide Directed Basic Research, Princeton, New Jersey 08543

Sharon C. Braunagel, Department of Entomology, Texas Agricultural Experiment Station, Texas A & M University, College Station, Texas 77843-2475

Lynn A. Brennan, American Cyanamid Company, Agricultural Products Research Division, Insecticide Directed Basic Research, Princeton, New Jersey 08543

Rollie J. Clem, Department of Molecular Microbiology and Immunology, Johns Hopkins School of Hygiene and Public Health, Baltimore, Maryland 21205

Jenny S. Cory, Ecology and Biocontrol Group, NERC Institute of Virology and Environmental Microbiology, Oxford OX1 3SR, England

Peter M. Dierks, American Cyanamid Company, Agricultural Products Research Division, Insecticide Directed Basic Research, Princeton, New Jersey 08543

Peter Faulkner, Department of Microbiology and Immunology, Queen's University, Kingston, Ontario K7L 3N6, Canada

Brian A. Federici, Department of Entomology and Interdepartmental Graduate Program in Genetics, University of California at Riverside, Riverside, California 92521

Paul D. Friesen, Institute for Molecular Virology and Department of Biochemistry, Graduate School and College of Agriculture and Life Sciences, University of Wisconsin, Madison, Wisconsin 53706

C. Joel Funk, Department of Microbiology, Oregon State University, Corvallis, Oregon 97331-7301

Ivan E. Gard, American Cyanamid Company, Agricultural Products Research Division, Insecticide Directed Basic Research, Princeton, New Jersey 08543

Rosemary S. Hails, Ecology and Biocontrol Group, NERC Institute of Virology and Environmental Microbiology, Oxford OX1 3SR, England

Donald L. Jarvis, Department of Entomology and Center for Advanced Invertebrate Molecular Sciences, Texas A & M University, College Station, Texas 77843

Peter J. Krell, Department of Microbiology, University of Guelph, Guelph, Ontario N1G 2W1, Canada

Albert Lu, DuPont Agricultural Products, DuPont Stine-Haskell Research Center, Newark, Delaware 19714-0030

Lois K. Miller, Departments of Entomology and Genetics, University of Georgia, Athens, Georgia 30602-2603

David R. O'Reilly, Department of Biology, Imperial College of Science, Technology, and Medicine, London SW7 2BB, England

Robert D. Possee, NERC Institute of Virology and Environmental Microbiology, Oxford OX1 3SR, England

George F. Rohrmann, Department of Microbiology, Oregon State University, Corvallis, Oregon 97331-7301

Steven M. Sait, Ecology and Biocontrol Group, NERC Institute of Virology and Environmental Microbiology, Oxford OX1 3SR, England

Just M. Vlak, Department of Virology, Wageningen Agricultural University, 6709 PD Wageningen, The Netherlands

Greg V. Williams, CSIRO, Australian Animal Health Laboratory, Geelong, Australia 3220

Preface

The past decade has witnessed an explosion of information on the molecular biology of insect viruses and a frenzy of activity in applying this information to medicine and agriculture. Baculovirus gene expression vectors are now found in the laboratories of most industrial and academic research institutions in the world. Genetically engineered baculoviruses are also being widely tested in agricultural fields in preparation for their commercial development as pesticides, and the study of baculoviruses is providing remarkable insights into basic virological and cellular processes such as apoptosis.

The purpose of this volume is to provide the reader with sufficient knowledge of current basic and applied baculovirology so that current literature in the field can be appreciated. The chapters of the volume are meant to be read in the order presented, especially by the novice, although each chapter can be read and understood individually. The choice of chapter topics was influenced by the molecular bias of the editor, but collectively the chapters provide a comprehensive view of the baculovirus field.

In addition to the authors, many people contributed to the development and quality of the volume. The following persons reviewed individual chapters of the books or provided valuable comments or information that have been incorporated into the volume: Dr. Gary Blissard (Boyce Thompson Institute at Cornell University), Dr. Greg Dwyer (University of Massachusetts, Amherst), Dr. Hans Flipsen (Wageningen), Dr. Robert R. Granados (Boyce Thompson Institute at Cornell University), Dr. Jeanne McLachlin (University of Georgia), Dr. Suzanne Thiem (Michigan State University), and Dr. Robert Weaver (University of Kansas). The contributions of Dr. Albert Lu and Dr. George F. Rohrmann extended far beyond the normal duties of authorship.

The volume is dedicated to the memory of Dr. Norman Crook, a leader in granulovirus research and a codiscoverer of the *iap* family of apoptotic inhibitors.

Lois K. Miller

Contents

Chapter 3

Baculovirus Pathogenesis

Brian A. Federici

Chapter 4

Cytological Changes and Viral Morphogenesis during Baculovirus Infection

Greg V. Williams and Peter Faulkner

Chapter 5

Baculovirus Genome Organization and Evolution

Robert D. Possee and George F. Rohrmann

Chapter 8

Regulation of Baculovirus Late and Very Late Gene Expression

Albert Lu and Lois K. Miller

Chapter 9

The Molecular Basis of Baculovirus Host Range

Lois K. Miller and Albert Lu

Chapter 10

Regulation of Programmed Cell Death by Baculoviruses

Rollie J. Clem

Chapter 11

Auxiliary Genes of Baculoviruses

David R. O'Reilly

Chapter 12

Baculovirus Ecology

Jenny S. Cory, Rosemary S. Hails, and Steven M. Sait

Chapter 13

Commercialization of Baculoviral Insecticides

Bruce C. Black, Lynn A. Brennan, Peter M. Dierks, and Ivan E. Gard

Chapter 14

Baculovirus Expression Vectors

Donald L. Jarvis

CHAPTER 1

Introduction to the Baculoviruses

Lois K. Miller

I. AN OVERVIEW OF BACULOVIROLOGY

A. General Considerations

Of all the viruses known to mankind, baculoviruses are the most beneficial from an anthropocentric viewpoint. While other viruses are studied because they cause problems, the basis of modern baculovirology was stimulated by the potential utility of baculoviruses to control insect pests (see Chapter 13, this volume). The current use of baculoviruses as gene expression vectors (see Chapter 14) evolved from the molecular biology of baculoviruses, a research area pursued while assessing the safety and improving the efficacy of baculovirus pesticides. Basic molecular studies also led to the use of baculoviruses and their genes to explore fundamental questions in biology such as the nature and function of apoptotic pathways (see Chapter 10). However, baculoviruses also cause disease in beneficial species, not just pest species, and understanding what controls the host range of these viruses (see Chapter 9) will become even more important if a balanced ecological perspective displaces anthropocentrism.

LOIS K. MILLER • Departments of Entomology and Genetics, University of Georgia, Athens, Georgia 30602-2603.

The Baculoviruses, edited by Lois K. Miller. Plenum Press, New York, 1997.

B. The Prevalence and Diversity of Baculoviruses

Baculoviruses are virtually ubiquitous in the environment. They infect arthropods that inhabit terrestrial and marine ecosytems. On land, baculoviruses have been identified in hundreds of insect species inhabiting forests, fields, rivers, and households (Martignoni and Iwai, 1986). Being specially designed to survive outside their host, they can reside in soil or in the crevices of plants or other refugia for years (Jacques, 1975). They are present in remarkable numbers in the food we eat and in the air we breathe (Heimpel *et al.*, 1973). But baculoviruses have never been found to cause disease in any organism outside the phylum *Arthropoda.* Their prevalence in the environment can be ascribed to the ubiquity of arthropod species. An understanding of their behavior in the environment will be important in predicting their prevalence and impact in various ecological niches (see Chapter 12).

Based on the diversity of arthropod species, the diversity of baculovirus "species" is probably enormous. Existing literature contains descriptions of baculovirus diseases in over 500 different species of insects. However, solid information on the nature and breadth of genetic diversity of baculoviruses is just emerging. Currently, the entire sequence of the genomes of three different nuclear polyhedrosis viruses (NPVs), belonging to the genus nucleopolyhedroviruses, are known (see Chapter 5). The basic organization of the genes in these NPV genomes is remarkably conserved, although rearrangements of genomic segments and the presence or absence of auxiliary genes (see Chapter 11) contributes to viral diversity.

Only limited sequence information is currently available for viruses belonging to the other major genus of the family *Baculoviridae,* the granuloviruses (GVs). The sequences that are available suggest that there will be fundamental genes common to both NPV and GV genomes (see, for example, Chapters 2, 10, and 11) but almost infinite potential for genetic variation on this basic theme. The full extent of the diversity of baculoviruses will not be known until more is learned about the granuloviruses and baculoviruses that infect marine arthropods. Understanding the evolution of the baculovirus family will also require a broader knowledge of insect viruses in general, many of which remain unclassified and probably even undiscovered. One group of insect viruses that are not covered in this book are viruses that were formerly called the "nonoccluded" baculoviruses, or nudiviruses, and were recently removed from the *Baculoviridae* family until further information is available concerning them.

II. BRIEF HISTORY

Although baculoviruses do not cause human disease, people have been aware of their existence for hundreds of years. According to baculovirus lore, the earliest written accounts of baculovirus infections are found in ancient Chinese literature describing silkworm culture. The first description of baculovirus disease in Western literature is credited (Benz, 1986) to a poem by Marco Vida of Cremona, an Italian bishop of the 16th century. Originally written in Latin, in perfect dactylic hexameter (the style of Virgil's *Aeneid*), Vida described the disease of silkworms in "De Bombyce" as follows (translated in utilitarian English):

> All at once, in the weak ones, the skin appears yellow. Then they swell up and a foul (connotation of smell) inactivity (connotation of paralysis) comes in the bodies of those who have fallen down. Finally they break open and everything is infected with repulsive putrid gore; diseased blood, from all sides flows from the bodies. (Translation provided by N. A. Miller)

A modern description and more sophisticated view of the pathology of baculovirus disease is provided in Chapter 3. Vida's putrid gore and diseased blood flowing from the larvae is likely to be a description of a sign of certain baculovirus diseases that is often referred to in the literature as "melting" or "wilting." Liquefaction of the diseased larvae occurs in the process of converting insect biomass into progeny virus. It was not until the 19th century that polyhedral crystals could be observed in the "gore" by microscopy and correlated with the "wilting disease" of insects (reviewed by Benz, 1986).

Research during the first half of the 20th century established that virus particles were present in the polyhedral crystals, and the view that baculoviruses were important in the natural control of insect populations was advanced. Also during this period, the GVs were described. In contrast to the NPVs, the environmentally stable or "occluded form" of GVs is small and granular in appearance, since they contain only a single virion per occlusion body (see Chapter 2). In the 1930s and 1940s, Bergold discovered rod-shaped virions within the crystalline polyhedra and initiated his landmark studies on the biochemical properties of NPVs (Bergold, 1953). During the same period, baculoviruses were observed to be effective biological control agents of an insect pest (Balch and Bird, 1944). In this study, the European spruce sawfly, a pest that was accidentally introduced into North America, was found to be effectively controlled by the subsequent introduction of a baculovirus.

From 1950 to 1975, the development of baculoviruses as biological control agents of insect pests was championed by Steinhaus and his students (Steinhaus, 1963). In 1975, the first baculovirus was registered as a pesticide in the United States (Ignoffo, 1981), but the product, Elcar®, was a commercial failure for a variety of reasons. A notable success of a baculovirus pest control agent, however, was the use of an NPV to control the Douglas fir tussock moth by the US Forest Service (Martignoni, 1984). Significantly, the development of baculoviruses as biological pesticides stimulated efforts to understand the molecular biology of baculoviruses, and this, in turn, led to renewed industrial interest in baculovirus pesticide development in the 1990s (see Chapter 13).

From 1970 to 1985, several important advances were made in understand the pathology and genetics of baculoviruses. The realization that there were two different forms of baculoviruses, a budded virus (BV) form as well as the occluded virus (OV) form, was important to understanding the behavior of the virus in cell culture and its pathology in the insect host. Whereas OV are infectious only in the midgut of the insect host, the BV are responsible for the spread of infection to other tissues of the insect and in cell culture. The ability to plaque-purify BV in cell culture led to the isolation of mutants and the ability to confidently characterize clonal viral isolates. *Autographa californica* nuclear polyhedrosis virus (AcMNPV) became the focus of much of the research because of the ease of propagating it in cell culture, its relative stability in a cell culture environment, and its relatively broad range of pest insect hosts. When the National Institutes of Health revised their guidelines for recombinant DNA research in late 1979, it became possible to clone defined segments of baculovirus DNA into *Escherichia coli* and analyze individual genes and their function in detail.

Baculovirology emerged as a dynamic and technologically important field in the last two decades of the 20th century. The use of baculoviruses as vectors for the expression of heterologous genes (see Chapter 14) led to their extensive use throughout the biomedical research community and the use of baculoviruses for human gene therapy has been proposed recently (see Chapter 9). Although much of the success in bringing baculoviruses into the forefront of biotechnology can be ascribed to progress made in the areas of molecular biology and genetics, advances in insect cell culture were critical to this success. Cell cultures provided a basis not only for studying baculovirus genetics, but also allowed the production of valuable proteins and pesticides in mass-scale cell culture facilities that were conducive to industrial development.

III. THE CURRENT STATUS OF BACULOVIROLOGY

This volume presents current, fundamental information regarding both basic and applied baculovirology. Information regarding the nature, regulation, and function of baculovirus genes is reviewed. Considerable effort in the past decade was devoted to defining the viral *cis*- and *trans*-acting factors involved in DNA replication (Chapter 7) and the cascade of early, late, and very late gene expression (Chapters 6 and 8). Progress was also made in defining the genes encoding the structural proteins of both BV and OV (Chapter 2). The study of cytological changes that occur during virus infection and the mechanisms by which viruses rearrange the cellular environment in the process of virus replication remains primarily observational (see Chapter 4), but this field is certain to be one of the more exciting ones as the viral genes responsible for these events are defined and functionally characterized. Several of the viral genes involved in manipulating cellular pathways have been identified and characterized to varying degrees (see Chapters 10 and 11).

The choice of focus areas was difficult given space limitations, and some areas are not treated as fully as they deserve. For example, the effects of serial passage of the virus are not explicitly covered, although some coverage of spontaneous insertions and deletions is found in Chapters 5, 7, and 11. The nature of transposable element insertion into the baculovirus genome, the second-most evident effect of serial passage, is a fascinating subject that was not covered in detail due to lack of space. The reader is referred to the early history of this phenomenon (Miller, 1986), to a more recent review (Friesen, 1993), and by current literature (Bauser *et al.*, 1996; Elick *et al.*, 1996a, b; Harrison *et al.*, 1996). Overall, the volume is biased toward molecular biology, and the chapters regarding ecology, viral pesticides, and expression vectors (Chapters 12, 13, and 14, respectively) each represent a vast body of literature compared to most of the other chapters. Collectively, the chapters reflect the enormous progress of baculovirus research in the 20th century.

Although the knowledge base was extended enormously this century, many of the same fundamental problems faced by baculovirologists in the 19th century still exist today. A significant problem in the late 1800s was diseases of beneficial insects such as silkworms, which eventually led to the demise of the silk industry in Europe. Today, the emerging shrimp aquaculture industry is threatened by baculovirus diseases. The use of NPVs to control pest insect populations was also an important motivation for research in the 19th century. Insect pest problems remain as one of the most important problems facing the 21st century. Considering the pressure of a rapidly expanding human population, the contamination of the environment with chemical pesticides, and the decline of biodiversity, the development of biological pesticides is becoming increasingly important. The need to learn more about baculoviruses, so that they and their hosts can be managed in an intelligent manner, should be a priority of the future. It is also likely, given the recent history of baculovirus research, that future baculovirus research will continue to contribute novel insights into basic biological processes and generate new technologies of broad utility.

REFERENCES

Balch, R. E., and Bird, F. T., 1944, A disease of the European spruce sawfly, *Gilpinia hercyniae* (Htg.) and its place in natural control. *Sci. Agric.* **25:**65.

Bauser, C. A., Elick, T. A., and Fraser, M. J., 1996, Characterization of hitchhiker, a transposon insertion frequently associated with baculovirus FP mutants derived upon passage in the TN-368 cell line, *Virology* **216:**235.

Benz, G. A., 1986, Introduction: Historical perspective, in: *The Biology of Baculoviruses* (R. R. Granados, and B. A. Federici, eds.), Vol. I: *Biological Properties and Molecular Biology*, pp. 1–35, CRC Press, Boca Raton, FL.

Bergold, G., 1953, Insect viruses, *Adv. Virus Res.* **1:**91.

Elick, T. A., Bauser, C. A., and Fraser, M. J., 1996a, Excision of the piggyBac transposable element *in vitro* is a precise event that is enhanced by the expression of its encoded transposase, *Genetica* **98:**33.

Elick, T. A., Bauser, C. A., Principe, N. M., and Fraser, M. J., 1996b, PCR analysis of insertion site specificity, transcription, and structural uniformity of the Lepidopteran transposable element IFP2 in the TN-368 cell genome, *Genetica* **97:**127.

Friesen, P. D., 1993, Invertebrate transposable elements in the baculovirus genome: Characterization and significance in: *Parasites and Pathogens of Insects* (N. E. Beckage, S. N. Thompson, and B. A. Federici, eds.), pp. 147–178, Academic Press, San Diego.

Harrison, R. L., Jarvis, D. L., and Summers, M. D., 1996, The role of the AcMNPV 25K gene, "FP25," in baculovirus polh and p10 expression, *Virology* **226:**34.

Heimpel, A. M., Thomas, E. O., Adams, J. R., and Smith, L. J., 1973, The presence of nuclear polyhedrosis viruses of *Trichoplusia ni* on cabbage from the market shelf, *Environment. Entomol.* **2:**173.

Ignoffo, C. M., 1981, Living microbial insecticides in: *Essays in Applied Microbiology* (J. R. Norris and M. H. Richmond, eds.), Vol. 7, pp. 2–31, John Wiley, New York.

Jacques, R. P., 1975, Persistence, accumulation, and denaturation of nuclear polyhedrosis and granulosis viruses, in: *Baculoviruses for Insect Pest Control: Safety Considerations* (M. D. Summers, R. Engler, L. A. Falcon, and P. V. Vail, eds.), pp. 90–99, American Society of Microbiology, Washington, DC.

Martignoni, M. E., 1984, Baculovirus: An attractive biological alternative, in: *Chemical and Biological Controls in Forestry* (W. Y. Garner and J. Harvey, eds.), pp. 55–67, American Chemical Society, Washington, DC.

Martignoni, M. E., and Iwai, P. J., 1986, *A Catalog of Viral Diseases of Insects, Mites, and Ticks,* 4th ed, General Technical Report PNW-195, US Department of Agriculture, Forest Service, Pacific Northwest Research Station, Portland, OR.

Miller, L. K., 1986, The genetics of baculoviruses, in: *The Biology of Baculoviruses* (R. R. Granados and B. A. Federici, eds.), Vol. I: *Biological Properties and Molecular Biology,* pp. 217–238, CRC Press, Boca Raton, FL.

Steinhaus, E. A., 1963, *Insect Pathology, An Advanced Treatise* (E. A. Steinhaus, ed.), Vol. 2, Academic Press, New York.

CHAPTER 2

Baculovirus Structure

C. JOEL FUNK, SHARON C. BRAUNAGEL, AND GEORGE F. ROHRMANN

I. INTRODUCTION

The *Baculoviridae* is a large family of occluded viruses that is composed of two genera that are differentiated by the size of their occlusion bodies. The nuclear polyhedrosis viruses (NPVs) produce large polyhedron-shaped structures called *polyhedra,* which contain many virions, while the granulosis viruses (GVs) have smaller occlusion bodies called *granules,* which normally contain a single virion. NPVs have been more extensively investigated than GVs, and therefore this chapter will mainly present information on NPVs. However, much of this information is likely to be pertinent to GVs because of the close relatedness of these two viral genera.

The evolution of the baculovirus structure appears to be in response to the unique features of the life cycles of their invertebrate hosts. Baculovirus host insects are often characterized by population expansion due to seasonal (e.g., winter–summer, or wet–dry seasons) or to ecological factors that can lead to cumulative increases in population over a number of years. In forest insects, such population explosions can involve millions of hectares of forest land and culminate in a collapse often caused by the spread of viral disease through the insect population. Critical to the ability of baculoviruses to efficiently replicate within insects and then to spread the infection throughout a population is the structure of their virions, which are present

C. JOEL FUNK AND GEORGE F. ROHRMANN • Department of Microbiology, Oregon State University, Corvallis, Oregon 97331-7301. SHARON C. BRAUNAGEL • Department of Entomology, Texas Agricultural Experiment Station, Texas A & M University, College Station, Texas 77843-2475.

The Baculoviruses, edited by Lois K. Miller. Plenum Press, New York, 1997.

in two forms during a single infection cycle. The occlusion-derived virus (ODV) is encapsulated in a protein matrix composed predominantly of a single protein called polyhedrin (or granulin in GVs), whereas the budded virus (BV) is not occluded. Infection is initiated when the alkali-soluble occlusion body is ingested by a susceptible host and dissolved by the high pH of the insect midgut, thereby releasing virions. Although occluded virions are highly efficient at infecting insects, they do not spread the infection within insect tissues. Early in infection BVs are produced, which are not occluded and which efficiently spread the infection from cell to cell within insects. Nucleocapsids destined to become BV are transported from the nucleus through the cytoplasm and acquire an envelope when they bud through the plasma membrane which has been modified by a viral-encoded glycoprotein called *gp64* (see Section VII). At later stages of infection, virions remain within insect cell nuclei where they become enveloped and occluded. These occluded virions are released upon disintegration of dead insects and contaminate foliage, which is subsequently ingested by other susceptible insects.

Although their nucleocapsids are similar in structure, the two viral forms differ in the composition of their envelopes (Fig. 1). These differences likely reflect the different functions required of each form. The BV envelope is adapted for movement and infection of tissues within the insect, whereas the ODV envelope is adapted for interacting with polyhedron structures in the occlusion process and for facilitating the infection of midgut epithelium in the harsh environment of the insect gut. The occluded viral form is highly stable in the environment, which is a major factor in allowing persistence of the virus and subsequent initiation of sporadic infections of insect populations at intervals of 1 year or more.

The structural composition of baculovirions was previously reviewed (Rohrmann, 1992) and extensive information on relatedness of capsid and polyhedron-associated proteins was presented. However, there have been numerous recent advances in the identification of virions components, particularly those associated with ODV envelopes. Therefore, this review will focus on three areas including the composition of polyhedra, the composition and organization of nucleocapsids, and the characterization of BV and ODV envelope components.

II. BACULOVIRUS STRUCTURAL PROTEINS

Baculoviruses have large genomes with the potential to encode over 150 proteins (Ahrens *et al.*, 1997; Ayres *et al.*, 1994; S. Maeda, personal communication). Because transfection of susceptible cultured insect cells with purified virus DNA results in the production of viable virus (Burand *et al.*, 1980; Carstens *et al.*, 1980; Kelly and Wang, 1981; Potter and Miller, 1980), it is apparent that under these conditions no virion-associated proteins are essential for virus replication. This is in contrast to the complexity of some viruses that are unable to productively infect cells via DNA transfection. For example, poxvirus virions carry an extensive array of enzymes that are essential for early gene transcription and other functions. Although the production of virus from transfected DNA suggests that the virion structure functions solely as a delivery system for the introduction of the virus genome into cells, it is likely that a variety of proteins with enzymatic activities are associated with the virion envelope, or are present as nucleocapsid structural components or reside within polyhedra. The adventitious association of some proteins within occlusion bodies and with the membrane of the BV is also likely to occur.

Although several genes encoding *Autographa californica* MNP (AcMNPV) structural proteins are located in close proximity, most are distributed throughout the genome with no obvious pattern to their location. In addition, the location of homologues of several of these genes have been identified in genomes of other baculoviruses, and the surrounding genes often vary among viruses, indicating that they can be expressed irrespective of the exact location in the genome or proximity to other specific genes.

III. STRUCTURAL PROTEINS OF POLYHEDRA

A. Polyhedrin (AcMNPV orf8, BmNPV orf1, OpMNPV orf3)/Granulin

Polyhedrin and granulin are proteins of about 245 amino acids (29K) and are the major component of polyhedra and granules, respectively. Polyhedrin/granulin has been reviewed previously (Rohrmann, 1986, 1992), and so will not be discussed extensively here. Based on complete genome sequences, polyhedrin appears to be the most highly conserved baculovirus protein (89% amino acid identity) between AcMNPV and the *Orgyia pseudotsugata* MNPV (OpMNPV) (Ahrens *et al.*, 1997). Despite this conservation, there are extensive differences between these proteins from various virus strains. At last count, 20 lepidopteran polyhedrin genes had been sequenced and those showing the most variation are about 70% identical in amino acid sequence, whereas the most closely related sequences are 97–99% identical (Chou *et al.*, 1996). In addition, lepidopteran polyhedrins show about 50% amino acid identity with granulins and 40% identity with a hymenopteran NPV polyhedrin (*Neodiprion sertifer* SNPV) (Rohrmann, 1986, 1992). An N-terminal sequence of polyhedrin from a dipteran (*Tipula paludosa*) NPV demonstrated little sequence identity to lepidopteran polyhedrins (Rohrmann *et al.*, 1981). However, the N-terminus is the most variable region, so this sequence may not be representative of the complete molecule. Although the polyhedrin of a NPV pathogenic for the pink shrimp, *Penaeus duorarum*, appears to be almost twice the size (50 kDa) of insect baculovirus polyhedrins, it was reported to react with antisera against

AcMNPV polyhedrin and *Trichoplusia ni* granulosis virus (TnGV) granulin antisera (Summers, 1977).

B. The Polyhedron Envelope/Calyx Protein (AcMNPV orf131, BmNPV orf108, OpMNPV orf129)

Polyhedra are surrounded by an electron-dense envelope that has been termed the polyhedron envelope (PE) or polyhedron calyx. Although the PE was originally reported to be composed of carbohydrate (Minion *et al.*, 1979), a protein component of the PE of AcMNPV was identified (Whitt and Manning, 1988). Subsequently, the gene encoding this protein (PEP, pp34) was identified in OpMNPV, AcMNPV (Gombart *et al.*, 1989b), and *Lymantria dispar* MNPV (LdMNPV) (Bjornson and Rohrmann, 1992b). Immunoelectron microscopic examination of cross-sections of polyhedra showed that the PE protein may be a major component of the PE of OpMNPV and AcMNPV polyhedra (Russell and Rohrmann, 1990b; van Lent *et al.*, 1990). The PE has been shown to be sensitive to protease, which suggests that protein is an integral component of this structure and it is not composed exclusively of carbohydrate. The PE protein localizes to the margins of p10 fibrillar structure (see Section III.C) and appears to be a component of structures called electron-dense spacers (Russell *et al.*, 1991). Interruption of the PE gene from OpMNPV with a *GUS* reporter gene resulted in the production of polyhedra that retained an electron-dense structure surrounding polyhedra, but they were of small size and highly pitted, indicative of cavities left by dislodged virions (Gross *et al.*, 1994).

C. p10 (AcMNPV orf137, BmNPV orf114, OpMNPV orf133)

In addition to polyhedrin, the p10 protein is highly expressed late in the infection cycle, but in contrast to polyhedrin it is poorly conserved (Zuidema *et al.*, 1993). However, alignment of p10 from six viruses indicates that they all have similar amphipathic alpha-helical structures at the N-termini with a characteristic pattern of units of seven amino acids with hydrophobic residues at position "a" and "d" and aligning on one face of the peptide (Wilson *et al.*, 1995). Such domains are involved in the formation of coiled-coil structures, which are associated with protein oligomerization domains. p10 appears to be a component of extensive fibrillar structures (Russell *et al.*, 1991; van der Wilk *et al.*, 1987; Williams *et al.*, 1989) located in both the nucleus and cytoplasm of infected cells, and it is likely that the coiled-coil domain is involved in the aggregation of p10 into these structures. It has been suggested that phosphorylation of p10 induces microtubule elongation and is involved in process formation in infected cells (Cheley *et al.*, 1992).

Although it is not clear whether p10 is an essential component of polyhedra or is simply trapped during the occlusion process, it is commonly found associated with polyhedra (Quant-Russell *et al.*, 1987) and may have a role in the formation of the polyhedron envelope. As described above, the p10 fibrillar structures show a close association with the developing PE and may be integrally involved in the application of the PE to the surface of polyhedra (Rohrmann, 1992). In addition, immunoelectron microscopic evidence shows a direct association of the PE protein with the periphery of p10-containing fibrillar structures (Russell *et al.*, 1991). Studies in which the p10 gene has been interrupted with the *lacZ* or *GUS* genes are not consistent in the effect on the PE. Williams *et al.* (1989) showed that deletion of the p10 gene resulted in the absence of p10-associated fibrillar material and in polyhedra that lack or had fragmented PEs. They also showed that polyhedra from a mutant virus lacking p10 are fragile and are not released from nuclei late in infection, suggesting that p10 may be involved in cell lysis. However, van Oers *et al.* (1993) reported that a variety of mutants in which the p10 gene was either partially or completely deleted showed evidence of a polyhedron envelope. In addition, they found that virus with C-terminal deletions past amino acid number 79 showed an inability to release polyhedra from cells late in infection, indicating that the disintegration of nuclei did not occur. These data suggest that the p10 gene may have a role in the stability and efficient dissemination of polyhedra.

Infection of *O. pseudotsugata* larvae with recombinant OpMNPV with a p10 gene interrupted by a reporter gene yielded polyhedra with partial PE-like structures (Gross *et al.*, 1994). However, scanning electron micrographs of polyhedra from mutant virus showed rough and highly pitted surfaces that did not resemble the surface of wild-type polyhedra, which indicated that they may lack a PE. This suggests that the PE-like structure observed by transmission electron microscopy (TEM) may be an artifact of the fixation protocol or is an improperly assembled PE. In addition, polyhedra produced by a OpMNPV double-mutant *PEP*$^-$–*GUS*$^+$/*p10*$^-$ *LacZ*$^+$ were observed to aggregate during the purification process (Gross *et al.*, 1994), which is another indication of a surface alteration on mutant polyhedra. One possible function of a properly assembled PE could be to prevent the fusion or aggregation of networks of polyhedra. Such a function would be important in ensuring their maximal dissemination.

IV. OTHER PROTEINS ASSOCIATED WITH OCCLUSION BODIES

A. Viral Enhancing Factor

The viral enhancing factor (VEF) is a 104-kDa protein that forms about 5% of the mass of occlusion bodies of the TnGV (Hashimoto *et al.*, 1991). Although it has been shown to be present in several GV genomes, homologues are not present in AcMNPV (Ayres *et al.*, 1994) or OpMNPV (Ahrens *et al.*, 1997). However a homologue has been reported from the LdMNPV genome (J. Slavicek, unpublished data). The insect gut is lined with a chitin-rich structure called the peritrophic membrane, which may function as a barrier to invasion of gut cells by insect pathogens. The VEF protein appears to facilitate GV infection by disrupting the peritrophic membrane, thereby allowing virions access to the surface of gut cells (Derksen and Granados, 1988). It is not clear whether the VEF is an integral component of a virus structure or simply trapped in granules during the occlusion process.

B. Alkaline Proteases

The presence of alkaline proteases in polyhedra (Kozlov *et al.*, 1975; Yamafuji *et al.*, 1958) greatly hindered the initial characterization of baculovirus structural proteins. However, the observation that it could be heat inactivated (Kozlov *et al.*, 1975) led to the sequencing of polyhedrin protein (Kozlov *et al.*, 1981; Serebryani *et al.*, 1977). Subsequently, it was demonstrated that the proteases associated with baculoviruses have properties similar to those isolated from the insect gut (Rubinstein and Polson, 1983). These data, along with the finding that polyhedra produced in cell culture lack protease activity (McCarthy and DiCapua, 1979; Wood, 1980; Zummer and Faulkner, 1979), indicate that the protease is likely to be a contaminant derived from the insect gut or bacteria, and it becomes associated with polyhedra when the insect dies and disintegrates.

V. STRUCTURAL PROTEINS OF THE NUCLEOCAPSID

The nucleocapsids of BV and ODV appear to have a similar structure (Fig. 1), and generally proteins that have been identified to be a component of the capsid are present in both BV and ODV. However, Braunagel and Summers (1994) showed differences in the structural protein composition between BV and ODV, using sodium dodecyl sulfate–polyacrylamide gel electrophoresis (SDS-PAGE) analysis. At least one protein, ODV-EC27, has since been identified that is differentially incorporated into ODV capsid. Preliminary data indicate that if this protein is incorporated into BV, it is incorporated in a different form than ODV (Braunagel *et al.*, 1996b). It remains to be determined how extensive the differences are between the capsid compositions of BV and ODV.

A. p6.9, a DNA-Binding Protein (AcMNPV orf100, BmNPV orf84, OpMNPV orf101)

Baculoviruses have large genomes that must become highly condensed in order to be efficiently packaged within a nucleocapsid. Although histones neutralize the electrostatic repulsion of neighboring DNA sequences in cel-

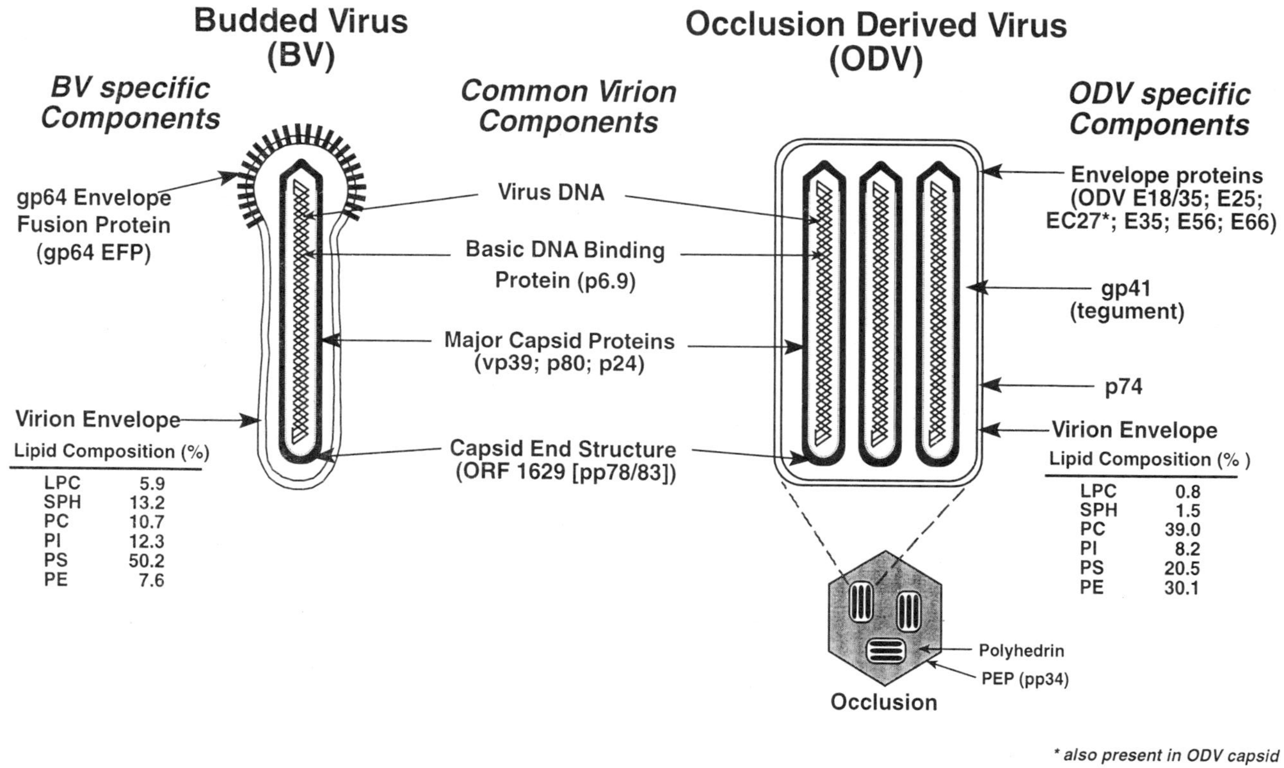

Figure 1. The location of virally encoded proteins in budded and occlusion-derived virus. References for structural proteins are p6.9 (Wilson *et al.*, 1987); vp39 (Blissard *et al.*, 1989; Pearson *et al.*, 1988; Russell *et al.*, 1991; Thiem and Miller, 1989); p80 (Lu and Carstens, 1992; Müller *et al.*, 1990); pp 78/83 (Russell *et al.*, 1997; Vialard and Richardson, 1993); polyhedrin (Hooft van Iddekinge *et al.*, 1983); PEP (PP34) (Gombart *et al.*, 1989b; Russell and Rohrmann, 1990b); ODV-E25 (Russell and Rohrmann, 1993); ODVE66 (Hong *et al.*, 1994); ODVE56 (Braungel *et al.*, 1996a; Theilmann *et al.*, 1996), ODVE18, E35, and EC27 (Braunagell *et al.*, 1996b): gp41 (Whitford and Faulkner, 1992a, b), p74 (Kuzio *et al.*, 1989) and gp64 (Blissard and Rohrmann, 1989; Whitford *et al.*, 1989). Lipid composition of BV and ODV envelopes (Braunagel and Summers, 1994). LPC, lysophosphatidycholine; SPH; sphingomyelin; PC, phosphatidycholine; PI, phosphatidylinositol; PS, phosphatidylserine; PE, phosphatidylethanolamine. (Adapted from Blissard, 1996).

lular DNA, they do not appear to be associated with DNA packaged within nucleocapsids (Wilson and Miller, 1986). However, an arginine–serine–threonine-rich basic protein was shown to be present in virions of the GV of *Plodia interpunctella* GV (PiGV) (Tweeten *et al.*, 1980) and was found to bind DNA (Wilson and Consigli, 1985b). Subsequently, a small gene encoding 54 amino acids (termed p6.9) was identified in AcMNPV that had a similar amino acid composition to the putative DNA-binding protein of PiGV (Wilson *et al.*, 1987). Homologues to the AcMNPV p6.9 genes have been isolated from both OpMNPV (Russell and Rohrmann, 1990a) and the NPV of *Bombyx mori* (BmNPV) (Maeda *et al.*, 1991) and a GV of *Cryptophlebia leucotreta* (ClGV) (Jehle and Backhaus, 1994). The predicted amino acid sequences of the NPV proteins are closely related (over 76% identity), but the BmMNPV gene contains an additional 12 amino acid arginine–serine-rich sequence, whereas the ClGV sequence is about 45% identical to the AcMNPV p6.9 amino acid sequence. All four of these NPV proteins contain about 40% arginine and approximately 30% serine or threonine residues. The high concentration of arginine, serine, and threonine in these proteins is similar to that of protamines, which are a class of proteins involved in the production of highly condensed DNA found in many fish, avian, and mammalian sperm nuclei. Protamines are small molecules (44–65 amino acids) composed of 55–70% arginine (Balhorn, 1982; Nakano *et al.*, 1989). Arginine has the highest affinity of all amino acids for interaction with the phosphate backbone of DNA. It is thought that the central polyarginine segment in protamines binds in the minor groove of DNA, cross-linking and neutralizing the phosphodiester backbone of DNA, while the N- and C-terminal ends of protamine forms bonds with other protamine molecules (Balhorn, 1982). Such an arrangement generates a neutral, insoluble chromatin complex of compact volume, which is to a large degree biochemically inert. In a sense, viral nucleocapsids and sperm have a similar function: the transport of genetic information in a highly compact form to a recipient cell. In baculoviruses, it has been suggested that the basic arginine residues of the DNA-binding protein neutralizes the acidic residues of the viral DNA. Phosphorylation appears to play a key role in the regulation of this basic protein. Evidence suggests that the basic protein is phosphorylated prior to assembly, but dephosphorylated within the capsid (Funk and Consigli, 1993). Upon entry into an insect cell, the DNA-binding protein may become phosphorylated by a protein kinase, which results in the unpackaging of the viral DNA. This theory is supported by the presence of a protein kinase associated with purified capsids of granulaosis viruses (Wilson and Consigli, 1985a, b) and with both BV and ODV of AcMNPV (Miller *et al.*, 1983).

B. vp39-Capsid (AcMNPV orf89, BmNPV orf72, OpMNPV orf90)

A major capsid protein of 39 kDa has been characterized from both OpMNPV and AcMNPV (Blissard *et al.*, 1989; Pearson *et al.*, 1988; Thiem and Miller, 1989). Immunoelectron microscopy confirmed that the vp39 protein is a component of the capsid (Russell *et al.*, 1991) and showed that it is randomly distributed over the surface of the nucleocapsid. This indicates that vp39 is a component of the capsid rather than a part of a more specialized structure (such as an end structure of the nucleocapsid). In addition to AcMNPV and OpMNPV, the vp39-capsid gene has been identified in LdMNPV (Bjornson and Rohrmann, 1992a). The AcMNPV and OpMNPV vp39 amino acid sequences are 59% identical and they are 39 and 47% identical, respectively, to the LdMNPV vp39 sequence. However, the carboxyl-terminal 65 amino acids of the LdMNPV vp39 sequence showed only 11 and 22% sequence identity with AcMNPV and OpMNPV vp39, respectively. This region in AcMNPV and OpMNPV vp39 is also variable (38% homology). Another feature of the vp39 sequence is that all eight cysteines are conserved. These likely form disulfide bonds that may play a major role in the structure of vp39.

Recently it has been demonstrated (Braunagel *et al.*, 1996a) that this capsid protein also locates to empty elongated capsid structures within the infected cell nucleus. It is not known if these structures represent aberrant capsids or function as precursor forms of the mature capsid.

C. p87-Capsid (AcMNPV orf104, BmNPV orf88, OpMNPV orf105)

In addition to vp39, another protein, p87, has been found to be associated with capsids of OpMNPV (Müller *et al.*, 1990) and its homologue, p80 is associated with AcMNPV capsids (Lu and Carstens, 1992). Although the AcMNPV homologue is found in a similar genome position and the C-terminal one third of the amino acids are well conserved (59% identity compared to OpMNPV), the overall amino acid sequence identity is rather low (34%). Monospecific antiserum produced against the OpMNPV protein binds to a protein of 87 kDa on Western blots of both ODV, BV, and purified capsids. Similar observations have been reported for AcMNPV (Lu and Carstens, 1992). The OpMNPV p87 reading frame encodes a protein of predicted mass of 71 kDa. No evidence of N-linked glycosylation has been found, and the cause of the difference between the predicted size of the sequenced ORF and the size calculated from Western blots has not been determined. Immunofluorescence microscopy demonstrated that p87 was concentrated in the nucleus late in OpMNPV infection similar to the vp39-capsid protein.

D. ORF1 (p24) (AcMNPV orf129, BmNPV orf106, OpMNPV orf127)

A gene originally referred to as ORF1 (now called orf129) is one of five contiguous open reading frames (ORFs) oriented in the same direction (Gombart *et al.*, 1989a; Oellig *et al.*, 1987) (the third ORF encodes the polyhedron envelope/calyx protein). The ORF129 locus is highly variable; in one strain of AcMNPV it contains a transposable element (Gombart *et al.*, 1989a, b; Schetter *et al.*, 1990), and in the LdMNPV genome only the carboxyl-terminal half of the gene appears to be present (Bjornson and Rohrmann, 1992b). Monospecific antiserum produced against a fusion protein containing the ORF129 protein of OpMNPV reacts with a protein of 24 kDa associated with both ODV and BV (Wolgamot *et al.*, 1993).

E. ORF1629; pp78/83 (AcMNPV orf9, BmNPV orf2, OpMNPV orf2)

AcMNPV encodes a late gene called orf9 (Ayres *et al.*, 1994) or orf1629 (Possee *et al.*, 1991) of 543 amino acids and a predicted molecular weight of 61 kDa that is located adjacent to and in the opposite orientation to the polyhedrin gene. Insertional mutagenesis at the carboxyl-terminal region indicated that orf9 is essential for AcMNPV viability (Possee *et al.*, 1991). This observation was exploited to develop a protocol for the selection of a high frequency of AcMNPV recombinants (Kitts and Possee, 1993). Subsequently, Vialard and Richardson (1993) characterized orf9 expression and found that it was present as both unphosphorylated and phosphorylated forms with apparent molecular masses of 78 and 83 kDa, respectively. Furthermore, they demonstrated by Western blot analysis that it was associated with both budded and occluded virions. Examination by immunoelectron microscopy indicated that it was localized to end structures of mature nucleocapsids. Surprisingly, although orf9 appears to be an essential component of a highly conserved virion structure, it appears to be poorly conserved with the orf9 homologue from *Heliothis zea* SNPV (HzSNPV) and OpMNPV, showing less than 30% amino acid sequence identity with the AcMNPV protein (Cowan *et al.*, 1994; Russell *et al.*, 1997).

The virogenic stroma, and electron-dense structure present in the nuclei of baculovirus-infected cells, contains intrastromal spaces and a convoluted periphery and has been implicated as being the site of nucleocapsid assembly. It is composed of a homogeneous electron-dense structure that is highly sensitive to RNase (Young *et al.*, 1993) and contains discrete compartments containing DNA. It has been suggested that the virogenic stroma is the site of DNA synthesis. Electron micrographic studies suggest that empty capsids align with one end pointed toward the stroma, and DNA is transferred from the stroma into the capsid. A caplike structure called the *apical cap* is observed in such capsids and it may be involved in the movement of the genomic DNA into the capsid. The other end of the capsid contains a structure termed the *basal structure* (Fraser, 1986). Recent evidence from electron microscopy on the OpMNPV orf1629 homologue demonstrate that it is present at only one end of the capsid, and this end is oriented away from the virogenic stroma. In addition, these studies indicate that the orf1629-containing end is opposite the end that interacts with membranes enveloping capsids in nuclei and is also opposite the end involved in the penetration of the nuclear membrane as nucleocapsids move into the cytoplasm on the way to becoming budded virus (Russell *et al.*, 1997). Collectively, these data suggest that the orf1629 protein is associated with the basal structure of the capsid.

F. Tegument-gp41 (AcMNPV orf80, BmNPV orf66, OpMNPV orf83)

An ODV structural glycoprotein of 41 kDa (called gp41) has been characterized from AcMNPV (Whitford and Faulkner, 1992a, b). It is expressed as a late gene with the mRNA initiating from two TAAG promoter sequences and is present from 12 to 36 hr postinfection (p.i.). Glycoprotein analysis of gp41 indicated that it contains single residues of *O*-linked *N*-acetylglucosamine. Virion fractionation studies indicated that gp41 is located between the virion envelope and the capsid. This structural region has been termed the tegument.

G. Protein Tyrosine Phosphatase (AcMNPV ORF1, BmNPV orf130, OpMNPV orf10)

Antiserum was made against a bacterially expressed fusion of a gene encoding a protein tyrosine/serine phosphatases (PTP). Western blot analysis of extracts of both BV and ODV indicated that PTP was associated with these structures. It was also found to localize to concentrated specific regions within fibrillar structures in the nucleus and cytoplasm (Li and Miller, 1995a). Deletion of this gene had no observable effect on the properties of AcMNPV infection in *Spodoptera frugiperda* larvae or cultured *Trichoplusia ni* cells. However, a portion of a population of *S. frugiperda* cells infected by the mutant were defective in occlusion body production (Li and Miller, 1995b).

H. Protein Kinase

Evidence suggests that a protein kinase may be associated with NPV and GV virons and that it may be involved in unpackaging virion DNA (Wilson and Consigli, 1985a, b; Miller *et al.*, 1983). See Section V.A, for more detail.

VI. ENVELOPES OF OCCLUDED VIRIONS

ODV are surrounded by a unique envelope structure that is likely to play an integral role in the association of ODV with polyhedrin as it crystallizes into occlusion bodies and may also play a major role in the infection process in the midguts of susceptible insects. Because of the harsh environment likely encountered in the gut, the ODV envelope may also contribute to virion stability prior to entry and infection of midgut cells.

Unlike other viruses that mature in the nucleus such as herpesvirus or plant nucleorhabdovirus, there is no evidence that baculovirus nucleocapsids destined to mature as ODV associate with the nuclear envelope. Instead, ODV obtains its envelope within the nucleoplasm. The source of the ODV envelope is unclear; however, a number of virus-encoded proteins present in the mature ODV envelope are also found in viral-induced intranuclear microvesicles (see Section VI.B.3). This strongly suggest that intranuclear microvesicles play an important role, either as a direct precursor or as an assembly focus for the ODV envelope (Braunagel *et al.*, 1996a, b; Hong *et al.*, 1994). The source of the membranes that comprise the microvesicles is unknown. Based on an apparent lack of continuity of microvesicles with the inner nuclear membrane and differences in morphology between these two membrane structures, *de novo* membrane morphogenesis has been proposed (Stoltz *et al.*, 1973). However, baculovirus infection also appears to induce an inward invagination of the inner nuclear membrane (Summers and Arnott, 1969; Tanada and Hess, 1976). Evidence using antibodies developed to ODV-E66 and -E56 to probe the transport pathway of ODV envelope proteins suggests that the nuclear envelope itself could function as the source of the microvesicles. This hypothesis proposes that the inner nuclear membrane expands, invaginates into the nucleus, and buds, to produce microvesicles which then migrate within the nucleoplasm (Braunagel *et al.*, 1996a, b; Hong *et al.*, 1994). It should be noted that these two hypothesis are not mutually exclusive. The inner nuclear membrane could serve as the assembly site for newly synthesized lipids (i.e., *de novo* membrane morphogenesis), and then as the membrane increases in size it invaginates and buds into the nucleoplasm.

A. Lipid Composition of the Envelope of BV and ODV

The envelope of ODV appears as a lipid bilayer membrane and becomes tightly associated with the nucleocapsids upon occlusion. Typically, lipid bilayer membranes contain phospholipids as major components and include a variety of proteins located within the membrane or bound to the surface. The two sides of the membrane are usually different in the distribution and orientation of proteins, oligosaccharides, and phospholipids, and this reflects their differing roles in membrane transport and other functions. A comparison of phospholipid content of AcMNPV membranes reveals that the composition of every lipid class except phosphatidylinositol is statistically different between BV and ODV (Braunagel and Summers, 1994). Phosphatidylcholine (39%) and phosphatidylserine (50%) are the major phospholipids of ODV and BV, respectively. The major fatty acid of both BV and ODV has 18 carbons and one double bond (18 : 1) and it comprises 31% of the fatty acids in ODV and 42% in the BV envelope. The ODV envelope contains more saturated fatty acids than BV and this is reflected in significantly different saturated–unsaturated fatty acid ratios (1.05 vs. 0.68, respectively). The cholesterol–phosphate ratio indicates that ODV contains somewhat more cholesterol than BV (0.58 vs. 0.42, respectively) and the ODV envelope is more densely packed with protein than BV as determined by both the phosphate–protein ratio (0.11 vs. 0.98 mole : mg) and the cholesterol–protein ratio (0.02 vs. 0.15 mg : mg respectively).

Biological membranes are not rigid structures, but the lipid and protein components are in constant motion. Such movement, or fluidity, depends on the temperature and lipid composition. A curious feature of the lipid composition of both BV and ODV is that it does not correspond with what would be predicted based on the apparent source of the membrane. The composition of BV, which obtains its envelope from the host cell plasma membrane, is more fluid than would be expected when compared to that of other mammalian cell plasma membranes. The composition of ODV predicts that its envelope is much less fluid than would be expected from an envelope that was obtained from an intracellular membrane source (Braunagel and Summers, 1994). These data suggest that baculovirus may have the ability to selectively modify host cell membrane composition; however, direct determination of the lipid composition of insect cellular membranes is needed to confirm this hypothesis.

The differences in both protein and lipid composition of BV and ODV envelopes likely reflect the different functions required of each form. A number of viral-encoded proteins have been found to be associated with ODV envelopes (see Section VI.B) and it is likely that these proteins provide modifications that assist interactions with polyhedrin in the crystallization process and facilitate fusion of the ODV envelope with the microvillar membranes of the insect gut during primary infection. In contrast, BVs are adapted for movement and infection of tissues within the insect. At least one BV envelope protein, gp64, has been identified to be essential in this process (see Section VII).

B. Proteins Associated with Occluded Viral Envelopes

1. ODV-E66 (AcMNPV orf46, BmNPV orf37, OpMNPV orf50)

A number of the structural proteins described below employ nomenclature used by Braunagel *et al.* (1996a). In this terminology, the source of the protein is indicated (ODV or BV), followed by the location, E (envelope) or C (capsid), and the apparent molecular weight.

ODV-E66 was initially identified by N-terminal amino acid sequencing of a protein that was unique to ODV and was not present in BV envelope. Transcription of the gene encoding ODV-E66 is initiated from two conserved TAAG motifs (-15 and -37) with transcripts detected from 12 to 72 hr p.i. The protein is detected in infected Sf9 cell extracts from 24 to 72 hr p.i. by Western blot analysis. Immunogold localization of ODV-E66 shows that the protein is present in the viral-induced microvesicles and ODV envelope of both nonoccluded and occluded virus within the cell nucleus (Hong *et al.*, 1994). AcMNPV and OpMNPV ODV-E66 share 75% amino acid sequence identity, while AcMNPV and LdMNPV ODV-E66 are 40% identical (Bjornson and Rohrmann, 1992b).

N-terminal amino acid sequencing revealed two forms of ODV-E66 in the mature virus. One form initiates from the first Met and shows that the N-terminal region is uncleaved, while a second determined sequence initiates 69 amino acids interior to the predicted N-terminus. It is unclear if the truncated version of the protein is a degradation product formed in the purification of the virus or if it represents a specific subset of ODV-E66. The smaller 60-kDa protein is truncated at a trypsin consensus site after a single arginine at amino acid 69, and this cleavage site resembles that of other viral proteins including Vp8 of rotavirus and HA_0 protein of influenza virus (Hong *et al.*, 1994).

The uncleaved N-terminus predicts a 23-amino-acid hydrophobic domain. Hong *et al.* (1997) demonstrated that reporter proteins fused to this domain [β-galactosidasae and green fluorescence protein (GFP)] are transported to the viral-induced intranuclear microvesicles and the ODV envelope. These data suggest that in an AcMNPV-infected cell, this amino acid sequence is sufficient to target, retain, and/or transport proteins to the ODV envelope. ODV-E66 is not *N*-glycosylated and preliminary data also suggest that it does not undergo *O*-glycosylation (Hong *et al.*, 1997).

2. ODV-E25,p25 (AcMNPV orf94, BmNPV orf77, OpMNPV orf95)

This protein was originally identified in OpMNPV and called p25 (Russell and Rohrmann, 1993). Antiserum produced against preoccluded virus of OpMNPV reacted with a fusion protein in an OpMNPV λgt11 expression library. One of the immunoreactive clones contained an insert that was subsequently identified to lie within the p25 ORF. Monospecific antiserum produced against a bacterial fusion containing this ORF reacted with a doublet of approximately 25 kDa on Western blots of SDS-PAGE-separated OpMNPV-infected *Lymantria dispar* cell extracts and AcMNPV-infected Sf9 cell extracts (Russell and Rohrmann, 1993). Immunoelectron microscopy showed that p25 was present in the nuclei of OpMNPV-infected cells and localized to the envelopes surrounding polyhedron-derived virus. No evidence of N- or O-linked glycosylation of the OpMNPV p25 was found. In addition, repeated attempts to produce a recombinant OpMNPV lacking this gene were unsuccessful, suggesting that it may be an essential gene (G. F. Rohrmann, unpublished data).

Like ODV-E66, ODV-E25 also locates to viral-induced microvesicles in AcMNPV-infected Sf9 cells (Braunagel *et al.*, 1996b, and unpublished data). The ODV-E25 amino acid sequence contains an N-terminal hydrophobic domain similar to that observed for ODV-E66; however, the N-terminus is uncleaved as determined by N-terminal amino acid sequencing (Hong *et al.*, 1997).

3. ODV-E56, ODVP-6E (AcMNPV orf148, BmNPV orf124, OpMNPV orf146)

The AcMNPV gene encoding ODV-E56 was identified from three independent λgt11 expression library clones that were screened using antisera developed to ODV or BV envelope proteins (Braunagel *et al.*, 1996a), while the OpMNPV gene was identified when the *ie-1* gene and adjoining regions were sequenced (Theilmann *et al.*, 1996). In both OpMNPV and AcMNPV, transcription of the gene encoding ODV-E56 initiates from a conserved late promoter TAAG sequence, with transcripts detected between 16 and 72 hr p.i. The protein is detected in AcMNPV-infected Sf9 cell extracts from 36 through 72 hr p.i., and is still present in OpMNPV-infected *Lymantria dispar* cell extracts at 96 hr p.i. One of the most distinguishing features of ODV-E56 is that it is detected at a variety of molecular weights when various baculoviruses are analyzed, and even within the same baculovirus it appears as multiple forms (Braunagel *et al.*, 1996a; Theilmann *et al.*, 1996). Although the molecular weight of the protein is different between baculoviruses, each baculovirus shows one predominant form in purified ODV. ODV-E56 from AcMNPV-infected cells is not N-glycosylated (Braunagel *et al.*, 1996a), and the source of the protein heterogeneity is unknown. The primary amino acid sequence predicted by the different baculovirus genes encoding ODV-E56 does not explain the apparent protein heterogeneity between baculoviruses. Homologues of the ODV-E56 have been sequenced from OpMNPV, *Choristoneura fumiferana* MNPV (CfMNPV), HzSNPV, and *Cydnia pomonella GV* (CpGV) and show 70, 69, 52, and 46% amino acid identity with AcMNPV, respectively, and comparisons of hydropathy profiles predict that the secondary structure of this protein would also be highly conserved (Theilmann *et al.*, 1996).

Immunoelectron microscopy showed that ODV-E56 associates with ODV envelopes (Braunagel *et al.*, 1996a; Theilmann *et al.*, 1996), virus-induced intranuclear microvesicles, and the outer and inner nuclear membrane (Braunagel *et al.*, 1996a). In cells infected with a recombinant AcMNPV in which the β-galactosidase gene was inserted at amino acid 151 of the ODV-E56 gene (E56-β-gal), the mutant E56 protein no longer localized to the viral-induced microvesicles or ODV envelope, but rather associated with nucleocapsids within the nucleus. Additionally, E56-β-gal does not associate with the elongated empty capsid structures as does vp39, the major capsid protein (Braunagel *et al.*, 1996a). This indicates that the manner of association or incorporation of E56-β-gal into nucleocapsids may differ from vp39 incorporation.

4. ODV-E18/35 (AcMNPV orf143/144, BmNPV orf119/120, OpMNPV 140/141)

ODV-E18 was identified by N-terminal sequencing of a protein of approximately 18 kDa, which was purified from ODV envelope preparations. The gene encoding this protein was located within the IE0 intron (Braunagel *et al.*, 1996b). Transcripts for ODV-E18 initiate from three conserved TAAG motifs, and transcripts are detected from 16 through 72 hr p.i. Antisera generated to a bacterial-expressed glutathione-*S*-transferase (GST)–E18 fusion protein detects a band of 18 kDa on Western blots of extracts from AcMNPV-infected Sf9 cells from 24 through 72 hr p.i., and in purified ODV envelope. A curious feature of the mobility of ODV-E18 is that the protein detected in purified ODV envelope has a diffuse banding pattern relative to the pattern visualized from infected cell extracts. The gene predicts a protein of 9.5 kDa, and when the protein is translated *in vitro,* only a 9.5-kDa protein is identified. This difference in size could be due to the anomalous migration on SDS-PAGE gels, the formation of an SDS-resistant dimer, or posttranslational modifications of ODV-E18. The ODV-E18 gene has been sequenced in AcMNPV, BmNPV, OpMNPV, and HzSNPV, and the AcMNPV protein shares 86, 68, and 52% amino acid identity to the predicted proteins respectively (Ahrens *et al.*, 1997; Braunagel *et al.*, 1996b; D. E. Tribe, unpublished data).

The nature of ODV-E35 is still unclear. The N-terminal amino acid sequence of ODV-E35 is identical to that of ODV-E18. ODV-E35 is detected in purified viral envelopes using antisera to ODV-E18 and antisera generated to an ORF immediately downstream that codes for ODV-EC27. Western blot analysis predicts that ODV-E35 is a structural protein of the ODV envelope. While the data are inconclusive at present, it is believed that ODV-E35 is translated from a transcript that includes both ORF 143 (ODV-E18) and 144 (ODV-EC27). Such a translation product would most likely be derived by ribosomal frameshifting. RNA structural analysis predicts a stem–loop secondary structure consistent with ribosomal frameshifting, and the stem–loop is associated with a cluster of "slippery sequences" that has been shown to promote frameshifting events at high levels (Braunagel *et al.*, 1996b).

Because antisera to ODV-E18 also recognize ODV-E35, immunoelectron microscopy shows that one or both of these proteins reside in the viral-induced microvesicles and ODV envelope. Unlike ODV-E66 and ODV-E56, however, ODV-E18/35 does not appear to associate with the nuclear envelope or cytoplasmic membranes.

5. ODV-EC27 (AcMNPV orf144, BmNPV orf120, OpMNPV orf141)

ODV-EC27 is the first structural protein of AcMNPV that appears to be present in both ODV nucleocapsids and envelopes. Additionally, it is the only nucleocapsid structural protein that appears to be specific for ODV and is not present in BV (Braunagel *et al.*, 1996b). ODV-EC27 is coded from ORF 144 and transcription initiates from a conserved TAAG motif. Transcripts are detected from 16 through 72 hr p.i., with protein detected in AcMNPV-infected cell extracts from 24 through 72 hr p.i. Antisera to ODV-EC27 also detect ODV-E35, and thus localization by immunoelectron microscopy is difficult; however, Western blot analysis detects ODV-EC27 in both the nucleocapsid and ODV envelope. The predicted AcMNPV ODV-EC27 protein demonstrates 95, 67, and 53% amino acid sequence identity with BmNPV, OpMNPV, and HzSNPV, respectively (Ahrens *et al.*, 1997; Braunagel *et al.*, 1996b; D. E. Tribe, unpublished data).

6. p74 (AcMNPV orf138; BmNPV orf115, OpMNPV orf134)

The p74 gene is located downstream and in the opposite orientation to the p10 gene in the AcMNPV (Kuzio *et al.*, 1989), OpMNPV (Leisy *et al.*, 1986), and CfMNPV (Hill *et al.*, 1993) genomes, and AcMNPV p74 is 77–79% identical to the OpMNPV and CfMNPV homologues. p74 is not glycosylated (Roberts, 1989) and it is expressed at low levels late in infection, even though a conventional late promoter sequence is not found near the mRNA start sites. Mutant virus with a partial p74 deletion are capable of replicating in cultured cells; however, the resulting occluded virus are noninfectious when fed to insects (Kuzio *et al.*, 1989). This indicates that p74 is required for infectivity when ingested by insects and also suggests that the protein must be associated with ODV. It has been suggested that this protein is associated with ODV envelopes (Hill *et al.*, 1993).

VII. BUDDED VIRUS

A. BV Entry into Cells

BVs appear to enter cells predominantly by endocytosis, since their infectivity can be neutralized by chloroquine and ammonium chloride, which buffer endosomal pH (Vòlkman and Goldsmith, 1985). It is believed that the process of endocytosis involves interaction of the virion with a host receptor, which leads to the invagination of the plasma membrane and formation of an endocytic vesicle containing the enveloped virion. The endosome is then acidified, which activates fusion of the viral and endosomal membrane, thereby releasing the nucleocapsid into the cytoplasm.

B. Proteins of the Budded Virus Envelope

1. gp64; Envelope Fusion Protein (AcMNPV orf128; BmNPV 105, OpMNPV orf126)

To date, only one viral-encoded protein has been found to be associated with the BV envelope. This protein is called gp64 (gp64 envelope fusion protein, or gp67), and most studies have focused on the gp64 proteins from AcMNPV and OpMNPV. Gp64 is encoded by an ORF predicted to express a protein of about 59 kDa (Blissard and Rohrmann, 1989; Whitford *et al.*, 1989), but the mature protein is heavily glycosylated and acylated (Roberts, 1989). The gp64 gene is regulated by both early and late promoters, and this combination of early plus late expression appears to be conserved among several baculoviruses (Blissard and Rohrmann 1989; Hill and Faulkner, 1994; Whitford *et al.*, 1989). The gp64 early promoter has been used as a model system for studies of early transcriptional regulation (Blissard *et al.*, 1992; Blissard and Rohrmann, 1991; Kogan *et al.*, 1995; Kogan and Blissard, 1994). In addition, the gp64 protein was also recently engineered for biotechnological applications. Chimeric membrane proteins containing the AcMNPV gp64 protein can be generated, and it is suggested that such constructs may be useful for displaying eukaryotic membrane proteins on the surface of baculovirus virions (Boublik *et al.*, 1995).

Immunologic and biochemical studies show that gp64 accumulates at the plasma membrane during the early and late phases of infection (Blissard and Rohrmann, 1989; Monsma *et al.*, 1996; Monsma and Blissard, 1995; Volkman, 1986). Nucleocapsids become enveloped within the gp64-modified plasma membrane by budding from the cell surface in the late phase. Electron micrographs indicate that in budded virions, nucleocapsids are surrounded by a loosely associated lipid-containing membrane. Spikelike structures (called peplomers) that contain gp64 are concentrated at one end of the virion (Volkman, 1986; Volkman and Goldsmith, 1984). Recent studies have shown that gp64 is a membrane fusion protein that is activated at low pH and involved in viral entry through the endocytic pathway (Blissard and Wenz, 1992; Monsma and Blissard, 1995). In addition, genetic studies demonstrated that gp64 is an essential structural protein, required for cell-to-cell propagation of the virus in cell culture and in the insect hemocoel (Monsma *et al.*, 1996).

The gp64 protein contains an N-terminal signal peptide and a C-terminal anchor domain. In the *Baculoviridae,* the gp64 protein appears to be highly conserved, with amino acid identities of 78–82% among AcMNPV, OpMNPV, and CfMNPV (Blissard and Rohrmann, 1989; Hill and Faulkner 1994). Recent studies of the kinetics of OpMNPV gp64 oligomerization and processing suggest that of the gp64 protein synthesized, only approximately 25% is oligomerized and transported to the cell surface, with the remainder apparently degraded as misfolded protein (Oomens *et al.*, 1995). As with other proteins that enter the secretory pathway, the extensive glycosylation of gp64 may play a role in folding and subsequent transport. In addition to glycosylation, palmitic acid is esterlinked to the AcMNPV gp64 peptide, making it the major acylated protein associated with BV (Roberts and Faulkner, 1989). Although its precise role in gp64 is unknown, it is believed that acylation of proteins may be involved in anchoring to membranes, membrane mobility membrane fusion, or regulation of intracellular transport.

The mature gp64 peplomer is found in the virion envelope as a disulfide-linked oligomer. Based on their migration in nonreducing SDS-PAGE gels, gp64 oligomers were initially thought to consist of a combination of tetramers and trimers. However, using a purified soluble form of the OpMNPV gp64 protein, Oomens *et al.*, (1995) detected only trimers by mass spectrometry. This indicated that the mature oligomer of gp64 is trimeric, and suggested that some heterogeneity may occur among trimers. Kinetic studies of gp64 synthesis and trimerization indicate that oligomerization to the trimeric form occurs rapidly and is complete or nearly complete within 15 min after synthesis (Oomens *et al.*, 1995). Sequence analysis suggests that the gp64 protein contains a centrally located amphipathic alpha-helix, or leucine zipper, a structure that is frequently associated with dimerization or oligomerization domains of proteins. Substitution mutations that reduce the hydrophobic character of one face of the predicted amphipathic helix resulted in inefficient or defective oligomerization. Similar effects were observed when helix-breaking amino acid substitutions were introduced into the predicted helix (Monsma and Blissard, 1995). Thus, gp64 trimerization appears to be mediated by the leucine zipper (amphipathic alpha-helix) domain in the central portion of the molecule. Studies of gp64 in isolation demonstrated that gp64 expressed alone in insect cells was capable of mediating pH-activated membrane fusion (Blissard *et al.*, 1992). Mediation of membrane fusion by proteins is believed to involve the interaction of hydrophobic regions of membrane fusion proteins with the target lipid bilayer. A small hydrophobic domain of gp64, composed of a cluster of six hydrophobic amino acids, has been implicated in membrane fusion by site-directed mutagenesis (Monsma and Blissard, 1995). Substitutions that reduce the local hydrophobicity of that small domain result in gp64 trimers defective for membrane fusion, while substitutions with hydrophobic amino acids had little or no effect. In addition, an antipeptide antiserum generated against this domain inhibits membrane fusion in syncytium formation assays (Monsma and Blissard, 1995). Fusion of virion cell membranes is a multistep reaction involving the triggering of a possible conformational change in the fusion protein. Triggering may result from reduction in pH or binding to a receptor(s). Models of membrane fusion suggest that the initial triggering of the fusion protein leads to the merger of the outer leaflet of the two bilayers.

This is followed by merger of the inner leaflets and the formation of a fusion pore between the two membranes. Recent studies of gp64-mediated membrane fusion have demonstrated that the triggering event can be separated from the initial merger of membranes (Chernomordik *et al.*, 1995).

The baculovirus gp64 protein has remarkable sequence similarity with the envelope glycoproteins from two orthomyxoviruslike arboviruses (Morse *et al.*, 1992) that are vectored by ticks. In these viruses the gp64 homologue appears to mediate both membrane fusion and hemagglutination, suggesting that like the influenza virus hemagglutinin protein, these gp64 homologues may be required for both host receptor binding and membrane fusion. The provocative similarity between the envelope proteins of these unrelated viruses poses fascinating questions regarding the evolutionary origins or transfer of viral genes.

C. Other Proteins Associated with Budded Virus

A number of proteins are found associated with budded virions; however, it is unclear whether they are randomly incorporated during the budding process or have a specific role in virion structure or function. In one instance, antiserum was generated against sucrose-gradient-purified OpMNPV BV grown in *L. dispar* cells and then used to examine uninfected *L. dispar* cells by immunofluorescence microscopy. A very strong fluorescence response was observed, suggesting that the BVs were associated with substantial amounts of cellular proteins (G. F. Rohrmann, unpublished data). IE-1 has been found associated with OpMNPV BV (Theilmann and Stewart, 1993). AcMNPV was found to contain a mixture of viral- and host-encoded ubiquitin on the inner surface of the BV envelope. It appeared to be attached by a novel type of phospholipid anchor (Guarino *et al.*, 1995), but it was not clear if it played a role in BV structure or function. Cellular actin may also associate with BV, since it was observed that BV nucleocapsids have actin-binding activity (Lanier *et al.*, 1996). A cathepsin-L-like cysteine proteinase is encoded by AcMNPV. Whereas the internal organs of larvae infected with wild-type virus liquefy during the final stages of infection, mutants lacking the cathepsinlike gene retained the integrity of their tissues during the final stages of infection (Slack *et al.*, 1995). Therefore, it has been suggested that this proteinase is involved in the release of the virus from the insect host. Subsequently, it was shown that wild-type BV were capable of degrading actin, whereas mutant virus lacking the viral cathepsinlike gene did not degrade actin (Lanier *et al.*, 1996). These data indicate that the viral cathepsin is present in association with BV.

VIII. CONCLUSIONS

Considerable progress has been achieved in the identification of baculovirus genes that are likely to encode virion structural proteins. In particular, the recent identification of proteins associated with the ODV envelope provide a foundation for the elucidation of the contribution of these proteins to the organization and function of this novel structure. These studies will also likely shed light on the role of the ODV envelope in polyhedron morphogenesis and infectivity. Proteins associated with the end of virions that interact with the virogenic stroma and nuclear membranes have not been identified.

Although gp64 is the most well-characterized baculovirus structural protein, the makeup of the BV envelope is still unclear. In particular, the role that host-derived molecules play in this structure is unknown. Future investigations on the capsid proteins will likely establish how they interact with one another and other virion components in producing the complete structure. Understanding the role of these proteins and their contribution to virion architecture is essential for the eventual understanding of the role of the virion in the infection cycle.

ACKNOWLEDGMENTS. We thank Gary Blissard for reviewing portions of this text and for allowing us to use Fig. 1. This review was supported by USDA grant 94-37302-0729 to G.R. and PHS postdoctoral fellowship F2AI09049A to C.J.F., NIH grant 2R01 GM47552 to S.C.B., and Texas Higher Education Board ARP/ATP #9902-011 to S.C.B.

REFERENCES

Ahrens, C. H., Russell, R., Funk, C. J., Evans, J. T., Harwood, S. H., and Rohrmann, G. F., 1997, The sequence of the *Orgyia pseudotsugata* multinucleocapsid nuclear polyhedrosis virus genome. *Virology* **229:**381.

Ayres, M. D., Howard, S. C., Kuzio, J., Lopez-Ferber, M., and Possee, R. D., 1994, The complete DNA sequence of *Autographa californica* nuclear polyhedrosis virus, *Virology* **202:**586.

Balhorn, R., 1982, A model for the structure of chromatin in mammalian sperm, *J. Cell Biol.* **93:**298.

Bjornson, R. M., and Rohrmann, G. F., 1992a, Nucleotide sequence of the vp39-capsid gene region of the *Lymantria dispar* nuclear polyhedrosis virus, *J. Gen. Virol.* **73:**1505.

Bjornson, R. M., and Rohrmann, G. F., 1992b, Nucleotide sequence of the polyhedron envelope protein gene region of the *Lymantria dispar* nuclear polyhedrosis virus, *J. Gen. Virol.* **73:**1499. (Author's correction, 1994, *J. Gen. Virol.* **75:**3687.)

Blissard, G. W., 1996, Baculovirus–insect cell interactions, *Cytotechnology,* **20:**73.

Blissard, G. W., Kogan, P. H., Wei, R., and Rohrmann, G. F., 1992, A synthetic early promoter from a baculovirus: Roles of the TATA box and conserved mRNA start site CAGT sequence in basal levels of transcription, *Virology* **190:**783.

Blissard, G. W., Quant-Russell, R. L., Rohrmann, G. F., and Beaudreau, G. S., 1989, Nucleotide sequence, transcriptional mapping, and temporal expression of the gene encoding vp39, a major structural protein of the multicapsid nuclear polyhedrosis virus of *Orgyia pseudotsugata, Virology* **168:**354.

Blissard, G. W., and Rohrmann, G. F., 1989, Location, sequence, transcriptional mapping, and temporal expression of the gp64 envelope glycoprotein gene of the *Orgyia pseudotsugata* multicapsid nuclear polyhedrosis virus, *Virology* **170:**537.

Blissard, G. W., and Rohrmann, G. F., 1991, Baculovirus gp64 gene expression: Analysis of sequences modulating immediate early transcription and transactivation by IE1, *J. Virol.* **65:**5820.

Blissard, G. W., and Wenz, J. R., 1992, Baculovirus gp64 envelope glycoprotein is sufficient to mediate pH-dependent membrane fusion, *J. Virol.* **66:**6829.

Boublik, Y., Di-Bonito, P., and Jones, I. M. J., 1995, Eukaryotic virus display: Engineering the major surface glycoprotein of the *Autographa californica* nuclear polyhedrosis virus (AcNPV) for the presentation of foreign proteins on the virus surface, *Bio-Technology* **13:**1079.

Braunagel, S. C., Elton, D. M., Ma, H., and Summers, M. D., 1996a, Identification and analysis of an *Autographa californica* nuclear polyhedrosis virus structural protein of the occlusion-derived virus envelope: ODV-E56, *Virology* **217:**97.

Braunagel, S. C., He, H., Ramamurthy, P., and Summers, M. D., 1996b, Transcription, translation, and cellular localization of three Autographa californica nuclear polyhedrosis virus structural proteins: ODV-E18, ODV-E35 and ODV-EC27, *Virology* **222:**100.

Braunagel, S. C., and Summers, M. D., 1994, *Autographa californica* nuclear polyhedrosis virus, PDV, and ECV viral envelopes and nucleopcapsids: Structural proteins, antigens, lipid and fatty acid profiles, *Virology* **202:**315.

Burand, J. P., Summers, M. D., and Smith, G. E., 1980, Transfection with baculovirus DNA, *Virology* **101:**286.

Carstens, E. B., Tjia, S. T., and Doerfler, W., 1980, Infectious DNA from *Autographa californica* nuclear polyhedrosis virus, *Virology* **101:**311.

Cheley, S., Kosik, K. S., Paskevich, P., Bakalis, S., and Bayley, H., 1992, Phosphorylated baculovirus p10 is a heat-stable microtubule-associated protein associated with process formation in Sf9 cells, *J. Cell Sci.* **102:**739.

Chernomordik, L., Leikina, E., Cho, M.-S., and Zimmerberg, J., 1995, Control of baculovirus gp64-induced syncytium formation by membrane lipid composition, *J. Virol.* **69:**3049.

Chou, C.-M., Huang, C.-J., Lo, C.-F., Kuo, G.-H., and Wang, C.-H., 1996, Characterization of *Perina nuda* nucleopolyhedrovirus (PenuNPV) polyhedrin gene, *J. Invertebr. Pathol.* **67:**256.

Cowan, P., Bulach, D., Goodge, K., Robertson, A., and Tribe, D. E., 1994, Nucleotide sequence of the polyhedrin gene region of *Helicoverpa zea* single nucleocapsid nuclear polyhedrosis virus: Placement of the virus in lepidopteran nuclear polyhedrosis virus group II, *J. Gen. Virol.* **75:**3211.

Derksen, A. C. G., and Granados, R. R., 1988, Alteration of a lepidopteran peritrophic membrane by baculoviruses and enhancement of viral infectivity, *Virology* **167:**242.

Fraser, M., J., 1986, Ultrastructural observations of virion maturation in *Autographa californica* nuclear polyhedrosis virus infected *Spodoptera frugiperda* cell cultures, *J. Ultrastruct. Mol. Struct. Res.* **95:**195.

Funk, C. J., and Consigli, R. A., 1993, Phosphate cycling on the basic protein of *Plodia interpunctella* granulosis virus, **193:**396.

Gombart, A. F., Blissard, G. W., and Rohrmann, G. F., 1989a, Characterization of the geneticorganization of the HindIII-M region of the multicapsid nuclear polyhedrosis virus of *Orgyia pseudotsugata* reveals major differences among baculoviruses, *J. Gen. Virol.* **70:**1815.

Gombart, A. F., Pearson, M. N., Rohrmann, G. F., and Beaudreau, G. S., 1989b, A baculovirus polyhedral envelope-associated protein: Genetic location, nucleotide sequence, and immunocytochemical characterization, *Virology* **169:**182.

Gross, C. H., Russell, R. L. Q., and Rohrmann, G. F., 1994, The *Orgyia pseudotsugata* baculovirus p10 and polyhedron envelope protein genes: Analysis of their relative expression levels and role in polyhedron structure, *J. Gen. Virol.* **75:**1115.

Guarino, L. A., Smith, G., and Dong, W., 1995, Ubiquitin is attached to membranes of baculovirus particles by a novel type of phospholipid anchor, *Cell* **80:**301.

Hashimoto, Y., Cosaro, B. G., and Granados, R. R., 1991, Location and nucleotide sequence of the gene encoding the viral enhancing factor of the *Trichoplusia ni* granulosis virus, *J. Gen. Virol.* **72:**2645.

Hill, J. E., and Faulkner, P., 1994, Identification of the gp67 gene of a baculovirus pathogenic to the spruce budworm, *Choristoneura fumiferana* multinucleocapsid nuclear polyhedrosis virus, *J. Gen. Virol.* **75:**1811.

Hill, J. E., Kuzio, J., Wilson, J. A., MacKinnon, E. A., and Faulkner, P., 1993, Nucleotide sequence of the p74 gene of a baculovirus pathogenic to the spruce budworm, *Choristoneura fumiferana* multicapsid nuclear polyhedrosis virus, *Biochim. Biophys, Acta* **1172:**187.

Hong, T., Braunagel, S. C., and Summers, M. D., 1994, Transcription, translation, and cellular localization of PDV-E66: A structural protein of the PDV envelope of *Autographa californica* nuclear polyhedrosis virus, *Virology* **204:**210.

Hong, T., Summers, M. D., and Braunagel, S. C., 1997, N-terminal sequences from *Autographa californica* nuclear polyhedrosis virus envelope proteins ODV-E66 and OV-E25 are sufficient to direct reporter proteins to the nuclear envelope, intranuclear microvesicles and the envelope of occlusion derived virus. *Proc. Natl. Acad. Sci. USA* **94:**4050.

Hooft van Iddekinge, B. J. L., Smith, G. E., and Summers, M. D., 1983, Nucleotide sequence of the polyhedrin gene of *Autographa californica* nuclear polyhedrosis virus. *Virology* **131:**561.

Jehle, J. A., and Backhaus, H., 1994, Genome organization of the DNA-binding protein gene region of *Cryptophlebia leucotreta* granulosis virus is closely related to that of nuclear polyhedrosis viruses, *J. Gen. Virol.* **75:**1815.

Kelly, D. C., and Wang, X., 1981, The infectivity of nuclear polyhedrosis virus DNA, *Ann. Virol.* **132E:**247.

Kitts, P. A., and Possee, R. D., 1993, A method for producing recombinant baculovirus expression vectors at high frequency, *BioTechniques* **14:**810.

Kogan, P. H., and Blissard, G. W., 1994, A baculovirus gp64 early promoter is activated by host transcription factor binding to CACGTG and GATA elements, *J. Virol.* **68:**813.

Kogan, P. H., Chen, X., and Blissard, G. W., 1995, Overlapping TATA-dependent and TATA-independent early promoter activities in the baculovirus gp64 envelope fusion protein gene, *J. Virol.* **69:**1452.

Kozlov, E. A., Sidorova, N. M., and Serebryani, S. B., 1975, Proteolytic cleavage of polyhedral protein during dissolution of inclusion bodies of the nuclear polyhedrosis viruses of *Bombyx mori* and *Galleria mellonella* under alkaline condition, *J. Invertebr. Pathol.* **25:**97.

Kozlov, E. A., Levitina, T. L., Gusak, N. M., and Serebryani, S. B., 1981, Comparison of the amino acid sequences of inclusions body proteins of *Bombyx mori, Porthetria dispar,* and *Galleria mellonella, Bioorg. Khim.* **7:**1008.

Kuzio, J., Jaques, R., and Faulkner, P., 1989, Identification of p74, a gene essential for virulence of baculovirus occlusion bodies, *Virology* **173:**759.

Lanier, L. M., Slack, J. M., and Volkman, L. E., 1996, Actin binding and proteolysis by the baculovirus AcMNPV: The role of virion-associated V-cath, *Virology* **216:**380.

Leisy, D. J., Rohrmann, G. F., Nesson, M., and Beaudreau, G. S., 1986, Nucleotide sequence and transcriptional mapping of the p10 gene region of the *Orgyia pseudotsugata* multicapsid nuclear polyhedrosis virus, *Virology* **153:**157.

Li, Y., and Miller, L. K., 1995a, Expression and localization of a baculovirus protein phosphatase, *J. Gen. Virol.* **76:**2941.

Li, Y., and Miller, L. K., 1995b, Properties of a baculovirus mutant defective in the protein phosphatase gene, *J. Virol.* **69:**4533.

Lu, A., and Carstens, E. B., 1992, Nucleotide sequence and transcriptional analysis of the p80 gene of *Autographa californica* nuclear polyhedrosis virus: A homologue of the *Orgyia pseudotsugata* nuclear polyhedrosis virus capsid-associated gene, *Virology* **190:**201.

Maeda, S., Kamita, S. G., and Kataoka, H., 1991, The basic DNA-binding protein of *Bombyx mori* nuclear polyhedrosis virus: The existence of an additional arginine repeat, *Virology* **180:**807.

McCarthy, W. J., and DiCapua, R. A., 1979, Characterization of solubilized proteins from tissue culture and host-derived nuclear polyhedra of *Lymantria dispar* and *Autographa californica, Intervirology* **11:**174.

Miller, L. K., Adang, M. J., and Browne, D., 1983, Protein kinase activity associated with the extracellular and occluded forms of the baculovirus *Autographa californica* NPV, *J. Virol.* **46:**275.

Minion, F. C., Coons, L. B., and Broome, J. R., 1979, Characterization of the polyhedral envelope of the nuclear polyhedrosis virus of *Heliothis virescens, J. Invertebr. Pathol.* **34:**303.

Monsma, S. A., and Blissard, G. W., 1995, Identification of a membrane fusion domain and an oligomerization domain in the baculovirus GP64 envelope fusion protein, *J. Virol.* **69:**2583.

Monsma, S. A., Oomens, A. G. P., and Blissard, G. W., 1996, The gp64 envelope fusion protein is an essential baculovirus protein required for cell-to-cell transmission of infection, *J. Virol.* **70:**4607.

Morse, M. A., Marriott, A. C., and Nuttall, P. A., 1992, The glycoprotein of Thogoto virus (a tick-borne orthomyxo-like virus) is related to the baculovirus glycoprotein gp64, *Virology* **186:**640.

Müller, R., Pearson, M. N., Russell, R. L. Q., and Rohrmann, G. F., 1990, A capsid-associated protein of the multicapsid nuclear polyhedrosis virus of *Orgyia pseudotsugata:* Genetic location, sequence, transcriptional mapping, and immunocytochemical characterization, *Virology* **176:**133.

Nakano, M., Kasai, K., Yoshida, K., Tanimoto, T., Tamaki, Y., and Tobita, T., 1989, Conformation of the fowl protamine, galline, and its binding properties to DNA, *J. Biochem.* **105:** 133.

Oellig, C., Happ, B., Muller, T., and Doerfler, W., 1987, Overlapping sets of viral RNAs reflect the array of polypeptides in the EcoR1 J and N Fragments (map positions 81.2 to 85.0) of the *Autographa californica* nuclear polyhedrosis virus, *J. Virol.* **61:**3048.

Oomens, A. G. P., Monsma, S. A., and Blissard, G. W., 1995, The baculovirus gp64 envelope fusion protein: Synthesis, oligomerization, and processing, *Virology* **209:**592.

Pearson, M. N., Russell, R. L. Q., Rohrmann, G. F., and Beaudreau, G. S., 1988, p39, a major baculovirus structural protein: Immunocytochemical characterization and genetic location, *Virology* **167:**407.

Potter, K. N., and Miller, L. K., 1980, Transfection of two invertebrate cell lines with DNA of *Autographa californica* nuclear polyhedrosis virus, *J. Invertebr. Pathol.* **36:**431.

Possee, R. D., Sun, T.-P., Howard, S. C., Ayres, M. D., Hill-Perkins, M., and Gearing, K. L., 1991, Nucleotide sequence of the *Autographa californica* nuclear polyhedrosis 9.4 kbp EcoRI-I and -R (polyhedrin gene) region, *Virology* **185:**229.

Quant-Russell, R. L., Pearson, M. N., Rohrmann, G. F., and Beaudreau, G. S., 1987, Characterization of baculovirus p10 synthesis using monoclonal antibodies, *Virology* **160:**9.

Roberts, T. E., and Faulkner, P., 1989, Fatty acid acylation of the 67K envelope glycoprotein of a baculovirus: *Autographa californica* nuclear polyhedrosis virus, *Virology* **172:**377.

Rohrmann, G. F., 1986, Polyhedrin structure, *J. Gen. Virol.* **67:**1499.

Rohrmann, G. F., 1992, Baculovirus structural proteins, *J. Gen. Virol.* **73:**749.

Rohrmann, G. F., Pearson, M. N., Bailey, T. J., Becker, R. R., and Beaudreau, G. S., 1981, N-terminal polyhedrin sequences and occluded baculovirus evolution, *J. Mol. Evol.* **17:**329.

Rubinstein, R., and Polson, A., 1983, Midgut and viral associated proteases of *Heliothis armigera, Intervirology* **19:**16.

Russell, R., Funk, C. J., and Rohrmann, G. F., 1997, Association of a baculovirus encoded protein with the capsid basal region, *Virology* **227:**142.

Russell, R. L. Q., Pearson, M. N., and Rohrmann, G. F., 1991, Immunoelectron microscopic examination of *Orgyia pseudotsugata* multicapsid nuclear polyhedrosis virus-infected *Lymantria dispar* cells: Time course and localization of major polyhedron-associated proteins, *J. Gen. Virol.* **72:**275.

Russell, R. L. Q., and Rohrmann, G. F., 1990a, The p6.5 gene region of a nuclear polyhedrosis virus of *Orgyia pseudotsugata:* DNA sequence and transcriptional analysis of four late genes, *J. Gen. Virol.* **71:**551.

Russell, R. L. Q., and Rohrmann, G. F., 1990b, A baculovirus polyhedron envelope protein: Immunogold localization in infected cells and mature polyhedra, *Virology* **174:**177.

Russell, R. L. Q., and Rohrmann, G. F., 1993, A 25 kilodalton protein is associated with the envelopes of occluded baculovirus virions, *Virology* **195:**532.

Schetter, C., Oellig, C,. and Doerfler, W., 1990, An insertion of insect cell DNA in the 81-map-unit segment of *Autographa californica* nuclear polyhedrosis virus DNA, *J. Virol.* **64:**1844.

Serebryani, S. B., Levitina, T. L., Kautsman, M. L., Radavski, Y. L., Gusak, N. M., Ovander, M. N., Sucharenko, N. V., and Kozlov, E. A., 1977, The primary structure of the polyhedral protein of nuclear polyhedrosis virus (NPV) of *Bombyx mori, J. Invertebr. Pathol.* **30:**442.

Slack, J. M., Kuzio, J., and Faulkner, P., 1995, Characterization of *v-cath,* a cathepsin-L-like proteinase expressed by the baculovirus *Authographa californica* multiple nuclear polyhedrosis virus, *J. Gen. Virol.* **76:**1091.

Stoltz, D. B., Pavan, C., and DaCunha, A. B., 1973, Nuclear polyhedrosis virus: A possible example of *de novo* intranuclear membrane morphogenesis, *J. Gen. Virol.* **19:**145.

Summers, M. D., 1977, Characterization of shrimp baculovirus. EPA Ecological Research Series November 1977, EPA-600/3-77-130, EPA, Gulf Breeze, FL.

Summers, M. D., and Arnott, H. J., 1969, Ultrastructural studies on inclusion formation and virus occlusion in nuclear polyhedrosis and granulosis virus-infected cells of *Trichoplusia ni, J. Ultrastruct. Res.* **28:**462.

Tanada, Y., and Hess, R., 1976, Development of a nuclear polyhedrosis virus in midgut cells and penetration of the virus into the hemocoel of the armyworm, *Pseudoletia unipuncta, J. Invertebr. Pathol.* **28:**67.

Theilmann, D. A., Chantler, J. K., Stewart, S., Flipsen, H. T. M., Vlak, J. M., and Crook, N. E., 1996, Characterization of a highly conserved baculovirus structural protein that is specific for occlusion-derived virions, *Virology* **218:**148.

Theilmann, D. A., and Stewart, S., 1993, Analysis of the *Orgyia pseudotsugata* multicapsid nuclear polyhedrosis virus *trans*-activators IE-1 and IE-2 using monoclonal antibodies, *J. Gen. Virol.* **74:**1819.

Thiem, S. M., and Miller, L. K., 1989, Identification, sequence, and transcriptional mapping of the major capsid protein gene of the baculovirus *Autographa californica* nuclear polyhedrosis virus, *J. Virol.* **63:**2008.

Tweeten, K. A., Bulla, L. A., and Consigli, R. A., 1980, Characterization of an extremely basic protein derived from granulosis virus nucleocapsid, *J. Virol.* **33:**866.

van der Wilk, F., van Lent, J. W. M., and Vlak, J. M., 1987, Immunogold detection of polyhedrin, p10 and virion antigens in *Autographa california* nuclear polyhedrosis virus-infected *Spodoptera frugiperda* cells, *J. Gen. Virol.* **68:**2615.

van Lent, J. W. M., Groenen, J. T. M., Klinge-Roode, E. C., Rohrmann, G. F., Zuidema, D., and Vlak, J. M., 1990, Localization of the 34kDa polyhedron envelope protein in *Spodoptera frugiperda* cells infected with *Autographa californica* nuclear polyhedrosis virus, *Arch. Virol.* **111:**103.

van Oers, M. M., Flipsen, J. T. M., Reusken, C. B. E. M., Sliwinsky, E. L., Goldbach, R. W., and Valk, J. M., 1993, Functional domains of the p10 protein of *Autographa californica* nuclear polyhedrosis virus, *J. Gen. Virol.* **74:**563.

Vialard, J. E., and Richardson, C. D., 1993, The 1,629-nucleotide open reading frame located downstream of the *Autographa californica* nuclear polyhedrosis virus polyhedrin gene encodes a nucleocapsid-associated phosphoprotein, *J. Virol.* **67:**5859.

Volkman, L. E., 1986, The 64K envelope protein of budded *Autographa californica* nuclear polyhedrosis virus, *Curr. Top. Microbiol. Immunol.* **131:**103.

Volkman, L. E., and Goldsmith, P. A., 1984, Budded *Autographa californica* NPV 64K protein: Further biochemical analysis and effects of postimmunoprecipitation sample preparation conditions, *Virology* **139:**295.

Volkman, L. E., and Goldsmith, P. A., 1985, Mechanism of neutralization of budded *Autographa californica* nuclear polyhedrosis virus by a monoclonal antibody: Inhibition of entry by adsorptive endocytosis, *Virology* **143:**185.

Whitford, M., and Faulkner, P., 1992a, Nucleotide sequence and transcriptional analysis of a gene encoding gp41, a structural glycoprotein of the baculovirus *Autographa californica* nuclear polyhedrosis virus, *J. Virol.* **66:**4763.

Whitford, M., and Faulkner, P., 1992b, A structural polypeptide of the baculovirus *Autographa californica* nuclear polyhedrosis virus contains *0*-linked *N*-acetylglucosamine, *J. Virol.* **66:**3324.

Whitford, M., Stewart, S., Kuzio, J., and Faulkner, P., 1989, Identification and sequence analysis of a gene encoding gp67 an abundant envelope glycoprotein of the baculovirus, *Autographa californica* nuclear polyhedrosis virus, *J. Virol.* **63:**1393.

Whitt, M. A., and Manning, J. S., 1988, A phosphorylated 34-kDa protein and a subpopulation of polyhedrin are thiol-linked to the carbohydrate layer surrounding a baculovirus occlusion body, *Virology* **163:**33.

Williams, G. V., Rohel, D. Z., Kuzio, J., and Faulkner, P., 1989, A cytopathological investigation of *Autograph californica* nuclear polyhedrosis virus p10 gene function using insertion/deletion mutants, *J. Gen. Virol.* **70:**187.

Wilson, J. A., Hill, J. E., Kuzio, J., and Faulkner, P., 1995, Characterization of the baculovirus *Choristoneura fumiferana* multicapsid nuclear polyhedrosis virus p10 gene indicates that the polypeptide contains a coiled-coil domain. *J. Gen. Virol.* **76:**2923.

Wilson, M. E., and Consigli, R. A., 1985a, Characterization of a protein kinase activity associated with purified capsids of the granulosis virus infection *Plodia interpunctella. Virology* **143:**516.

Wilson, M. E., and Consigli, R. A., 1985b, Functions of a protein kinase activity associated with purified capsids of the granulosis virus infecting *Plodia interpunctella. Virology* **143:**526.

Wilson, M. E., and Miller, L. K., 1986, Changes in the nucleoprotein complexes of a baculovirus DNA during infection. *Virology* **151:**315.

Wilson, M. E., Mainprize, T. H., Friesen, P. D., and Miller, L. K., 1987, Location, transcription, and sequence of a baculovirus gene encoding a small arginine-rich polypeptide. *J. Virol.* **61:**661.

Wolgamot, G. M., Gross, C. H., Russell, R. L. Q., and Rohrmann, G. F., 1993, Immunocytochemical characterization of p24, a baculovirus capsid-associated protein. *J. Gen. Virol.* **74:**103.

Wood, H. A., 1980, *Autographa californica* nuclear polyhedrosis virus-induced proteins in tissue culture. *Virology* **102:**21.

Yamafuji, K., Yoshihara, F., and Hirayama, K., 1958, Protease and deoxyribonuclease in viral polyhedral crystals. *Enzymologia* **19:**53.

Young, J. C., MacKinnon, E. A., and Faulkner, P., 1993, The architecture of the virogenic stroma in isolated nuclei of *Spodoptera frugiperda* cells in vitro infected by *Autographa californiica* nuclear polyhedrosis virus. *J. Struct. Biol.* **110:**141.

Zuidema, D., van Oers, M. M., van Strien, E. A., Caballero, P. C., Klok, E. -J., Goldbach, R. W., and Valk, J. M., 1993, Nucleotide sequence and transcriptional analysis of the p10 gene of *Spodoptera exigua* nuclear polyhedrosis virus. *J. Gen. Virol.* **74:**1017.

Zummer, M., and Faulkner, P., 1979, Absence of protease in baculovirus polyhedral bodies propagated *in vitro. J. Invert. Pathol.* **33:**383.

CHAPTER 3

Baculovirus Pathogenesis

BRIAN A. FEDERICI

I. INTRODUCTION

Humans have been aware of diseases caused by baculoviruses for over 2000 years. The earliest historical accounts originated with descriptions of silkworm "jaundice," a disease of *Bombyx mori* that we now know is caused by a nuclear polyhedrosis virus. In addition, people in various cultures throughout history have witnessed, without knowing their cause, spectacular epizootics and subsequent population declines brought about by nuclear polyhedrosis and granulosis viruses in caterpillars and sawfly larvae that feed in the forests and on field and vegetable crops (Benz, 1986). However, it was not until this century, and especially since the end of World War II, that the etiologic agents that cause these diseases were identified as a unique family of viruses, now known as the baculoviruses, which are largely restricted to insects. Moreover, although the literature from the first half of this century contains good descriptions of the diseases caused by nuclear polyhedrosis and granulosis viruses, it is only recently that we have begun to understand the progression of these diseases in their hosts and to identify the genes and gene products that underlie their various pathologies.

My purpose in this chapter is to review the different pathologies caused by the two major types of baculoviruses—nuclear polyhedrosis viruses (NPVs) and granulosis viruses (GVs)—and show how they invade, develop, and spread throughout their insect hosts, eventually resulting in host death. I will begin with an overview of the different types of NPVs and GVs, their phylogenetic distribution, and their tissue tropisms, as there are interesting

---

BRIAN A. FEDERICI • Department of Entomology and Interdepartmental Graduate Program in Genetics, University of California at Riverside, Riverside, California 92521.

The Baculoviruses, edited by Lois K. Miller. Plenum Press, New York, 1997.

and evolutionarily important correlations among these. I will then discuss representative viruses and the disease that each causes, relating the gross pathology observed in the whole insect with events that occur in infected cells and tissues, identifying the causative genes and gene products where known.

The best-studied baculovirus is the multinucleocapsid NPV of the alfalfa looper, *Autographa californica,* the so-called AcMNPV, which attacks most tissues in its lepidopteran hosts, replicating particularly well in the fat body, tracheal matrix, and epidermis. Owing to our considerable knowledge of the AcMNPV, the discussion of baculovirus pathogenesis will initially focus on this virus. Our understanding of AcMNPV pathogenesis will then serve as the basis for a discussion of the diseases caused by other NPVs of Lepidoptera and the GVs. Members of the latter group also typically attack two or more tissues, but are only known from lepidopteran hosts. Finally, NPVs attacking insects of other orders, such as those of sawflies (Hymenoptera) and mosquitoes (Diptera), where replication is generally restricted to the midgut epithelium, will be discussed.

A. General Properties of Nuclear Polyhedrosis Viruses

The NPVs are large dsDNA viruses in which rod-shaped nucleocapsids are enveloped singly (SNPV type) or in groups (MNPV type) in a membranous envelope and subsequently are occluded in large protein crystals known as polyhedra (see Chapter 2, this volume). These typically range from 1 to 5 μm in diameter and form in the nuclei of infected cells.

NPVs are the most common and widely distributed baculovirus phylogenetically. They have been reported from more than 400 insect species distributed among seven insect orders (Volkman *et al.*, 1995). They occur by far most commonly among species of Lepidoptera. For this order alone, several hundred species have been reported as hosts, and with more than 200,000 species of lepidopterans described, it is likely there are at least several thousand species of NPVs that attack members of this order. Though much less common, NPVs have also been reported from members of the Hymenoptera, Diptera, Coleoptera, Thysanura, and Trichoptera (Table 1). Several NPVs are also known from shrimp (class Crustacea, order Decapoda).

An interesting aspect of NPV tissue tropism is that, with the important exception of the viruses attacking lepidopterous insects and a single disease of the cranefly *Tipula paludosa,* virus replication in all these arthropod groups is restricted to the midgut epithelium (Federici, 1993). Thus, in mosquito, sawfly, and caddisfly larvae, their NPVs invade and produce occlusion bodies only in midgut epithelial nuclei. In the lepidopterans, the virus establishes a transient infection in the midgut, without producing occlusion bodies, and then invades the other tissues.

TABLE I. Tissue Tropisms of Baculoviruses

Virus type	Major tissues[a]				
	Midgut epithelium[b]	Hemocytes	Fat body	Epidermis	Tracheal matrix
Nuclear polyhedrosis viruses					
Lepidoptera (caterpillars)	*	+	+	+	+
Hymenoptera (sawflies)	+	−	−	−	−
Diptera (flies)	+	−	−	−	−
Type 1 (*Tipula paludosa*)	+	+	−	−	−
Type 2 (*Aedes triseriatus*)	+	−	−	−	−
Thysanura (silverfish)	+	−	−	−	−
Trichoptera (caddisflies)	+	−	−	−	−
Crustacea (shrimp)	+	−	−	−	−
Granulosis viruses					
Lepidoptera					
Type 1 (*T. ni* GV)	*	−	+	−	−
Type 2 (*C. pomonella* GV)	*	+	+	+	+
Type 3 (*H. brillians* GV)	+	−	−	−	−

[a]Only the major insect tissues are listed here. However, viruses that infect the fat body, epidermis, and tracheal matrix typically have a very broad tissue tropism, and thus will be found replicating in most other tissues including the hemocytes, muscle, nerve, malpighian tubules, and reproductive and glandular tissues, though the amount of virus produced in these tissues is usually very limited.

[b]Viral reproductive strategies vary the most with respect to the extent of replication and pathogenesis in the midgut epithelium. For the nuclear polyhedrosis viruses (NPVs) those of lepidopterous insects undergo an initial colonizing phase of replication in the midgut, but typically form few if any occlusion bodies, using this tissue as a "beach head" prior to invading most other tissues in the body. In these tissues replication leads to the formation of occlusion bodies. This pattern of infection, with a transient phase in the midgut epithelium, is indicated by an asterisk. In the NPVs attacking insects of other orders, with the exception of the NPV of the cranefly, *Tipula paludosa,* the midgut epithelium is the only site of replication, and this is where the occlusion bodies form. Granulosis viruses (GVs) are only known from insects of the order Lepidoptera. With the exception of the GV of the Western grapeleaf skeleonizer, *Harrisina brillians,* these viruses also have only a transient phase of replication in the midgut epithelium, after which they invade other tissues and produce occlusion bodies. In the GV of *H. brillians,* the infection is restricted to the midgut epithelium, where replication and occlusion body formation occurs. The NPVs of shrimp infect the hepatopancreas, a digestive organ comparable to the midgut of insects.

B. General Properties of Granulosis Viruses

The granulosis viruses (GVs) produce virions biochemically and structurally similar to those of NPVs, but the virions are occluded individually in small occlusion bodies referred to as granules. These granules average about 150 nm in diameter by 400–600 nm in length, the length depending on the length of the occluded virion, which can vary among different viruses (Federici, 1986; Huger *et al.*, 1963; Tanada and Hess, 1991; Tweeten *et al.*, 1981). GVs have only been reported from members of the order Lepidoptera, and presently slightly more than 100 lepidopteran species are known to be attacked by this virus type (Volkman *et al.*, 1995). In all but one of the known GVs, the reproductive strategy is the same as that of the lepidopteran NPVs. The virus establishes a transient infection in the midgut epithelium, and then, depending on the GV type, either the fat body or the fat body and other major tissues are invaded, where the virus replicates first in the nucleus and then in the cytoplasm after rupture of the nuclear envelope (Table 1). The exception to this typical strategy is the GV of the Western grapeleaf skeletonizer, *Harrisina brillians*. This virus, like the NPVs of mosquitoes and sawflies, replicates and produces occlusion bodies only in the midgut epithelium, though the cytopathology in this tissue is similar to that which occurs in the fat body and other tissues of the typical GV.

II. BACULOVIRUSES OF LEPIDOPTERAN INSECTS

A. Nuclear Polyhedrosis Viruses

1. Gross Pathology

Baculovirus diseases are primarily diseases of the larval stages, and the progression and signs of disease depend on several factors including the instar in which NPV infection becomes apparent, infective dose, nutrition, temperature, degree of compatibility of the virus with its host, and the physical characteristics of the larva. When infected by ingestion of several hundred to several thousand polyhedra during the first few instars (1–3), death can occur within 24 to 72 hr, with few gross signs of disease before death because of the small size of the larvae. However, if infected with the same amount of virus during the fourth or early fifth instar, the disease generally runs its course over a period of 5–10 days, at temperatures from 25–30°C. Death results at the end of this period, followed by liquefaction of the larval carcass.

In typical NPV infections, such as the diseases caused by the NPVs of *A. californica* (AcMNPV), *Trichoplusia ni* (TnSNPV), and *Helicoverpa zea* (HzMNPV) in noctuid larvae, there are few gross signs of disease during the first 3 days of infection. Then, at about the 4th day, infected larvae begin to respond much more slowly than healthy larvae to tactile stimuli such as touching. Their feeding also begins to slow, virtually ceasing by day 6 or 7 (Aizawa, 1963; Benz, 1963; Tanada and Kaya, 1993). At day 4 or 5, the larva will begin to appear swollen and the cuticle can appear glossy, with small melanotic spots occurring in some species. In species in which the larval cuticle is translucent or lightly pigmented, such as *T. ni*, the larva will take on a white to yellow, creamy coloration (Fig. 1). This is due to the presence of polyhedra (which are white) accumulating in epidermal and fat body nuclei. In larvae with a translucent cuticle, such as the avocado leafroller, *Amorbia cuneana*, swollen nuclei packed with polyhedra can be observed as white spherules at 25× with the aid of a dissecting microscope. The hemolymph of infected larvae at this stage is also cloudy owing to the circulation of large numbers of infected hemocytes and polyhedra released into the

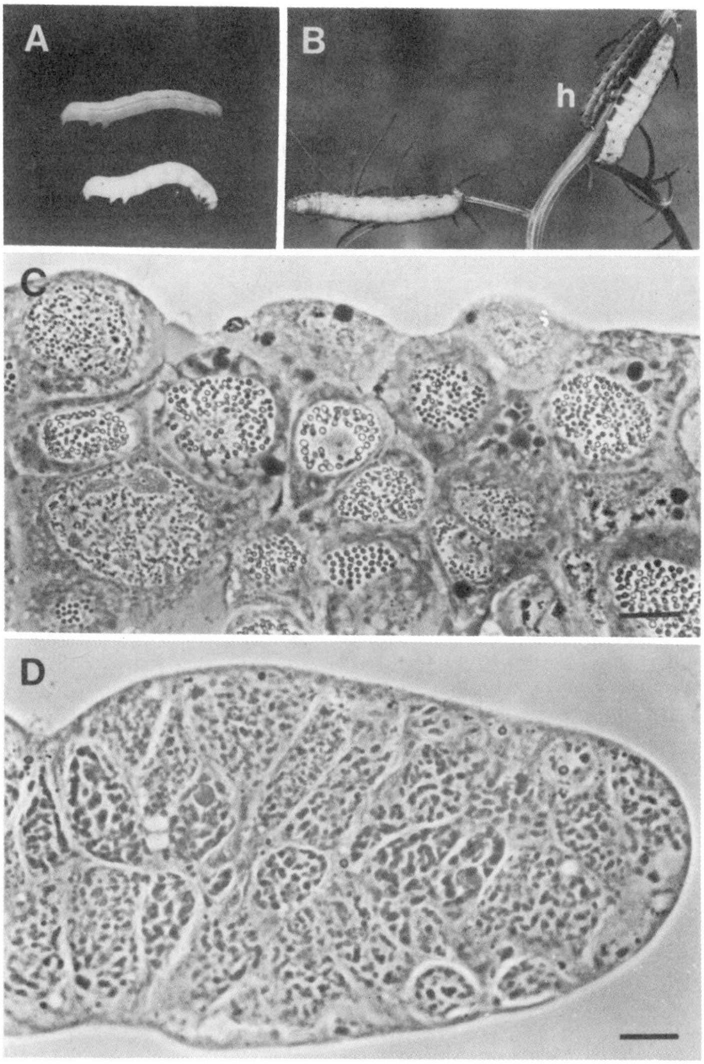

FIGURE 1. General characteristics of the pathology caused by nuclear polyhedrosis viruses (NPVs) and granulosis viruses (GVs). (A) Healthy fifth-instar larva (top) and AcMNPV-infected larva (bottom) of *Trichoplusia ni*, the latter at 6 days postinfection. (B) Healthy (h) and GV-infected larva of the clover cutworm, *Scotogramma trifolii*. Note the discoloration and hypertrophy of the diseased larvae in A and B. (C, D) Phase-contrast micrographs of typical fat body lobes of fifth-instar noctuid larvae infected with, respectively, NPV (C) and GV (D) at 6 days postinfection. Polyhedra in NPV infections are restricted to the nuclei, whereas masses of granules (gray clusters) in GV infections are in cells that no longer have nuclei. Larvae in A and B are illustrated at about half life-size; bars in C and D, 25 μm.

blood as a result of lysis of cells in various tissues at an advanced stage of disease.

Larvae at this stage of disease will die within 1 or 2 days. Just prior to this stage, larvae of many lepidopteran species will crawl to the top of the vegetation on which they were feeding and then die. After death, the larvae lose their turgor, become flaccid, and are frequently observed anchored to the vegetation by their proleg crochets, hanging head down or in an inverted V position, with the head and rear portion of the abdomen facing downward. The cuticle is very fragile at this point due to its almost complete destruction by the virus. Within a few hours, the cuticle ruptures, releasing billions of polyhedra that have been liberated from lysed cells of the fat body, tracheal matrix, epidermis, and other tissues. Rupture of the cuticle at the end of disease results from infection of the cuticular epidermis and production of viral-encoded chitinase and protease in infected tissues (Hawtin *et al.*, 1995).

In species such as the gypsy moth, *Lymantria dispar,* in which the larval cuticle is dark, heavily pigmented, and covered with cuticular hairs, the behavioral changes are the same, but it is not usually possible to observe the changes in color to internal tissues. However, the ventrum of these larvae, where the cuticle often is not so heavily pigmented, may take on a creamy coloration.

One aspect of gross pathology caused by at least some NPVs, and which is not easily observed, is failure of the larva to molt after infection. This is because of the production by the NPV of an ecdysteroid UDP–glucosyl transferase that glucosylates the molting hormone, ecdysone, thus preventing further molts (O'Reilly and Miller, 1989). The energy and resources that normally go into molting are instead directed to viral reproduction, thereby increasing the amount of progeny virus (O'Reilly and Miller, 1991).

2. Pathogenesis

The gross pathology described above is of course the final manifestation of several cycles of viral replication in a variety of tissues that began soon after the larva ingested polyhedra. Here the principal stages of host invasion and viral spread to the major tissues, where secondary infection and occlusion body production occurs, are reviewed and discussed.

a. Initiation of Midgut Infection

Although NPVs can be transmitted via the egg (Tanada and Kaya, 1993), or can initiate infection through the spiracles (Kirkpatrick *et al.*, 1994), the predominant route of infection is by ingestion of polyhedra and entry of the virus through the midgut epithelium (Fig. 2). After ingestion, the occlusion bodies/polyhedra dissolve quickly, typically within seconds to at most a few minutes, in the alkaline juices at the anterior end of the midgut. Dissolution

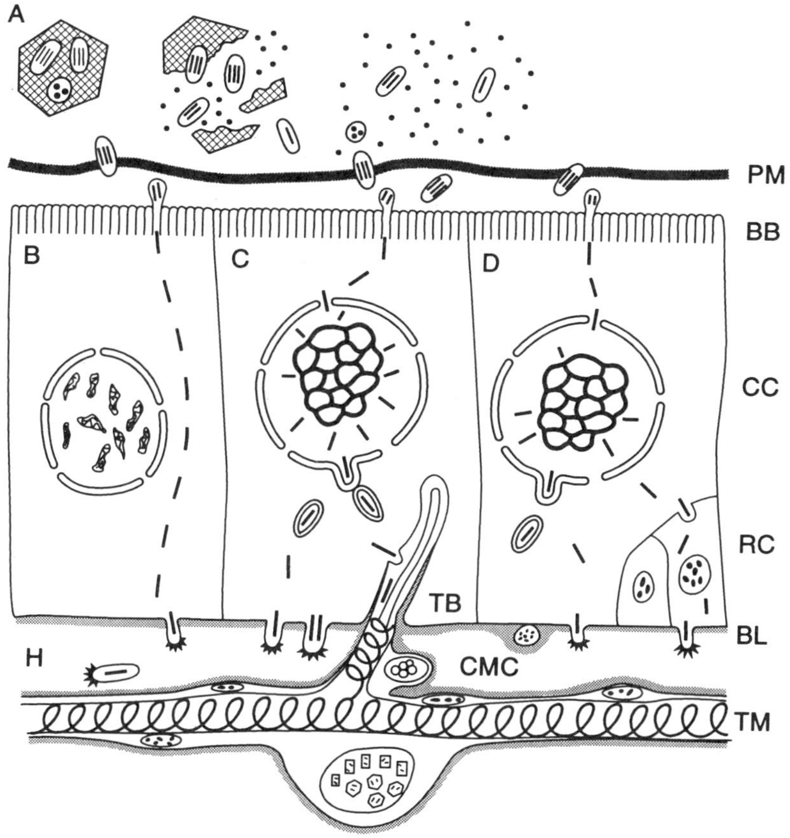

FIGURE 2. Schematic illustration of the initial baculovirus infection process in the midgut epithelium of an infected caterpillar. (A) Dissolution of polyhedra in the midgut lumen and passage of virions through the peritrophic membrane (PM). (B–D) Different infection pathways. (B) The virion envelope fuses with the microvillar membrane and the nucleocapsid passes directly through the cell to the basal lamina, followed by budding through this matrix into the hemolymph. (C) An initial cycle of replication occurs in a midgut epithelial cell nucleus, after which progeny-budded virions pass directly into the hemolymph or into the tracheal matrix via tracheoblasts. (D) After the initial cycle of replication, midgut regenerative cells are invaded, and progeny BV from these invade the hemolymph. PM, peritrophic membrane; BB, microvillar brush border; CC, columnar cell; RC, regenerative cell; BL, basal lamina; TM, tracheal matrix; CMC, circular muscle; H, hemolymph.

is aided by both the high pH (8–11) characteristic of most lepidopterans and the presence of proteases (Granados, 1978; Granados and Lawler, 1981; Granados and Williams, 1986; Harrap, 1970; Harrap and Robertson, 1968; Horton and Burand, 1993). Thus, polyhedra have apparently evolved to persist and protect the virions outside the host, but also to dissolve quickly under midgut conditions, thereby facilitating rapid infection of a host.

The virions released from occlusion bodies—occlusion-derived virions (ODV)—then pass through the peritrophic membrane, a protective lining secreted by the midgut that surrounds food and other ingested materials. Passage of virions through the peritrophic membrane is neither well understood nor is there agreement on how this occurs. In some lepidopterans, the peritrophic membrane is fibrous, such as in the larvae of the Douglas fir tussock moth, *Orgyia pseudotsugata,* allowing particles less than 800 nm to pass, whereas in others, like *T. ni,* the membrane is multilayered and channeled, though the channels are not continuous from one side of the membrane to the other (Adang and Spence, 1981, 1983). Thus, it is not clear how the virions traverse the peritrophic membrane in many species commonly infected by baculoviruses. One possibility is that these viruses produce enzymes that facilitate localized digestion of the membrane to permit virion passage. And in fact, though originally identified in GVs, Granados and coworkers have described a group of proteins, referred to as enhancins, that appear to serve this function (Gijzen *et al.,* 1995; Hashimoto *et al.,* 1991; Lepore *et al.,* 1996; Wang *et al.,* 1994). Studies of one of these, an enhancin isolated from the *Trichoplusia ni* GV (TnGV), suggest that it is a metalloprotease with a specific affinity for mucins in the peritrophic membrane of *T. ni* (Lepore *et al.,* 1996). This enzyme is capable of markedly disrupting the structural integrity of the peritrophic membrane, creating lesions through which virions can pass.

Other evidence suggests that the peritrophic membrane is not a significant barrier to infection because it is shed during molting, permitting virions to come into direct contact with midgut microvilli (Washburn *et al.,* 1995). However, larvae probably ingest most virus during the intermolt period. As the peritrophic membrane in many lepidopteran species may not contain channels sufficiently large for virion passage, the existence of the enhancins and their enzymatic properties suggests these aid virion passage through the membrane. Though current evidence indicates that enhancins aid viral infection, it is not known whether they are a feature of all baculoviruses.

b. Entry into Midgut Cells

After crossing the peritrophic membrane, the virion envelope comes in contact with the microvillar membrane of columnar midgut epithelial cells. Specific binding proteins on the ODV envelope have not been identified, nor have receptors on microvilli, although binding studies suggest they exist (Horton and Burand, 1993). After contact, the virion envelope fuses with the microvillar membrane, allowing the nucleocapsids to enter the microvilli (Fig. 2). Virions have also been observed to enter cells by viropexis, but electron microscopic studies indicate the primary route of entry is via fusion of the virion envelope with the microvillar membrane (Granados and Lawler, 1981).

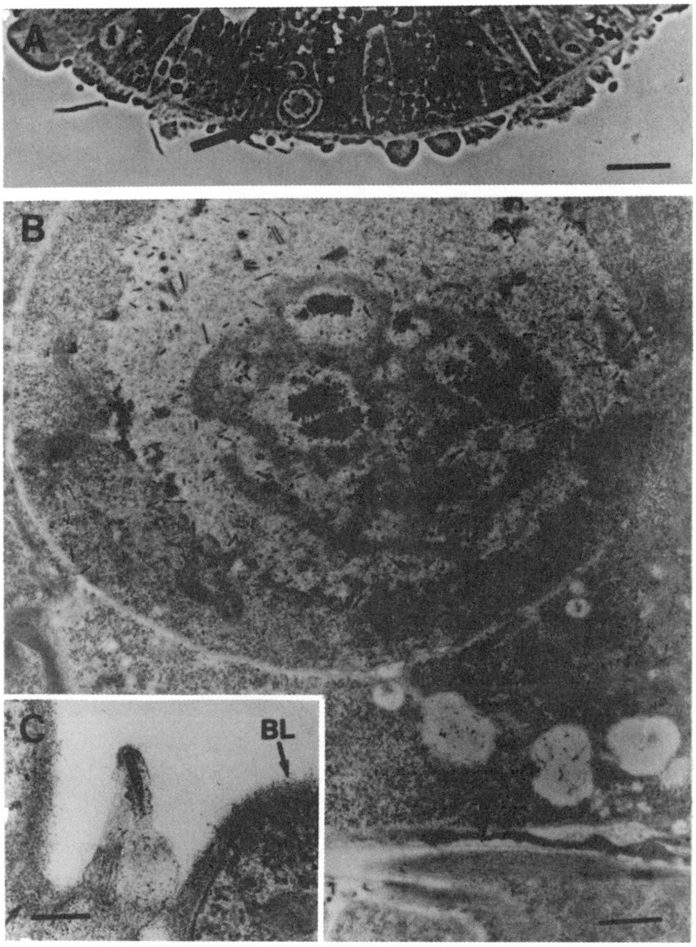

FIGURE 3. Light and electron micrographs illustrating aspects of the initial baculovirus colonization phase in the midgut epithelium. (A) Phase-contrast light micrograph illustrating an infected regenerative cell (arrow). (B) Electron micrograph of an infected regenerative cell showing the virogenic stroma and production of progeny nucleocapsids. Arrow indicates basal lamina at proximal end of the cell. (C) Electron micrograph of progeny-budded virion passing out of cell into hemolymph via the basal lamina. Bar, 10 μm in A, 1 μm in B, and 200 nm in C.

After entry, the nucleocapsids travel along the microvillus, apparently with the aid of microtubles, to the nuclear pore and then into the nucleus where the nucleocapsids uncoat, releasing the viral DNA to initiate replication (Charlton and Volkman, 1993; Granados and Lawler, 1981). Within 8 hr, the nucleus hypertrophies, the nucleoli enlarge, move toward the nuclear membrane, and then decrease in size as a large virogenic stroma forms within the center of the nucleus (Fig. 3B). As the stroma develops, nucleocapsids begin to assemble around the periphery and in the stroma lacunae

(Figs. 2 and 3). They leave the stroma and pass through the nuclear membrane, often acquiring an envelope from this membrane. The nucleocapsids shed this membrane before aligning end-on with the plasmalemma at the base of the midgut cell. Here they become budded virions (BV) by emerging from the cell. In this process, the nucleocapsids acquire a new envelope derived from the plasmalemma. This envelope includes virion-encoded proteins, such as the gp64 envelope fusion protein, essential for the infection of other tissues (Adams *et al.*, 1977; Blissard and Rohrmann, 1990; Blissard and Wenz, 1992; Granados and Lawler, 1981; Monsma *et al.*, 1996; also see Chapters 2 and 4, this volume).

An interesting and important aspect of this phase of pathogenesis is that, though polyhedrin crystals may form in midgut epithelial cells, occlusion of virions in this tissue does not occur in most lepidopterous insects, or where it does occur, occurs only rarely (Flipsen *et al.*, 1993). Instead, infected cells are cast off as the larva molts and are replaced by new cells as the midgut regenerates, either immediately or as part of the molting process (Flipsen *et al.*, 1993). Of course, the new cells can be infected by fresh inoculum.

A slightly different invasion pathway has been reported in larvae of the beet armyworm, *Spodoptera exigua,* infected with the AcMNPV. Using a recombinant AcMNPV with two reporter genes under control of constitutive (*Drosophila melanogaster* hsp70) and very late viral (AcMNPV p10) promoters, it was shown that the virus first invades the columnar cells and then the underlying regenerative cells before passing on into the hemolymph (Flipsen *et al.*, 1995; Keddie *et al.*, 1989).

c. *Spread of Virions to the Hemolymph*

Supporting the midgut epithelium at its base is the basal lamina, a fibrous matrix of glycoproteins secreted by the epithelial cells. As BVs pass out of the midgut cells, they either traverse this matrix directly into the hemolymph or accumulate between the basal lamina and midgut cells, remaining there or possibly traversing the matrix into the hemolymph later (Figs. 2 and 3). The hemolymph then circulates the BVs throughout the body, facilitating the rapid transmission to and infection of other tissues (Adams *et al.*, 1977; Benz, 1963; Granados and Lawler, 1981). Distribution of BV by the hemolymph appears to be the primary mode by which the virus is spread to other tissues.

Though the route of infection outlined above has been generally accepted (Granados and Williams, 1986), several other routes for viral spread to the hemolymph have been proposed. For example, Granados and Lawler (1981), using early fifth instars of *T. ni* and high doses of AcMNPV, provided evidence that the nucleocapsids could circumvent the midgut nucleus, attach to the plasmalemma along the base of the midgut epithelial cell, and bud directly into the hemolymph. They fed larvae a dose of 10^6 polyhedra/larva (a high dose that might be encountered during epizootics) and found infectious hemolymph as early as 30 min postfeeding. This is much too soon for progeny nucleocapsids to have been produced, but sufficient time to obtain very early expression of *gp64*, which encodes the BV gp64 envelope fusion glycoprotein in both AcMNPV (Jarvis and Garcia, 1994) and OpMNPV (Blissard and Rohrmann, 1989, 1990). Thus, a plausible explanation for the early appearance in the hemolymph is direct passage of nucleocapsids derived from infecting polyhedra into the hemolymph by circumvention of the nucleus (Fig. 2). It is also possible that ODV pass between midgut cells and through the basal lamina to the hemolymph. This route was suggested in an early study of the ODV infection pathway (Tanada and Leutenegger, 1968), but now seems unlikely given the demonstration of viral replication in midgut epithelial cells (Granados and Williams, 1986).

A more recent and interesting suggestion for passage of the virus to other tissues is that the BV spreads via the tracheal matrix, using this tissue as a conduit for the systemic spread of infection throughout the host (Engelhard *et al.*, 1994). Using fourth instars of *T. ni* and a recombinant AcMNPV that produces β-glactosidase under control of a constitutive promoter, these authors showed that tracheoblasts infiltrating midgut epithelial cells were markedly blue in the presence of 5-bromo-4-chloro-3-indolyl-β-D-galactopyranoside (X-gal), a chromogenic indicator for *lacZ* expression, and thus were infected within 16 hr postfeeding with very low doses of virus. Moreover, the tracheal matrix was heavily infected by 24 to 48 hr postfeeding, the period during which infection became marked in several other tissues.

To explain the rapid spread of the virus, Engelhard *et al.* (1994) postulated that BVs were incapable of traversing the basal lamina of the midgut because the pores within it, which exclude particles larger than 15 nm in diameter (Reddy and Locke, 1990), were too small to allow virions to pass. To cross this barrier, they proposed that the BVs traveled from the midgut via the tracheoles, either directly into the hemolymph and then to other tissues, but more probably to other tissues laterally along the tracheal matrix through the tracheae-specific lymph system. While interesting, available evidence does not support this hypothesis. For example, the tracheoles themselves are surrounded with a basal lamina as they extend from the midgut to the main tracheal trunks. This basal lamina would have to be traversed for the BV to enter the hemolymph. Thus, if the BV has to cross a basal lamina, it is not clear why, if the pores throughout this matrix are of equal size, it is easier for them to pass the basal lamina of the tracheoles than that of the midgut. Second, if the BVs are to travel along the tracheal matrix to other tissues via the tracheal lymph system, they will have to run a "receptor gauntlet" along the way, repeatedly encountering the plasmalemma of susceptible tracheal matrix cells on their way to other tissues. Third, other studies have shown that AcMNPV BVs can traverse the basal lamina of the midgut epithelium (Fig. 3) (Flipsen *et al.*, 1993, 1995; Granados and Lawler, 1981). Since BV can cross this matrix, it is possible that this is an enzymatic process, rather than just a physical one, as proposed by Engelhard *et al.* (1994), and that the virion envelope contains as yet unknown enzyme(s) on its surface.

In summary, analysis of available evidence indicates that BV spreads the virus from the midgut to most tissues via the hemolymph. The tracheal matrix is infected quickly and directly from the midgut owing to its intimate physical association with this tissue (Fig. 2), but can also be infected at sites not associated with the midgut by infection of BV transported by the hemolymph. Aside from the tracheal matrix, the hemocytes, fat body, and muscles closely associated with the midgut are also very quickly infected (Benz, 1963; Granados and Lawler, 1981). Early infection of these tissues would be unlikely if BVs had to first pass solely through the tracheal matrix, especially if one or more cycles of replication were necessary to boost the number of infectious virions for spread to these tissues along the tracheal lymph system.

d. Infection of Other Tissues by Budded Virions

The budded virions initiate the second phase of replication in which occluded virions and polyhedra/occlusion bodies are produced in most tissues of infected lepidopterans. After the midgut, the first tissues reported by most investigators to show signs of infection in susceptible hosts are the hemocytes, tracheal matrix, and fat body cells, with infection of the latter two tissues being most obvious in cells in direct physical contact with the midgut (Adams *et al.*, 1977; Aizawa, 1963; Benz, 1963; Engelhard *et al.*, 1994; Granados and Lawler, 1981; Granados and Williams, 1986; Keddie *et al.*, 1989). The hemocytes are the easiest in which to detect infection because they are free-floating in the hemolymph, being readily collected from larvae, have no basal lamina, and show a marked cytopathic effect, i.e., nuclear and cellular hypertrophy, followed by the presence of polyhedra within the nucleus. At high viral doses, infected hemocytes containing polyhedra can be detected with light microscopy as early as 15 hr postfeeding of polyhedra to larvae, with over 80% of the hemocytes producing polyhedra by 4 days postfeeding (Granados and Lawler, 1981). But infected tracheal matrix cells near the midgut (Adams *et al.*, 1977; Engelhard *et al.*, 1994), and in some cases fat body cells adhering closely to the midgut (Benz, 1963), also exhibit infection by the presence of polyhedra within 24–48 hr postingestion of polyhedra. Subsequently, infection is observed in nerve, muscle, pericardial cells, reproductive tissues, and glandular tissues. Some tissues, such as the Malpighian tubules and salivary glands, are typically not well infected (Aizawa, 1963). The virus gains entry to these tissues and immediate early genes but not late genes are expressed (Knebel-Mörsdorf *et al.*, 1996). Interestingly, in hosts of low susceptibility, hemocytes are capable of clearing infection via the encapsulation process (Washburn *et al.*, 1996).

The mechanism by which BV gains access to the secondary sites of viral replication is not well understood. For the tracheal matrix, infection appears to be direct. Within the midgut epithelial cells, the tracheoblast and midgut plasmalemmas are in direct contact where the fingerlike projections of tracheoblasts penetrate these cells. Thus, BV can pass from a midgut cell directly into the tracheoblast. Infection of hemocytes is likely initiated by BVs circulating in the hemolymph that adhere to the hemocyte plasmalemma, with endocytosis, and then the fusion event, whereby nucleocapsids are released into the hemocyte cytoplasm, being mediated by gp64, the envelope fusion glycoprotein (Blissard and Wenz, 1992).

Infection of tissues surrounded with a basal lamina probably occurs in two ways. Where the basal lamina is thin, e.g., less than 20 nm across, such as along the fat body, epidermis, and tracheal matrix epidermis in early instars, BV may be able to penetrate directly into the tissues from the hemolymph. If BV can penetrate through the midgut basal lamina to the hemolymph (Fig. 3C), it may be capable of moving from the hemolymph through this matrix to gain access to other tissues. A second possible mechanism for cases where the basal lamina is thick, e.g., >30 nm, is by entry into a tissue via the tracheoles. The basal lamina surrounding the tracheoles suspended in the hemolymph is often quite thin (<30 nm), and thus in this case the tracheoles may provide a pathway into other tissues. Evidence for this comes from the studies of Engelhard *et al.* (1994), where it was found that the tracheal matrix was infected at the sites where trachea branched into tissues such as the Malpighian tubules.

Keddie *et al.* (1989) proposed that virus could also spread to secondary tissues via hemocytes. They observed infected hemocytes adhering to the epidermis, and suggested that virus was passing from the hemocytes into this tissue. It is known that uninfected hemocytes often adhere to tissues, so the relevance of adherence by infected hemocytes is not clear. Though it is unlikely that hemocytes serve as a major mechanism for the direct transmission of virions to tissues, since infected hemocytes would contain gp64 in the plasmalemma, it is possible virus is occasionally spread in this manner, particularly where the basal lamina lining a tissue is thin.

e. *Spread of the Virus within Tissues*

Once within a secondary site of infection, such as the fat body, tracheal matrix, or epidermis, two different cycles of virion production occur. The first corresponds to the infection cycle in the midgut and leads to the production of BV that migrate from the cell and infect nearby cells. This spread of the infection leads to concentric patterns of infected cells, much like plaques in a cell culture monolayer, which can be observed histologically during the early phases of disease, usually 3 to 4 days after initial infection. Within individual cells, BV production peaks at about 12–16 hr. Then viral replication shifts to the occlusion phase, in which occluded virions and polyhedra are produced (Granados and Williams, 1986; Volkman and Knudson, 1986; see also Chapters 4, 7, and 8, this volume).

f. Occlusion Phase of Viral Pathogenesis

The pathology associated with the occlusion phase of virion production is one of the most dramatic that occurs in insect viral diseases. The high level of polyhedrin production brought about by the *polh* gene and the formation of hundreds of polyhedra filling each nucleus result in enormous hypertrophy of infected nuclei. These become from 5 to 10 times the diameter of those in healthy cells in all NPV infections (Figs. 1C and 4). The nuclear hypertrophy leads to hypertrophy of the infected cells and tissues, and this ultimately accounts for the swollen appearance of larvae, which occurs toward the end of disease. The tissues that produce the heaviest infections and largest quantities of polyhedra are the fat body, epidermis, and tracheal matrix (Aizawa, 1963; Tanada and Kaya, 1993).

As virus replication advances, physical distension and weakening of the plasmalemma brought about by occlusion body formation and concomitant cellular hypertrophy, and possibly the production of viral protease, cause the nuclei and cells to lyse. The viral p10 protein also assists in nuclear lysis (van Oers *et al.*, 1993), and cell lysis is probably aided by cathepsins. As more and more cells lyse, millions of free polyhedra accumulate in tissues (Fig. 4E). With most of the cells infected, the basal lamina, which may also be weakened by viral enzymes, loses its integrity and ruptures, releasing polyhedra into the hemolymph. This increases the cloudiness already present due to infected hemocytes, which also lyse. At about this point, the larva dies. During the occlusion phase, the viral-encoded chitinase (Hawtin *et al.*, 1995) is produced, and after larval death likely aids in disruption of the chitin-rich cuticle, releasing polyhedra from the body of the dead caterpillar.

B. Granulosis Viruses

1. Gross Pathology

As noted above and in Table 1, GVs are only known from lepidopterous insects, and in these there are three types of GV disease, each characteristic of a different type of tissue tropism encoded by the virus. In type 1 GVs, such as the TnGV and many other but not all GVs of noctuid larvae, the virus invades the host through the midgut epithelium, much like the typical NPV, but subsequently only infects the fat body tissue (Federici, 1993; Hamm, 1968; Huger, 1963). This type of tissue tropism has several consequences for the gross pathology. Because other important tissues such as the

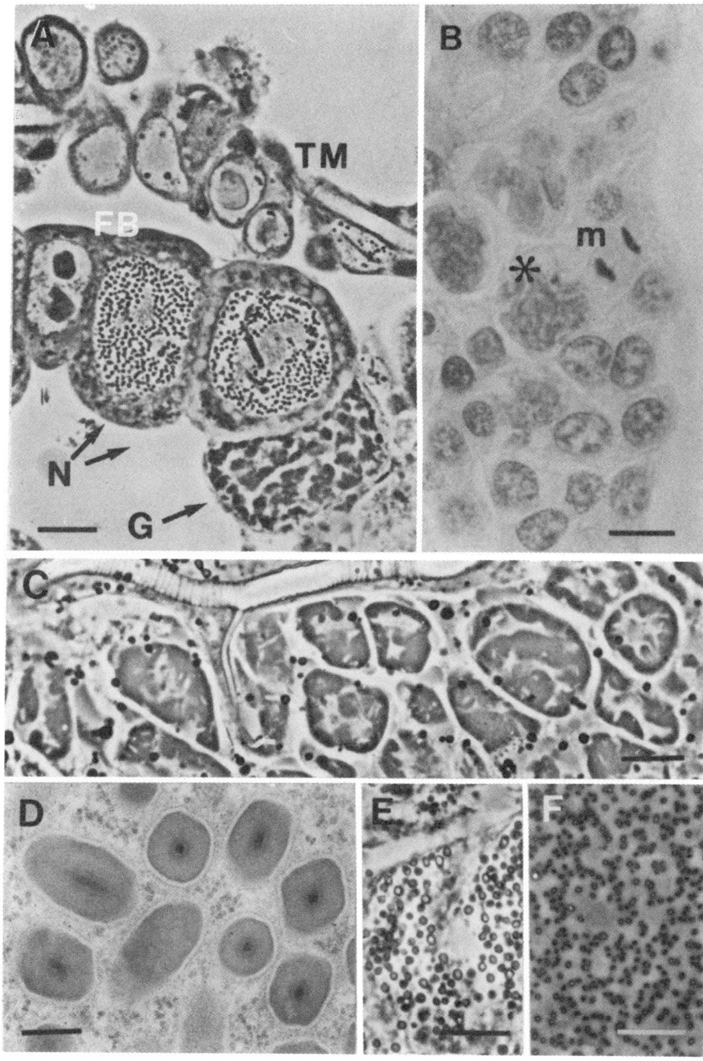

FIGURE 4. Micrographs illustrating different aspects of NPV and GV pathogenesis. (A) Phase-contrast light micrograph of NPV-infected tracheal matrix (TM) and NPV- and GV-infected fat body (FB) in larva of *Trichoplusia ni.* N indicates NPV infected cells, G a GV-infected cell. Note the difference in cytopathology; no nucleus is apparent in the GV-infected cell. (B) Cell undergoing mitosis (m) in GV-infected fat body stained with the Feulgen technique. The asterisk indicates a cell in which most of the chromatin is now viral. (C) Phase-contrast light micrograph of GV-infected fat body at an advanced stage of infection. (D) Each of the round granular clusters are cells packed with thousands of granules. (E) Wet-mount preparation of polyhedra released from infected cells in a moribund NPV-infected larva. (F) Granules in a moribund GV-infected larva. Bar, 15 μm in A, 4 μm in B, 2 μm in C, 150 nm in D, 30 μm in E, and 7 μm in F.

tracheal matrix and epidermis are not attacked, the larva may live as much as a week longer than a larva of the same stage infected with a similar amount of NPV. When infected in the fourth instar, it is not unusual for larvae of *T. ni* to live 10–14 days after infection. Unlike NPV-infected larvae, these larvae will typically maintain their appetites and grow much larger than healthy larvae, only becoming lethargic and slow to respond to tactile stimuli within a day or two of death.

After infection, as the disease progresses, the infected fat body proliferates and larvae become markedly swollen, developing a creamy yellow appearance (Fig. 1B). The latter discoloration is due to the accumulation in the fat body of large numbers of golden-colored infected cells packed with viral granules (Fig. 4C), the color of the infected fat body being easily observed because of the light pigmentation of the *T. ni* larval cuticle. Soon after this stage is reached, the larvae die. Typically, there is little or no liquefaction of the body, probably because the epidermis is not infected. More often, the larvae turn dark brown to black because of invasion of the body by gut flora and then they dry up or disintegrate.

In the type 2 GV, such as the GV of the codling moth, *Cydia pomonella* (CpGV), the viral infection and gross pathology parallel that of the typical lepidopteran NPV disease. This is probably because the tissue tropism is similar to that of NPVs. After invasion of the midgut, the virus attacks most of the major body tissues, including the tracheal matrix, epidermis, and fat body (Huger, 1963; Tanada and Leutenegger, 1968; Tanada and Kaya, 1993; Tweeten *et al.*, 1981). The disease is more acute than that caused by type 1 GV, typically lasting only 5–10 days in larvae infected during the fourth instar. As the disease progresses, larvae swell and distend slightly and develop irregular white to yellow patches beneath the cuticle. These are easily observed in larvae with little cuticular pigmentation, such as codling moth and cabbage worm (*Pieris brassicae*). After death, the body liquefies, probably due to infection of the epidermis. It is not known whether this type of GV also produces chitinase and other enzymes that foster cuticle rupture and liquefaction as in the NPVs.

The third type of GV disease is unique to the HbGV. This virus attacks the Western grapeleaf skeletonizer, *Harrisina brillians*, and is characterized by a tissue tropism restricted to the midgut epithelium (Federici and Stern, 1990; Smith *et al.*, 1956; Steinhaus and Hughes, 1952). As far as is known, HbGV replicates only in this tissue, producing virions and occlusion bodies in both larvae and adults.

The disease caused by HbGV is an acute one, and larvae infected in the third or fourth instar usually die within 4 to 7 days. After ingestion, the virus invades the midgut epithelium and rapidly spreads from cell to cell over a period of a few days. The larvae, which are normally brightly colored with alternating purple and yellow stripes that circle the body, develop a noticeable diarrhea a few days after infection. The liquid diarrhetic discharge and feces contain infectious granules and cells sloughed from the infected gut. As the disease progresses, the bright yellow stripes change to a brownish yellow. The larvae typically begin to shrivel up during the 4th or 5th day of disease, apparently due to their inability to maintain water within the body, and then die. There is no evidence that the virus invades other tissues.

2. Pathogenesis

The pathogenesis of GV diseases has not been studied nearly as well as that of NPVs, though the studies that have been done indicate that the initial stages of host infection are very similar to those that occur in NPVs. There are, however, major differences between the GVs and NPVs in tissue tropism, already discussed, and in cytopathology and histopathology. The summary below for GVs is a composite, and it should be realized that many of the details of these processes remain to be determined for individual GVs. Thus, the features described may not be universal, even within the same GV pathotype.

a. Infection of Midgut Epithelium

As in NPVs, GV granules dissolve in the alkaline midgut juices of lepidopteran larvae within minutes of ingestion (Summers, 1969, 1971). The virions then pass through the peritrophic membrane, in some cases with the aid of an enhancin (Derksen and Granados, 1988; Wang *et al.*, 1994). The virion envelope then fuses with the apical microvillar membrane, and the nucleocapsid passes through the cell to the nuclear pore. A difference between NPVs and GVs occurs at this stage, in that rather than pass into the nucleus as in NPVs, the GV nucleocapsid injects the DNA into the nucleus at the nuclear pore (Summers, 1971).

b. Midgut Cytopathology and Initial Viral Replication

The initial phase of GV replication is similar to that of NPVs and includes enlargement and margination of the nucleoli, development of a virogenic stroma, hypertrophy of the nucleus, and production of nucleocapsids (Huger, 1963; Walker *et al.*, 1982). At this point, GV cytopathology begins to differ considerably from that of NPVs. In GV diseases, as the nucleus enlarges, the nuclear membrane ruptures and fragments and the nucleoplasm and cytoplasm become intermixed. Nucleocapsids continue to form and pass out of the cell, forming budded virus along the basal plasmalemma. It appears that peplomers similar to those that occur on the envelope of NPV BVs exist, but these have not been clearly identified, nor has a gp64 analogue. At this point, it is assumed that BVs occur in type 1 and type 2 GVs, and that these enter the hemolymph by traversing the basal lamina much as do the BVs of NPVs. They then spread to other tissues, entering by direct penetration through the basal lamina, or via tracheal junctions with these tissues. In GVs that infect the tracheal matrix, such as CpGV, the virus may gain access to this tissue by direct penetration of BV from the midgut cells into tracheoblasts or via the hemolymph.

As in NPVs of lepidopterous insects, no occlusion bodies or virion occlusion occurs in midgut epithelial cells, with the sole exception of the HbGV. The mechanism by which the HbGV spreads from cell to cell in the midgut epithelium of *H. brillians* is not known.

c. Cytopathology and Histopathology of the Occlusion Phase

The cytopathology that occurs in the occlusion phase in secondary sites of GV infection differs markedly from that of NPVs, but parallels events that occur in GV-infected midgut cells. The following description is based on reports of infections caused by all three GV types (Federici and Stern, 1990; Hamm, 1968; Huger, 1963; Tanada and Leutenegger, 1968; Walker *et al.*, 1982). Regardless of the tissue tropism, the phase of replication leading to occluded virions and granules is similar in all types.

After cell infection, the nucleus hypertrophies as the nucleoli enlarge and move toward the nuclear membrane along with the chromatin. As the nucleus enlarges, it ruptures and the cytoplasm and nucleoplasm mix (Figs. 1D and 4). At about this stage, numerous small virogenic stromata begin to form as dense granular areas distributed throughout the cell, with the stromata interconnected by thin nucleoprotein strands. Along the edges of these, nucleocapsids begin to assemble and shortly thereafter are enveloped by membranes lying outside the stroma. In the same region, as mature virions begin to accumulate, masses of granulin appear, often as large crystals. The virions are quickly occluded and initially form distinct masses distributed throughout the cell. As more granules form, the masses coalesce and the cell greatly enlarges. Finally, the cell becomes completely filled with thousands of granules.

This cytopathology leads to a distinctive histopathology quite different than that which occurs in NPVs. After lysis of the nuclei, infected cells disassociate from one another as the gap junctions, desmosomes, and hemidesmosomes degenerate, apparently due to the loss of nuclear functions. More or less simultaneously, the infected cells also separate from the basal lamina and, in a tissue such as the fat body, move to and accumulate in the center of the lobe (Fig. 1D). In HbGV, this same pathology results in large numbers of infected midgut epithelial cells being sloughed into the midgut lumen.

In all infected tissues, the separation of infected cells from the basal lamina results in proliferation of regenerative cells situated on the internal surface of this matrix. Cell proliferation is apparently a natural response of the infected tissue as it attempts to replace cells lost to disease. Thus, in the midst of GV infection, mitotic regenerative cells are found along the basal lamina in diseased tissues, and these continue to occur until death (Fig. 4B).

As the regenerative cells begin to differentiate, they all develop GV infections. It is tempting to think that the GV may have transformed these cells in the sense of a malignancy, which the pathology resembles, but no evidence for this exists. More likely, differentiating cells are infected by BV derived from nearby cells more advanced in disease. Interestingly, the mitotic activity in infected tissues was recognized in the early days of insect virology by Paillot (1935), the discoverer of the GVs. He also recognized the enlarged fat body cells filled with granules as a major characteristic of GV disease that differentiates these from NPV diseases.

III. BACULOVIRUSES OF OTHER INSECT ORDERS AND CRUSTACEA

The baculoviruses that occur in nonlepidopterous insects and crustaceans (shrimp) are all NPVs, and in most of these the infection is restricted to the midgut epithelium. Moreover, all have virions of the SNPV type. While there are differences in the gross pathology of these diseases, the cytopathology and histopathology are similar to those characteristic of the occlusion phase of lepidopteran NPVs. Most of these viruses and the diseases they cause have received relatively little study, especially in comparison to NPVs that attack lepidopterous insects. The process of infection is not well understood, though infection where it has been studied has been shown to occur by ingestion of polyhedra. Moreover, little is known about the biochemistry and molecular biology of these viruses and their pathogenesis.

The focus in this section therefore will be on a description of representative NPVs that occur in sawflies (Hymenoptera), mosquitoes and the cranefly, *Tipula paludosa* (Diptera), and shrimp (class Crustacea). the NPVs of caddisfly larvae (Trichoptera) and silverfish (Thysanura) (Larsson, 1984) will not be discussed because the cytopathology and histopathology of these is similar to that observed in sawfly larvae.

A. Nuclear Polyhedrosis Viruses of Hymenoptera

Among hymenopterous insects, NPVs are known to cause disease in about 20 different species of sawflies (Cunningham and Entwistle, 1981). Sawflies are primitive hymenopterans belonging to the suborder Symphyta. The larvae resemble caterpillars, and many sawfly species are important forest pests. Interestingly, there is some molecular evidence, based on similarities of polyhedrin sequences, that the sawfly NPVs originated evolutionarily from the NPVs that attack the Lepidoptera (Rohrmann, 1986).

The diseases caused by the sawfly NPVs are quite similar, and the disease of the European spruce sawfly, *Gilpinia hercyniae,* will be used to illustrate this group. Generally, sawfly NPVs cause acute diseases, with the larvae dying within 4 to 7 days after being infected (Steinhaus, 1949). In *G. hercyniae,* the larvae are green, but within 2 to 3 days of infection, a yellowish-white discoloration appears in the midabdominal segments, and this becomes marked over the next day or two. As in other NPVs, as the disease progresses, the larvae lose appetite and may decrease in size due to water loss. The larvae die in 4 to 7 days after infection, turning brown, and after death, depending on environmental conditions, either dry up or putrefy.

The gross signs of disease are due to infection of the midgut epithelium where the virus replicates in the nuclei. As polyhedra accumulate in the midgut nuclei, the tissue becomes opaque white, and this accounts for the yellow discoloration that can be observed through the green cuticle in the abdomen of infected larvae.

As in the midgut GV infection of *H. brillians,* infection of the sawfly larva midgut results in an infectious diarrhea and to some extent a proliferation of midgut epithelial cells to replace those sloughed at an advanced stage of cellular disease. Otherwise, the histopathology and cytopathology are similar to that observed in the cuticular epidermis, for example, of the lepidopteran NPVs.

B. Nuclear Polyhedrosis Viruses of Diptera

Two different types of NPV diseases occur in the Diptera: those restricted to the midgut, as exemplified by mosquito NPV diseases, and the NPV hemocyte disease of larvae of the cranefly, *T. paludosa.*

1. Mosquito Nuclear Polyhedrosis Viruses

There are now about six NPVs reported from mosquito larvae, primarily from species of *Aedes* and *Culex,* and all of these attack the midgut epithelium. The best-studied mosquito NPV is one originally isolated from the saltmarsh mosquito, *Aedes sollicitans* (AsSNPV), but studied primarily in larvae of the treehole mosquito, *Aedes triseriatus.* This NPV will serve to illustrate the typical mosquito NPV pathology.

Larval development in *Aedes* mosquitoes is quite rapid, typically taking only 4–6 days from eclosion to pupation. Thus, the NPV disease is an acute one, usually lasting only 3–4 days in larvae infected during the third or early fourth instar (Federici and Lowe, 1972). The gross signs of disease caused by the AsSNPV are a shortening of the larva, cessation of feeding, and the development of white spherules, which are infected nuclei packed with occlusion bodies throughout the midgut. These can be observed at about 25× with a dissecting microscope by examining the ventral larval surface and are particularly noticeable through the translucent intersegmental ab-

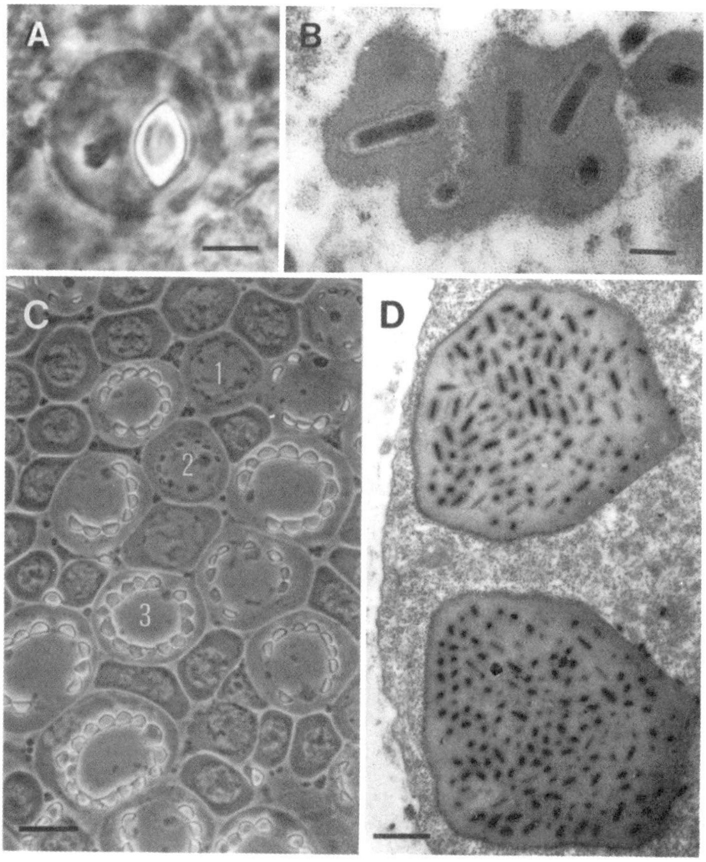

FIGURE 5. Light and electron micrographs illustrating NPV diseases of dipterous larvae. (A) Nucleus of NPV-infected mosquito (*Aedes triseriatus*) larva showing a single large spindle-shaped occlusion body. (B) Virions of the mosquito NPV in occlusion bodies at an early stage of occlusion body formation. (C) Phase-contrast light micrograph of a wet mount of cranefly (*Tipula paludosa*) hemocytes infected with the *T. paludosa* NPV. (D) Polyhedra of the *T. paludosa* NPV developing on the nuclear membrane of a hemocyte. Note that in both of these NPVs, the virions are of the SNPV type. Bar, 10 μm in A, 100 nm in B, 6 μm in C, and 500 nm in D.

dominal cuticle. Larvae die within 4 days of infection and are usually quickly devoured by other larvae and microcrustacea.

The cytopathology, i.e., hypertrophy of the nucleus, formation of a virogenic stroma, and production of nucleocapsids and virions, is similar to that which occurs in lepidopteran NPVs. The AsSNPV differs from these, however, in occlusion body formation. Here, the occlusions begin to coalesce as they grow and eventually fuse together to form only one or a few large spindle-shaped occlusions in the nucleus (Fig. 5A) (Federici and Anthony, 1972).

2. Nuclear Polyhedrosis Virus of *Tipula paludosa*

The larvae of craneflies live just beneath the soil or along the edges of streams, where they feed on vegetation such as the roots of grasses. This NPV, originally discovered in the 1920s and then again in the 1950s, is unique in that the primary site of viral replication is the larval hemocytes (Smith, 1955). The virus appears to initiate infection through the midgut, but the actual pathway of infection is unknown. In terms of gross pathology, the disease lasts for 10 to 14 days before larvae die. Healthy larvae are normally brown. But within 4 days of infection, a white to yellow color develops in diseased larvae, and this becomes more pronounced as the disease advances. The discoloration is due to the accumulation of thousands of infected hemocytes in the hemolymph.

The histopathology and cytopathology of this disease are also unique. NPV infection results in rapid proliferation of the hemocytes, all of which eventually develop the disease (Fig. 5C). With respect to cytopathology, the early stages are characteristic of other NPVs, but occlusion formation is again different. In this disease, the occlusion bodies are shaped like an orange segment and form by the direct deposition of polyhedrin and virions on the nuclear membrane (Smith, 1955).

C. Nuclear Polyhedrosis Viruses of Crustacea

NPVs that attack crustacea such as shrimp are becoming increasingly important because they are capable of causing significant disease outbreaks in commercial shrimp production. There are now more than eight species of shrimp reported to be hosts to NPVs, and these have been reported from around the world where shrimp are produced commercially (Couch, 1991).

One of the best-studied shrimp baculoviruses is the NPV of *Penaeus monodon*. The disease caused by this virus is not easily detected because viral replication is restricted to the hepatopancreas. Thus, there is little observable gross pathology. At an advanced stage of disease, the shrimp are lethargic and can develop a foul odor.

Histologically, as the disease progresses and occlusion bodies form, these can be observed in fresh preparations of the hepatopancreas with the aid of a phase contrast microscope. Typically the occlusions are large and tetrahedral in shape, being triangular in outline. Otherwise, the cytopathology resembles that of lepidopteran NPVs (Couch, 1991).

IV. BACULOVIRUS PATHOGENESIS, DIVERSITY, AND EVOLUTION

The studies reviewed in this chapter show that we have a reasonably good understanding of how baculovirus pathogenesis at the level of the cell and tissue results in the signs of disease exhibited by many insect species. Yet, we are only now beginning to explore and understand the molecular basis of pathogenesis. Site-directed mutagenesis, particularly the use of deletion mutants and chromogenic marker enzyme systems that enable different stages of replication to be monitored *in vivo,* have been very useful in studying pathogenesis recently and will prove even more so in the future.

Most of what we know about the molecular basis of baculovirus pathogenesis is based on studies of the AcMNPV, and to some extent, studies of the NPVs of *Orgyia pseudotsugata, Bombyx mori,* and *Spodoptera exigua.* Whereas much remains to be learned about the molecular basis of lepidopteran NPV pathogenesis, we know even less about the GVs and the NPVs of nonlepidopteran hosts. Most of these viruses differ considerably in their pathology from the AcMNPV, and thus study of them would prove useful in defining the molecular basis for different baculovirus pathologies. The following list of questions regarding the AcMNPV and other viruses will suffice to show just how much remains to be learned in a few key areas:

1. What controls the lack of occlusion body formation in the midgut epithelium by lepidopteran NPVs?
2. Why is AcMNPV replication in the epidermis so extensive, while being highly limited in an active tissue such as the salivary glands?
3. What enzymes, if any, are involved in the passage of BV through the basal lamina?
4. To what extent does the insect defense system limit baculovirus host range?
5. What is the molecular basis for the various tissue tropisms observed in the different types of GVs?
6. Which genes in GVs control early lysis of the nuclear envelope?
7. How do viruses such as the HbGV and the mosquito and sawfly NPVs spread from cell to cell?
8. Do they produce BV and gp64-like proteins, or is another mechanism used for cell-to-cell spread?

Studies of baculovirus pathogenesis carried out to date have dealt largely with the progression of disease in different tissues of a limited range of viruses and their hosts. Nevertheless, these studies provide insight into the possible ancestral origin and evolution of baculoviruses when the phylogenetic distribution of these viruses is correlated with the apparent evolutionary success of different viral replication strategies and pathologies.

If we assume that baculoviruses have a common evolutionary origin, which of course may not be the case, then two possibilities emerge. The first, and most probable, is that the baculovirus ancestor was a virus that attacked the midgut or hepatopancreas of an ancient arthropod. This ancestral virus then would have evolved and diversified along with the arthropods and, based on our knowledge of the current phylogenetic distribution of baculoviruses (Table 1), the virus group it gave rise to would have "blossomed" in parallel with species radiation in the order Lepidoptera. The large number of NPVs and GVs known today from lepidopterous insects, in comparison to the relatively few baculoviruses (all NPVs) reported from other arthropods, serve as documentation of an extensive viral radiation that accompanied lepidopteran species divergence.

The reasons for the evolutionary success of lepidopteran baculoviruses can only be speculated on at this point, but would appear at least in part to be due to the dominant replication strategy and pathogenesis that evolved. The NPVs of nonlepidopteran insects and crustacea attack and produce polyhedra in midgut epithelial tissues, and this type of tissue tropism and pathology, especially in gregarious insects, leads to a rapid spread of the virus throughout a population, for example, as occurs in sawfly larvae. Such a replication strategy is only seen in a single baculovirus of a lepidopteran, the GV of the grape leaf skeletonizer, *H. brillians,* which interestingly is considered a primitive lepidopteran. Thus, it would appear that in baculoviruses that attack lepidopterous insects, there was rapid selection against the production of infectious virus and formation of occlusion bodies in the midgut epithelium. Concomitantly, a replication strategy evolved that enabled the lepidopteran baculoviruses to invade the internal tissues of the host. There, horizontal transmission was limited, but this loss was more than compensated for by the larger numbers of progeny virions that could be produced in a range of tissues, as opposed to just the midgut, that includes the fat body, epidermis, and tracheal matrix.

The second possibility for the origin of the baculoviruses is that they originated within the order Lepidoptera and then radiated to other orders via horizontal transmission. Little evidence exists for this possibility, but analysis of polyhedrin sequence data by Rohrmann (1986) suggests that the sawfly NPVs may have originated by cross-infection from lepidopteran NPVs. These data, however, are far too few to give this possibility much credence at this time.

Whether dealing with the mechanisms of pathogenesis at the level of the cell, tissue, or whole insect, or studies of the origin and evolution of baculoviruses, it is clear that there are many fruitful avenues for future research on baculovirus pathogenesis.

ACKNOWLEDGMENTS. The author thanks Robert R. Granados, Johannes T. M. Flipsen, and Lois K. Miller for constructively critical reviews of the manuscript, and Max Bergoin for the light and electron micrographs of the *T. paludosa* NPV.

REFERENCES

Adams, J. R., Goodwin, R. H., and Wilcox, T. A., 1977, Electron microscope investigations on invasion and replication of insect baculoviruses *in vivo* and *in vitro*. *Biol. Cellulaire* **28:**261.

Adang, M. J., and Spence, K. D., 1981, Surface morphology of peritrophic membrane formation in the cabbage looper, *Trichoplusia ni, Cell Tiss. Res.* **218:**214.

Adang, M. J., and Spence, K. D., 1983, Permeability of the peritrophic membrane of the Douglas fir tussock moth (*Orgyia pseudotsugata*), *Comp. Biochem. Physiol. A* **74:**233.

Aizawa, K., 1963, The nature of infections caused by nuclear-polyhedrosis viruses, in: *Insect Pathology, An Advanced Treatise,* Vol. 1 (E. A. Steinhaus, ed.), pp. 381–412, Academic Press, New York.

Benz, G., 1963, A nuclear polyhedrosis of *Malacosoma alpicola* (Staudinger), *J. Insect Pathol.* **5:**215.

Benz, G. A., 1986, Introduction: Historical perspectives, in: *The Biology of Baculoviruses,* Vol. I, *Biological Properties and Molecular Biology* (R. R. Granados and B. A. Federici, eds.), pp. 1–35, CRC Press, Boca Raton, FL.

Blissard, G. W., and Rohrmann, G. F., 1989, Location, sequence, transcriptional mapping, and temporal expression of the gp64 envelope glycoprotein gene of the *Orgyia pseudotsugata* multinucleocapsid nuclear polyhedrosis virus, *Virology* **170:**537.

Blissard, G. W., and Rohrmann, G. F., 1990, Baculovirus diversity and molecular biology, *Annu. Rev. Entomol.* **35:**127.

Blissard, G. W., and Wenz, J. R., 1992, Baculovirus gp64 envelope glycoprotein is sufficient to mediate pH-dependent membrane fusion, *J. Virol.* **66:**6829.

Charlton, C. A., and Volkman, L. E., 1993, Baculoviruses, vertebrate viruses, and cytoskeletons, in: *Parasites and Pathogens of Insects,* Vol. 2. *Pathogens* (N. E. Beckage, S. N. Thompson, and B. A. Federici, eds.), pp. 103–125, Academic Press, San Diego.

Couch, J. A., 1991, Baculoviridae. Part 2. Nuclear polyhedrosis viruses of invertebrates other than insects, in: *Atlas of Invertebrate Viruses* (J. R. Adams and J. R. Bonami, eds.), pp. 205–226, CRC Press, Boca Raton, FL.

Cunningham, J. C., and Entwistle, P. F., 1981, Control of sawflies by baculovirus, in: *Microbial Control of Pests and Plant Diseases, 1970–1980* (H. D. Burges, ed.), pp. 379–407, Academic Press, London.

Derksen, A. C. G., and Granados, R. R., 1988, Alteration of a lepidopteran peritrophic membrane by baculoviruses and enhancement of viral infectivity, *Virology* **167:**241.

Engelhard, E. K., Kam-Morgan, L. N. W., Washburn, J. O., and Volkman, L. E., 1994, The insect tracheal system: A conduit for the systemic spread of *Autographa californica* M nuclear polyhedrosis virus, *Proc. Natl. Acad. Sci. USA* **91:**3224.

Federici, B. A., 1986, Ultrastructure of baculoviruses, in: *The Biology of Baculoviruses,* Vol. I, *Biological Properties and Molecular Biology* (R. R. Granados and B. A. Federici, eds.), pp. 61–88, CRC Press, Boca Raton, FL.

Federici, B. A., 1993, Viral pathobiology in relation to insect control, in: *Parasites and Pathogens of Insects,* Vol. 2, *Pathogens* (N. E. Beckage, S. N. Thompson, and B. A. Federici, eds.), pp. 81–101, Academic Press, San Diego.

Federici, B. A., and Anthony, D. W., 1972, Formation of virion-occluding protenic spindles in a baculovirus disease of *Aedes triseriatus, J. Invertebr. Pathol.* **20:**129.

Federici, B. A., and Lowe, R. E., 1972, Studies on the pathology of a baculovirus in *Aedes triseriatus, J. Invertebr. Pathol.* **20:**14.

Federici, B. A., and Stern, V. M., 1990, Replication and occlusion of a granulosis virus in larval and adult midgut epithelium of the western grapeleaf skeletonizer, *Harrisina brillians, J. Invertebr. Pathol.* **56:**401.

Flipsen, J. T. M., van Lent, J. W. M., Goldbach, R. W., and Vlak, J. M., 1993, Expression of polyhedrin and p10 in the midgut of AcMNPV-infected *Spodoptera exigua* larvae: An immunoelectron microscope investigation, *J. Invertebr. Pathol.* **61:**17.

Flipsen, J. T. M., Martens, J. W. M., Van Oers, M. M ., Vlak, J. M., and Van Lent, J. W. M., 1995, Passage of *Autographa californica* nuclear polyhedrosis virus through the midgut epithelium of *Spodoptera exigua* larvae, *Virology* **208:**328.

Gijzen, M., Roelvink, P., and Granados, R. R., 1995, Characterization of a viral enhancing protein from *Trichoplusia ni* granulosis virus, *J. Invertebr. Pathol.* **65:**289.

Granados, R. R., 1978, Early events in the infection of *Heliothis zea* midgut cells by a baculovirus, *Virology* **90:**170.

Granados, R. R., and Lawler, K. A., 1981, *In vivo* pathway of *Autographa californica* baculovirus invasion and infection, *Virology* **108:**297.

Granados, R. R., and Williams, K. A., 1986, *In vivo* infection and replication of baculoviruses, in: *The Biology of Baculoviruses,* Vol. I, *Biological Properties and Molecular Biology* (R. R. Granados and B. A. Federici, eds.), pp. 89–108, CRC Press, Boca Raton, FL.

Hamm, J. J., 1968, Comparative histopathology of a granulosis and a nuclear polylhedrosis of *Spodoptera frugiperda, J. Invertebr. Pathol.* **10:**320.

Harrap, K. A., 1970, Cell infection by a nuclear polyhedrosis virus, *Virology* **42:**311.

Harrap, K. A., and Robertson, J. S., 1968, A possible infection pathway in the development of a nuclear polyhedrosis virus, *J. Gen. Virol.* **3:**221.

Hashimoto, Y., Corsaro, B. G., and Granados, R. R., 1991, Location and nucleotide sequence of the gene encoding the viral enhancing factor of the *Trichoplusia ni* granulosis virus, *J. Gen. Virol.* **72:**2645.

Hawtin, R. E., Arnold, K., Ayres, M. D., de A. Zanotto, P. M., Howard, S. C., Gooday, G. W., Chappell, L. H., Kitts, P. A., King, L. A., and Possee, R. D., 1995, Identification and preliminary characterization of a chitinase gene in the *Autographa californica* nuclear polyhdrosis virus genome, *Virology* **212:**673.

Horton, H. M., and Burand, J. P., 1993, Saturable attachment sites for polyhedron-derived baculovirus on insect cells and evidence for entry via direct membranc fusion, *J. Virol.* **67:**1860.

Huger, A., 1963, Granuloses of insects, in: *Insect Pathology, An Advanced Treatise,* Vol. 1 (E. A. Steinhaus, ed.), pp. 531–575, Academic Press, New York.

Jarvis, D. L., and Garcia, Jr., A., 1994, Biosynthesis and processing of the *Autographa californica* nuclear polyhedrosis virus gp64 protein, *Virology* **205:**300.

Keddie, B. A., Aponte, G. W., and Volkman, L. E., 1989, The pathway of infection of *Autographa californica* nuclear polyhedrosis virus in an insect host, *Science* **243:**1729.

Kirkpatrick, B. A., Washburn, J. O., Engelhard, E. K., and Volkman, L. E., 1994, Primary infection of insect tracheae by *Autographa californica* M nuclear polyhedrosis virus, *Virology* **203:**184.

Larsson, R., 1984, Insect pathological investigations on Swedish Thysanura: A nuclear polyhedrosis virus of the Bristletail *Dilta hibernica, J. Invertebr. Pathol.* **44:**172.

Lepore, L. S., Roelvink, P. R., and Granados, R. R., 1996, Enhancin, the granulosis virus protein that facilitates nucleopolyhedrosis (NPV) infections, is a metalloprotease, *J. Invertebr. Pathol.* **68:**131.

Monsma, S. A., Oomens, A. G. P., and Blissard, G. W., 1996, The gp64 envelope fusion protein is an essential baculovirus protein required for cell-to-cell transmission and infection, *J. Virol.* **70:**4607.

Knebel-Mörsdorf, D., Flipsen, J. T. M., Roncarati, R., Jahnel, F., Kleefsmen, A. W. F., and Vlak, J. M., 1996, Baculovirus infection of *Spodoptera exigua* larvae: *lac*Z expression driven by promoters of early genes pe38 and me53 in larval tissue, *J. Gen. Virol.* **77:**815.

O'Reilly, D. R., and Miller, L. K., 1989, A baculovirus blocks insect molting by producing ecdysteroid UDP-glucosyl transferase, *Science* **245:**1110.

O'Reilly, D. R., and Miller, L. K., 1991, Improvement of a baculovirus pesticide by deletion of the EGT gene, *Bio/technology* **9:**1086.

Paillot, A., 1935, Nouvel ultravirus parasite d'*Agrotis segetum* provoquant une proliferation des tissues infectes, *C R Hebd. Seances Acad. Sci.* **201:**1062.

Reddy, J. T., and Locke, M., 1990, The size limited penetration of gold particles through insect basal laminae, *J. Insect Physiol.* **36:**397.

Rohrmann, G. F., 1986, Evolution of occluded baculoviruses, in: *The Biology of Baculoviruses,* Vol. I, *Biological Properties and Molecular Biology* (R. R. Granados and B. A. Federici, eds.), pp. 203–215, CRC Press, Boca Raton, FL.

Smith, K. A., 1955, Intranuclear changes in the polyhedrosis of *Tipula paludosa* (Diptera), *Nature* **176:**255.

Smith, O. J., Hughes, K. M., Dunn, P. H., and Hall, I. M., 1956, A granulosis virus disease of the Western grapeleaf skeletonizer and its transmission, *Can. Entomol.* **88:**507.

Steinhaus, E. A., 1949, *Principles of Insect Pathology,* McGraw-Hill, New York.

Steinhaus, E. A., and Hughes, K. M., 1952, A granulosis virus of the Western grapeleaf skeletonizer, *J. Econ. Entomol.* **24:**744.

Summers, M. D., 1969, Apparent *in vivo* pathway of granulosis virus invasion and infection, *J. Virol.* **4:**188.

Summers, M. D., 1971, Electron microscopic observations on granulosis virus entry, uncoating, and replication process during infection of the midgut cells of *Trichoplusia ni, J. Ultrastruct. Res.* **35:**606.

Tanada, Y., and Hess, R. T., 1991, Baculoviridae. Granulosis viruses, in: *Atlas of Invertebrate Viruses* (J. R. Adams and J. R. Bonami, eds.), pp. 227–257, CRC Press, Boca Raton, FL.

Tanada, Y., and Kaya, H. K., 1993, *Insect Pathology,* Academic Press, San Diego.

Tanada, Y., and Leutenegger, R., 1968, Histopathology of a granulosis virus disease of the codling moth, *Carpocapsa pomonella, J. Invertebr. Pathol.* **10:**39.

Tweeten, K. A., Bulla, L. A., and Consigli, R. A., 1981, Applied and molecular aspects of insect granuloses viruses, *Microbiol. Rev.***45:**379.

van Oers, M. M., Flipsen, J. T. M., Reusken, C. B. E. M., Sliwinsky, E. L., Goldbach, R. W., and Vlak, J. M., 1993, Functional domains of the p10 protein of *Autographa californica* nuclear polyhedrosis virus, *J. Gen. Virol.* **74:**563.

Volkman, L. E., and Knudson, D. L., 1986, *In vitro* replication of baculoviruses, in: *The Biology of Baculoviruses,* Vol. I, *Biological Properties and Molecular Biology* (R. R. Granados and B. A. Federici, eds.), pp. 109–120, CRC Press, Boca Raton, FL.

Volkman, L. E., Blissard, G. W., Friesen, P., Keddie, B. A., Possee R., and Theilmann, D. A., 1995, Family *Baculoviridae,* in: *Virus Taxonomy* (F. A. Murphy, C. M. Fauquet, D. H. L. Bishop, S. A. Ghabrial, A. W. Jarvis, G. P. Martelli, M. A. Mayo, and M. D. Summers, eds.), pp. 104–113, Springer-Verlag, New York.

Walker, S., Kawanishi, C. Y., and Hamm, J. J., 1982, Cellular pathology of a granulosis virus infection, *J. Ultrastr. Res.* **80:**163.

Wang, P., Hammer, D. A., and Granados, R. R., 1994, Interaction of *Trichoplusia ni* granulosis virus-encoded enhancin with the midgut epithelium and peritrophic membrane of four lepidopteran insects, *J. Virol.* **75:**1961.

Washburn, J. O., Kirkpatrick, B. A., and Volkman, L. E., 1995, Comparative pathogenesis of *Autographa californica* M nuclear polyhedrosis virus in larvae of *Trichoplusia ni* and *Heliothis virescens, Virology* **209:**561.

Washburn, J. O., Kirkpatrick, B. A., and Volkman, L. E., 1996, Insect protection against viruses, *Nature* **383:**767.

CHAPTER 4

Cytological Changes and Viral Morphogenesis during Baculovirus Infection

GREG V. WILLIAMS AND PETER FAULKNER

I. INTRODUCTION

A. General

The family *Baculoviridae* encompasses a group of arthropod-specific DNA viruses that are characterized by a circular DNA genome packaged within a rod-shaped capsid and enclosed by a lipid envelope (Volkman *et al.*, 1995). At some stage of their life cycle all baculoviruses exist as an occluded form in which virions are embedded within a proteinaceous matrix (Miller, 1996). The family is divided into the genera nucleopolyhedrovirus [nuclear polyhedrosis viruses (NPV)] and granulovirus [granulosis viruses (GV)], which differ in the morphology of their occluded virion form as well as the cytopathology they induce. The occlusion matrix proteins polyhedrin (NPV) and granulin (GV) are closely related and highly conserved, and have been used to assist in the taxonomy of baculoviruses (Rohrmann, 1986a,b; Zanotto *et al.*, 1993). NPV have polyhedral-shaped occlusion bodies, 0.5–15μm in diameter, which contain many virions embedded either as single nucleocapsids per envelope (SNPV) or multiple nucleocapsids per envelope (MNPV), and morphogenesis occurs within the nucleus of infected cells. In contrast,

GREG V. WILLIAMS • CSIRO, Australian Animal Health Laboratory, Geelong, Australia 3220. PETER FAULKNER • Department of Microbiology and Immunology, Queens University, Kingston, Ontario K7L 3N6 Canada.

The Baculoviruses, edited by Lois K. Miller. Plenum Press, New York, 1997.

GV produce much smaller ovicylindrical occlusions (capsules, 0.16–0.30 μm × 0.30–0.50 μm) that contain one or occasionally two virions. Replication of GV begins in the cell nucleus but induces apparent breakdown of the nuclear membrane, such that morphogenesis is completed in the cytoplasm or hybrid cellular compartment (Crook, 1991; Granados and Williams, 1986). Thus, GV cytopathology is distinct from that of NPV. Attempts to produce a satisfactory tissue culture system for the propagation and study of GVs have had limited success (Kelly, 1982; Tanada and Hess, 1991), and consequently most studies describing the structure and morphogenesis of granulosis viruses have been done using tissues derived from infected insects. Winstanley and Crook (1993) have recently established new cell lines that, when maintained 21°C, are permissive for *in vitro* replication of *Cydia pomonella* granulosis virus (CpGV). Serial passage of the cells at higher temperature, as is the standard for insect tissue culture, caused gradual and permanent loss of susceptibility to CpGV infection.

Many continuous insect cell lines are permissive for MNPV replication, and their use has facilitated detailed investigations of virus molecular biology and morphogenesis (Bilimoria, 1991; Blissard, 1996; Faulkner and Carstens, 1986; Vaughn and Dougherty, 1985; Volkman and Knudson, 1986). This review is centered on the morphogenesis and cytopathology of the type species of NPV, *Autographa californica* MNPV, in *Spodoptera frugiperda* and *Trichoplusia ni* continuous cell lines, since this is the best-characterized baculovirus system reported on to date and it represents the standard to which other baculoviruses are compared.

B. NPV Molecular Biology

The genome of baculoviruses is composed of a single covalently closed circular dsDNA molecule (reviewed by Faulkner, 1981; Miller, 1986) 90–160 kilobasepair (kbp) in size with a G+C content ranging from 28 to 59%. There is no correlation of base composition with morphological subdivisions (MNPV, SNPV, GV) (Harrap and Payne, 1979). AcMNPV (C6 strain) was fully sequenced (Ayres *et al.*, 1994) and found to be 133,894 bp in length with an overall A+T content of 59%. Genome analysis identified 337 open reading frames (ORFs) initiated with a methionine codon and encoding polypeptides of 50 or more amino acids (Ayres *et al.*, 1994). Of these, 154 were chosen for further analysis based on a criteria of the ORF being a single, contiguous, nonoverlapping coding region. Although many of the coding regions that are actually utilized during viral replication remain to be determined, the coding potential of the NPV genome underscores the complexity of their replication, which results in the production of two genotypically identical but phenotypically distinct forms of virion.

Several strains of AcMNPV are under study, including E-2 (Smith and Summers, 1979), L-1 (Miller and Dawes, 1979), D (Lubbert *et al.*, 1981), E (Tjia *et al.*, 1979), C6 (Possee *et al.*, 1991; Possee, 1986), and HR3 (Brown *et al.*, 1979; Cochran *et al.*, 1982). A standard restriction map and genome orientation was proposed by Vlak and Smith (1982) to facilitate comparison of strains. Genes generally appear to lack introns, with the exception of one known immediate early gene that contains an intron and is processed by cellular transcription and splicing machinery (Chisholm and Henner, 1988) (see Chapter 5, this volume).

Transcription begins immediately after release of viral DNA into the cell nucleoplasm and these very early transcripts are grouped in a class called immediate early (IE) or α class genes (see Chapter 6, this volume). Later transcripts appear in a temporal fashion, as products of each progressive phase are required for the expression of the next phase. Classically, the gene expression phases recognized for NPV are α (immediate early), β (delayed early), γ (structural or late), and δ (temporally delayed and hyperexpressed late genes) (Blissard and Rohrmann, 1990; Friesen and Miller, 1986). Delayed early genes are transcribed in the presence of α gene products, and together these constitute the early phase of infection, until the start of viral DNA replication and γ class expression at about 6 hr postinfection (p.i.) (Tjia *et al.*, 1979). Significant overlap of temporal expression occurs (Nissen and Friesen, 1989; Tomalski *et al.*, 1988), and it is during the γ phase (6–18 hr p.i.) that host protein synthesis is significantly curtailed (Friesen and Miller, 1986). The δ phase genes are distinct in that transcription of this class is maximal by 24 hr p.i. and continues until cell death, well after γ class expression has curtailed. The predominant δ gene products are the occlusion matrix protein polyhedrin (Rohel *et al.*, 1983) and the nonstructural polypeptide designated p10 (Kuzio *et al.*, 1984; Rohel *et al.*, 1983), and these may comprise a very high proportion of cell protein mass; polyhedrin alone constitutes about 20% of total cell protein by the end of the infection cycle (Blissard and Rohrmann, 1990; Rohrmann, 1986a). The γ and δ genes are transcribed from a consensus late promoter start site, the TAAG motif (Blissard and Rohrmann, 1990) (see Chapter 8, this volume).

C. Overview of Baculovirus Replication and Morphogenesis

1. *In Vitro*

The replication and morphogenesis of baculoviruses have been studied most intensely in cell culture systems. Baculoviruses have a biphasic replication cycle that produces two different forms (Fig. 1), which have a common nucleocapsid structure and carry identical genetic information (reviewed by Blissard, 1996; Faulkner, 1981; Rohrmann, 1992). The form responsible for horizontal transmission between insect hosts is the occluded virus (OV), whereas systemic spread through the insect host and propagation in tissue culture are dependent on a budded virus (BV) form, which acquires its envelope at the host cell plasma membrane.

Nucleocapsids gain access to the cytoplasm proper and then migrate to

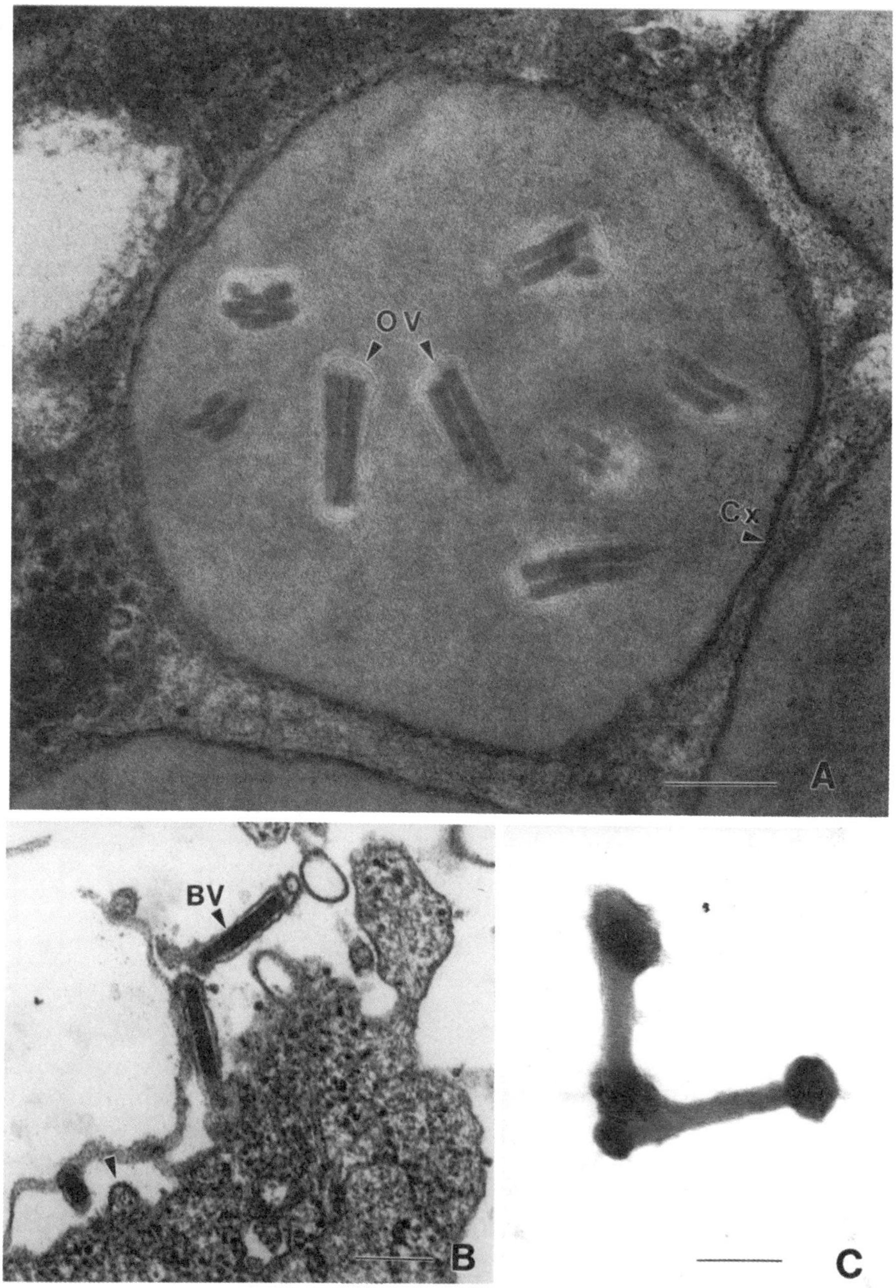
OV
Cx
A
BV
B
C

the nuclear membrane by an unknown mechanism; it is generally accepted that uncoating occurs at nuclear pore complexes or just inside the nuclear membrane (Bilimoria, 1991). A complex cascade of transcription, protein synthesis, and the initiation of viral DNA replication occurs within the first 6–8 hr following NPV infection. Early cytopathology that begins during this period includes nuclear hypertrophy (swelling) followed by progressive alteration of cell shape (rounding), with coincident changes to heterochromatin distribution and nucleolar morphology. The switchover from early to late viral gene expression (Blissard, 1996) coincides with the start of viral DNA replication and intranuclear development of the viral replication center, called the virogenic stroma (Knudson and Harrap, 1976). The virogenic stroma provides the microenvironment necessary for assembly of progeny nucleocapsids, which initially egress from the cell nucleus to bud at the plasma membrane (BV). Later in infection, BV production is curtailed and nucleocapsids are diverted by intracellular envelopment and incorporation within the matrix of OV. Cellular infection culminates in cytolysis and the release of mature OV (Van Oers *et al.*, 1993; Williams *et al.*, 1989).

Several parameters that may alter the outcome of infection influence the progression of NPV infection in cell culture. For example, there is substantial variability in permissiveness for NPV replication among cell lines (Carpenter and Bilimoria, 1983; Knudson, 1978; Lynn and Hink, 1980), even among those derived from the same insect species (Vaughn and Dougherty, 1985). Both *T. ni* Tn368 and *S. frugiperda* Sf21 cells are permissive for replication of AcMNPV, but plaque assays in each cell line yielded an apparent three- to fivefold difference in titer (Brown and Faulkner, 1978). Lynn and Hink (1980) compared the growth of AcMNPV in five insect cell lines from different insect species. Although AcMNPV infection was productive in all cell lines, yield and relative virulence of progeny OV tested *in vivo* varied significantly. Some cell lines may support only one NPV species, while other lines support several viruses. Even with the same cell lines, factors local to a particular laboratory such as cell incubation temperature (Dougherty *et al.*, 1981; Kelly, 1981; Knudson and Tinsley, 1974), composition of the culture medium (nutrients, inorganic components) (Goodwin *et*

FIGURE 1. The occluded and budded forms of baculoviruses. (A) Thin section of AcMNPV occluded virus (OV) in Sf21 cells at 38 hr p.i. Enveloped nucleocapsid bundles (arrowheads by OV) are dispersed throughout the paracrystalline matrix of the occlusion body, which in turn is encapsulated within a sheathlike structure, the calyx (Cx). (B) AcMNPV nucleocapsid budding at the plasma membrane occurs at outpockets enriched with peplomer structures (arrowhead), and these are present on a cap structure that remains localized to the apical end of the budded virus (BV) particle. Arrays of microtubules and microfilaments are present along the cell periphery adjacent to regions of virion budding, but have an uncertain affiliation with the process. A small vesiclelike structure devoid of peplomers is frequently seen on the basal end of the virion particle, and these loose-fitting end structures can be seen by EM examination of whole virions from tissue culture supernatant (panel C). Scale bars, 200 nm (panels A, B) and 100 nm (panel C).

al., 1973; Goodwin and Adams, 1980), and starting cell density and cell growth rate (Wood *et al.*, 1982) may affect the efficiency and temporal details of NPV replication (reviewed by Bilimoria, 1991; Volkman and Knudson, 1986). Similarily, the multiplicity of infection determines the proportion of cells infected (Dougherty *et al.*, 1981) and affects the timing of replicative events (Knudson and Tinsley, 1974) and relative yields of BV and OV (Brown and Faulkner, 1975).

2. *In Vivo*

During the natural life cycle of baculoviruses, infection of a suitable host insect is initiated by the ingestion of OV and subsequent dissolution of the occlusion matrix by the highly alkaline midgut environment, thereby releasing occlusion-derived virions (ODVs). Virions must traverse the peritrophic membrane (PM), a tubular barrier that surrounds the food bolus, to gain access to midgut cells (Derksen and Granados, 1988; Romoser, 1981; Stoltz and Summers, 1971). Factors released by dissolution of OV transiently caused major alterations to the integrity and elasticity of the PM (Derksen and Granados, 1988), presumably to enhance virion access to the midgut tissue. ODVs bind to the brush border microvilli of epithelial cells to establish the primary infection and release nucleocapsids into the cytoplasm proper: Penetration is insensitive to inhibitors of adsorptive endocytosis (Horton and Burand, 1993), and ultrastructural studies suggest direct fusion of viral and cellular membranes as the mechanism of ODV entry (Granados and Lawler, 1981; Harrap, 1970; Summers, 1971). Nucleocapsids may proceed to the nucleus and uncoat, although there is evidence that some nucleocapsids from the original inoculum may pass directly through the cell to infect underlying cells or gain entry to the hemocoel (Granados and Lawler, 1981; Keddie *et al.*, 1989) (see Chapter 3, this volume). Nucleocapsids that pass directly through cells by this mechanism appear to acquire a standard BV envelope, complete with apical spikes (peplomers): A possible explanation for this phenomenon has been postulated in a recent review (Blissard, 1996). Virion penetration and uncoating within the midgut cells initiate the cascade of transcription and protein synthesis as described above for replication *in vitro*. Progeny virions emerge shortly after formation of the virogenic stroma by budding from the basolateral surface of the midgut cells at sites enriched with the major envelope glycoprotein GP64 (Keddie *et al.*, 1989).

Infection throughout the host insect is facilitated by the spread of progeny BV to neighboring cells and tissues. Several potential invasion pathways to effect systemic spread have been recognized, although their relative importance remains to be determined. Direct penetration of BV through the midgut basal lamina was observed by electron microscope (EM) when high doses of OV ($>5 \times 10^6$/larva) were used to initiate infection (Granados and Lawler, 1981); however, systemic spread along trachioles appeared to be the major conduit for virus when insects were inoculated *per os* with lower doses (Engelhard *et al.*, 1994). Once infection is established at secondary sites, further spread within the target tissue is facilitated wholly or in part by direct cell-to-cell transmission (Engelhard *et al.*, 1994; Keddie *et al.*, 1989). Additional perspectives on *in vivo* pathogenesis are provided in Chapter 3 (this volume).

There is a pronounced variation in tissue tropism among baculoviruses. Infection by the European spruce sawfly NPV is restricted to cells of the midgut epithelium (monoorganotrophic) (Benz, 1986; see Chapter 3), whereas many other NPV infections progress to involve most host tissues (polyorganotrophic) (Faulkner, 1981; Granados and Williams, 1986). Little is known about the underlying mechanisms that determine baculovirus tropism. A recent study utilized recombinant AcMNPV strains that carried reporter genes under control of early versus late viral promoters and showed that, in the case of the midgut, virus was able to gain entry into most or all cell types of *Spodoptera exigua* larvae, but infection did not progress in nonpermissive cell types such as goblet cells (Knebel-Mörsdorf *et al.*, 1996). The implications of these findings focuses attention on the virus–cell interactions following attachment and penetration, and recent studies of baculovirus replication *in vitro* are beginning to eludicate principles of these interactions on the molecular level (Clem *et al.*, 1991; Crook *et al.*, 1993; Kamita and Maeda, 1993; Maeda *et al.*, 1993). Progression of polyorganotrophic NPV infection in the insect host indicates a further level of complexity to viral pathogenesis. In the case of AcMNPV, infected cells are cleared from midgut tissue by shedding into the gut lumen after systemic spread has occurred, and by 60 hr p.i., the tissue has almost fully recovered (Englehard *et al.*, 1994; Keddie *et al.*, 1989). Clearing infection from this site may benefit the virus by maintaining the gut in a functional state and extending insect feeding (Volkman and Keddie, 1990), and therefore provide metabolites that would not be available if the tissue was heavily or chronically infected.

II. CELL INFECTION AND EARLY CYTOPATHOLOGY

A. Virus Attachment, Entry, and Uncoating

BVs and ODVs have unique envelope compositions and appear to enter cells by different mechanisms (Rohrmann, 1992; see Chapter 2, this volume). Ultrastructural studies have shown that ODV attach to the brush border microvilli of midgut epithelial cells and probably gain entry into the cytoplasm by direct fusion of membranes (Granados and Lawler, 1981; Harrap, 1970; Horton and Burand, 1993; Kawanishi *et al.*, 1972; Summers, 1971). The interaction of ODV with the brush border membrane probably involves a specific cellular ligand (receptor), because virus binding was saturable in an *in vitro* test system and reactivity was abrogated by pretreatment of target cells with proteinases (Horton and Burand, 1993). Virus attachment to target membranes and subsequent membrane fusion was demonstrated and found to be optimal at alkaline pH, coinciding with the native midgut environment. In contrast, BV attach to and enter cells in culture primarily by adsorptive endocytosis (Blissard and Wenz, 1992; Volkman and Goldsmith, 1985). Viral particles appear to be internalized into clathrin-coated vesicles by invagination of the plasma membrane: Subsequent fusion of the viral envelope with the endosome membrane is thought to release the viral nucleocapsid into the cytoplasm proper (Blissard, 1996). Some BV may enter cells by membrane fusion at the cell surface (Volkman *et al.*, 1986), although the significance of this mechanism is unknown. The major envelope glycoprotein of BV is designated GP64 EFP (envelope fusion protein) (Blissard, 1996; Volkman and Goldsmith, 1985; Volkman *et al.*, 1984; Whitford *et al.*, 1989). This protein can mediate membrane fusion at low pH (Monsma and Blissard, 1995), but it is not known whether GP64 EFP is responsible for binding to the cell surface. Kinetic studies suggest that BVs also bind to specific receptor structures on the target membrane in a saturable manner (Wickham *et al.*, 1992), but the molecules involved in these interactions have not been identified for either phenotype.

Nucleocapsids traverse the cytoplasm and align with nuclear pores within a 1 to 4 hr window following inoculation both *in vivo* and *in vitro* (Granados, 1978; Knudson and Harrap, 1976; Vaughn and Dougherty, 1985). A capsid-associated kinase (Miller *et al.*, 1983; Wilson and Consigli, 1985a) may be involved in the release of the viral genome at or just inside a nuclear pore by phosphorylation of the highly basic core protein p6.9 (Funk and Consigli, 1993; Wilson and Consigli, 1985b).

Release of the viral genome into the nucleoplasm of a susceptible cell is rapidly followed by transcription of immediate early (α class) viral genes, most likely through utilization of host cell RNA *pol*II and associated cellular machinery (Fuchs *et al.*, 1983; Hoopes and Rohrmann, 1991; Huh and Weaver, 1990a,b). Virus-specific RNA can be detected in cells by 30 min postinoculation (Chisholm and Henner, 1988). Because transfection of viral DNA into insect cells yields productive infection *in vitro* (Burand *et al.*, 1980; Carstens *et al.*, 1980; Kelly and Wang, 1981; Potter and Miller, 1980), it is generally accepted that productive infection does not require virion-associated proteins to initiate transcription of immediate early genes or to alter host replication machinery. The presence of the transcriptional activator IE1 (immediate early 1) has been reported in the BV phenotype of *Orgyia pseudotsugata* MNPV (OpMNPV) (Theilmann and Stewart, 1992), although the significance of this finding is currently unknown.

B. Correlation of Cytopathology and Molecular Biology of Early Infection

The entry of NPV into insect cells induces formation of F-actin bundles within 30 min after inoculation in a process that does not require new protein synthesis (Charlton and Volkman, 1991, 1993a). Single nucleocapsids were usually associated with one end of an F-actin tract, and the number of tracts formed in the cytoplasm was proportional to the multiplicity of infection used in the experiment. However, addition of cytochalasin D to depolymerize F-actin prior to infection did not cause obvious differences relative to controls (Volkman *et al.*, 1987). In a study of NPV infection in midgut epithelial cells, Granados (1978) reported possible association between nucleocapsids traversing the cytoplasm and microtubules, perhaps indicating tissue-specific cytoskeletal interactions. Immobilized F-actin was reported to bind, directly or indirectly, the major capsid protein p39 and a 67-kDa polypeptide from infected cell extracts (Charlton and Volkman, 1991; Lanier *et al.*, 1996).

Within the first few hours following infection, the virus must seize control of the host cell's metabolic machinery, and on a molecular level the virus undergoes a frenetic cascade of activity to achieve this goal. In addition to expressing the early genes required to initiate viral DNA replication, the virus must counteract host defense mechanisms such as apoptosis (see Chapter 10, this volume). Considering the numerous biochemical changes that are occurring during this period, there are remarkably few gross cytological changes observable.

III. THE ASSEMBLY AND STRUCTURE OF THE REPLICATION COMPLEX

A. Development of the Virogenic Stroma and Concomitant Changes to Nuclear Structure

The onset of late viral gene expression at about 6 hr p.i. (AcMNPV) is correlated with nuclear hypertrophy (swelling) and progressive cell rounding. The latter is attributed to significant reorganization of the cytoskeleton, including changes to the microtubules (Volkman and Zaal, 1990) and microfilaments (Charlton and Volkman, 1991), which may influence the state of the host cell nucleus (Ben-Ze'ev, 1983). Within the cell nucleus, clumps of heterochromatin normally seen dispersed throughout the nucleoplasm are reduced or redistributed and may be displaced so that the heterochromatin generally becomes progressively marginated along the inner nuclear membrane. Nucleolar morphology is also affected (Benz, 1986; Tanada and Hess, 1976), and there is a transient increase in the abundance of RNA detected by histochemistry (Benz, 1986). It is not known whether nucleolar changes result from alterations to the organelle-associated heterochromatin (nucleolar organizing region), or if other cytopathological changes affect the structure. Coincident with these changes is the development of the intranuclear viral replication center, the virogenic stroma (Vaughn and Dougherty, 1985; Volkman and Knudson, 1986), which is first seen as a loose granular material dispersed throughout the nucleoplasm (Fig. 2A.)

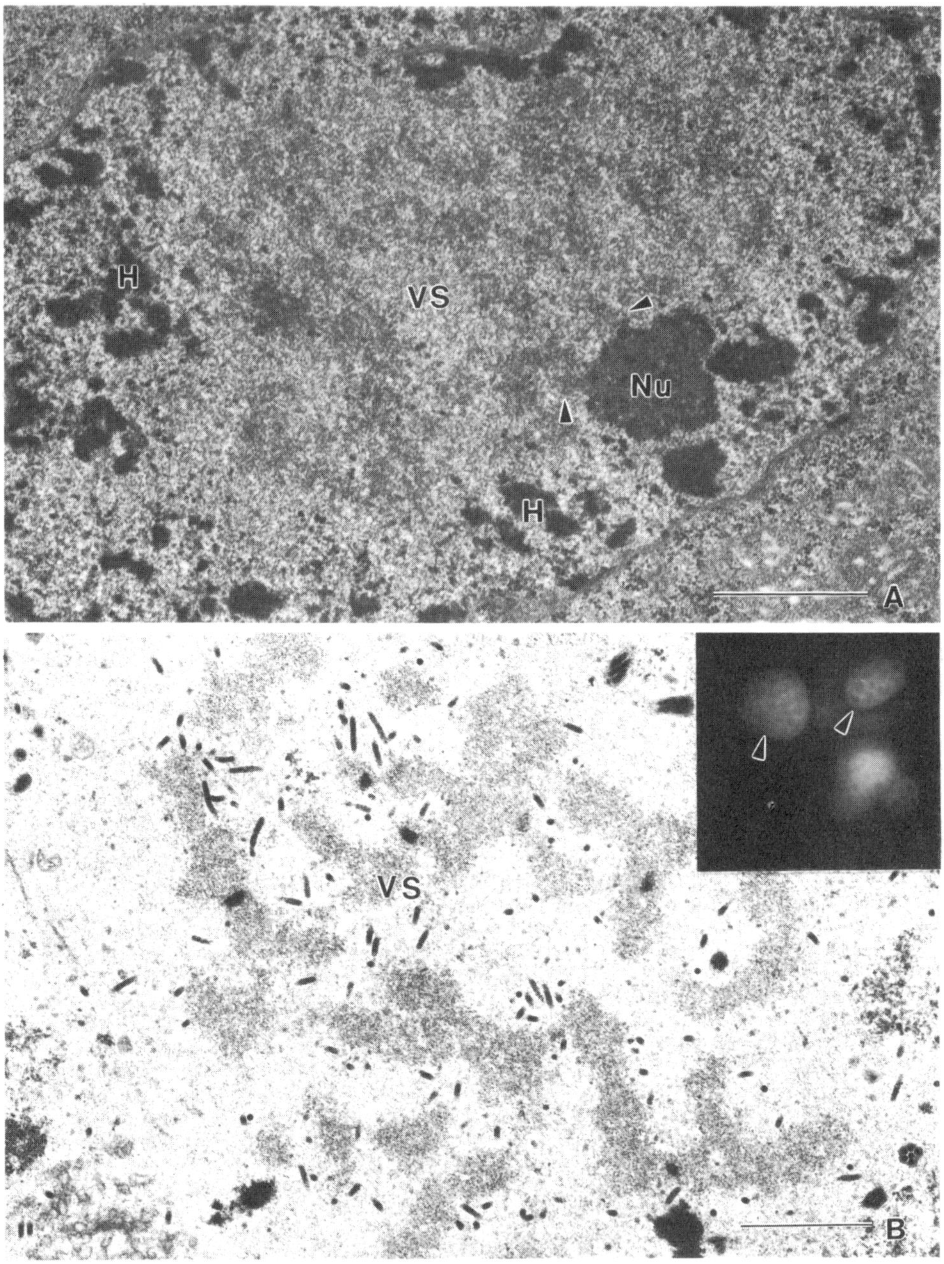

FIGURE 2. Formation of the virogenic stroma. (A) TEM of early virogenic stroma (8 hr p.i.). The virogenic stroma (VS) is first manifest as a loose (dispersed) granular matrix within the cell nucleus, often in association with nucleoli (Nu, arrowheads). Clumps of heterochromatin (H) are displaced peripherally by the forming stroma. There is usually no evidence of capsid formation at this time. (B) By 22 hr, the intrastromal spaces of the virogenic stroma are enriched in developing nucleocapsids. Maturation of the stroma coincides with its condensation into a dense latticelike structure, which is DNA-rich by DAPI staining (B, inset). Scale bars, 1 μm.

The virogenic stroma is considered a *de novo* product of baculovirus infection in which progeny virions are assembled. The stroma forms around the time viral DNA replication is initiated (6–8 hr p.i.), and progeny nucleocapsids appear in the stromal network shortly thereafter (about 10 hr p.i.). The onset of these morphological events coincides with the switch over from early to late viral gene expression (Blissard, 1996; Faulkner, 1981; Knudson and Harrap, 1976), involving the appearance of a novel virus-induced RNA polymerase activity (Beniya *et al.*, 1996; Yang *et al.*, 1991) (see Chapter 8, this volume). Viral DNA synthesis continues for at least 12 hr, with the level of intracellular viral DNA plateauing by 18 hr p.i. (Erlandson and Carstens, 1983; Tjia *et al.*, 1979). The virogenic stroma persists throughout infection, although the structure appears to regress and contains few if any nucleocapsids after OV formation.

By 16–18 hr p.i., the virogenic stroma condenses into a dense (mature) structure, usually in the central region of the nucleus, and contains a large quantity of nucleocapsids (Fig. 2B). When observed in the electron microscope the mature stroma has two distinct regions: a fibrillar electron-dense matte forms a prominent "chromatic mass" interspersed with electron-lucent (granular) intrastromal spaces in which progeny nucleocapsids are assembled (Harrap, 1972b; Summers, 1971; Young *et al.*, 1993). The chromatic mass stains intensely with DNA-specific fluorescent dyes such as 4′, 6-diamidino-2-phenylindole (DAPI) and propidium iodide from very early in infection (6 hr p.i.), and the staining pattern changes from dispersed or amorphous early in stromal development to a lattice or cagelike appearance as the stroma condenses into the mature structure (Fig. 2B, inset). The distribution of staining in the mature stroma also indicates that the concentration of DNA occurs within the matte and not the intrastromal spaces. Maturation of the virogenic stroma together with progressive nuclear hypertrophy yield a significant and morphologically distinct peristromal compartment of nucleoplasm, called the ring zone (Benz, 1986). It is within this compartment that intranuclear envelopment of virions and occlusion body formation take place.

The molecular basis for changes in nuclear architecture and factors governing the assembly of the virogenic stroma are not known. In general, the degree of condensation of chromatin is principally regulated by modification of histone proteins and possibly alterations to the nuclear scaffold. Wilson and Miller (1986) examined the changes in nucleoprotein complexes of host and viral DNA at several times during infection using a micrococcal nuclease assay. Viral DNA appeared to adopt a chromatinlike nucleosome structure initially, but adopted a unique nucleoprotein structure between 10 and 24 hr p.i. Host chromatin, however, remained in a nucleosomal structure. However, histones were not examined for chemical modifications that may have influenced the condensation of the host chromatin. Margination of the host cell chromatin occurs during the late phase and roughly coincides with down-regulation of the steady-state levels of host transcripts between 12 and 18 hr p.i. (Ooi and Miller, 1988), although a causal relationship has not been established. Within a similar time frame, cellular DNA and protein synthesis are also inhibited (Bilimoria, 1991; Vaughn and Dougherty, 1985), and by 8 to 12 hr p.i., only a small percentage of DNA synthesis is cellular. Host cell protein synthesis is reduced beginning at about 10 hr p.i., and its shutoff is complete at 24 hr p.i. (Carstens *et al.*, 1979; Kelly and Lescott, 1982).

The decline in the steady-state levels of host transcripts and protein synthesis does not occur when viral DNA synthesis is inhibited by aphidicolin treatment (Ooi and Miller, 1988; Rice and Miller, 1986). Similarly, studies using temperature-sensitive mutants have demonstrated that in the absence of viral DNA replication, both host DNA synthesis (Brown *et al.*, 1979) and protein synthesis (Gordon and Carstens, 1984) are maintained. Because the shutoff of host gene expression does not occur when DNA replication and subsequent late gene expression is blocked, it is thought that viral early genes do not play a direct role in host synthesis shutoff. As in the case of host shutoff, formation of the virogenic stroma is dependent on viral DNA synthesis and late gene expression. Two temperature-sensitive mutants have been described that are defective in late gene expression and cause disappearance of heterochromatin and nucleoli but fail to form a defined stroma within most cells (Carstens *et al.*, 1994).

B. Substructure of the Virogenic Stroma

Little is known about the fine structure and molecular components of the virogenic stroma or the degree to which the host cell contributes to its structure or function. The constituents of progeny nucleocapsids are known to accumulate within the stromal matte, but it is generally accepted that capsid shell formation and nucleocapsid maturation occur in the intrastromal spaces (Fraser, 1986; Harrap, 1972b). A recent study provided evidence of specialized substructures within the stromal mat (chromatic mass) (Young *et al.*, 1993). DNase-I-sensitive filament bundles were revealed by purification of nuclei in the presence of 0.5% Nonidet P40 and 250 m$\underline{M}$ sucrose. These structures localized throughout the stromal matte and were intimately associated with nascent capsids in the adjacent instrastromal spaces (lacunae); hence, the filament bundles were postulated to be precondensed viral genomes. The distribution of filament bundles was consistent with the intrastromal distribution of the basic DNA-binding protein p6.9 (Williams, 1995), which is associated with viral genomes within the capsid core (Maeda *et al.*, 1991; Russell and Rohrmann, 1990b; Tweeten *et al.*, 1980; Wilson *et al.*, 1987). Another major viral protein associated principally with the stromal matte is the 39K early gene phosphoprotein, pp31 (Guarino *et al.*, 1992). The protein has four internal basic regions that were attributed to nuclear localization and accumulation in the virogenic stroma and to binding to dsDNA (Broussard *et al.*, 1996a). The intranuclear localization and DNA-binding properties of pp31 may be modulated by dynamic phosphorylation (Broussard *et al.*, 1996b). These recent studies on pp31 are the first to recognize specific motifs required for partitioning within the virogenic stroma. Although immunolocalization of pp31 suggests that it is a major constituent of the stromal matte in infected cells, its constitutive expression in transformed Sf9 cells was manifest as an amorphous intranuclear distribution, without gross evidence of higher-order assembly or association (G. V. Williams, unpublished observations).

The nuclear matrix may be defined as the major underlying scaffold structure of the eukaryotic nucleus, and by definition it is the fraction resistant to extraction by successive treatments with nucleases (RNase, DNAse), nonionic detergent, and high salt (2 M NaCl). A large amount of viral protein remains associated with the nuclear matrix fraction following extraction procedures (Wilson and Price, 1988), including the stroma-associated proteins p6.9 and 39K (pp31). The morphology of the virogenic stroma is severely disrupted by nucleases or moderate salt (0.4 M) treatment (Young *et al.*, 1993), and thus is sensitive to considerably milder conditions than treatments used to reveal the nuclear matrix.

The structural foundation of the virogenic stroma may be built on a preexisting scaffold provided by the nuclear matrix, but it is structurally distinct and should not be considered an extension of the nuclear scaffold. For NPV morphogenesis, the nuclear matrix may be utilized for several intranuclear events, including OV formation. The occlusion matrix protein polyhedrin is found in the nuclear matrix fraction (Wilson and Price, 1988), and disruption of whole nuclei late in infection by gentle procedures causes release of virogenic stroma ringed by OV than remain spatially intact (Fig. 3A). Other intranuclear viral structures, such as preoccluded virions or intranuclear membrane patches, also appear to be aligned on "strings" of unidentified material (Fig. 3B,C). Although these structures are not seen in morphologically intact preparations, similar stringlike structures can be revealed by immunogold localization (Fig. 3D), and most likely reflect a *de facto* scaffold utilized for transport or assembly of virion components rather than an artifact of preparation.

The virogenic stroma is considered the most probable site of viral DNA replication and late transcription (Volkman and Keddie, 1990), and its assembly parallels the onset of these processes (Knudson and Harrap, 1976). Recent attempts to identify components required for viral DNA replication and late gene expression by transient complementation assay have shown that pp31, which localizes to the stroma, is necessary for transcription or transcript processing or stability (Lu and Miller, 1995; Todd *et al.*, 1995), and is consistent with the hypothesis that late viral transcription occurs within the stroma. In addition, extraction of infected cell nuclei by 0.4 M salt solubilizes a pool of proteins that includes the baculovirus-induced RNA *pol* activity (Beniya *et al.*, 1996), and these conditions also cause severe disruption and extraction of the virogenic stroma (Young *et al.*, 1993).

VS
OB
A
NM
C
D
Ph
B

RNase treatment of NPV-infected nuclei caused severe disruption of the stromal matrix, suggesting that this structure contains a significant amount of RNA (Young *et al.*, 1993) in addition to protein and DNA. However, sites of DNA replication and transcription have not been mapped by *in situ* techniques, so their localization to the virogenic stroma has not been demonstrated directly.

Studies utilizing transient complementation assays have demonstrated the requirement of 18 viral genes for replication of plasmids transfected into insect cells and for transcription from late viral promoters (Lu and Miller, 1995; Todd *et al.*, 1995, 1996). A total of nine viral genes were necessary for plasmid replication in Sf21 cells, including the putative DNA polymerase (*dnapol*) and helicase (*p143*) genes, and an additional nine viral genes were required for transcription from late and very late viral promoters (see Chapter 7 and 8, this volume). There is currently little information regarding the intracellular localization of any of the gene products involved, but association with the virogenic stroma is likely. Structural and functional analyses of the virogenic stroma have been hampered by the difficulties associated with cell fractionation and isolation of pure stroma, and the likelihood for selective extraction or contamination during cell fractionation is high. An alternate approach that may yield higher-quality data on the assembly, cellular associations, and fine structure of the viral replication complex may be the use of subviral transient expression (complementation) systems as described above to map and dissect morphological changes in the host cell nucleus.

Potential roles of the homologous repeat sequences (HRS) in the accumulation or positioning of viral DNA within the virogenic stroma have yet to be assessed. The HRS are interspersed throughout baculovirus genomes. In transient expression assays, HRS act as enhancer elements, especially of delayed early mRNA transcription (Guarino and Summers, 1986a,b; see Chapter 6, this volume). HRS also act as origins of DNA replication in plasmid replication assays (Kool *et al.*, 1993a,b; Leisy and Rohrmann, 1993; Pearson *et al.*, 1992; see Chapter 7, this volume).

←

FIGURE 3. Nuclear substructure utilized by baculoviruses during replication and morphogenesis. (A) Scanning EM micrograph of intact nuclear contents from a AcMNPV-infected Sf21 cell (36 hr p.i.). Lysis of purified nuclei directly onto specimen stubs before aldehyde fixation preserves the spatial relationship of OV situated around the virogenic stroma (VS). (B, C) Postfixation extraction of whole cell preparations by TEM (36 hr p.i.) indicates association of viral structures such as *de novo* membrane patches (arrowheads) and polyhedrin condensations (Ph) within a matrix of fine fibers that extend throughout the nucleus. (C) Intranuclear microvesicles are often observed extending from the nuclear membrane (NM) to patches of *de novo* membrane. (D) Immunogold staining using monoclonal antibodies directed against select NPV nonstructural proteins demonstrates the presence of tracts of viral protein throughout the nucleus of infected cells and perhaps indicates distribution along the scaffold structure revealed by extraction. Scale bars, 5 μm (A), 1 μm (B, D), and 500 nm (C).

IV. NUCLEOCAPSID ASSEMBLY, MIGRATION, AND BUDDING AT THE PLASMA MEMBRANE

A. Nucleocapsid Morphogenesis within the Virogenic Stroma

The virogenic stroma is central to the replication of baculoviruses, as it provides the environment for assembly of progeny nucleocapsids. Baculovirus nucleocapsids are composed of a tubular capsid shell with distinct cap structures on each end, enclosing an inner nucleoprotein core in which the genome is condensed about 100-fold (Bud and Kelly, 1980). The supercoiled viral genome is complexed with a highly basic protein (pI 12.8–12.9) that has been identified in other baculoviruses (Maeda *et al.*, 1991; Russell and Rohrmann, 1990b; Tweeten *et al.*, 1980; Wilson *et al.*, 1987). Morphological and other data suggest that viral genomes are prepackaged with the highly basic DNA-binding protein designated p6.9 (NPV) or VP12 (GV), perhaps at specialized regions within the stromal matte (Young *et al.*, 1993). These nucleoprotein complexes are subsequently inserted into the preformed capsid shells that dock with or perhaps are assembled along the edges of the stromal matte within the intrastromal spaces (Fraser, 1986; Mackinnon *et al.*, 1974). Assembly of the nucleoprotein core may be regulated by the state of phosphorylation of p6.9, which modulates its affinity for the viral DNA (Funk and Consigli, 1993; Oppenheimer and Volkman, 1995); this polypeptide is arginine- and serine–threonine-rich and may condense with the viral genome by ionic interactions between arginine residues and the phophopentose backbone of the DNA and with tyrosine residues intercalated among the stacked base pairs (Kelly *et al.*, 1983). The nucleoprotein core may also contain the highly basic polyamines putrescine and spermidine, which were found associated with *Spodoptera littoralis* MNPV OV (Elliot and Kelly, 1977) and were present in sufficient quantity to neutralize 16.4% of phosphates on the vDNA, and so may aid p6.9 with packaging of the genome. However, localization within the virion was not determined, and the polyamines might instead be associated with the virus envelope, as is the case with herpes simplex virus (Gibson and Roizman, 1971).

Specialized end structures found on baculovirus nucleocapsids are distinct, with a flat disk on one end called a base plate (basal end) and a nipple-shaped structure on the other (apical) end (Federici, 1986; Kawamoto *et al.*, 1977a; Summers, 1971); hence, the nucleocapsids display polarity. The structures may contribute to assembly of nucleocapsids, and it has been speculated that the apical end cap of the virus may mediate packing of the nucleoprotein into the capsid (Fraser, 1986). The empty capsid shell appears to form in association with and perhaps by sequential addition to the basal plate structure (Fraser, 1986). A structural comparison of baculovirus capsids by optical diffraction could not distinguish between GV and NPV, and suggested an organization of 4.5-nm stacked rings (Beaton and Filshie, 1976), indicating a highly conserved capsid structure. Molecules involved with the end structures have not been identified, although Vialard and Richardson (1993) identified a 78-to 83-kDa NPV phosphoprotein, encoded by ORF 9 (also known as ORF 1629), that associates with the nucleocapsid in a polarized fashion and may be a component of one of the end structures. Interestingly, the protein localizes to the ring zone of infected cells at the periphery of the virogenic stroma. It is uncertain whether p78 association with forming nucleocapsids occurs within the intrastomal spaces or as the nascent nucleocapsids migrate from the stroma into the surrounding nucleoplasm of the ring zone. Young *et al.*, (1993) observed that DNase treatment of purified nuclei caused degradation of the electron-dense core of nucleocapsids within the stroma, but did not affect the morphology of nucleocapsids in the ring zone; hence, nucleocapsids may be immature structures until or after they leave the microenvironment of the virogenic stroma.

The capsid shell may be composed of nine or more protein species (Kelly, 1985), although the predominant species is a 39-kDa protein (designated p39), which has been characterized for several NPVs (Bjornson and Rohrmann, 1992a; Blissard *et al.*, 1989; Pearson *et al.*, 1988; Russell *et al.*, 1991; Thiem and Miller, 1989). Western blots of OV and BV phenotypes (Pearson *et al.*, 1988) and immunocytochemistry (Russell *et al.*, 1991) showed p39 to be a major structural component present uniformly along the capsid. In infected cells, p39 is present mainly in the mat of virogenic stroma, yet capsid assembly occurs in the intrastromal spaces. The timing and regulation of p39 multimer formation, movement from the stromal matte to the intrastromal space, and regulation of capsid length are not known. No structural motifs have been recognized on p39, either for association with molecules within the stromal matte or for capsid assembly. A 24-kDa protein is also present in the capsid, and appears to be evenly distributed through the nucleocapsid in both the OV and BV phenotypes (Wolgamot *et al.*, 1993), but appears to be nonessential for virus replication (Schetter *et al.*, 1990). Other virus structural proteins have been identified (see Chapter 2, this volume), but their localization with respect to the virogenic stroma is not known.

The process of nucleocapsid maturation within the virogenic stroma was shown to sensitive to the actin-depolymerizing drug cytochalasin D (CD) (Charlton and Volkman, 1991). By immunofluorescence staining, the distribution of p39 in infected cells is principally intranuclear, with bright patches of staining also found along the plasma membrane. A modest dose of CD (0.5 μg/ml) applied to cells through infection altered the distribution of p39 (Volkman, 1988); in particular, CD abrogated the localization along the plasma membrane and caused foci of cytoplasmic staining with staining along the nuclear membrane. Within the nuclei of CD-treated cells, p39 was immunogold localized to bundles of aberrant tubular (capsid) structures juxtaposed to the inner nuclear membrane, which apparently lacked a nucleoprotein core (Volkman, 1988), although viral DNA synthesis was not inhibited (Volkman *et al.*, 1987). Actin was found to have a peristromal distribution in infected cells during the time of most active nucleocapsid morphogenesis (Charlton and Volkman, 1991). In the presence of CD, packaging of viral DNA into the capsid shells was disrupted, and cells released particles enriched in GP64 EFP but lacking nucleocapsids (Volkman *et al.*, 1987). Removal of the drug restored the morphogenic process even in the absence of new protein synthesis (Volkman *et al.*, 1992); thus, reassembly of microfilaments was apparently sufficient to reestablish nucleocapsid assembly. CD treatment was subsequently found to induce proteolytic degradation of the unphosphorylated form of p6.9, thus preventing nucleocapsid maturation (Oppenheimer and Volkman, 1995). The rate of p6.9 synthesis, phosphorylation, and dephosophorylation were not affected. The role of actin in the assembly of nucleocapsids is not clear, except that it appears to provide a scaffold structure required by the process. The involvement of p6.9 is unclear: Degradation may occur because nucleoprotein cores cannot be packaged within capsid shells, or conversely, the degradation of p6.9 may occur as it is dephosphorylated and thereby preclude packaging.

B. Routes of Nucleocapsid Egress from Nuclear to Cytoplasmic Compartments

Intrastromal spaces communicate with the surrounding nucleoplasm of the ring zone and mature nucleocapsids are able to migrate freely into this peristromal compartment, where they could follow one of two possible morphogenic pathways. In early stages of the infection cycle, the progeny nucleocapsids are destined to be components of BV, so they egress from the nucleus for exit at the plasma membrane (Fraser, 1986; Kawamoto *et al.*, 1977a). In the case of AcMNPV, maximum titers of BV are achieved by about 24 hr p.i., at which time BV production declines as intranuclear envelopment and occlusion begins (Faulkner, 1981). Several methods of nucleocapsid exit from the nucleus have been suggested, including via nuclear pores (Hess and Falcon, 1978; Summers, 1971), migration into or through the endoplasmic reticulum (E. A. Mackinnon, personal communication), or passage through discontinuities in the nuclear membrane (Adams *et al.*, 1977). However, the most common method of egress observed in EM studies of NPV involves a budding process (synhymenosis) in which nucleocapsids acquire a double membrane vesicle derived from the nuclear membrane (the transport vesicle) (Fig. 4A).

FIGURE 4. Alterations to nuclear membrane structure induced by NPV infection. (A) TEM micrograph of nucleocapsid egress through the nuclear membrane and acquisition of a double membrane transport vesicle (arrowhead). (B,D) The nuclear membrane becomes highly convoluted, with electron dense patches evident along the inner membrane (arrowheads, panel B). (C) Freeze fracture of infected cells demonstrates unique cagelike structures (arrowheads) that may correspond to sites of nuclear membrane budding and vesiculation seen by standard thin section methods (panel D). (D) Cytoskeletal-like elements similar in diameter to F-actin (microfilaments) are located on both sides of the nuclear membrane (NM) and are associated with vesicles budding into the cytoplasm (arrowheads; N = nucleus). Scale bars, 0.1 μm (A, B, D) and 0.5 μm (C).

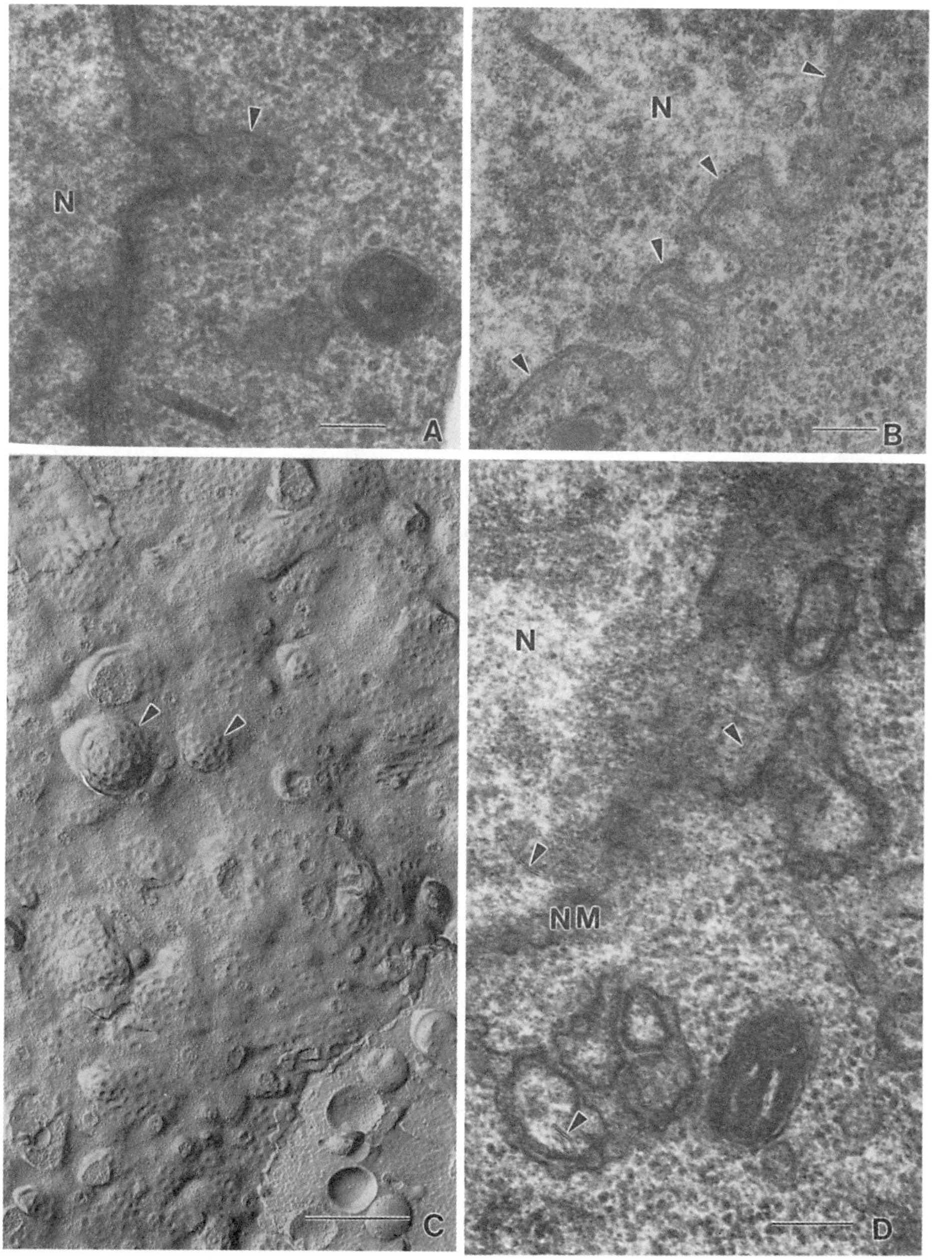
N
A
N
B
C
N
NM
D

Numerous changes occur to the nuclear membrane during the course of NPV infection. Electron-dense patches are often observed along the nucleoplasmic face of the inner nuclear membrane (Fig. 4B). As well, nuclear cisternae become progressively swollen and can be found in direct communication with similarly swollen pockets of endoplasmic reticulum. There may be partial disruption of regions along the inner nuclear membrane, with pockets of infolding. By freeze fracture, nuclear membranes develop cagelike outpockets during peak nucleocapsid egress (Fig. 4C), and these may correspond to regions enriched with a meshwork of cytoskeletonlike filaments seen in thin sections of specimens that have been stained *en bloc* with tannic acid and uranyl acetate (Fig. 4D). Based on morphology, the nuclear membrane is heavily utilized and extensively modified by NPV during the course of replication. A fundamental difference in replication strategy between NPV and GV is the complete breakdown of the nuclear membrane induced by the latter (Tanada and Hess, 1991; Winstanley and Crook, 1993).

Nucleocapsids of both budded and occluded phenotypes are seen to align with and attach to membranes at the apical (nipple) end of the capsid (Fraser, 1986; Kawamoto *et al.*, 1977a,b), and this is thought to trigger envelopment of nucleocapsid particles. Electron micrographs have shown an association of nucleocapsid apical ends with nuclear pores (Granados and Lawler, 1981; Summers, 1971). A 16-kDa nonstructural late glycoprotein of OpMNPV appears to be associated with lamellarlike structures near the nuclear membrane and with nucleocapsid-containing transport vesicles (Gross *et al.*, 1993): this polypeptide is highly conserved relative to AcMNPV (Roberts, 1989) and has been designated as ORF 130 (Ayres *et al.*, 1994).

The intracytoplasmic nucleocapsids en route to the plasma membrane appear to lose their nuclear membrane-derived vesicle in transit, as naked capsids are predominant at sites of budding along the plasma membrane (Fig. 1). No model for the release of capsids from transport vesicles has been proposed, and the process may involve fusion and/or disintegration of membranes. Multivesicular aggregates that contain many nucleocapsids are often seen in the cytoplasm, and this would suggest a series of successive membrane fusions. Simple fusion of the inner and outer nuclear membrane-derived membranes should also release nucleocapsids from the transport vesicle directly into the cytosol. However, disintegration of the transport vesicle also cannot be dismissed as a possible mechanism of nucleocapsid release. It will be important to utilize immunocytochemistry to determine whether the double membrane transport vesicles derived from the nuclear membrane are lined by lamin proteins. The nuclear lamina lines the inner surface of the nuclear envelope with a cagelike organization and is composed of intermediate filament-class polypeptides (lamins) that form a complex filamentous meshwork important for the structural integrity of nuclei (Aebi *et al.*, 1986; Gerace *et al.*, 1984). Phosphorylation of lamins during mitosis promotes filament disassembly, thus causing breakdown of the nuclear envelope (Dessev *et al.*, 1988). A Ca^{2+} and cAMP-independent protein kinase copurifies with the nuclear lamina of mammalian cells (Dessev *et al.*, 1988); thus, a similar-acting kinase packaged within the transport vesicle or perhaps nucleocapsid-associated may facilitate disintegration of the transport vesicle by phosphorylating lamins within the vesicle. Whatever the mechanism(s) of nucleocapsid release from transport vesicles, it is clearly established that unenveloped nucleocapsids migrate to and bud through the plasma membrane.

Nucleocapsids situated just inside the plasma membrane align their apical (nipple) end with the plasma membrane in the process of budding from the host cell, and this is seen both *in vivo* (Kawamoto *et al.*, 1977a) and in cell culture (Adams *et al.*, 1977). Contact with the capsid may induce a localized thickening of the membrane at the site of contact (Kawamoto *et al.*, 1977a), although sites of budding are probably pre-enriched for BV envelope proteins prior to contact (Adams *et al.*, 1977; Volkman, 1986; Volkman *et al.*, 1984). Usually single nucleocapsids bud at the plasma membrane, acquiring a loose-fitting envelope with prominent peplomers (spikes) (Fig. 1). These peplomer projections are present on the apical end of the budding virions only and are composed of GP64 EFP (Volkman, 1986; Volkman *et al.*, 1984). A modified form of ubiquitin bearing a phospholipid anchor is also present and is localized to the inner surface of the BV envelope (Guarino *et al.*, 1995). The function of ubiquitin is not known, and both cellular and viral forms are present in the budded virions (Guarino *et al.*, 1995; Guarino, 1990).

C. Changes in Cytoplasmic Organization

Prominent changes occur to the cytoplasm during NPV infection, including alterations to cytoskeleton and endomembrane systems and progressive development of large fibrillar structures throughout the cytoplasm (Fig. 5). Microfilament and microtubule distribution are both altered by infection (Charlton and Volkman, 1993b). The distribution of filamentous actin (F-actin) changes as viral replication proceeds, and after initial bundle formation induced by virus, entry is manifest as transient cytoplasmic agregates on the basal side of infected cells. This occurs prior to viral DNA replication and is dependent on early gene expression (Charlton and Volkman, 1991). After the switch over to viral DNA replication and late gene expression, the distribution of F-actin becomes intranuclear and is concentrated around the virogenic stroma (Charlton and Volkman, 1991). Microtubules undergo significant changes as a result of baculovirus infection and may account for the rounding of cells normally observed (Volkman and Zaal, 1990), and is one of the first signs of viral cytopathic effect (CPE) evident by light microscopy (Vaughn and Dougherty, 1985). Changes to the

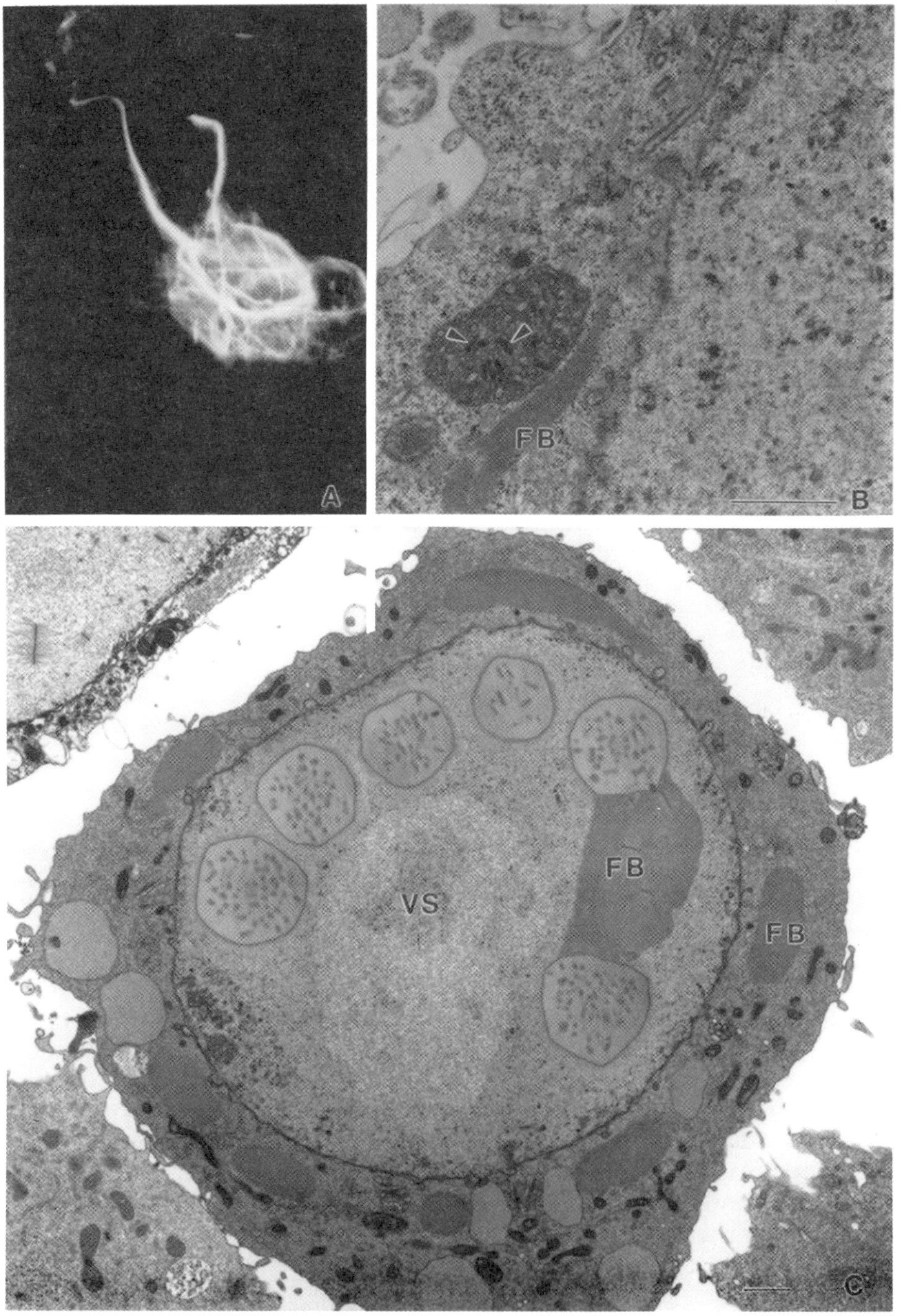
A
FB
B
VS
FB
FB
C

microtubule cytoskeleton are attributed to the action of both early and late viral genes (Volkman and Zaal, 1990).

Following the onset of late gene expression, very late gene (δ phase) expression begins. The primary products of this phase are the occlusion matrix protein (polyhedrin in NPV; granulin in GV) (see Chapter 2, this volume) and a small nonstructural polypeptide designated p10 (Kuzio *et al.*, 1984; Williams *et al.*, 1989). In cell culture, transcripts from the polyhedrin and p10 genes are present in approximately equal amounts, and together account for 90% of the viral RNA in insect cells at 48 hr p.i. (Smith *et al.*, 1983); the corresponding polypeptides may constitute 50% of cell protein. These genes have similar promoter structures (see, Chapter 8, this volume), and both proteins are synthesized in copious amounts (Rohel and Faulkner, 1984; Smith *et al.*, 1983). Their transcription probably involves a unique subset of factors (McLachlin and Miller, 1994; Roelvink *et al.*, 1992; see Chapter 8, this volume). Regulation of p10 and polyhedrin transcription is slightly different but not necessarily independent (Chaabihi *et al.*, 1993).

Both p10 and polyhedrin proteins initially accumulate in the cytoplasm and can be demonstrated immunocytochemically by 10–12 hr p.i. (p10) and 14–16 hr p.i. (polyhedrin). Later in infection (after about 20 hr p.i.), both proteins also localize to within the nucleus. Polyhedrin distribution becomes predominantly intranuclear, whereas p10 forms fibrillar masses in both cytoplasm and nucleus of infected cells (Vlak *et al.*, 1988; Williams *et al.*, 1989). In the cytoplasm, these associate with microtubules, but are not attributed as the cause of the cytoskeletal changes (Volkman and Zaal, 1990). The cytoplasmic distribution of p10 is maintained throughout infection, but its function at this location is not known. During early morphogenesis (before OV formation) the cytoplasmic fibrillar structures have an intermediate filamentlike distribution. Coexpression of both wild-type and a heterologous p10 indicated that only p10 molecules of the same type were present in individual fibrillar structures (van Oers *et al.*, 1994).

Microtubules are associated with p10 fibrillar structures but are not an integral component of them, and although there is significant interaction between the structures, structure and dynamics of each is viewed as autonomous. Depolymerization of microtubules by colchicine (Fig. 6A) induces cell rounding but does not prevent formation of p10 fibrillar tracts; thus, microtubules are not essential for development of p10 structures. Changes

FIGURE 5. Distribution of p10 in early and late infection. (A) Immunofluorescence localization of AcMNPV p10 in *T. ni* cells at 18 hr p.i. demonstrates a predominantly cytoplasmic accumulation with an intermediate filamentlike distribution that extends down cell processes. (B) By TEM examination, the p10-rich fibrillar bodies (FB) course through the cytoplasm and are occasionally seen associated with vesicular aggregates containing nucleocapsids (arrowheads). At times very late in infection (48 hr p.i.) p10 bodies are located in both the cytoplasm and nucleus. Intranuclear p10 is intimately associated with maturing OV in the ring zone. Note that at late times postinfection, few free nucleocapsids are present in the virogenic stroma (VS). Scale bars, 1 μm.

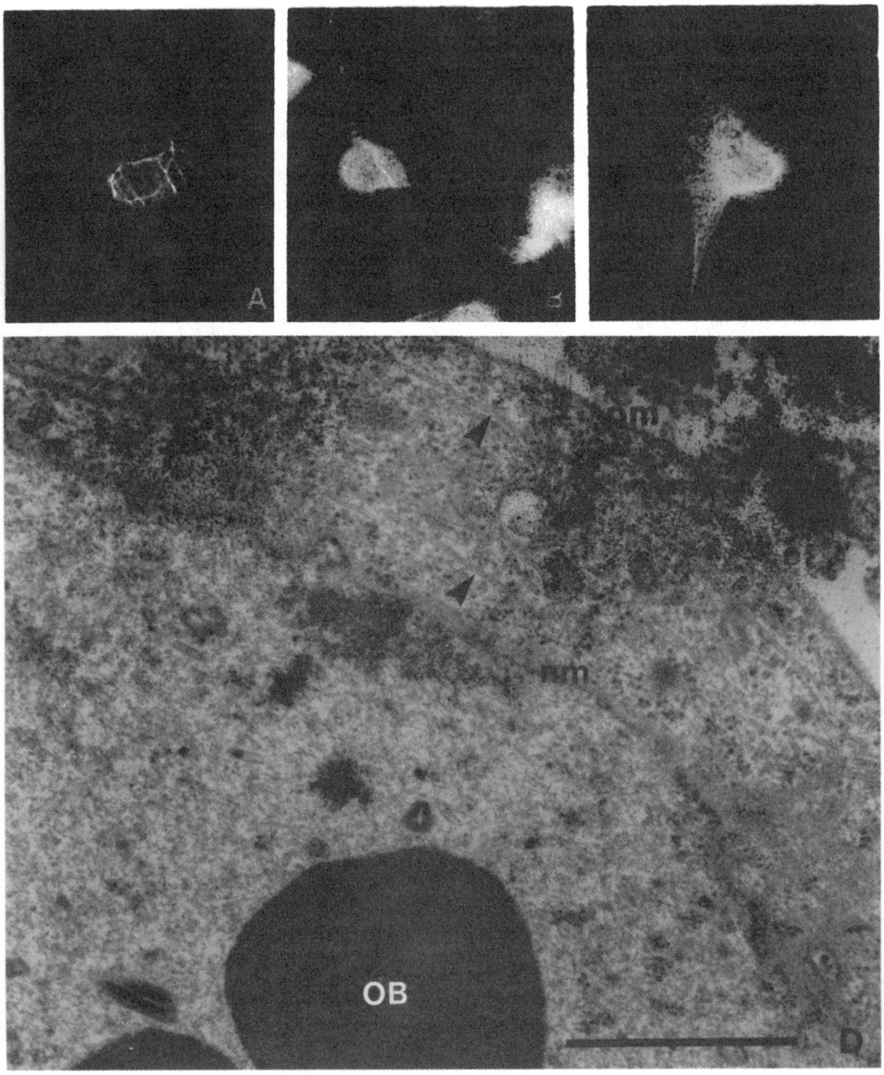

FIGURE 6. Persistence of microtubules through infection. (A) Immunofluorescence staining of AcMNPV-infected *T. ni* cells that have been treated with colchicine (0.1 μg/ml) from 6 hr p.i. until harvest (18 hr p.i.) form cytoplasmic fibrillar p10 structures in the absence of microtubules. (B, C) Immunofluorescence staining of tubulin in *T. ni* cells infected with AcMNPV (33 hr p.i., panel B), or mock-infected (C) indicates that microtubules were present in infected cells at late times postinfection (B) but lacked the latticelike complexity of microtubule distribution in uninfected cells (C). (D) By TEM study, microtubules persisted in Sf cells infected with AcMNPV until at least 30 hr p.i. (nm, nuclear membrane; pm, plasma membrane; OV, occlusion body). Scale bar, 1 μm.

to the organization and stability of microtubules are induced by AcMNPV infection (Fig. 6B,C) independent of p10 and are manifest as a lack of mesh or latticelike complexity. In contrast to findings in an earlier study (Volkman and Zaal, 1990), which had suggested that microtubules are depolymerized in *S. frugiperda* cells infected by AcMNPV by 24 hr p.i., we have observed microtubules to persist in infected *S. frugiperda* and *T. ni* cells throughout infection (Fig. 6B,D). Intact microtubules could be seen as late as 68 hr p.i. by immunofluorescence staining and EM. In the study of Volkman and Zaal (1990), microtubules may have been obscured by nuclear hypertrophy or physically disrupted during tissue processing.

Domains important for the association of p10 with microtubules may reside partially or completely within its C-terminal half (Quant-Russell *et al.*, 1987). Antibodies specific to a peptide domain of OpMNPV ($_{77}$LPEIPDVP$_{84}$) cross-reacted with a host cell cytoskeletal element (Quant-Russell *et al.*, 1987), and part of this sequence is conserved among all p10 species examined thus far (van Oers *et al.*, 1994; Zuidema *et al.*, 1993). The highly basic C-terminus of AcMNPV p10 has also been a focus of study because overexpression of a mammalian cAMP-dependent protein kinase altered the morphology of infected cells, caused phosphorylation of two serine residues within the p10 sequence $_{87}$RRGKR*SS*K$_{94}$, and appeared to affect p10 affinity for microtubules (Cheley *et al.*, 1992). The C-terminus of p10 may be functionally similar to the basic C-terminus of microtubule-associated proteins (MAP) such as τ (tau), MAP-1B and MAP-2, which interact with the acidic C-terminus of tubulin (Noble *et al.*, 1989; Serrano *et al.*, 1985; Serrano *et al.*, 1984) by charge complementarity (Hagestedt *et al.*, 1989). However, the primary amino acid sequence of the p10 microtubule-binding domain might be quite distinct from those of cellular microtubule-binding proteins, since there is wide variation of binding domains among cellular MAPs (Noble *et al.*, 1989). Phosphorylation of τ at the microtubule-binding domain is thought to decrease affinity for binding to microtubules (Hagestedt *et al.*, 1989; Hirokawa, 1994), and site-specific phosphorylation of other cytoskeletal elements, for example, the intermediate filament (IF) class polypeptides vimentin (Inagaki *et al.*, 1987) and nuclear lamins (Dessev *et al.*, 1988; Gerace, 1986), induces disassembly of filaments. It is possible that phosphorylation of p10 at the C-terminal serines may similarly alter the dynamics of association with fibrillar structures and/or microtubules. Site-directed mutagenesis of p10, which substituted one or both C-terminal serine residues with alanine, failed to affect fibrillar structure formation or nuclear accumulation (van Oers, 1994), although the kinetics of p10 association–dissociation from fibrillar bodies or microtubules was not examined. Deletion of the entire eight amino acid C-terminal domain abrogated fibrillar structure and prevented aggregation of p10 in the nucleus and cytoplasm of infected cells (van Oers *et al.*, 1993); hence, this sequence appears essential for the formation of higher-order fibrillar structures and possibly for association with microtubules, but it is not required for nuclear transport.

Large quantities of polyhedrin are required for OV morphogenesis; thus, the high rate of synthesis can by reconciled with polypeptide function. In contrast, p10 is not essential for virus replication *in vitro* or *in vivo* (Kuzio, 1990) and is not a structural component of BV, ODV, or OV (Vlak *et al.*, 1981). The p10 gene has been identified in all baculoviruses characterized to date and is likely to provide some survival advantage in the field. Studies on p10 mutants (Vlak *et al.*, 1988; Williams *et al.*, 1989) or immunolocalization of p10 in wild-type infected cells (Croizier *et al.*, Quant-Russell *et al.*, 1987; van der Wilk *et al.*, 1987) led to the correlation of p10 with cytoplasmic and nuclear fibrillar masses formed during virus replication. Although studies have demonstrated p10-dependent events related to OV morphogenesis and cytolysis (Williams *et al.*, 1989), these studies failed to establish a functional significance for the highly ordered cytoplasmic distribution of p10.

V. LATE EVENTS IN MORPHOGENESIS

A. Switchover to Occluded Virus Production

Late in AcMNPV infection (about 22–24 hr p.i.), BV production is curtailed by an undefined mechanism in favor of intranuclear maturation of virions destined for occlusion (Stoltz *et al.*, 1973; Tanada and Hess, 1976; Volkman *et al.*, 1976). The intranuclear ring zone is the site of several morphogenic processes closely associated with OV development, and is marked by the appearance of intranuclear membrane profiles that envelop nucleocapsids to form preoccluded virions (POV). Patches of membranous profiles and vesicles that develop within the ring zone are precursor structures utilized for intranuclear envelopment of virions. These membranes have trilamellar morphology and dimensions distinct from cellular membranes (Stoltz *et al.*, 1973). Because of these observations and the apparent lack of continuity with the nuclear membrane, intranuclear membranous profiles were postulated to be the result of *de novo* synthesis. In other studies, baculovirus infection has been observed to induce an invagination of the inner nuclear membrane (Summers and Arnott, 1969; Tanada and Hess, 1976) (Fig. 7), and this may be the source material for intranuclear membrane vesicles. Although BV production is reduced substantially in favor of intranuclear envelopment at late times postinfection, some virion budding at the plasma membrane can still be observed very late in infection (after 48 hr p.i.); thus, the phenotype switchover is not absolute. Polyhedrin condensation subsequently occludes POV, and morphogenesis of the OV is completed by the attachment of a distinct outer layer: the calyx (OV membrane, polyhedron membrane). The onset of OV morphogenesis roughly

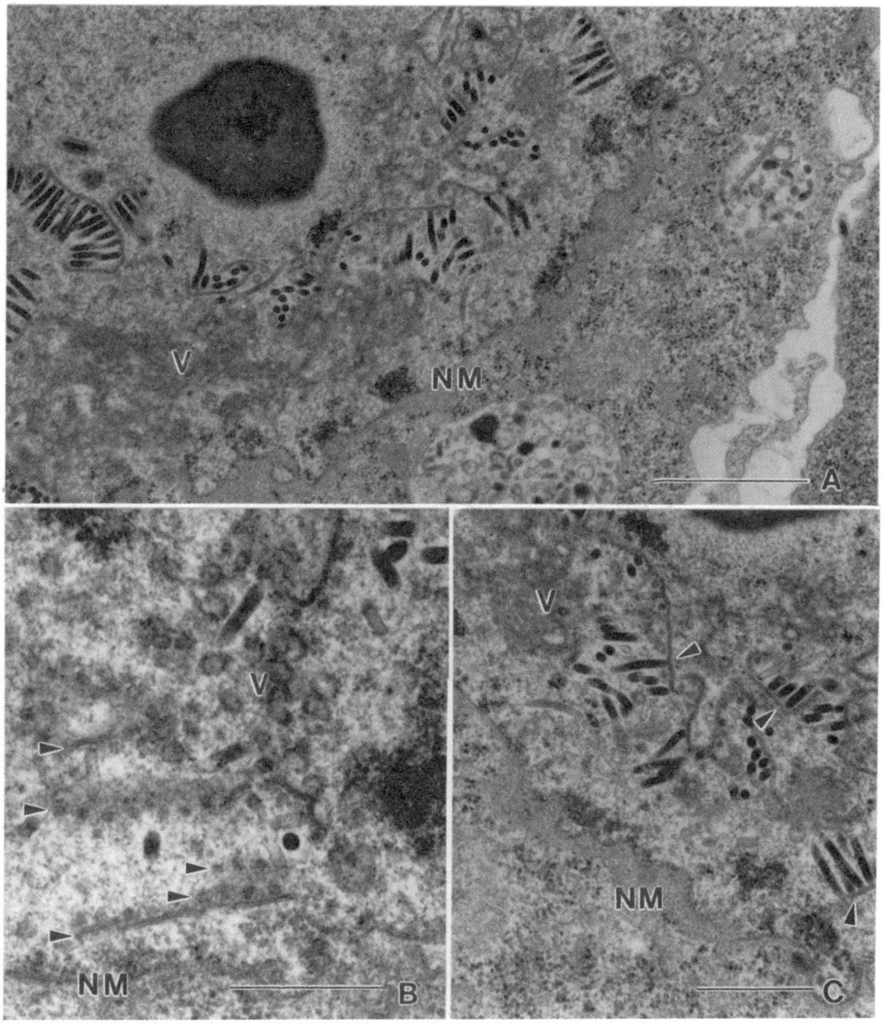

FIGURE 7. Membrane formation and intranuclear envelopment of nucleocapsids. (A) Large aggregates of membrane vesicles and profiles (V) form in the ring zone of NPV-infected cells and are utilized for the envelopment of nucleocapsids. (B) Profiles of membrane material juxtaposed to the nuclear membrane (NM) appear to break down into microvesicles (arrowheads), and distal to the nuclear membrane reform into distinct trilamellar membrane profiles (refer to Fig. 3 B, C). (C) Nucleocapsids align with *de novo* membranous profiles end-on (arrowheads) and acquire their envelope. Scale bars, 1 μm (A) and 0.5 μm (B, C).

coincides with intranuclear accumulation of p10 and polyhedrin, although deletion of one or both of these genes does not abrogate the switchover from budded to occluded phenotype virions, and intranuclear morphogenesis of other OV-related structures such as calyx precursor is similarly maintained in their absence (Williams, 1995).

B. Morphogenesis of Intranuclear Membrane and Envelopment of Nucleocapsids

Morphogenesis of POV occurs in the peristromal ring zone in infected cells by interception and envelopment of nucleocapsids as they leave the virogenic stroma (Fraser, 1986; Kawamoto *et al.*, 1977a,b). Prerequisite to this process is the formation of viral envelope precursors in the form of microvesicles and unit membranelike profiles (Hong *et al.*, 1994). We have done extensive analysis of the processes of morphogenesis of this membrane (Fig. 7) using fixation parameters and *en bloc* staining regimens to maximally stabilize membrane components. Under these processing conditions, there is strong evidence to support nuclear membrane involvement in this morphogenic process. However, this process is complex. Lipid composition of the OV envelope is distinct (Braunagel and Summers, 1994) and envelope association of a recently identified 66-kDa OV structural protein (PDV-E66) appears to occur concomitant with or following formation of the microvesicles and membranous profiles (Hong *et al.*, 1994). Immunocytochemical localization indicated that PDV-E66 colocalized to these intranuclear membrane patches but not to the adjacent nuclear membrane. In addition, a 25-kDa ODV envelope protein of OpMNPV was found to localize to virions within the ring zone and within mature OV (Russell and Rohrmann, 1993) and was also identified at the periphery of the virogenic stroma, suggesting late insertion in the membrane and perhaps indicating dynamic modification or reassortment of membrane-associated species following association with nucleocapsids. The capsid end structure-associated protein (p78) identified in both virion phenotypes has a similar distribution to the 25-kDa species (Vialard and Richardson, 1993), and the most likely explanation for these observed distributions of virion components is that nucleocapsids leaving the virogenic stroma are intercepted by molecules necessary for completion of morphogenesis. Regulation of peristromal accumulation of POV-specific components is one possible mechanism that may govern the switch over from budded to occluded virion production. Kawamoto *et al.* (1977b) observed interaction of nucleocapsids leaving the virogenic stroma with electron-dense particles that were intimately involved in the envelopment process and appeared to form the tegument layer following envelopment; this material most likely contained gp41, the major tegument glycoprotein (Whitford and Faulkner, 1992a,b).

Although morphological changes to the nuclear membrane (Fig. 4) and the development of membrane patches within the ring zone (Fig. 7) outwardly appear dramatic, they may be due to very subtle changes within the host cell. For example, up-regulation of a single nuclear pore complex protein in BHK cells (Nup 153) (Bastos *et al.*, 1996) induced both nuclear membrane-associated lamellar profiles and intranuclear membrane arrays. The lamellar profiles, although intranuclear rather than cytoplasmic, are grossly similar to those reported to be associated with a late-expressed 16-kDa glycoprotein of OpMNPV (Gross *et al.*, 1993), and intranuclear membrane profiles are similar to those observed in NPV infection where late gene expression is limited (Carstens *et al.*, 1994). Further, cell-cycle-dependent action of a nuclear-membrane-associated kinase can cause nuclear membrane disintegration, grossly similar to what is observed during GV morphogenesis, by phosphorylation of nuclear lamins (Dessev *et al.*, 1988). The molecular details of nuclear membrane-associated cytopathology induced by NPV and GV replication have not been determined.

VI. OCCLUSION BODY MORPHOGENESIS

A. Polyhedrin Condensation and the Occlusion of Virions

The NPV occlusion matrix protein, polyhedrin, is highly conserved 29-kDa polypeptide (Rohrmann, 1986a,b, 1992) and accounts for about 95% of the OV mass (Faulkner, 1981). Polyhedrin is related to the GV matrix protein granulin (Kozlov *et al.*, 1992; Vlak and Rohrmann, 1985) but is very different from the matrix proteins of entomopoxvirus (EPV), spheroids (110–115 kDa) (Arif, 1995; Hall and Moyer, 1993; Harrap and Payne, 1979) and cytoplasmic polyhedrosis virus (CPV) occlusions (25–30 kDa) (Vlak and Rohrmann, 1985).

The incorporation of POVs into the matrix of condensed polyhedrin is well underway by 24 hr p.i. in cells infected with AcMNPV (Volkman and Knudson, 1986). It appears that only enveloped virus particles become incorporated into OV (Vaughn and Dougherty, 1985), but the regulatory mechanism governing the process has not been identified. Mutations exhibiting a few polyhedra (FP) phenotype arise spontaneously upon serial passage of AcMNPV in tissue culture, and the resulting OV contain few if any occluded virions (Potter *et al.*, 1976; Ramoska and Hink, 1974). The genetic defect is usually attributed to loss of a 25-kDa protein (Beames and Summers, 1989; Fraser *et al.*, 1983). Deletion of the gene, FP-25 (ORF 61), encoding this protein results in reduced OV formation, reduced intranuclear envelopment of nucleocapsids, and elevated levels of BV production (Beames and Summers, 1989; Fraser *et al.*, 1983; Harrison and Summers, 1995a,b). It has been speculated that the 25K protein could be an important factor for initiation of OV formation, perhaps by causing nucleation of polyhedrin (Wood, 1980). However, FP-25 does not appear to associate with forming OV and may in fact be predominantly cytoplasmic (Harrison and Summers, 1995b).

The size and shape of baculovirus OVs are governed to some degree by the occlusion matrix protein, polyhedrin or granulin. Several morphology mutants of AcMNPV with altered polyhedrin coding regions have been characterized at the nucleotide level. Cells infected with the mutant M5 synthesize one large cubic OV per cell that has a paracrystalline lattice and outer calyx, but occludes very few virions (Brown *et al.*, 1980). In contrast, cells infected with the mutant M29 produce copious amounts of small intranuclear polyhedrin condensations that lack an obvious lattice structure and do not occlude virions or associate with calyx material (Duncan and Faulkner, 1982; Duncan *et al.*, 1983). A single point mutation results in the observed phenotype of both viruses (Carstens *et al.*, 1986, 1987). Two additional mutants, M276 and M934 (Partington *et al.*, 1990), also have defects in the polyhedrin coding region that result in large amorphous condensations within infected cells (Carstens *et al.*, 1992). These mutants all have alterations in highly conserved regions of polyhedrin, based on amino acid comparisons of at least four NPV polyhedrin sequences (Kozlov *et al.*, 1986), which disrupt the paracrystalline lattice and abrogate or greatly reduce occlusion of virions.

B. Assembly of Calyx Precursors within the Intranuclear Ring Zone

Calyx precursor structures, also called bilamellar fibrous sheets (Williams *et al.*, 1989) or spacers (Mackinnon *et al.*, 1974), form in the ring zone of cells infected with NPVs (Figs. 8, 9). The origin of these structures is unknown and is considered a *de novo* process (Mackinnon *et al.*, 1974; van der Wilk *et al.*, 1987). A 34-kDa phosphoprotein (pp34) (Gombart *et al.*, 1989) has been immunolocalized to the precursors and to mature calyx (Russell and Rohrmann, 1990a; van Lent *et al.*, 1990), and its presence is essential for calyx formation (Zuidema *et al.*, 1989). Other protein components of the calyx material have not been identified, and the structure appears to be predominantly carbohydrate (Minion *et al.*, 1979). Fibrillar structures enriched in p10 are commonly found associated with the calyx precursors and with maturing OV (Croizier *et al.*, 1987; van der Wilk *et al.*, 1987; Vlak *et al.*, 1988; Williams *et al.*, 1989). The p10 polypeptide, however, is not required for formation of calyx precursor material (van Oers *et al.*, 1993, 1994; Williams *et al.*, 1989) or OV (van der Wilk *et al.*, 1987). Morphogenesis of p10-rich fibrillar bodies, OV, and the OV calyx seems to occur independently, as deletion mutants that affect one of the structures do not affect formation of the other structures (van der Wilk *et al.*, 1987; Zuidema *et al.*, 1989).

FIGURE 8. TEM analysis of bilamellar fibrous sheet formation. (A) Nuclei of Sf cells infected with AcMNPV (24 hr p.i.) contain extensive profiles of bilamellar fibrous sheet material (s) associated with p10-rich fibrillar structures (f) situated adjacent to the nuclear membrane (nm). Bilamellar sheet profiles often abut the inner nuclear membrane (inset, arrowhead). (B) Profiles of sheet material appeared to form at or near the nuclear membrane, and are morphologically complete with spikelike projections (arrowhead) before incorporation into fibrillar structures. An undefined RNP-like matrix (r) is usually found at or near sites of sheet formation. Scale bars, 1 μm (A), 0.2 μm (A, inset), and 0.5 μm (B).

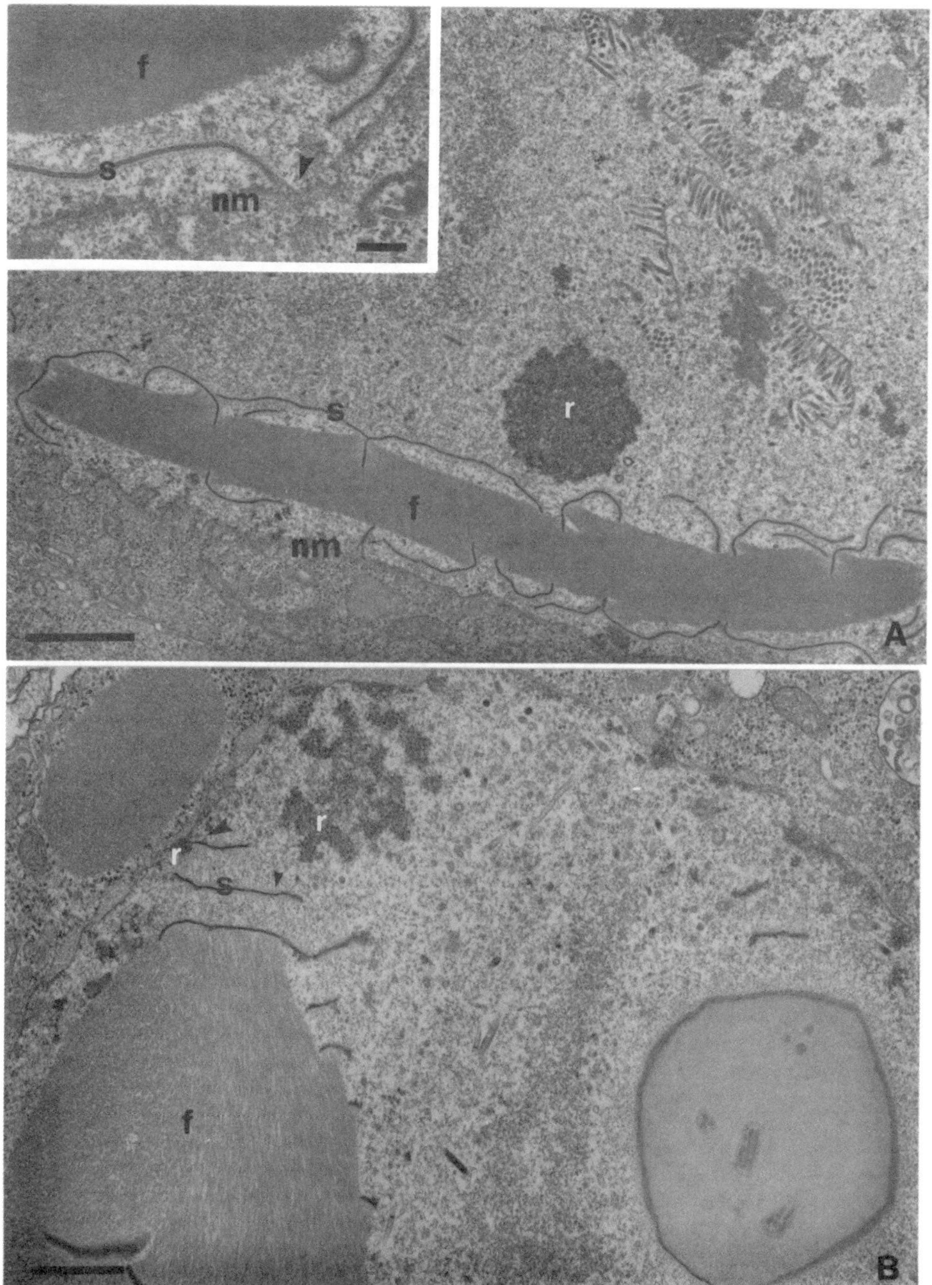
f
s
nm
r
A
B

Mature calyx precursor is usually associated with fibrillar structures (Fig. 8) and is often aligned close to the nuclear membrane, sometimes abutting the inner membrane (Fig. 8, inset). Parallel arrays of fibrous sheet with spiked outer surfaces also occur in the nucleoplasm near p10 bodies (Fig. 8B), with shorter profiles occurring near the nuclear membrane. Adjacent regions of the inner nuclear membrane stain intensely and are lined with beadlike structures predominantly on the cisternal side of the membrane. Calyx precursor synthesis may occur at specialized regions on or near the inner nuclear membrane. Looped end structures joining bilamellar sheets have been reported (Mackinnon *et al.*, 1974), and would be consistent with a model involving the involution and collapse of unilamellar sheets to form the observed bilamellar profiles. Interestingly, an electron-dense ribonucleoprotein (RNP)-like material is usually associated with the sites where precursor formation seems to occur, although the composition and function of this material is not known.

C. Attachment of Calyx to the Outer Surface of Developing Occlusion Bodies

Structurally, the calyx is composed of a regular bilammellar array distinct from the polyhedrin lattice, with 15-nm holes or hollows evenly distributed across the structure (Harrap, 1972a). The structure appears to be porous, and occluded virions can be stained within intact OV by fluorescent DNA-intercalating dyes such as DAPI. Dissolution of OV does not affect the calyx, which remains as a residual baglike structure (Harrap, 1972a). A 34-kDa phosphoprotein (pp34) (Gombart *et al.*, 1989) has been immunolocalized to the calyx (Russell and Rohrmann, 1990a; van Lent *et al.*, 1990) and is covalently associated by thiol bonds (Whitt and Manning, 1988). A subpopulation of polyhedrin is also thiol linked to the calyx (Whitt and Manning, 1988), which may anchor this structure to the surface of the OV. Deletion mutants lacking the pp34 protein fail to produce a calyx or the bilamellar fibrous sheet (precursor) structure; OVs of these mutants have an increased sensitivity to alkali (Vlak *et al.*, 1988; Zuidema *et al.*, 1989). A 34.8-kDa spheroidinlike glycoprotein (SLP) (Vialard *et al.*, 1990) has also been localized to the OV of AcMNPV and may be a component of the calyx (Rohrmann, 1992), but its role in calyx morphogenesis or structure is not known, and in OpMNPV the putative SLP does not seem to be calyx-associated (Rohrmann, 1992). By EM of thin section specimens, the calyx has distinct spikelike projections on the outer surface (Fig. 9). Similar structures are present on the bilamellar fibrous sheet precursor material, and attachment onto the surface of OV appears to be facilitated by unfolding of these sheets around the polyhedrin matrix to form a unilamellar structure. This process is intimately associated with but is not dependent on p10-rich fibrillar masses, and it appears to be important for the structural integrity of the OV (Williams *et al.*, 1989). Some mutations in the

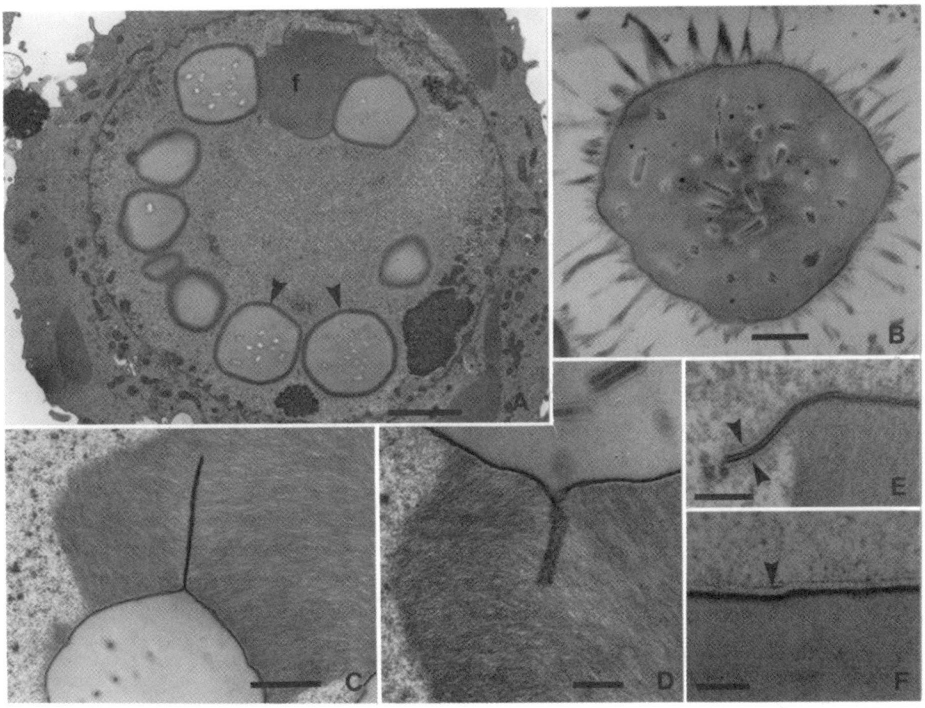

FIGURE 9. Calyx attachment to OV. Sf21 cells infected with AcMNPV were harvested between 30 and 48 hr p.i. and processed for TEM study with tannic acid and uranyl acetate *en bloc* staining to stabilize and enhance membrane and other structures. (A) Fibrillar bodies (f) within the nucleus at 30 hr p.i. are intimately associated with forming OV, and form a shell of 100–120 nm around their surface (arrowheads) just prior to or during calyx attachment. (B) Mature OV seen in partially extracted Sf cells (48 hr p.i.) have p10 fibrillar structures intimately associated with the OV surface after removal of most nucleoplasm. (C, D) OV calyx and p10-associated bilamellar fibrous sheet structures are contiguous. (D) Attachment of calyx to the surface of maturing OV appears to involve separation of fibrous sheet lamellae at the OV surface (38 hr p.i.). (E) Bilamellar fibrous sheets (30 hr p.i.) associated with fibrillar body have spikelike projections along the outer surfaces (arrowheads). (F) Corresponding structures are seen on the outer surface of the unilamellar calyx structure attached to the OV surface (48 hr p.i.). Scale bars, 2 μm (A), 0.5 μm (B, C), 0.2 μm (D, E), and 0.1 μm (F).

polyhedrin gene cause an unusual affinity between polyhedrin and calyx material (Carstens *et al.*, 1992), and it is possible that a shift in polyhedrin tertiary structure at the surface of a growing occlusion condensation may signal the attachment of the calyx.

D. Fine Structure of the Occlusion Body Lattice

The paracrystalline matrix of OV is a face-centered cubic lattice comprised of 4 to 5-nm subunits packed in a rhombic arrangement (Federici,

1986; Harrap, 1972a; Kelly, 1985). Virions within the matrix are randomly distributed and do not interfere with the regularity of the lattice (Bergold, 1963; Harrap, 1972a; Harrap and Payne, 1979). Harrap (1972a) examined polyhedrin crystals of partially dissolved OV by transmission electron microscopy, using a negative stain procedure, and determined that polyhedrin oligomers form a six-nodal structure that presumably associates in an ordered array to form the matrix. OV are resistant to putrefaction and disruption by a variety of chemical agents (Benz, 1986) or physical treatments such as freezing and desiccation (Jacques, 1985), but are dissolved by weak alkali (pH 9.5–11.5) in the midgut of the host insect (Granados and Williams, 1986), which this facilitates release of the occluded virions. Reducing conditions are additionally required to dissolve the polyhedrin crystal and release embedded virions of the *T. paludosa* NPV (Guelpa *et al.*, 1977). In addition to containing virions, the OV matrix of GVs and possibly some NPVs also contains a viral-enhancing factor that acts on the insect peritrophic membrane to aid OV access to the ectoperitrophic space (Derksen and Granados, 1988). Also, OV derived from insects contains a significant amount of silicon (Stairs, 1968), which may be incorporated into the paracrystalline lattice, but its distribution has not been determined nor have tissue-culture-derived OV been examined for inorganic components. There are also several accounts of the presence of RNA in purified OV (Aizawa and Tida, 1963; Faulkner, 1962; Himeno *et al.*, 1969) in approximately equal quantities relative to viral DNA (Faulkner, 1962). These molecules and inorganic components have not been localized within the OV structure and may be associated with embedded virions, dispersed throughout the occlusion matrix, or copurified as contaminants at the surface of the OV.

VII. CYTOLYSIS AND RELEASE OF OCCLUSION BODIES

The end of NPV infection in tissue culture is marked by cytolysis and release of OV (Williams, 1995; Williams *et al.*, 1989). This phenomenon is dependent on viral and cellular factors and may not be seen in some *in vitro* systems. Ultrastructurally, release is correlated with the development of large cytoplasmic vacuoles that become associated with both plasma membrane and the outer nuclear membrane (Fig. 10A–E). The p10 protein of AcMNPV is essential to OV release from the nucleus (Williams *et al.*, 1989), and this function has been mapped to a segment residing between amino acids 52 and 79 (van Oers *et al.*, 1993). Immunocytochemically, p10 is intimately associated at the periphery of the cytoplasmic vesicle (Fig. 10). Regulation of the process is complex and cells appear to commit to the event at least 6 hr prior to OV release. The action of a currently uncharacterized kinase as well as the presence of both cellular microfilaments and the virus-encoded cathepsin are also essential to OV release (Williams, 1995). In unrelated experiments, expression of *bcl*-2 in infected cells was found to delay the internucleosomal DNA cleavage, which normally occurs late in infec-

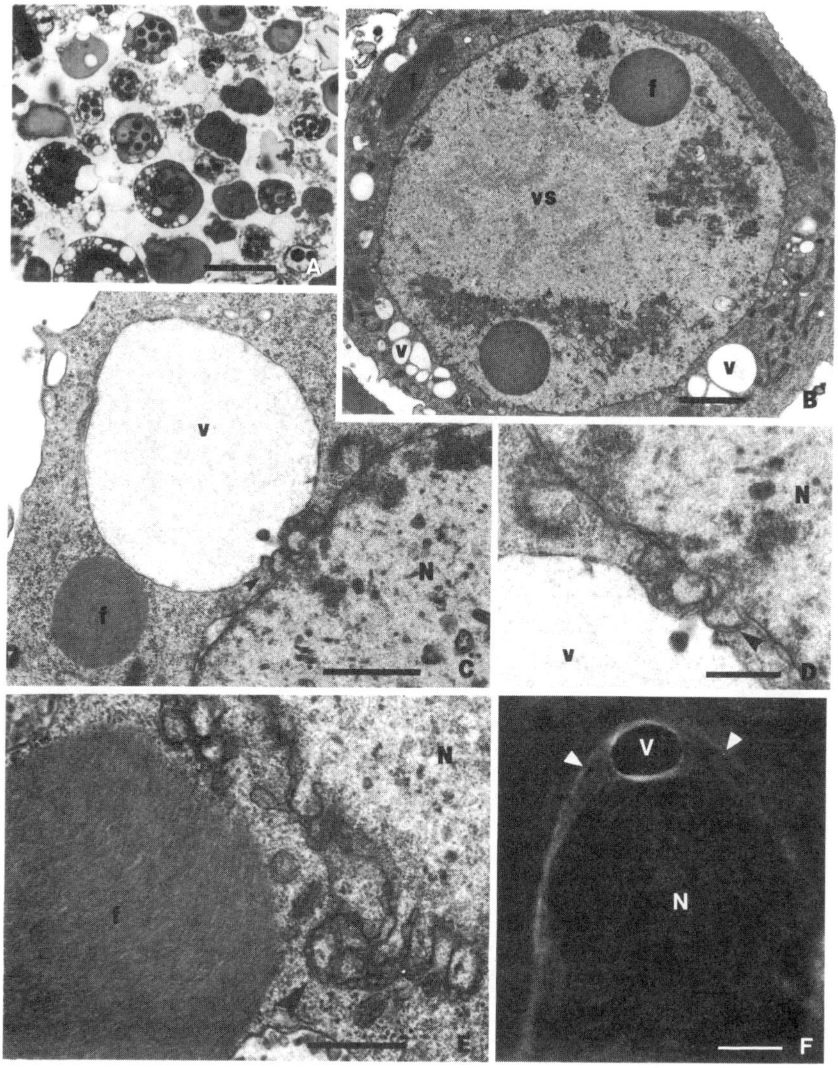

FIGURE 10. Light and electron microscopic examination of cytoplasmic vesicles associated with cytolysis and OV release. (A) Toluidine blue-stained section of Sf cells infected with AcMNPV at 68 hr p.i. Large cytoplasmic vesicles are prominent features of infected cells and appear to fuse with the plasma membrane, exposing a portion of nuclear membrane immediately adjacent to the extracellular space (arrowheads). (B) by TEM examination, multiple vesicles (v) are present in the cytoplasm of infected cells at 22 hr p.i. (C) Vesicles become larger as infection progresses, and associate with p10-rich fibrillar bodies (f) and nuclear membrane (arrowhead). (D) Narrow channels connect the vesicle membrane with the outer nuclear membrane (arrowhead), and may anchor the vesicle membrane when fusion with the plasma membrane occurs. Severe vesiculation of the nuclear membrane occurs near cytoplasmic vesicles (C, D) and in areas of close contact with cytoplasmic fibrillar bodies (E, arrowhead). (F) Immunofluorescent staining of p10 in *T. ni* cells infected with wild-type virus (moi = 10, 22 hr p.i.) labeled cytoplasmic tracts (arrowheads) that appear to be concentrated around cytoplasmic vesicles (v). Scale bars, 20 μm (A), 2 μm (B), 1 μm (C), 0.5 μm (D, E), and 10 μm (F).

tion (Alnemri *et al.*, 1992), and this is probably related to the observed nuclear disintegration that precedes or is concomitant with OV release (van Oers *et al.*, 1993). Disintegration and/or reorganization of intranuclear components may in some way affect the release of OV.

VIII. SUMMARY

Baculovirus infection results in a number of cytological changes and, in the process of their morphogenesis, baculoviruses induce the formation of a several novel structures including fibrillar structures, intranuclear trilamellar membranes, bilamellar fibrous sheets, and paracrystalline occlusions. The basis for these remarkable changes are just beginning to be understood at the molecular level, and these studies are being aided by immunocytochemistry, genetics, and electron microscopy. Unraveling the mechanisms by which these unique cytological and morphological events occur promises to be an exciting avenue for future baculovirus research.

REFERENCES

Adams, J. R., Goodwin, R. H., and Wilcox, T. A., 1977, Electron microscopic investigations on invasion and replication of insect baculoviruses *in vivo* and *in vitro*, *Biologie Cellulaire* **28:**261–268.

Aebi, U., Cohn, J., Buhle, L., and Gerace, L., 1986, The nuclear lamina is a meshwork of intermediate-type filaments, *Nature* **323:**560–564.

Aizawa, K., and Tida, S., 1963, Nucleic acids extracted from the virus polyhedra of the silkworm, *Bombyx mori* (Linnaeus), *J. Insect Pathol.* **5:**344–348.

Alnemri, E. S., Robertson, N. M., Fernandes, T. F., Croce, C. M., and Litwack, G., 1992, Overexpressed full-length human BCL2 extends the survival of baculovirus-infected Sf9 insect cells, *Proc. Natl. Acad. Sci. USA* **89:**7295–7299.

Arif, B. M., 1995, Recent advances in the molecular biology of entomopoxviruses, *J. Gen. Virol.* **76:**1–13.

Ayres, M. D., Howard, S. C., Kuzio, J., Lopez-Ferber, M., and Possee, R. D., 1994, The complete DNA sequence of *Autographa californica* nuclear polyhedrosis virus, *Virology* **202:**586–605.

Bastos, R., Lin, A., Enarson, M., and Burke, B., 1996, Targeting and function in mRNA export of nuclear pore complex protein Nup 153, *J. Cell Biol.* **134:**1141–1156.

Beames, B., and Summers, M. D., 1988, Location and nucleotide sequence of the 25K protein missing from baculovirus few polyhedra (FP) mutants, *Virology* **168:**344–353.

Beaton, C. D., and Filshie, B. K., 1976, Comparative ultrastructural studies of insect granulosis and nuclear polyhedrosis viruses, *J. Gen. Virol.* **31:**151–161.

Beniya, H., Funk, C. J., Rohrmann, G. F., and Weaver, R. F., 1996, Purification of a virus-induced RNA polymerase from *Autographa californica* nuclear polyhedrosis virus-infected Spodoptera frugiperda cells that accurately initiates late and very late transcription *in vitro*, *Virology* **216:**12–19.

Benz, G. A., 1986, Introduction: Historical perspectives, in: *The Biology of Baculoviruses* (R. R. Granados and B. A. Federici, eds.) pp. 1–35, CRC Press, Boca Raton, FL.

Ben-Ze'ev, A., 1983, Virus replication in infected epithelial cells is coupled to cell shape-responsive metabolic controls, *J. Cell Physiol.* **114:**145–152.

Bergold, G. H., 1963, The nature of nuclear-polyhedrosis viruses, in: *Insect Pathology, An Advanced Treatise.* (E. A. Steinhaus, ed.), Vol. I, pp. 413–456. Academic Press, New York.

Bilimoria S. L., 1991, The biology of nuclear polyhedrosis viruses, in: *Viruses of Invertebrates* (E. Kurstak, ed.), pp. 1–72, Marcel Dekker, New York.

Bjornson, R. M., and Rohrmann, G. F., 1992a, Nucleotide sequence of the p39-capsid gene region of the *Lymantria dispar* nuclear polyhedrosis virus, *J. Gen. Virol.* **73:**1505–1508.

Blissard, G. W., 1996, Baculovirus–insect cell interactions, *Cytotechnology* **20:**73–93.

Blissard, G. W., and Rohrmann, G. F., 1990, Baculovirus diversity and molecular biology, *Annu. Rev. Entomol.* **35:**127–155.

Blissard, G. W., and Wenz, J. R., 1992, Baculovirus gp64 envelope glycoprotein is sufficient to mediate pH dependent membrane fusion, *J. Virol.* **66:**6829–6835.

Blissard, G. W., Quant-Russell, R. L., Rohrmann, G. F., and Beaudreau, G. S., 1989, Nucleotide sequence, transcriptional mapping, and temporal expression of the gene encoding p39, a major structural protein of the multicapsid nuclear polyhedrosis virus of *Orgyia pseudotsugata, Virology* **168:**354–362.

Braunagel, S. C., and Summers, M. D., 1994, *Autographa californica* nuclear polyhedrosis virus, PDV, and ECV viral envelopes and nucleocapsids: Structural proteins, antigens, lipid and fatty acid profiles, *Virology* **202:**315–328.

Broussard, D. R., Guarino, L. A., and Jarvis, D. L., 1996a, Mapping functional domains in AcMNPV pp31, *Virology* **222:**318–331.

Broussard, D. R., Guarino, L. A., and Jarvis, D. L., 1996b, Dynamic phosphorylation of *Autographa californica* nuclear polyhedrosis virus pp31, *J. Virol.* **70:**6767–6774.

Brown, M., and Faulkner, P., 1978, Plaque assay of nuclear polyhedrosis viruses in cell culture, *Appl. Environ. Microbiol.* **36:**31–35.

Brown, M., and Faulkner, P., 1975, Factors affecting the yield of virus in a cloned cell line of *Trichoplusia ni* infected with a nuclear polyhedrosis virus, *J. Invertbr. Pathol.* **26:**251–257.

Brown, M., Crawford, A. M., and Faulkner, P., 1979, Genetic analysis of a baculovirus, *Autograph californica* nuclear polyhedrosis virus. 1. Isolation of temperature sensitive mutants and assortment into complementation groups, *J. Virol.* **31:**190–198.

Brown, M., Faulkner, P., Cochran, M. A., and Chung, K. L., 1980, Characterization of two morphology mutants of *Autographa californica* nuclear polyhedrosis virus with large cuboidal inclusion bodies, *J. Gen. Virol.* **50:**309–316.

Bud, H. M., and Kelly, D. C., 1980, An electron microscope study of partially lysed baculovirus nucleocapsids: The intrancleocapsid packaging of viral DNA, *J. Ultrastruc. Res.* **73:**361–368.

Burand, J. P., Summers, M. D., and Smith, G., 1980, Transfection with baculovirus DNA, *Virology* **101:**286–290.

Carpenter, W. M., and Bilimoria, S. L., 1983, A semipermissive nuclear polyhedrosis virus infection: Characterization of infection kinetics and morphogenesis, *Virology* **130:**227–231.

Carstens, E. B., Tjia, S. T., and Doerfler, W., 1979, Infection of *Spodoptera frugiperda* cells with *Autographa californica* nuclear polyhedrosis virus. I. Synthesis of intracellular proteins after virus infection, *Virology* **99:**386–398.

Carstens, E. B., Tjia, S. T., and Doerfler, W., 1980, Infectious DNA from *Autographa californica* nuclear polyhedrosis virus, *Virology* **101:**311–314.

Carstens, E. B., Krebs, A., and Gallerneault, C. E., 1986, Identification of an amino acid essential to the normal assembly of *Autographa californica* nuclear polyhedrosis virus polyhedra, *J. Virol.* **58:**684–688.

Carstens, E. B., Ye, L. -B., and Faulkner, P., 1987, A point mutation in the polyhedrin gene of a baculovirus, *Autographa californica* MNPV, prevents crystallization of occlusion bodies, *J. Gen. Virol.* **68:**901–905.

Carstens, E. B., Williams, G. V., Faulkner, P., and Partington, S., 1992, Analysis of polyhedra morphology mutants of *Autographa californica* nuclear polyhedrosis virus: Molecular and utranstructural features, *J. Gen. Virol.* **71:**3035–3040.

Carstens, E. B., Chan H., Yu, H., Williams G. V., and Casselman, R., 1994, Genetic analysis of temperature-sensitive mutations in baculovirus late expression factors, *Virology* **204:**323–337.

Chaabihi, H., Ogliastro, M. H., Martin, M., Giraud, C., Devauchelle, G., and Cerutti, M., 1993, Competition between baculovirus polyhedrin and p10 gene expression during infection of insect cells, *J. Virol.* **67:**2664–2671.

Charlton, C. A., and Volkman, L. E., 1991, Sequential rearrangement and nuclear polymerization of actin in baculovirus-infected *Spodoptera frugiperda* cells, *J. Virol.* **65:**1219–1227.

Charlton, C. A., and Volkman, L. E., 1993a, Penetration of *Autographa californica* nuclear polyhedrosis virus nucleocapsids into IPLB Sf-21 cells induces actin cable formation, *Virology* **197:**245–254.

Charlton, C. A., and Volkman, L. E., 1993b, Baculoviruses, vertebrate viruses and cytoskeletons, in: *Parasites and Pathogens of Insects* (N. E. Beckage, S. N. Thompson, and B. A. Federici, eds.), pp. 103–125, Academic Press, New York.

Cheley, S., Kosik, K. S., Paskevich, P., Bakalis, S., and Bayley, H., 1992, Phosphorylated baculovirus p10 is a heat-stable microtubule-associated protein associated with process formation in Sf-9 cells, *J. Cell Sci.* **102:**739–752.

Chisholm, G. E., and Henner, D. J., 1988, Multiple early transcripts and splicing of the *Autographa californica* nuclear polyhedrosis virus IE-1 gene, *J. Virol.* **62:**3193–3200.

Clem, R. J., Fechheimer, M., and Miller, L. K., 1991, Prevention of apoptosis by a baculovirus gene during infection of insect cells, *Science* **254:**1388–1390.

Cochran, M. A., Carstens, E. B., Eaton, B. E., and Faulkner, P., 1982, Molecular cloning and physical mapping of restriction endonuclease fragments of *Autographa californica* nuclear polyhedrosis virus, *J. Virol.* **41:**940–946.

Croizier, G., Gonnet, P., and Devauchelle, G., 1987, Localization cytologique de la proteine non structurale p10 du baculovirus de la polyedrose nucleaire du Lepidoptere *Galleria mellonella* L, *C R Acad. Sci. IV* **305:**677–681.

Crook, N. E., 1991, Granulosis viruses, in: *Viruses of Invertebrates* (E. Kurstak, ed.), pp. 73–110, Marcel Dekker, New York.

Crook N. E., Clem, R., and Miller, L. K., 1993, An apoptosis-inhibiting baculovirus gene with a zinc finger-like motif, *J. Virol.* **67:**2168–2174.

Derksen, A. C. G., and Granados, R. R., 1988, Alteration of a lepidopteran peritrophic membrane by baculoviruses and enhancement of viral infectivity, *Virology* **167:**242–250.

Dessev, G., Iovcheva, C., Tasheva, B., and Goldman, R., 1988, Protein kinase activity associated with the nuclear lamina, *Proc. Natl. Acad. Sci. USA* **85:**2994–2998.

Dougherty, E. M., Weiner, R., Vaughn, J., and Reichelderfer, C., 1981, Physical factors that affect *in vitro Autographa californica* nuclear polyhedrosis virus infection, *Appl. Environ. Microbiol.* **41:**1166–1172.

Duncan, R., and Faulkner, P., 1982, Bromodeoxyuridine-induced mutants of *Autographa californica* nuclear polyhedrosis virus defective in occlusion body formation, *J. Gen. Virol.* **62:**369–373.

Duncan, R., Chung, K. L., and Faulkner, P., 1983, Analysis of a mutant of *Autographa californica* nuclear polyhedrosis virus with a defect in the morphogenesis of the occlusion body macromolecular lattice, *J. Gen. Virol.* **64:**1531–1542.

Elliot, R. M., and Kelly, D. C., 1977, The polyamine content of a nuclear polyhedrosis virus from *Spodoptera littoralis, Virology* **76:**472–474.

Engelhard, E. K., Kam-Morgan, L. N. W., Washburn, J. O., and Volkman, L. E., 1994, The insect tracheal system: A conduit for the systemic spread of *Autographa californica* M nuclear polyhedrosis virus, *Proc. Natl. Acad. Sci. USA* **91:**3224–3227.

Erlandson, M. A., and Carstens, E. B., 1983, Mapping early transcription products of *Autographa californica* nuclear polyhedrosis virus, *Virology* **126:**398–402.

Faulkner, P., 1962, Isolation and analysis of ribonucleic acid from inclusion bodies of the nuclear polyhedrosis of the silkworm, *Virology* **16:**479–484.

Faulkner, P., 1981, Baculovirus, in: *Pathogenesis of Invertebrate Microbial Diseases* (E. W. Davidson, ed.), pp. 3–33, Allanheld, Osmun, Totowa, NJ.

Faulkner, P., and Carstens, E. B., 1986, An overview of the structure and replication of baculoviruses, *Curr. Top. Microbiol. Immunol.* **131:**1–19.

Federici, B. A., 1986, Ultrastructure of baculoviruses, in: *The Biology of Baculoviruses,* Vol. I. Biological Properties and Molecular Biology, (R. R. Granados and B. A. Federici, eds.), pp. 61–88, CRC Press, Boca Raton, FL.

Fraser, M. J., 1986, Ultrastructural observations of virion maturation in *Autographa californica* nuclear polyhedrosis virus infected *Spodoptera frugiperda* cell cultures, *J. Ultrastr. Mol. Struct. Res.* **95:**189–195.

Fraser, M. J., Smith, G. E., and Summers, M. D., 1983, Acquisition of host cell DNA sequences by baculoviruses: Relationship between host DNA insertions and FP mutants of *Autographa californica* and *Galleria mellonella* nuclear polyhedrosis viruses, *J. Virol.* **47:**287–300.

Friesen, P. D., and Miller, L. K., 1986, The regulation of baculovirus gene expression, *Curr Top Microbiol Immunol.* **131:**31–49.

Fuchs, L. Y., Woods, M. S., and Weaver, R. F., 1983, Viral transcription during *Autographa californica* nuclear polyhedrosis virus infection: A novel RNA polymerase induced in infected *Spodoptera frugiperda* cells, *J. Virol.* **48:**641–646.

Funk, C. J., and Consigli, R. A., 1993, Phosphate cycling on the basic protein of *Plodia interpunctella* granulosis virus, *Virology* **193:**396–402.

Gerace, L., 1986, Nuclear lamina and organization of nuclear architecture, *Trends Biochem. Sci.* **11:**443–446.

Gerace, L., Comeau, C., and Benson, M., 1984, Organization and modulation of nuclear lamina structure, *J. Cell Sci.* Suppl. **1:**137–160.

Gibson, W., and Roizman, B., 1971, Compartmentalization of spermine and spermidine in the herpes simplex virion, *Proc. Natl. Acad. Sci. USA* **68:**2818–2821.

Gombart, A. F., Pearson, M. N., Rohrmann, G. F., and Beaudreau, G. S., 1989, A baculovirus polyhedral envelope-associated protein: Genetic location, nucleotide sequence, and immunocytochemical characterization, *Virology* **169:**182–193.

Goodwin, R. H., and Adams, J. R., 1980, Nutrient factors influencing viral replication in serum free insect cell line culture, in: *Invertebrate Systems in Vitro* (E. Kurstak, K. Maramorosch, and A. Dubendorfer, eds.), pp. 493–514, Elsevier/North-Holland Biomedical Press, Amsterdam.

Goodwin, R. H., Vaughn, J. L., Adams, J. R., and Louloudes, S. J., 1973, The influence of insect cell lines and tissue-culture media on baculovirus polyhedra production, *Misc. Publ. Entomol. Soc. Am.* **9:**66–72.

Gordon, J. D., and Carstens, E. B., 1984, Phenotypic characterization and physical mapping of a temperature sensitive mutant of *Autographa californica* nuclear polyhedrosis virus defective in DNA synthesis, *Virology* **138:**69–81.

Granados, R. R., 1978, Early events in the infection of *Heliothis zea* midgut cells by a baculovirus, *Virology* **90:**170–174.

Granados, R. R., and Lawler, K. A., 1981, *In vivo* pathway of *Autographa californica* baculovirus invasion and infection, *Virology* **108:**297–308.

Granados, R. R., and William, K. A., 1986, *In vivo* Infection and replication of baculoviruses, in: *The Biology of Baculoviruses,* (R. R. Granados and B. A. Federici, eds.), Vol. I. Biological Properties and Molecular Biology. pp. 90–108. CRC Press, Boca Raton, FL.

Gross, C. H., Wolgamot, G. M., Russell, R. L. Q., Pearson, M. N., and Rohrmann, G. F., 1993, A baculovirus encoded 10-kDa glycoprotein localizes near the nuclear membrane of infected cells, *Virology* **192:**386–390.

Guarino, L. A., 1990, Identification of a viral gene encoding a ubiquitin-like protein, *Proc. Natl. Acad. Sci. USA* **87:**409–413.

Guarino, L. A., and Summers, M. D., 1986a, Functional mapping of a *trans*-activating gene required for expression of a baculovirus delayed early gene, *J. Virol.* **57:**563–571.

Guarino, L. A., and Summers, M. D., 1986b, Interspersed homologous DNA of *Autographa californica* nuclear polyhedrosis virus enhances delayed early gene expression, *J. Virol.* **60:**215–223.

Guarino, L. A., Dong, W., Xu, B., Broussard, D. R., Davis, R. W., and Jarvis, D. L., 1992, Baculovirus phosphoprotein pp31 is associated with virogenic stroma, *J. Virol.* **66:**7113–7120.

Guarino, L. A., Smith, G., and Dong, W., 1995, Ubiquitin is attached to membranes of baculovirus particles by a novel type of phospholipid anchor, *Cell* **80:**301–309.

Guelpa, B., Bergoin, M., and Crozier, G., 1977, La protéine d'ínclusion et les protéines du virion du baculovirus du diptère *Tipula paludosa* (Meigen), *C R Acad. Sci Serie D* **284:**779–782.

Hagestedt, T., Lichtenberg, B., Wille, H., Mandelkow, E.-M., and Mandelkow, E., 1989, Tau protein becomes long and stiff upon phosphorylation: Correlation between paracrystalline structure and degree of phosphorylation, *J. Cell Biol.* **109:**1643–1651.

Hall, D. L., and Moyer, R. W., 1993, Identification of an *Amsacta* spheroidin-like protein within the occlusion bodies of *Choristoneura* entomopoxviruses, *Virology* **192:**179–187.

Harrap, K. A., 1970, Cell infection by a nuclear polyhedrosis virus, *Virology* **42:**311–318.

Harrap, K. A., 1972a, The structure of nuclear polyhedrosis viruses. I. The inclusion body, *Virology* **50:**114–123.

Harrap, K. A., 1972b, The structure of nuclear polyhedrosis viruses. III. Virus assembly, *Virology* **50:**133–139.

Harrap, K. A., and Payne, C. C., 1979, The structural properties and identification of insect viruses, *Ad. Virus Res.* **25:**273–355.

Harrison, R. L., and Summers, M. D., 1995a, Biosynthesis and localization of the *Autographa californica* nuclear polyhedrosis virus 25K gene product, *Virology* **208:**279–288.

Harrison, R. L., and Summers, M. D., 1995b, Mutations in the *Autographa californica* nuclear polyhedrosis virus 25K gene result in reduced virion occlusion, altered intranuclear envelopment and enhanced virus production, *J. Gen. Virol.* **76:**1451–1459.

Hess, R. T., and Falcon, L. A., 1978, Electron microscope observations of the membrane surrounding polyhdral inclusion bodies of insects, *Arch. Virol.* **56:**169–176.

Himeno, M., Yasuda, S., Kohsaka, T., and Onodera, K., 1968, The fine structure of a nuclear-polyhedrosis virus of the silkworm, *J. Invertbr. Pathol.* **11:**516–519.

Hong, T., Braunagel, S. C., and Summers, M. D., 1994, Transcription, translation, and cellular localization of PDV E-66: A structural protein of the PDV envelope of *Autographa californica* nuclear polyhedrosis virus, *Virology* **204:**210–222.

Hirokawa, N., 1994, Microtubule organization and dynamics dependent on microtubule-associated proteins, *Curr. Opin. Cell Biol.* **6:**74–81.

Hoopes, R. R., and Rohrmann, G. F., 1991, *In vitro* transcription of baculovirus immediate early genes: Accurate mRNA initiation by nuclear extracts from both insect and human cells, *Proc. Natl. Acad. Sci. USA* **88:**4513–4517.

Horton, H. M., and Burand, J. P., 1993, Saturable attachment sites for polyhedron-derived baculovirus on insect cells and evidence for entry via direct membrane fusion, *J. Virol.* **67:**1860–1868.

Huh, N. E., and Weaver, R. F., 1990a, Categorizing some early and late transcripts directed by the *Autographa californica* nuclear polyhedrosis virus, *J. Gen. Virol.* **71:**2195–2200.

Huh, N. E., and Weaver, R. F., 1990b, Identifying the RNA polymerases that synthesize specific transcripts of the *Autographa californica* nuclear polyhedrosis virus, *J. Gen. Virol.* **71:**195–201.

Inagaki, M., Nishi, Y., Nishizawa, K., Matsuyama, M., and Sato, C., 1987, Site-specific phosphorylation induces disassembly of vimentin filaments *in vitro*, *Nature* **328:**649–651.

Jacques, R. P., 1985, Stability of insect viruses in the environment, in: *Viral Insecticides for Biological Control* (K. Maramorosch and K. E. Sherman, eds.), pp. 285–360, Academic Press, New York.

Kamita, S. G., and Maeda, S., 1993, Inhibition of *Bombyx mori* nuclear polyhedrosis virus (NPV) replication by the putative DNA helicase gene of *Autographa californica* NPV, *J. Virol.* **67:**6239–6245.

Kawamoto, F., Suto, C., Kumada, N., and Kobayashi, M., 1977a, Cytoplasmic budding of a nuclear polyhedrosis virus and comparative ultrastructural studies of envelopes, *Microbiol. Immunol.* **21:**255–265.

Kawamoto, F., Kumada, N., and Kobayashi, M., 1977b, Envelopment of the nuclear polyhedrosis virus of the Oriental tussock moth, *Euproctis subflava*, *Virology* **77:**867–871.

Kawanishi, C., Summers, M. D., Stoltz, D. B., and Arnott, H. J., 1972, Entry of an insect virus *in vitro* by fusion of viral envelope and microvillus membrane, *J. Invertbr. Pathol.* **20:**104–108.

Keddie, B. A., Aponte, G. W., and Volkman, L. E., 1989, The pathway of infection of *Autographa californica* nuclear polyhedrosis virus in an insect host, *Science* **243:**1728–1730.

Kelly, D. C., 1981, Non-occluded viruses, in: *Pathogenesis of Invertebrate Microbial Diseases* (E. W. Davidson, ed.), pp. 39–60, Allanheld, Osmun, Totowa, NJ.

Kelly, D. C., 1985, The structure and physical characteristics of baculoviruses, in: *Viral Insecticides for Biological Control* (K. Maramorosch and K. E. Sherman, eds.), pp. 469–488, Academic Press, New York.

Kelly D. C., 1982, Baculovirus replication, *J. Gen. Virol.* **63:**1–13.

Kelly, D. C., and Lescott, T., 1981, Baculovirus replication: Protein synthesis in *Spodoptera frugiperda* cells infected with *Trichoplusia ni* nuclear polyhedrosis virus, *Microbiologica* **4:**35–57.

Kelly, D. C., and Wang, X., 1981, The infectivity of nuclear polyhedrosis virus DNA, *Ann. Virol.* **132E:**247–259.

Kelly, D. C., Brown, D. A., Ayres, M. D., Allen, C. J., and Walker, I. O., 1983, Properties of the major nucleocapsid protein of *Heliothis zea* singly enveloped nuclear polyhedrosis virus, *J. Gen. Virol.* **64:**399–408.

Knebel-Mörsdorf, D., Flipsen, J. T. M., Roncarati, R., Jahnel, F., Kleefsman, A. W. F., and Vlak, J. M., 1996, Baculovirus infection of *Spodoptera exigua* larvae: lacZ expression driven by promoters of early genes pe38 and me53 in larval tissues, *J. Gen. Virol.* **77:**815–824.

Knudson, D. L., 1978, Pathogenic-invertebrate viruses: *In vivo* specificity, in: *Viral Pesticides*, Symposium Proceedings (M. D. Summers and C. Y. Kawanishi, eds.), pp. 151–163, US Environmental Protection Agency, Washington, DC.

Knudson, D. L., and Harrap, K. A., 1976, Replication of a nuclear polyhedrosis virus in a continuous cell culture of *Spodoptera frugiperda:* Microscopy study of the sequence of events of the virus infection, *J. Virol.* **17:**254–268.

Knudson, D. L., and Tinsley, T. W., 1974, Replication of a nuclear polyhedrosis virus in a continuous cell line of *Spodoptera frugiperda:* Partial characterization of the viral DNA, comparative DNA-DNA hybridization, and patterns of DNA synthesis, *Virology* **87:**42–47.

Kool, M., van Den Berg, P. M. M. M., Tramper, J., Goldbach, R. W., and Vlak, J. M., 1993a, Location of two putative origins of DNA replication of *Autographa californica* nuclear polyhedrosis virus, *Virology* **192:**94–101.

Kool, M., Voeten, J. T. M., Goldbach, R. W., Tramper, J., and Vlak, J. M., 1993b, Identification of seven putative origins of *Autographa californica* multiple nucleocapsid nuclear polyhedrosis virus DNA replication, *J. Gen. Virol.* **74:**2661–2668.

Kozlov, E. A., Levitina, T. L., and Gusak, N. M., 1986, The primary structure of baculovirus inclusion body proteins. Evolution and structure-function aspects, *Curr. Top. Microbiol. Immunol.* **131:**135–164.

Kozlov, E. A., Rodnin, N. V., Levitina, T. L., Gusak, N. M., Radomsku, N. F., and Palchikovskaya, L. J., 1992, The amino acid sequence determination of a granulin and polyhedrin from two baculoviruses infecting *Agrotis segetum, Virology* **189:**320–323.

Kuzio, J., 1990, Characterization of two genes affecting virulence of a baculovirus, AcNPV: Polypeptides p10 and p74, PhD thesis, Queen's University.

Kuzio, J., and Faulkner, P., 1984, Regions of repeated DNA in the genome of *Choristoneura fumiferana* nuclear polyhedrosis virus, *Virology* **139:**185–188.

Kuzio, J., Rohel, D. Z., Curry, C. J., Krebs, A., Carstens, E. B., and Faulkner, P., 1984, Nucleotide sequence of the p10 polypeptide gene of *Autographa californica* nuclear polyhedrosis virus, *Virology* **139:**414–418.

Lanier, L. M., Slack, J. M., and Volkman, L. E., 1996, Actin binding and proteolysis by the baculovirus AcMNPV: The role of virion-associated V-cath, *Virology* **216:**380–388.

Leisy, D. J., and Rohrmann, G. F., 1993, Characterization of the replication of plasmids containing hr sequences in baculovirus-infected *Spodoptera frugiperda* cells, *Virology* **196:**722–730.

Lu, A., and Miller, L. K., 1995, The roles of eighteen baculovirus late expression factor genes in transcription and DNA replication, *J. Virol.* **69:**975–982.

Lubbert, H., Kruczekg, S., Tjia, S., and Doerfler, W., 1981, The cloned EcoRI fragments of *Autographa californica* nuclear polyhedrosis virus DNA, *Gene* **16:**343–345.

Lynn, D. E., and Hink, W. F., 1980, Comparison of nuclear polyhedrosis virus replication in five lepidopteran cell lines, *J. Invertbr. Pathol.* **35:**234–240.

Mackinnon, E. A., Henderson, J. F., Stoltz, D. B., and Faulkner, P., 1974, Morphogenesis of nuclear polyhedrosis virus under conditions of prolonged passage *in vitro, J. Ultrastruc. Res.* **49:**419–435.

Maeda, S., Kamita, S. G., and Kataska, H., 1991, The basic DNA-binding protein of *Bombyx mori* nuclear polyhedrosis virus: The existence of an additional arginine repeat, *Virology* **180:**807–810.

Maeda, S., Kamita, S. G., and Kondo, A., 1993, Host range expansion of *Autographa californica* nuclear polyhedrosis virus (NPV) following recombination of a 0.6-kilobase-pair DNA fragment originating from *Bombyx mori* NPV, *J. Virol.* **67:**6234–6238.

McLachlin, J. R., and Miller, L. K., 1994, Identification and characterization of vlf-1, a baculovirus gene involved in very late gene expression, *J. Virol.* **68:**7746–7756.

Miller, L. K., 1986, The genetics of baculoviruses, in: *The Biology of Baculoviruses,* (R. R. Granados and B. A. Federici, eds.), Vol. I, Biological Properties and Molecular Biology, pp. 217–238, CRC Press, Boca Raton, FL.

Miller, L. K., 1996, Insect viruses, in: *Field's Virology,* 3rd ed. (B. N. Fields, D. M. Knipe, P. M. Howley, R. M. Chanock, J. L. Melnick, T. P. Monath, B. Roizman, and S. E. Straus, eds.), pp. 533–556, Lippincott-Raven, Philadelphia.

Miller, L. K., and Dawes, K. P., 1979, Physical map of the DNA genome of *Autographa californica* nuclear polyhedrosis virus, *J. Virol.* **29:**1044–1055.

Miller, L. K., Adang, M. J., and Browne, D., 1983, Protein kinase activity associated with the extracellular and occluded forms of the baculovirus *Autographa californica* nuclear polyhedrosis virus, *J. Virol.* **46:**275–278.

Minion, F. C., Coons, L. B., and Broome, J. R., 1979, Characterization of the polyhedral envelope of the nuclear polyhedrosis virus of *Heliothis virescens, J. Invertbr. Pathol.* **34:**303–307.

Monsma, S. A., and Blissard, G. W., 1995, Identification of a membrane fusion domain and an oligomerization domain in the baculovirus GP64 envelope fusion protein, *J. Virol.* **69:** 2583–2595.

Nissen, M. S., and Friesen, P. D., 1989, Molecular analysis of the transcriptional regulatory region of an early baculovirus gene, *J. Virol.* **63:**493–503.

Noble, M., Lewis, S. A., and Cowan, N. J., 1989, The microtubule binding domain of microtubule-associated protein MAP1B contains a repeated sequence motif unrelated to that of MAP2 and *tau*, *J. Cell Biol.* **109:**3367–3376.

Ooi, B. G., and Miller, L. K., 1988, Regulation of host RNA levels during baculovirus infection, *Virology* **166:**515–523.

Oppenheimer, D. I., and Volkman, L. E., 1995, Proteolysis of p6.9 induced by cytochalasin D in *Autographa californica* M nuclear polyhedrosis virus-infected cells, *Virology* **207:**1–11.

Partington, S., Yu, H., Lu, A., and Carstens, E. B., 1990, Isolation of temperature sensitive mutants of *Autographa californica* nuclear polyhedrosis virus: Phenotype characterization of mutants defective in very late gene expression, *Virology* **175:**91–102.

Pearson, M., Bjornson, R., Pearson, G., and Rohrmann, G. F., 1992, The *Autographa californica* baculovirus genome: Evidence for multiple replication origins, *Science* **257:**1382–1384.

Pearson, M. N., Russell, R. L. Q., Rohrmann, G. F., and Beaudreau, G. S., 1988, P39, a major baculovirus structural protein: Immunocytochemical characterization and genetic location, *Virology* **167:**407–413.

Possee, R. D., 1986, Cell-surface expression of influenza virus haemagglutinin in insect cells using a baculovirus vector, *Virus Res.* **5:**43–59.

Possee, R. D., Sun, T. -P., Howard, S. C., Ayres, M. D., Hill-Perkins, M., and Gearing, K. L., 1991, Nucleotide sequence of the *Autographa californica* nuclear polyhedrosis 9.4 kbp EcoRI-I and -R (polyhedrin gene) region, *Virology* **185:**229–241.

Potter, K. N., and Miller, L. K., 1980, Transfection of two invertebrate cell lines with DNA of *Autographa californica* nuclear polyhedrosis virus, *J. Invertbr. Pathol.* **36:**431–439.

Potter, K. N., Faulkner, P., and Mackinnon, E. A., 1976, Strain selection during serial passage of *Trichoplusia ni* nuclear polyhedrosis virus, *J. Virol.* **18:**1040–1050.

Quant-Russell, R. L., Pearson, M. N., Rohrmann, G. F., and Beaudreau, G. S., 1987, Characterization of baculovirus p10 synthesis using monoclonal antibodies, *Virology* **160:**9–19.

Ramoska, W. A., and Hink, W. F., 1974, Electron microscope examination of two plaque variants from a nuclear polyhedrosis virus of the alfalfa looper, *Autographa californica*, *J. Invertbr. Pathol.* **23:**197–201.

Rice, W. C., and Miller, L. K., 1986, Baculovirus transcription in the presence of inhibitors and in nonpermissive *Drosophila* cells, *Virus Res.* **6:**155–172.

Roberts, T. E., 1989, Fatty acid acylated baculovirus polypeptides and aspects of intracellular protein processing, PhD thesis, Queen's University.

Roelvink, P. W., van Meer, M. M. M., de Kort, C. A. D., Possee, R. D., Hammock, B. D., and Vlak, J. M., 1992, Dissimilar expression of *Autographa californica* multiple nucleocapsid nuclear polyhedrosis virus polyhedrin and p10 genes, *J. Gen. Virol.* **73:**1481–1489.

Rohel, D. Z., and Faulkner, P., 1984, Time course analysis and mapping of *Autographa californica* nuclear polyhedrosis virus transcripts, *J. Virol.* **50:**739–747.

Rohel, D. Z., Cochran, M. A., and Faulkner, P., 1983, Characterization of two abundant mRNAs of *Autographa californica* nuclear polyhedrosis virus present late in infection, *Virology* **124:**357–365.

Rohrmann, G. F., 1986a, Polyhedrin structure, *J. Gen. Virol.* **67:**1499–1513.

Rohrmann, G. F., 1986b, Evolution of occluded baculoviruses, in: *The Biology of Baculoviruses,* (R. R. Granados and B. A. Federici, eds.), Vol. I. *Biological Properties and Molecular Biology,* pp. 203–216, CRC Press, Boca Raton, FL.

Rohrmann, G. F., 1992, Baculovirus structural proteins, *J. Gen. Virol.* **73:**749–761.

Romoser, W. S., 1981, *The Science of Entomology,* 2nd ed., Macmillan, New York.

Russell, R. L. Q., and Rohrmann, G. F., 1990a, A baculovirus polyhedron envelope protein: immunogold localization in infected cells and mature polyhedra, *Virology* **174:**177–184.

Russell, R. L. Q., and Rohrmann, G. F., 1990b, The p6.5 gene region of a nuclear polyhedrosis virus of *Orgyia pseudotsugata:* DNA sequence and transcriptional analysis of four late genes, *J. Gen. Virol.* **71:**551–560.

Russell, R. L. Q., and Rohrmann, G. F., 1993, A 25-kDa protein is associated with the envelopes of occluded baculovirus virions, *Virology* **195:**532–540.

Russell, R. L. Q., Pearson, M. N., and Rohrmann, G. F., 1991, Immunoelectron microscopic examination of *Orgyia pseudotsugata* multicapsid nuclear polyhedrosis virus-infected *Lymantria dispar* cells: Time course and localization of major polyhedron-associated proteins, *J. Gen. Virol.* **72:**275–283.

Schetter, C., Oellig, C., and Doerfler, W., 1990, An insertion of insect cell DNA in the 81-map-unit segment of *Autographa californica* nuclear polyhedrosis virus DNA, *J. Virol.* **64:**1844–1850.

Serrano, L., Avila, J., and Maccioni, R. B., 1984, Controlled proteolysis of tubulin by subtilisin: Localization of the site for MAP-2 interaction, *Biochemistry* **23:**4675–4681.

Serrano, L., Montejo, E., Hernandez, M. A., and Avila, J., 1985, Localization of the tubulin binding site for *tau* protein, *Eur. J. Biochem.* **153:**595–600.

Smith, G. E., and Summers, M. D., 1979, Restriction maps of five *Autographa californica* MNPV variants, *Trichoplusia ni* MNPV, and *Galleria mellonella* MNPV DNAs with endonucleases SmaI, KpnI, BamHI, *J. Virol.* **30:**828–838.

Smith, G. E., Vlak, J. M., and Summers, M. D., 1983, Physical analysis of *Autographa californica* nuclear polyhedrosis virus transcripts for polyhedrin and 10,000-molecular weight protein, *J. Virol.* **45:**215–225.

Stairs, G. R., 1968, Inclusion-type insect viruses, *Curr. Top. Microbiol. Immunol.* **42:**1–23.

Stoltz, D. B., and Summers, M. D., 1971, Pathway of infection of the mosquito iridescent virus. I. Preliminary observations on the fate of the ingested virus, *J. Virol.* **8:**900–909.

Stoltz, D. B., Pavan, C., and DaCunha, A. B., 1973, Nuclear polyhedrosis virus: A possible example of *de novo* intranuclear membrane morphogenesis, *J. Gen. Virol.* **19:**145–150.

Summers, M. D., 1971, Electron microscopic observations on granulosis virus entry, uncoating and replication processes during infection of the midgut cells of *Trichoplusia ni*, *J. Ultrastruc. Res.* **35:**606–625.

Summers, M. D., and Arnott, H. J., 1969, Ultrastructural studies on occlusion formation and virus occlusion in nuclear polyhedrosis and granulosis-infected cells of *Trichoplusia ni* (Hübner), *J. Ultrastruc. Res.* **28:**462–480.

Tanada, Y., and Hess, R. T., 1976, Development of a nuclear polyhedrosis virus in midgut cells and penetration of the virus into the hemocoel of the armyworm, *Pseudaletia unipuncta*, *J. Invertbr. Pathol.* **28:**67–76.

Tanada, Y., and Hess, R. T., 1991, *Baculoviridae.* Granulosis viruses, in: *Atlas of Invertebrate Viruses* (J. R. Adams and J. R. Bonami, eds.), pp. 227–257, CRC Press, Boca Raton, FL.

Theilmann, D. A., and Stewart, S., 1992, Tandemly repeated sequence at the 3′ end of the IE-2 gene of the baculovirus *Orgyia pseudotsugata* multicapsid nuclear polyhedrosis virus is an enhancer element, *Virology* **187:**97–106.

Thiem, S. M., and Miller, L. K., 1989, Identification, sequence, and transcriptional mapping of the major capsid protein gene of the baculovirus *Autographa californica* nuclear polyhedrosis virus, *J. Virol.* **63:**2008–2018.

Tjia, S. T., Carstens, E. B., and Doerfler, W., 1979, Infection of *Spodoptera frugiperda* cells with *Autographa californica* nuclear polyhedrosis virus. II. The viral DNA and the kinetics of its replication, *Virology* **99:**391–409.

Todd, J. W., Passarelli, A. L., and Miller, L. K., 1995, Eighteen baculovirus genes, including lef-11, p35, 39K, and p47, support late gene expression, *J. Virol.* **69:**968–974.

Todd, J. W., Passarelli, A. L., Lu, A., and Miller, L. K., 1996, Factors regulating baculovirus late and very late gene expression in transient-expression assay, *J. Virol.* **70:**2307–2317.

Tomalski, M. D., Wu, J., and Miller, L. K., 1988, The location, sequence, transcription, and regulation of a baculovirus DNA polymerase gene, *Virology* **167:**591–600.

Tweeten, K. A., Bulla, L. A., and Consigli, R. A., 1980, Characterization of an extremely basic protein derived from granulosis virus nucleocapsid, *J. Virol.* **33:**866–876.

van der Wilk, F. van Lent, J. W. M., and Vlak, J. M., 1987, Immunogold detection of polyhedrin, p10 and virion antigens in *Autographa californica* nuclear polyhedrosis virus-infected *Spodoptera frugiperda* cells, *J. Gen. Virol.* **68:**2615–2613.

van Lent, J. W. M., Groenen, J. T. M., Klinge-Roode, E. C., Rohrmann, G. F., Zuidema, D., and Vlak, J. M., 1990, Localization of the 34 kDa polyhedron envelope protein in *Spodoptera frugiperda* cells infected with *Autographa californica* nuclear polyhedrosis virus, *Arch. Virol.* **111:**103–114.

van Oers, M. M., 1994, Functional analysis of the baculovirus 10 kilodalton protein, PhD thesis, Agricultural University in Wageningen, The Netherlands.

van Oers, M. M., Flipsen, J. T. M., Reusken, C. B. E. M., Sliwinsky, E. L., Goldbach, R. W., and Vlak, J. M., 1993, Functional domains of the p10 protein of *Autographa californica* nuclear polyedrosis virus, *J. Gen. Virol.* **74:**563–574.

van Oers, M. M., Flipsen, J. T. M., Reusken, C. B. E. M., and Vlak, J. M., 1994, Specificity of baculovirus p10 functions, *Virology* **200:**513–523.

Vaughn, J. L., and Dougherty, E. M., 1985, The replication of baculoviruses, in: *Viral Insecticides for Biological Control* (K. Maramorosch and K. E. Sherman, eds.), pp. 569–633, Academic Press, New York.

Vialard, J. E., and Richardson, C. D., 1993, The 1,629-nucleotide open reading frame located downstream of the *Autographa californica* nuclear polyhedrosis virus polyhedrin gene encodes a nucleocapsid-associated phosphoprotein, *J. Virol.* **67:**5859–5866.

Vialard, J. E., Yuen, L., and Richardson, C. D., 1990, Identification and characterization of a baculovirus occlusion body glycoprotein which resembles spheroidin, and entomopoxvirus protein, *J. Virol.* **64:**5804–5811.

Vlak, J. M., and Rohrmann, G. F., 1985, The nature of polyhedrin, in: *Viral Insecticides for Biological control* (K. Maramorosch and K. Sherman, eds.), pp. 489–542, Academic Press, New York.

Vlak, J. M., and Smith, G. E., 1982, Orientation of the genome of *Autographa californica* nuclear polyhedrosis virus: A proposal, *J. Virol.* **41:**1118–1121.

Vlak, J. M., Smith, G. E., and Summers, M. D., 1981, Hybridization selection and *in vitro* translation of *Autographa californica* nuclear polyhedrosis virus mRNA, *J. Virol.* **40:**762–771.

Vlak, J. M., Klinkenberg, F. A., Zaal, K. J. M., Usmany, M., Klinge-Roode, E. C., Geervliet, J. B. F., Roosien, J., and van Lent, J. W. M., 1988, Functional studies on the p10 gene of *Autographa californica* nuclear polyhedrosis virus using a recombinant expression a p10-β-galactosidase fusion gene, *J. Gen. Virol.* **69:**765–776.

Volkman, L. E., 1986, The 64K envelope protein of budded *Autographa californica* nuclear polyhedrosis virus, *Curr. Top. Microbiol. Immunol.* **131:**103–118.

Volkman, L. E., 1988, *Autographa californica* MNPV nucleocapsid assembly: Inhibition by cytochalasin D., *Virology* **163:**547–553.

Volkman, L. E., and Goldsmith, P. A., 1985, Mechanism of neutralization of budded *Autographa californica* nuclear polyhedrosis virus by a monoclonal antibody: inhibition of entry by adsorptive endocytosis, *Virology* **143:**185–195.

Volkman, L. E., and Keddie, B. A., 1990, Nuclear polyhedrosis virus pathogenesis, *Semin. Virol.* **1:**249–256.

Volkman, L. E., and Knudson, D. L., 1986, *In vitro* replication of baculoviruses, in: *The Biology of Baculoviruses,* (R. R. Granados and B. A. Federici, eds.), Vol. I. *Biological Properties and Molecular Biology,* pp. 110–127, CRC Press, Boca Raton, FL.

Volkman, L. E., and Zaal, K. J. M., 1990, *Autographa californica* M nuclear polyhedrosis virus: Microtubules and replication, *Virology* **175:**292–302.

Volkman, L. E., Summers, M. D., and Hsieh, C. H., 1976, Occluded and non-occluded nuclear polyhedrosis virus grown in *Trichoplusia ni:* Comparative neutralization comparative infectivity, and *in vitro* growth studies, *J. Virol.* **19:**820–832.

Volkman, L. E., Goldsmith, P. A., Hess, R. T., and Faulkner, P., 1984, Neutralization of budded *Autographa californica* NPV by a monoclonal antibody: Identification of the target antigen, *Virology* **133:**354–362.

Volkman, L. E., Goldsmith, P. A., and Hess, R. T., 1986, Alternate pathway of entry of budded *Autographa californica* nuclear polyhedrosis virus: Fusion at the plasma membrane, *Virology* **148:**288–297.

Volkman, L. E., Goldsmith, P. A., and Hess, R. T., 1987, Evidence for microfilament involvement in budded *Autographa californica* nuclear polyhedrosis virus production, *Virology* **156:**32–39.

Volkman, L. E., Talhouk, S. N., Oppenheimer, D. I., and Charlton, C. A., 1992, Nuclear f-actin: A functional component of baculovirus-infected lepidopteran cells? *J. Cell Sci.* **103:**15–22.

Volkman, L. E., Blissard, G. W., Friesen, P., Keddie, B. A., Possee, R., and Theilmann, D. A., 1995, *Baculoviridae,* in: *Virus Taxonomy,* (F. A. Murphy, C. M. Fauquet, D. H. L. Bishop, S. A. Ghabrial, A. W. Jarvis, G. P. Martelli, M. A. Mayo, and M. D. Summers, eds.), Sixth report of the international committee on taxonomy of viruses, pp. 104–111, Springer-Verlag, New York.

Whitford, M., and Faulkner, P., 1992a, A structural polypeptide of the baculovirus *Autographa californica* nuclear polyhedrosis virus contains *O*-linked *N*-acetylglucosamine, *J. Virol.* **66:**3324–3329.

Whitford, M., and Faulkner, P., 1992b, Nucleotide sequence and transcriptional analysis of a gene encoding gp41, a structural glycoprotein of the baculovirus *Autographa californica* nuclear polyhedrosis virus, *J. Virol.* **66:**4763–4768. (Erratum, **67:**2427, 1993.)

Whitford, M., Stewart, S., Kuzio, J., and Faulkner, P., 1989, Identification and sequence analysis of a gene encoding gp67, an abundant envelope glycoprotein of the baculovirus, *Autographa californica* nuclear polyhedrosis virus, *J. Virol.* **63:**1393–1399.

Whitt, M. A., and Manning, J. S., 1988, A phosphorylated 34-kDa protein and a subpopulation of polyhedrin are thiol linked to the carbohydrate layer surrounding a baculovirus occlusion body, *Virology* **163:**33–42.

Wickham, T. J., Shuler, M. L., Hammer, D. A., Granados, R. R., and Wood, H. A., 1992, Equilibrium and kinetic analysis of *Autographa californica* nuclear polyhedrosis virus attachment to different insect cell lines, *J. Gen. Virol.* **73:**3185–3194.

Williams, G. V., 1995, Cytopathology of *Autographa californica* nuclear polyhedrosis virus infection: Morphogenic processes associated with the late viral polypeptides p10 and p74, PhD thesis, Queen's University.

Williams, G. V., Rohel, D. Z., Kuzio, J., and Faulkner, P., 1989, A cytopathological investigation of *Autographa californica* nuclear polyhedrosis virus p10 gene function using insertion/deletion mutants, *J. Gen. Virol.* **70:**187–202.

Wilson, M. E., and Consigli, R. A., 1985a, Characterization of a protein kinase associated with purified capsids of the granulosis virus infecting *Plodia interpunctella, Virology* **143:**516–525.

Wilson, M. E., and Consigli, R. A., 1985b, Functions of a protein kinase activity associated with purified capsids of the granulosis virus infecting *Plodia interpunctella, Virology* **143:**526–535.

Wilson, M. E., and Miller, L. K., 1986, Changes in the nucleoprotein complexes of a baculovirus DNA during infection, *Virology* **151:**315–328.

Wilson, M. E., and Price, K. H., 1988, Association of *Autographa californica* nuclear polyhedrosis virus (AcMNPV) with the nuclear matrix, *Virology* **167:**233–241.

Wilson, M. E., Mainprize, T. H., Friesen, P. D., and Miller, L. K., 1987, Location, transcription, and sequence of a baculovirus gene encoding a small arginine-rich polypeptide, *J. Virol.* **61:**661–666.

Winstanley, D., and Crook, N. E., 1993, Replication of *Cydia pomonella* granulosis virus in cell cultures, *J. Gen. Virol.* **74:**1599–1609.

Wolgamot, G. M., Gross, C. H., Russell, R. L. Q., and Rohrmann, G. F., 1993, Immunocytochemical characterization of p24, a baculovirus capsid-associated protein, *J. Gen. Virol.* **74:**103–107.

Wood, H. A., 1980, Isolation and replication of an occlusion body deficient mutant of the *Autographa californica* nuclear polyhedrosis virus, *Virology* **105:**338–344.

Wood, H. A., Johnston, L. B., and Burand, J. P., 1982, Inhibition of *Autographa californica* nuclear polyhedrosis virus replication in high-density *Trichoplusia ni* cell cultures, *Virology* **119:**245–254.

Yang, C. L., Stetler, D. A., and Weaver, R. F., 1991, Structural comparison of the *Autographa californica* nuclear polyhedrosis virus-induced RNA polymerase and the three nuclear RNA polymerases from the host, *Spodoptera frugiperda, Virus Res.* **20:**251–264.

Young, J. C., Mackinnon, E. A., and Faulkner, P., 1993, The architecture of the virogenic stroma in isolated nuclei of *Spodoptera frugiperda* cells *in vitro* infected by *Autographa californica* nuclear polyhedrosis virus, *J. Struct. Biol.* **110:**141–153.

Zanotto, P. M. A., Kessing, B. D., and Maruniak, J. E., 1993, Phylogenetic interrelationships among baculoviruses: Evolutionary rates and host associations, *J. Invertbr. Pathol.* **62:**147–164.

Zuideman, D., Klinge-Roode, E. C., van Lent, J. W. M., and Vlak, J. M., 1989, Construction and analysis of an *Autographa californica* nuclear polyhedrosis virus mutant lacking the polyhedral envelope, *Virology* **173:**98–108.

Zuidema, D., Van Oers, M. M., Van Strien, E. A., Caballero, P. C., Klok, E. J., Goldbach, R. W., and Vlak, J. M., 1993, Nucleotide sequence and transcriptional analysis of the p10 gene of *Spodoptera exigua* nuclear polyhedrosis virus, *J. Gen. Virol.* **74:**1017–1024.

CHAPTER 5

Baculovirus Genome Organization and Evolution

ROBERT D. POSSEE AND GEORGE F. ROHRMANN

I. INTRODUCTION

The development of DNA-sequencing technology and the advancement of computer-assisted sequence analysis have provided a powerful means of determining the genetic content of baculovirus genomes and investigating their gene organization and evolution. The entire nucleotide sequence of three baculovirus genomes are now available (Ahrens *et al.*, 1997; Ayres *et al.*, 1994; Maeda, 1994). Two of these genomes, those of *Autographa californica* nuclear polyhedrosis virus (AcMNPV) and *Bombyx mori* nuclear polyhedrosis virus (BmNPV), are closely related, whereas the third sequenced genome, that of *Orgyia pseudotsugata* nuclear polyhedrosis virus (OpMNPV), is more distantly related. Comparison of the overall gene content of divergent baculovirus genomes provides an initial view of what genes are retained in all genomes, and thus are likely to be fundamental for virus survival. Such genes constitute the "essence" of being a baculovirus, i.e., the minimal basis of membership in the baculovirus family, and are likely to provide insight into the ancient origins of this viral family. In contrast, characterization of the genes that are found in only some members of the family provide novelty to the virus and will influence individual phenotypic traits such as host or tissue tropism, virulence, and morphology

---

ROBERT D. POSSEE • NERC Institute of Virology and Environmental Microbiology, Oxford OX1 3SR, England. GEORGE F. ROHRMANN • Department of Microbiology, Oregon State University, Corvallis, Oregon 97331-7301.

The Baculoviruses, edited by Lois K. Miller. Plenum Press, New York, 1997.

[e.g., nuclear polyhedrosis viruses (NPVs) vs. granulosis viruses (GVs)]. The nature of these "individualistic" genes will probably reveal how baculoviruses are currently changing and acquiring new, distinctive properties. This chapter focuses on a comparison of the AcMNPV and OpMNPV genomes and highlights some of the fundamental as well as unique properties of these viruses.

II. THE BASICS OF BACULOVIRUS DNA

A. Early Analysis of Baculovirus Genomes

DNA was first identified in virus particles derived from the occlusion bodies of an NPV by Breindl and Jirovec (1935). Further information on the base composition of NPV and GV DNAs was provided by Wyatt (1953). Electron microscopy and biochemical and biophysical studies provided the first evidence that baculovirus genomes were large circular DNAs (Bud and Kelly, 1977; Burgess, 1977; Summers and Anderson, 1972a,b; reviewed by Harrap and Payne, 1979).

The introduction of restriction endonuclease analysis to the study of baculovirus genomes in the 1970s provided a powerful new means of characterizing baculovirus genomes. Initially, these analyses provided the means of estimating total genome sizes more accurately, assessing genomic heterogeneity, and distinguishing different baculoviruses (Miller and Dawes, 1978a,b; Rohrmann *et al.*, 1977; Smith and Summers, 1978). Coupled with Southern blot hybridization, restriction endonuclease analysis also suggested that some baculovirus DNAs exhibited considerable sequence divergence (Jewell and Miller, 1980). But most importantly, restriction endonucleases provided a means of producing discrete fragments of baculovirus DNA that could be ordered—to provide physical maps of the genomes—and cloned—to provide pure samples for genetic and sequence analyses.

In the 1980s and early 1990s, defined portions of baculovirus genomes were sequenced and functionally analyzed. The first complete baculovirus gene to be sequenced was the AcMNPV polyhedrin gene (Hooft van Iddekinge *et al.*, 1983). In the ensuing decade, numerous studies of individual baculovirus genes provided a tantalizing glimpse of the nature and complexity of the baculovirus genome (summarized in Kool and Vlak, 1993). Many of these studies included transcriptional mapping and/or functional analysis as well as DNA sequence analysis. Transcriptional analyses provided evidence for the interspersion of early, late, and very late gene classes throughout the genome; different transcriptional classes of genes are not clustered with respect to their temporal expression. By the time the entire sequence of the AcMNPV genome was published (Ayres *et al.*, 1994), approximately half of the AcMNPV genome had already been sequenced and approximately 40 of the genes had been functionally or transcriptionally characterized to some extent. The availability of the entire sequence, however, revolutionized the way genetic and molecular studies could be approached.

B. Heterogeneity of Baculovirus Genomes

The current statement on baculovirus taxonomy (Murphy *et al.*, 1995) suggests that the genome size for different members of this virus family may vary from 90 to 160 kilobasepairs (kbp). Such a wide variation in the size of baculovirus genomes indicates that some baculoviruses may lack many of the genes present in other members of the family. Because of their rod-shaped nucleocapsid structure (see Chapter 2, this volume), baculovirus nucleocapsids are able to expand to accommodate such natural variations in genome content, and they also can accept engineered insertions of foreign genetic material (see Chapter 14, this volume).

While most molecular genetic studies of baculoviruses have been performed with plaque-purified strains propagated in cell culture, field isolates of baculoviruses often show considerable genetic heterogeneity. Early studies with AcMNPV and its close relatives revealed sequence heterogeneity as monitored by restriction endonuclease digestion of plaque-purified isolates (Lee and Miller, 1978; Smith and Summers, 1978). Isolates of other baculoviruses such as *Spodoptera frugiperda* MNPV (Maruniak *et al.*, 1984) and *Heliothis zea* SNPV (Corsaro and Fraser, 1987) exhibited similar heterogeneity. Some baculoviruses exhibit a high degree of genetic instability when propagated in cell culture, and variants identified following passage in cell culture may not represent the constituents of the original isolate. For example, propagation of *Spodoptera exigua* MNPV in cell culture rapidly generates deletion mutants that cannot be amplified *in vivo* (Heldens *et al.*, 1996). AcMNPV, although comparatively stable in cell culture, may be genetically altered during serial passage in cell culture, even when passaged at high dilution (reviewed by Miller, 1986). At least some of these changes are due to insertion of host transposable elements into the viral genome (reviewed by Friesen, 1993), although specific types of deletions are also observed (Kumar and Miller, 1987). However, the sequence heterogeneity first reported for AcMNPV (Lee and Miller, 1978; Smith and Summers, 1978) is probably an accurate reflection of the heterogeneity of the original field isolate.

Passage of a baculovirus through a single permissive insect species may also influence (bias) the proportion of the variants within the population. For example, the proportion of two virus variants of *Panolis flammea* MNPV, PfMNPV A and B (Weitzmann *et al.*, 1992), changes according to the larval host in which the virus is propagated, but both genotypes are always present. The fact that minority genotypes are not eliminated indicates that this heterogeneity is important for virus survival and suggests that field isolates containing heterogeneous populations may be more valid models for studies of baculovirus host range and pathogenicity than cloned isolates. The realities of molecular genetic research, however, usually require the use of cloned isolates. The sequence of the C6 strain of AcMNPV is described here, but it should be remembered that evidence already exists for microheterogeneity (occasional nucleotide differences) among the isolates of AcMNPV used by different laboratories in the world, e.g., C6, L-1, HR3, and E2.

III. THE AcMNPV GENOME

A. Coding Potential

The AcMNPV C6 genome contains 133,894 bp. Analysis of such a complex set of information involves the use of computer programs that identify putative open reading frames (ORFs) that initiate with an ATG (methionine initiation codon) and terminate with a stop codon. For the analysis of AcMNPV, ORFs that were less than 150 nucleotides in length were excluded, which led to the identification of a total of 337 ORFs (Ayres *et al.*, 1994). A strong bias was introduced into these analyses by focusing on translatable genes and by the exclusion of gene products smaller than 50 amino acids. Such an approach, in which genomic DNA but not cDNA was characterized, excludes the identification of gene products of spliced mRNAs. However, since splicing of baculovirus mRNAs has only been identified in the transcript spanning IE-0 and IE-1 (Chisholm and Henner, 1988; Kovacs *et al.*, 1991) this parameter may not have a major effect on the prediction of baculovirus gene products. In addition, nontranslatable genes such as tRNAs would not be identified by this approach.

The selection of 337 ORFs in the AcMNPV genome still presents a mass of information that is difficult to interpret. Consequently, Ayres *et al.* (1994) reduced the number to 154 by assuming that ORFs are distributed as nonoverlapping, contiguous sequences. Some exceptions were made for previously well-defined baculovirus genes. For example, the 5′ end of *lef-2* (Passarelli and Miller, 1993a) starts within the 3′ region of ORF5 and the 3′ end of *lef-1* (Passarelli and Miller, 1993b) overlaps the start of ORF13. These additional criteria could have a significant impact on the prediction of baculovirus gene products, since baculoviruses have short intergenic regions and regulatory sequences often overlap with coding sequences. For example, the early promoters for CG-30, *lef-4* and *lef-6*, are located within the coding region of the gene immediately upstream of each gene (Passarelli and Miller, 1993c, 1994a,b). The *lef-10* gene, a small ORF overlapping extensively with a downstream gene (Lu and Miller, 1994), was eliminated by the ORF selection criteria.

The analysis of AcMNPV genes was assisted by the use of the neural-net ORF identification program, GRAIL (Uberbacher and Mural, 1991). This program searches for coding potential in the primary sequence data. Encouragingly, GRAIL selected 84% of the 154 ORFs as potential coding regions, providing some measure of confidence in the original, intuitive analysis.

A summary of the properties of the 154 AcMNPV ORFs selected by Ayres *et al.* (1994) is presented in Table 1. The direction of the ORFs with respect to their clockwise or counterclockwise position on the circular genome is indicated by rightward and leftward arrows, respectively; the ORFs are found on both strands of the DNA throughout the genome. The size of the predicted gene products of each of the ORFs is presented as both the predicted molecular weight and the number of amino acid residues. In those cases where some functional information has been determined for an ORF and/or some sequence homology has been determined by sequence analysis, the gene has been assigned a name and a reference is provided. The *lef-10* gene has also been added to the list of ORFs in Table 1 as ORF 53a and, to avoid confusion, future changes of this nature should follow the protocol of not altering the numbering of the original ORFs.

B. Transcription Signals

The recognition that baculovirus genes are expressed in a temporally regulated manner and that early and late genes possess distinctive transcription signals provides another method to identify potential coding regions. The early genes of AcMNPV utilize the host cell RNA polymerase II (RNA *polII*) and probably host transcription factors. In general, the promoters of early genes conform to those seen in other RNA *polII*-transcribed genes (see Chapter 6, this volume) and often include a TATA box 20 to 40 bps upstream of the transcriptional start point. Two additional early gene promoter motifs have been functionally identified in several early promoters: an initiator CAGT and an upstream regulatory element CGTGC (Friesen and Miller, 1987; Kogan and Blissard, 1994; Lu and Carstens, 1991; reviewed in Chapter 6, this volume). The presence of these three sequence motifs upstream of a potential ORF are likely to be suggestive of early transcription of that region. Those genes appearing to have early transcriptional motifs are indicated as E in the "trans" column (Table 1), while those with a CATG motif are indicated with a C in this column.

The search for baculovirus late transcription signals is easier, since most of these genes have a TAAG at the point of mRNA initiation. The 5′ leader sequences of most characterized late mRNAs are less than 80 nucleotides in length. A search for TAAGs within the 80 nucleotides upstream of each predicted ORF revealed that TAAG motifs are located in 71 of the 154 selected ORFS and only 11 of the 183 nonselected ORFS. The TAAG motif occurs less frequently than would be expected for a random sequence; it occurs 210 times on the clockwise (+) strand and 196 times on the counterclockwise (−) strand. For comparison, GAAT occurs 574 times on the + strand and 595 times on the − strand. The paucity of TAAG motifs may

TABLE I. The 154 Potentially Expressed ORFs of AcMNPV C6[a]

AcMNPV								OpMNPV			
ORF	Left	Dir	Right	M_r	AA	Trans	Name	ORF	AA	% ID	References
1	503	→	1007	19288	168	EL	*ptp*	10	220	60.1	Tilakaratne *et al.* (1991), Kim and Weaver (1993)
2	1041	←	2025	37769	328	L		115	88	40.9	
3	2084	←	2243	5590	53	L	*ctl*	136	53	75.5	Eldridge *et al.* (1992)
4	2295	←	2748	17577	151	E		8	146	64.4	
5	2779	→	3106	12435	109	L		7	77	49.9	
6	3089	→	3719	23926	210	L	*lef-2*	6	204	54.7	Passarelli and Miller (1993a)
7	3759	←	4362	23612	201	C	*orf603*				Gearing and Possee (1990)
8	4520	→	5255	28642	245	CL	*polh*	3	245	89.4	Hooft van Iddekinge *et al.* (1983)
9	5287	←	6916	60713	543	C	*orf1629*	2	473	28.1	Possee *et al.* (1991), Vialard and Richardson (1993)
10	6917	→	7733	31978	272	CL	*pk1*	1	274	66.5	Reilly and Guarino (1994)
11	7899	←	8919	40093	340	EC		11	331	43.9	
12	8958	→	9609	25412	217	EC					
13	9638	←	10619	38660	327	ECL		12	320	49.0	
14	10513	←	11311	30780	266	C	*lef-1*	13	243	58.1	Passarelli and Miller (1993b)
15	11426	→	12944	57033	506	C	*egt*	14	489	60.0	O'Reilly and Miller (1989, 1990)
16	13092	→	13767	25910	225			15	197	37.5	
17	13738	→	14230	18482	164	E		16	207	51.6	
18	14398	←	15457	40870	353	EC		17	355	47.2	
19	15461	→	15785	12162	108	EL		18	104	43.3	
20	16013	←	16220	7978	69	E					
21	16305	←	17262	36555	319	EC		19	298	35.1	
22	17301	→	18447	43777	382	CL		20	382	79.8	
23	18513	→	20583	79857	690	CL		21	627	40.1	
24	20634	←	21141	19193	169	L		44	166	41.8	
25	21183	←	22131	36554	316	C		43	300	43	
26	22209	→	22596	14644	129	EL		42	127	62.2	
27	22600	→	23458	33320	286	EC	*iap1*	41	275	61.9	Braunagel *et al.* (1992), Crook *et al.* (1993), Birnbaum *et al.* (1994)

28	23465	→	23984	20388	173	EC	*lef-6*	40	138	35.3	Passarelli and Miller (1994a)
29	24046	←	24259	8569	71			39	75	52.1	
30	24315	←	25704	54688	463	C		38	459	52.9	
31	25820	→	26273	16182	151	L	*sod*	29	152	72.2	Tomalski *et al.* (1991)
32	27041	←	27584	20619	181	C	*fgf*	27	205	39.7	Ayres *et al.* (1994)
33	27733	←	28279	20838	182	EL					
34	28294	←	28939	24885	215			26	209	51.9	
35	28962	→	29193	8653	77	L	*ubi*	25	93	84.4	Guarino (1990)
36	29242	←	30067	31282	275	C	*39k*	24	261	54.8	Guarino and Smith (1990)
37	30063	←	30399	13129	112		*lef-11*	23	125	48.6	Todd *et al.* (1995)
38	30364	←	31012	25257	216	L		22	209	72.7	
39	31078	←	32167	43490	363	E	*p43*				Carstens *et al.* (1993)
40	32177	←	33380	47530	401	E	*p47*	45	399	67.5	Carstens *et al.* (1993)
41	33381	→	33924	21058	181	C		46	171	46.7	
42	34010	→	35528	59058	506	C	*gta*	47	498	59.2	
43	35544	→	35775	8816	77	L		48	63	61.9	
44	35758	→	36151	15002	131	E		49	120	45.7	
45	36155	→	36731	22651	192	EC					
46	36718	→	38830	79075	704	L	*PDV-E66*	50	682	74.7	Hong *et al.* (1994)
47	38938	←	39202	10482	88	E	*ets*	51	81	31.2	Crawford and Miller (1988)
48	39278	←	39617	12873	113	EC	*etm*	52	109	44.0	Crawford and Miller (1988)
49	39643	←	40498	32121	285	E	*pcna*	53	250	30.1	Crawford and Miller (1988)
											O'Reilly *et al.* (1989)
50	40523	←	43151	101765	876	C	*lef-8*	54	884	69.3	Passarelli *et al.* (1994)
51	43180	→	44134	37532	318	E		55	318	28.7	
52	44339	←	44708	14866	123	EC					Fraser *et al.* (1995)
53	44712	→	45129	16995	139	L		56	146	62.5	
53a	45128	→	45364	8706	79	L	*lef-10*	57	81	42	Lu and Miller (1994)
54	45222	→	46317	42094	365			58	378	49.5	
55	46411	→	46630	8190	73	EC		59	68	54.9	
56	46634	→	46886	9858	84	C		60	82	47.6	
57	47073	→	47556	18994	161	EC		61	163	54.4	
58	47574	←	47745	6816	57						
59	47882	←	48089	8174	69	C		62	176	47.7	

(continued)

TABLE I. (*Continued*)

AcMNPV								OpMNPV			
ORF	Left	Dir	Right	M_r	AA	Trans	Name	ORF	AA	% ID	References
60	48103	←	48364	10148	87			63	90	60.5	
61	48513	←	49155	25228	214	CL	*25k*	64	208	72	Beames and Summers (1989)
62	49184	→	50732	59305	516		*lef-9*	65	489	76.1	Lu and Miller (1994)
63	50795	→	51260	18476	155	C					
64	51283	←	52189	34798	302	EL	*gp37*	69	321	65.5	Wu and Miller (1989)
65	52329	←	55281	114307	984	EC	*dnapol*	70	985	61.9	Tomalski *et al.* (1988)
66	55292	→	57716	93973	808	L		71	875	28.3	
67	57721	←	58876	44551	385	EC	*lef-3*	72	373	39.6	Li *et al.* (1993)
68	58720	→	59296	22333	192			73	131	61.5	
69	59276	→	60062	30355	262						
70	60110	→	60980	34408	290	EC	*hcf-1*				Lu and Miller (1995b)
71	61016	→	61763	28621	249	EL	*iap2*	74	259	55.1	Ayres *et al.* (1994)
72	61824	→	62004	7068	60	CL		75	89	53.3	
73	62015	←	62312	11526	99	EL		76	84	28.3	
74	62311	←	63106	30567	265	ECL		77	172	33.3	
75	63126	←	63525	15512	133	L		78	130	40.8	
76	63543	←	63795	9440	84	EL		79	84	81	
77	63813	←	64950	44363	379	L	*vlf-1*	80	374	79.6	McLachlin and Miller (1994)
78	64958	←	65285	12545	109	L		81	105	57.9	
79	65290	←	65602	12199	104			82	104	62.9	
80	65607	←	66834	45381	409	EL	*gp41*	83	367	64.9	Whitford and Faulkner (1992)
81	66826	←	67525	26920	233	L		84	218	66.5	
82	67376	←	67916	19752	180	L		85	155	31.0	
83	67884	→	70425	96209	847	L		86	819	60.3	
84	71165	→	71729	21720	188						
85	71934	→	72093	6367	53			87	55	23.9	
86	72131	←	74213	80759	694	C	*pnk/pn1*	M			
87	74356	→	74734	15048	126	EC		88	151	39.5	
88	74737	←	75529	30092	264	C	*cg30*	89	249	47.3	Thiem and Miller (1989b)
89	75534	←	76575	38951	347	L	*vp39*	90	351	59.3	Thiem and Miller (1989a)
90	76596	→	77988	53909	464		*lef-4*	91	457	50.9	Passarelli and Miller (1993c)

91	77987	←	78659	24138	224	CL		92	279	47.1	
92	78699	←	79476	30937	259	E		93	282	74.6	
93	79477	→	79960	18380	161	C		94	159	71.2	
94	79971	→	80655	25526	228	ECL	*p25*	95	229	62.4	Lu and Carstens (1991)
95	80694	←	84357	143213	1221	EL	*hel*	96	1223	58.1	Lu and Carstens (1991)
96	84346	→	84865	19840	173			97	172	68.6	
97	84839	→	85007	6535	56	EC					
98	85021	←	85981	38021	320	ECL	*38k*	99	313	57	
99	85918	→	86713	31010	265	C	*lef-5*	100	263	59.5	Lu and Carstens (1991) Passarelli and Miller (1993c)
100	86712	←	86877	6885	55	CL	*p6.9*	101	51	72.7	Wilson *et al.* (1987)
101	86921	←	88004	41537	361	ECL		102	354	61.5	
102	88026	←	88392	13335	122	EL		103	112	46.3	
103	88375	←	89536	45313	387	ECL		104	411	57.7	
104	89564	→	91637	79878	691	L	*vp80*	105	624	30.1	Lu and Carstens (1992)
105	91667	←	93326	65576	553	CL	*HE65*				Becker and Knebel-Mörsdorf (1993)
106	93873	→	94056	7098	61			106	256	76	
107	94059	→	94389	12547	110	E					
108	94392	←	94707	11840	105	L		108	108	53.8	
109	94721	←	95891	44802	390	EL		109	390	66.6	
110	95929	←	96097	6799	56	L		111	57	71.4	
111	96148	←	96349	8169	67	EC		112	72	56.4	
112	96521	→	96782	10460	87	EC					
113	96789	→	97296	20290	169	EC					
114	97886	←	99158	49292	424	EL		114	424	37.8	
115	99182	←	99794	23019	204	ECL		115	205	63.7	
116	99804	←	99972	6425	56	C					
117	99910	→	100195	10992	95	C		117	97	35.3	
118	100231	←	100702	18709	157	C					
119	100699	→	102289	59739	530	L		119	529	72.9	
120	102296	→	102542	9532	82			120	82	53.7	
121	102647	→	102821	6705	58	E					

(*continued*)

TABLE I. (*Continued*)

AcMNPV								OpMNPV			
ORF	Left	Dir	Right	M_r	AA	Trans	Name	ORF	AA	% ID	References
122	102713	←	102899	7216	62	EC		121	70	36.2	
123	102964	←	103609	24944	215	EC	*pk2*				Morris *et al.* (1994), Li and Miller (1995a)
124	103793	→	104534	28530	247	EL		122	243	44.0	
125	104553	←	105231	26612	226	C	*lef-7*	123	211	32.3	Morris *et al.* (1994), Lu and Miller (1995b)
126	105282	←	106935	61368	551	L	*chiA*	124	550	78.0	
127	106983	→	107952	36938	323	EL	*cath*	125	324	76.8	Rawlings *et al.* (1992)
128	108179	←	109769	60646	530	EL	*gp67*	126	509	79.2	Whitford *et al.* (1989)
129	109900	→	110494	22110	198	L		127	192	64.6	
130	110524	→	110842	12112	106	CL	*gp16*	128	103	71.8	
131	110903	→	111659	29079	252	EL	*pp34*	129	297	62.6	Oellig *et al.* (1987) Whit and Manning (1988)
132	111873	→	112530	25136	219	L		130	228	28.2	
133	112560	→	113817	48291	419	CL	*alk-exo*	131	424	53.4	Oellig *et al.* (1987) Ayres *et al.* (1994)
134	113870	←	116279	94540	803	E	*p94*				Friesen and Miller (1987) Clem *et al.* (1994)
135	116492	→	117389	34828	299	EL	*35k*				Friesen and Miller (1987) Clem *et al.* (1991)
136	118044	→	118764	27282	240		*p26*	132	230	46.1	Liu *et al.* (1986)
137	118839	→	119121	10310	94	L	*p10*	133	92	41.8	Kuzio *et al.* (1984)
138	119135	←	121070	73885	645	EL	*p74*	134	644	79.5	Kuzio *et al.* (1989)
139	121205	←	122552	52636	449	EC	*me53*	138	455	42.2	Knebel-Mörsdorf *et al.* (1993)

140	122625	→	122805	7096	60						
141	122832	→	123615	30109	261	CL	*ie-0*	139	245	51.8	
142	123632	→	125063	55417	477	EL		140	484	74.0	
143	125153	→	125339	6559	62		*ODV-E18*	141	85	68.4	Braunagel *et al.* (1996b)
144	125357	→	126227	33528	290	CL	*ODV-EC27*	142	297	67.1	Braunagel *et al.* (1996b)
145	126299	→	126530	8850	77			143	95	70.1	
146	126527	←	127130	22881	201	C		145	197	50.3	
147	127198	→	128944	66881	582	EC	*ie-1*	146	560	48.1	Guarino and Summers (1987) Chisolm and Henner (1988) Passarelli and Miller (1993a)
148	129008	←	130136	40863	376	CL	*ODV-E56*	147	374	69.8	Braunagel *et al.* (1996a)
149	130167	←	130488	12419	107	C					
150	130456	→	130753	11161	99						
151	130857	←	132081	47007	408	ECL	*ie-2*	152	405	38.8	Carson *et al.* (1991) Passarelli and Miller (1993a)
152	132109	←	132385	10829	92						
153	132526	→	133489	37425	321	ECL	*pe38*	153	307	37.1	Krappa and Knebel-Mörsdorf (1991)
154	133591	→	133834	9421	81	L					

[a]For AcMNPV, the selected ORFs are numbered sequentially, in order of appearance in the + strand of the genome. The left (column Left) and right (column Right) columns define the ends of the ORF irrespective of its encoding strand. The direction of the transcripts (column Dir) that could express the ORF is indicated by arrows. The number of amino acids encoded by the ORF (column AA) from the first ATG are listed. The transcription column (Trans) indicates if at least one early (E) or CATG (C) motif is present in the 160 nucleotides upstream of an ORF. The presence of a late promotor motif, TAAG (L), within 80 nucleotides upstream of an ATG is also shown in column Trans. Where an ORF has a common name defined by its function or by its sequence identity to other products in the protein databases, this is indicated (column Name). For OpMNPV, the ORFs are presented to indicate identity (column % ID) with the AcMNPV genome. The AcMNPV ORFs missing in OpMNPV are listed in Table 2; OpMNPV ORFs missing in AcMNPV are listed in Table 3. References are listed as a guide to the literature regarding previously published sequences or studies defining AcMNPV gene functions.

have evolved to minimize nonspecific initiation of late transcription within coding regions, since the TAAG site serves as the major promoter element for late and very late transcription (see Chapter 8, this volume).

C. Analysis of AcMNPV Gene Function

Computer-assisted predictions of potential AcMNPV genes and appropriate comparisons with databases can assist in predicting their function, but such predictions must be confirmed by appropriate experimentation. If a gene is nonessential for replication in cell culture, the gene can be functionally inactivated by inserting heterologous sequences into the coding region using homologous recombination methods (see Chapter 14, this volume). Usually the heterologous sequences are genes such as *lacZ* that confer a distinctive color to recombinant plaques and facilitate their identification (e.g., O'Reilly and Miller, 1989; Pennock *et al.*, 1984). Once the recombinant mutant is isolated, its phenotype can be characterized. This procedure is not applicable for essential genes, and the isolation of conditional lethal mutations is necessary for genetic confirmation of gene function. Two methods for isolating conditional baculovirus mutants have been used previously: the isolation of temperature-sensitive mutants (reviewed by Miller, 1986) and the isolation of a gene-disrupted recombinant in a cell line expressing the essential gene (Monsma *et al.*, 1995).

IV. COMPARISONS OF BACULOVIRUS GENOMES

The three baculovirus genomes that have been sequenced in their entirety are similar in size, but the AcMNPV genome (133,894 bp) is about 5.5 kbp larger than that of BmNPV (128,413 bp), whereas the OpMNPV genome (131,990 bp) is closer in size to that of AcMNPV. There are significant differences in the G+C contents of these viruses. The AcMNPV (41%) and BmNPV (40%) are very similar, but OpMNPV (55%) is much higher.

The availability of the complete sequences enables detailed comparisons to be made of the differences in their genetic content. The AcMNPV genome is used as the prototype sequence for these analyses. Although most of the AcMNPV and BmNPV ORFs are closely related with amino acid sequence identities in excess of 95%, OpMNPV is significantly different from these two viruses. The average amino acid identity between homologous ORFs of AcMNPV and OpMNPV is 56% (see Table 1). The most conserved ORF is polyhedrin (ORF3) (89% amino acid sequence identity), followed by the viral ubiquitin homologue (ORF25) (84%), although the latter has a 3′ extension of 16 amino acids in OpMNPV that is not included in this calculation. Several other OpMNPV ORFs show about 80% identity and include homologues of AcMNPV ORFs 22, 76, VLF-1 (ORF77), chitinase (ORF126), cathepsin (ORF127), gp64 (ORF127), and p74 (ORF 138).

In general, those genes that are known to be essential for AcMNPV are also found in the OpMNPV genome. All the genes implicated as encoding virion structural proteins (see Chapter 2, this volume) are present in both genomes. In addition, the set of AcMNPV genes required for transient DNA replication and late gene transcription (Kool *et al.*, 1994; Lu and Miller 1995a) are also present in OpMNPV except for *hcf-1* (Lu and Miller, 1995b) and the antiapoptotic gene *p35* (Clem *et al.*, 1991). The absence of the latter two genes is not unexpected, since *hcf-1* is a host cell-specific AcMNPV factor (see Chapter 9, this volume) and OpMNPV has a member of the *iap* family of antiapoptotic genes, which can functionally substitute for *p35* in the AcMNPV genome (Birnbaum *et al.*, 1994) and in replication and late gene transcriptional transient assays (Kool *et al.*, 1994; Lu and Miller, 1995a).

Table 2 lists the AcMNPV ORFs that are missing in BmNPV and OpMNPV. There are a number of AcMNPV ORFs that are not present in the genomes of both OpMNPV and BmNPV. Among these are ORF 20 and 58, which are present in both OpMNPV and BmNPV as fusions with a neighboring ORF. Although this suggests the possibility of a sequence error in the AcMNPV genome, there may be real differences in these sequences, and further sequence information and/or protein analysis will be necessary to resolve this question. Fourteen other ORFs appear to be completely missing in both OpMNPV and BmNPV genomes. Some of these ORFs are small (e.g., ORF 97, 112, 116, 121, 140, 152), which suggests that they may not actually be translated genes in AcMNPV. However, ORF 152 is the only one of these ORFs that lacks any of the transcriptional motifs; the other ORFs have at least one or two motifs associated with early transcription.

The AcMNPV ORFs that have been previously characterized to some degree and are missing in one or both of the genomes include: a conotoxinlike gene (*ctl;* AcMNPV ORF 3), missing only in BmNPV; ORF 7 (also known as ORF 603) (Gearing and Possee, 1990), which is located immediately upstream of the polyhedrin gene in AcMNPV; *p43* (AcMNPV ORF 39); proliferating cell nuclear antigen homolog (*pcna,* ORF 49, missing in BmNPV); host cell-specific factor, *hcf-1* (ORF70); a homologue of polynucleotide kinase/polynucleotide ligase (*pnk/pnl,* see Section V); ORF105 (*He65*), an early gene; protein kinase 2 (*pk2;* AcMNPV ORF 123); *p94* (AcMNPV ORF 134), and *p35* (AcMNPV ORF 135, lacking in OpMNPV). The latter two genes are located in the single large AcMNPV insert of 4.2 kb that is lacking in OpMNPV. In BmNPV, the *p94* gene appears to have been deleted, since a remnant of *p94* sequence is present adjacent to the *p35* gene. Of these eight genes, six are known by mutational analysis to be nonessential for AcMNPV replication, although disruption of *pcna, hcf-1,* and *p35* have demonstrable phenotypes in at least some cell lines and insect species (Clem *et al.*, 1991; Crawford and Miller, 1988; Lu and Miller, 1995b).

The BmNPV genome has at least two ORFs that are not found in AcMNPV. These ORFs are located between AcMNPV ORFs 96 and 98 and are

TABLE II. AcMNPV ORFs Missing in OpMNPV and BmNPV

OpMNPV	AA	BmNPV	Comments
	53	3 (*ctl*)	
7 (ORF603)	201	7 (ORF603)	
12	217	12	Absent in both
20	69	20	Fused to 21 in both
33	182	33	Absent in both
39 (*p43*)	363		
45	192		
	113	48	
	285	49 (*pcna*)	
52	123	52	Absent in both
58	57	58	Fused to 59 in both
63	155		
69	262		
70 (*hcf-1*)	290	70 (*hcf-1*)	Absent in both
84	188	84	Absent in both
	53	85	
86 (*pnk/pnl*)	694	86 (*pnk/pnl*)	Absent in both
97	56	97	Absent in both
105 (*HE65*)	553	105 (*HE65*)	Truncated in BmNPV
112	87	112	Absent in both
113	169	113	Absent in both
116	56	116	Absent in both
118	157	118	Absent in both
121	58	121	Absent in both
123 (*pk2*)	215		
134 (*p94*)	803	134	Truncated in BmNPV
135 (*p35*)	299		
140	60	140	Absent in both
149	107		
150	99		
152	92	152	Absent in both
154	81		

predicted to encode polypeptides of 239 and 318 amino acids. ORF 97 appears to be absent in the BmNPV genome, suggesting that it may have been disrupted by the insertion in this location.

Despite a general colinearity of their genomes, there are a number of genes present in OpMNPV that are lacking in AcMNPV (see Table 3). Furthermore, there are two major rearrangements and two major insertions in the OpMNPV genome relative to AcMNPV. These include inversions that encompass regions with homology to AcMNPV ORFs 1–10 and 24–38 and insertions of 6.8 and 2.7 kb located between OpMNPV homologues of AcMNPV ORFs 31 and 30 and 148 and 151, respectively. The latter region contains three genes that were previously described—*Opp32*, *Opp25*, and *Opp8.9*—and whose functions are unknown (Shippam *et al.*, 1993; Wu *et*

TABLE III. OpMNPV ORFs Missing in AcMNPV

OpMNPV	AA
PTP-2	160
CTL-2	52
dUTPase	317
RR R1	593
RR R2	349
IAP-3	268
Bm81FS	201
IAP-4	118
Opp32	285
Opp25	236
Opp8.9	75

al., 1993). The first major inversion involves the region surrounding the polyhedrin gene. In OpMNPV this region contains two homologues of protein tyrosine phosphatase (*ptp-1* and *ptp-2*), while in AcMNPV this region contains only *ptp-1* and *ctl-1;* the *ctl-1* gene is located elsewhere in the OpMNPV genome. However, there is also a second *ctl*-like gene (*ctl-2*) in OpMNPV located within the 6.8-kb insert (for information on *ctl* genes, see Chapter 11, this volume). In addition to *ctl-2,* this insert contains *iap-3,* which was previously known as *iap* (Birnbaum *et al.,* 1994) and has anti-apoptotic function. There are two other members of the *iap* family (*iap-1* and *iap-2*) also present in OpMNPV. In addition, there is also a fourth truncated *iap*-like gene (*iap-4*) in OpMNPV. The relationships between the full-length *iaps* of AcMNPV and OpMNPV are summarized in Table 4. The 6.8-kb OpMNPV insert also contains homologues of three genes involved in nucleotide metabolism in other organisms. These include a dUTPase homologue and two subunits of a putative ribonucleotide reductase. The large ribonucleotide reductase subunit, R, shares 29% amino acid sequence identity with the human enzyme and 24% identity with a vaccinia virus homologue. The small subunit, R2, shares 14 and 16% identity, respectively, with the same enzymes.

TABLE IV. Identity between AcMNPV and OpMNPV IAPs

		OpMNPV			
		IAP1	IAP2	IAP3	IAP4
AcMNPV	IAP1	62%	18%	28%	27%
	IAP2	20%	55%	20%	28%

The primary cellular nucleotide biosynthetic pathways flow into ribonucleotide triphosphate (rNTP) pools. Many DNA viruses have genes encoding enzymes that allow the virus to convert host cell rNTPs into dNTPs for use in DNA synthesis. These enzymes are thought to benefit the virus because they allow replication in nondividing cells in which dNTP pathways are inactive. Ribonucleotide reductase is the first enzyme in this pathway and catalyzes the reduction of rNDPs to dNDPs. These dNDPs are then phosphorylated to produce dATP, dGTP, dCTP, and dUTP. However, the production of dUTP can be deleterious because it is mutagenic if incorporated into DNA. Therefore, dUTPase is a critical enzyme because it converts dUTP to dUMP, thereby excluding dUTP from incorporation into DNA. In addition, dUMP is a starting nucleotide for the synthesis of dTTP, which is not directly produced by the ribonucleotide reductase pathway (reviewed in Mathews and van Holde, 1995).

The C-terminal 128 amino acids of OpMNPV ORF31 show about 30% amino acid sequence identity with a segment of human dUTPase and with a dUTPase derived from a polyprotein encoded by a subset of retroviruses. All the amino acids thought to comprise the active site of dUTPase (Wagaman *et al.*, 1993) are present in this region. Although a role for the dUTPase gene in OpMNPV infection has not been established, lack of a functional dUTPase gene is associated with increased mutation frequency in feline immunodeficiency virus (Lerner *et al.*, 1995) and other lentiviruses. In addition, such dUTPase mutations have been shown to delay replication in nondividing cells (Turelli *et al.*, 1995).

Contiguous with the OpMNPV dUTPase homologue is a gene, *r1*, with sequence similarity to the R1 subunit of ribonucleotide reductase, and in close proximity to *r1* is a gene, *r2*, with sequence similarity to R2, the small subunit of ribonucleotide reductase. OpMNPV R1 consists of 594 amino acids, which is much shorter than other R1 proteins (750–800 amino acids) and shows only about 25% amino acid sequence identity with R1s of herpes simplex virus, vaccinia virus, and *Homo sapiens* and 20% identity with *Escherichia coli* R1. Despite this low level of sequence conservation, amino acids shown to be essential for enzyme activity (Uhlin and Eklund, 1994) are present in OpMNPV R1. The R2 subunit is even less well-conserved, with 10–16% sequence identity with other R2 peptides. Although this level of identity is very low, homology was clustered in regions where major areas of conservation are present in other R2s. The *r1* gene (but not the *r2* gene) has also been identified in at least two other baculoviruses (E. A. van Strien, personal communication). In contrast to OpMNPV, the *r1* genes from these viruses are much more closely related (50–60% amino acid identity) to those from *Homo sapiens* and vaccinia virus and only distantly related to OpMNPV, herpes simplex virus, and *E. coli* R1 (25–27% identity). Therefore, at least two different lineages of R1 appear to be present in baculoviruses.

Deletion of vaccinia virus ribonucleotide reductase greatly reduces virulence *in vivo*, but has little effect on replication in cell culture (Child *et al.*, 1990). Therefore, possession of the enzyme could benefit OpMNPV during replication in insects. (For further discussion, see Section VI.)

V. BACULOVIRUS EVOLUTION AND THE ACQUISITION OF NEW GENES

Sequence analysis has shown that baculovirus carry many genes that are homologous to those found in the genomes of other categories of organisms. These include a number of genes related to those from eukaryotes, at least two genes that are similar to genes found in other virus families, and several genes that appear to be derived from prokaryotes. These genes are listed in Table 5. As new baculovirus genomes are sequenced and as the eukaryotic databases expand, this list of genes is likely to increase. The origin of these baculovirus genes and the mechanism(s) and frequency with which baculoviruses can acquire new genes are intriguing questions. Experimental studies have shown that baculovirus genomes can acquire host

TABLE V. Known Homologues of Baculovirus Genes

Gene	Homologue	Known locations
ptp	Protein tyrosine phosphatase	Eukaryotes, prokaryotes, poxviruses
ct1	Conotoxins	Cone snails, spiders
pk1	Protein kinase	Eukaryotes, prokaryotes, other virus families
iaps	Inhibitors of apoptosis	Mammals, insects
sod	Superoxide dismutase	Eukaryotes, prokaryotes, poxviruses
fgf	Fibroblast growth factor	Mammals
ubi	Ubiquitin	Eukaryotes, prokaryotes
gta	General transcription activator	Mammals
pcna	Proliferating cell nuclear antigen	Eukaryotes
dnapol	DNA polymerase	Eukaryotes, prokaryotes, other virus families
vlf-1	Integrase/resolvase	Yeast, prokaryotes
pnk/pnl	Polynucleotide kinase and ligase	Phage, yeast, plants
pk2	Protein kinase, elF2-alpha-like	Eukaryotes
chiA	Chitinase	Eukaryotes, prokaryotes
cath	Cathepsins	Eukaryotes
alk-exo	Alkaline exonuclease	Eukaryotes, prokaryotes
gp67	Viral glycoprotein	Thogoto virus
p34.8	Spindlin	Entomopoxviruses
dUTPase	dUTPase	Eukaryotes, prokaryotes, other virus families
RR 1, 2	Ribonucleotide reductase subunits	Eukaryotes, prokaryotes, other virus families

transposable elements (reviewed by Friesen, 1993), including retrotransposons (Miller and Miller, 1982), and this implies that genomic and cDNAs of arthropods might be acquired via transposition. For genes such as *egt,* the ecdysteroid UDP–glucosyl transferase gene, an insect origin seems likely, since ecdysteroid glucoside conjugates are found in some insects. For accurate phylogenetic analyses, multiple sequence alignments of homologues from a wide variety of organisms are necessary to construct an evolutionarily valid phylogenetic tree, so that only a few genes have been studied in this manner. In the case of *sod,* a homologue of super oxide dismutase, computer-assisted phylogenetic analysis suggests that the AcMNPV homologue is most closely related to insect *sods.* But one of these studies, involving the viral chitinase gene, has proposed a surprising origin for this gene that cannot be easily explained by known transposition mechanisms or simple recombination models.

The AcMNPV chitinase gene, *chiA,* has striking similarity to a bacterial chitinase homologue (Ayres *et al.,* 1994). The AcMNPV *chiA* encodes a functional enzyme with both endo- and exochitinase activities (Hawtin *et al.,* 1995) that appears to be involved in liquefaction of virus-infected larvae, suggesting that the chitinase plays a role in the efficient release of the polyhedra from the larvae (Hawtin *et al.,* submitted; see Chapter 3, this volume). A number of chitinase sequences from a variety of organisms were compared with the AcMNPV *chiA* sequence using a Monte Carlo method (Hawtin *et al.,* 1995) and by multiple sequence alignments to establish phyologenetic trees. Two bacterial enzymes, *Serratia marcescens* chitinase (Jones *et al.,* 1986) (61% identity; 56 SD above mean) and marine bacterium *Alteromonas* sp strain O-7 *chiA* (Tsujibo *et al.,* 1993) (48% identity; 50 SD above mean) showed the highest similarity.

Nucleotide composition provides an additional suggestion of common ancestry of the AcMNPV and *S. marcescens chiA.* The AcMNPV *chiA* has a G+C content (48%) higher than that expected from the rest of the virus genome (41%) (Hawtin *et al.,* 1995). The G+C content two genes flanking *chiA* in the AcMNPV genome were analyzed: the cathepsin gene (*cath*) has a G+C content of 40%, while *lef-7,* upstream of *chiA,* has a G+C content of 26%. The *S. marcescens chiA* has a G+C content of 59%. Therefore, the higher overall G+C composition of the AcMNPV *chiA* resembles the *S. marcescens chiA* G+C content.

Although it was originally proposed that a simple mutation of the bacterial gene upstream of the initiating methionine codon might be sufficient to provide a TAAG motif for late transcription from the AcMNPV genome, an alternate explanation seems more plausible at this time. Although residues important for the enzymatic activity as a chitinase are conserved, the AcMNPV and *S. marcescens* protein sequences diverge at their N-termini where the secretory signal sequence of the bacterial enzyme is replaced by a eukaryotic signal peptide (Ayres *et al.,* 1994). Furthermore, the carboxyl end of the AcMNPV chitinase has a KDEL motif that is characteristic of proteins that are retained in the endoplasmic reticulum of cells in widely divergent eukaryotic species (e.g., yeast, plants, and mammals) (Pelham, 1990). Indeed, the AcMNPV chitinase appears to localize within the endoplasmic reticulum (R. D. Possee, unpublished data). The KDEL motif is conserved in the predicted sequence of the OpMNPV chitinase and as RDEL in BmNPV, an acceptable conservative change in this motif (Pehlam, 1990), but is absent from the *S. marcescens* chitinase sequence. A comparison of the OpMNPV and AcMNPV chitinases, however, shows that the region immediately upstream of the KDEL motif is not well conserved. These data suggest that the ancestral chitinase sequences may have replaced most of the coding region of an original baculovirus late gene encoding a protein with a signal peptide and a KDEL endoplasmic reticulum localization motif.

VLF-1, the product of the very late factor-one gene (*vlf-1*), is also related to a family of genes—integrases and resolvases—that have been found in microorganisms but not invertebrates to date (McLachlin and Miller, 1994). However, in this case, members of this family are also found in lower eukaryotes such as yeast, suggesting that other members of this family might eventually be found in higher eukaryotes.

The origin of the AcMNPV *pnk/pnl* eventually may be traced to a prokaryotic source. This gene has the potential to encode multiple functional domains. It is very similar to two genes of T4 bacteriophage—*63* and *pseT*—which encode RNA ligase (Silber *et al.*, 19792; Snopek *et al.*, 1977) and polynucleotide kinase (Midgley and Murray, 1985) activities, respectively. The amino terminal portion of *pnk/pnl* is related to the T4 RNA ligase (31% identity, 72% similarity), while the carboxyl-terminal half appears to be related to the T4 polynucleotide kinase (26% identity, 66% similarity).

The T4 RNA ligase and polynucleotide kinase are not essential for T4 replication in common *E. coli* hosts, but are indispensible in a clinical isolate CT196 (Depew and Cozzarelli, 1974; Runnels *et al.*, 1982; Sirotkin *et al.*, 1978). The CT196 isolate has an additional gene, *ppr*, which encodes two physically associated restriction activities (Amitsur *et al.*, 1987, 1992; Morad *et al.*, 1993). These comprise the *Eco*pprI (Tyndal *et al.*, 1994) and the anticodon nuclease activity (ACNase) (Abdul-Jabbar and Synder, 1984), which is inhibited in normal conditions. After bacteriophage infection, the product of the T4 gene *Stp* inactivates the *Eco*prrI activity, but activates the ACNase (Kaufmann *et al.*, 1985; Penner *et al.*, 1995). The T4 polynucleotide kinase and ligase are thus involved in counteracting the reaction of the host in blocking the infection.

The relatively low sequence identity shared by baculovirus PNK/PNL and the bacterial gene products argues against a recent acquisition of these sequences. The higher similarity values, however, suggest that the baculovirus gene product may retain RNA ligase/kinase activity. The absence of this gene in BmNPV and OpMNPV suggests that it may not be required for replication of all baculoviruses. Further, it may be deleted from the AcMNPV genome without apparent effect on virus propagation in cell culture and is also absent in some natural isolates of AcMNPV (Durantel *et al.*, submitted).

Single proteins with both RNA ligase and polynucleotide kinase activities have also been described in wheat (Pick and Hurwitz, 1986; Pick *et al.*, 1986) and yeast (Xu *et al.*, 1990a,b). In yeast, the bifunctional protein is involved in the processing of tRNAs; mutants with these functions impaired are nonviable (Phizicky *et al.*, 1992). These studies suggest that chimeric RNA ligase/kinase proteins may have a role in higher eukaryotes, and eukaryotic homologues more closely related to PNK/PNL than the T4 phage genes may exist.

The suggestion that baculoviruses may have acquired a bacterial gene during their evolution poses many questions regarding how, where, and when such an event may have occurred. The presence of intracellular bacterial symbionts in many insect species could provide a source for such genes. It may be relevant that at least one other eukaryotic virus has been found to possess a gene that is most closely related to a T4 phage homologue, as determined by multiple sequence alignment and phylogenetic tree construction (Pellock *et al.*, 1996). This is the DNA polymerase gene of the African swine fever virus, a virus that replicates in both mammals and ticks. The symbiotic relationship that some invertebrates have with bacteria suggests a possible environment for gene exchange, although the existence of a virus that can replicate in both prokaryotes and eukaryotes would simplify the construction of models for such gene exchange.

Baculoviruses appear to have acquired or shared at least two genes with viruses of other families. The GP64 envelope fusion protein of AcMNPV is related to a glycoprotein of Thogoto virus, a tick-borne orthomyxolike virus (Morse *et al.*, 1992), while GP37 (derived from ORF 64, also known as p34.8) is homologous to spindlin of entomopoxviruses (Yuen *et al.*, 1990). The extent of similarity of these genes suggests that gene exchange may have occurred between viral families. The possibility that the spindlin gene may have been acquired by baculoviruses from entomopoxviruses was supported by the observation that the gp37 gene of AcMNPV has a different codon bias than other AcMNPV genes analyzed (Wu and Miller, 1989) and that this codon bias is similar to entomopoxvirus codon bias (Yuen *et al.*, 1990). However, baculoviruses differ dramatically in their G+C content as well as their codon bias (see Section VI). Thus, arguments of evolution based on codon biases or G+C content are weakened by the likely possibility that baculoviruses exchange genes among themselves. Indeed, the comparison of AcMNPV and OpMNPV genomes described above imply that baculoviruses might collectively be a rich source of genes, possibly acquired over long evolutionary history from many different organisms.

VI. CODON USAGE

The G+C contents of the genomes of OpMNPV and AcMNPV differ markedly (55 vs. 41%, respectively). The higher G+C content of the OpMNPV genome is reflected both in the codon usage as well as the amino acid frequency. The most frequently used codons in the predicted OpMNPV

ORFs all end in G or C and include GAC (Asp), GCG (Ala), AAC (Asn), GTG (Val), CGC (Arg), and CTG (Leu). Overall, the percent of codons that have a G or C in the third position is above 75% for 17 of the 25 codon sets. For OpMNPV ORFs, codon sets encoding amino acids that have more than one possible codon showed on average 25% more codons with G or C in the third position than AcMNPV. Furthermore, the first and second positions of the codon, which specify the amino acid for most codons, also varied between AcMNPV and OpMNPV. In OpMNPV, nearly an equal percent of codons had exclusively Gs and Cs in the first and second positions of the codon compared to those with only As and Ts (24% G and C vs. 29% A and T). In contrast, the AcMNPV-predicted codons had a much stronger bias for As and Ts in both these positions compared to Gs and Cs (40% A and T vs. 16% G and C). This is reflected in a higher number of A/T-rich codons specifying Asn, Ile, and Lys residues in AcMNPV-predicted ORFs compared to those of OpMNPV, which have a greater number of G/C-rich codons specifying Ala, Gly, and Arg residues.

A similar relationship has been found between two alphaherpesviruses: herpes simplex virus I (HSVI) (68% G+C) and varicella zoster virus (VZV) (46% G+C) (Schachtel *et al.*, 1991). Both the percent G+C at codon position 3 and amino acid usage were found to reflect the G+C content of these viruses. It was suggested that various selective events following speciation could evoke major changes in HSVI and VZV processes that are reflected in codon and amino acid divergence. These events could include residence in a new host species or cell type, replication at different times or in different cell types, or novel interactions with host factors.

In addition, major differences in G+C content could reduce the competition for nucleotides by closely related viruses that coinfect the same host or replicate simultaneously in the same cell (Schachtel *et al.*, 1991). OpMNPV was originally isolated as a mixture with OpSNPV, a virus with 44% G+C (Rohrmann *et al.*, 1978). Recent evidence suggests that T4 phage ribonucleotide reductase has a bias in its ability to reduce rNDPs, which reflects the phage G+C content (S. Hendricks, personal communication). *Orgyia pseudotsugata,* the host insect, has a G+C content of 36% (Rohrmann *et al.*, 1978). Because of its low G+C content, the host cell ribonucleotide reductase could have a bias toward dATP and dUTP production. Therefore, it may be to the advantage of viruses with higher G+C contents to alter the composition of dNTPs by providing their own dNTP synthetic enzymes.

VII. HOMOLOGOUS REGIONS

A novel feature of many baculovirus genomes is the presence of homologous regions (*hr*s) that are located throughout the genome (Ayres *et al.*, 1994; Cochran and Faulkner, 1983; Garcia-Maruniak *et al.*, 1996; Theilmann and Stewart, 1992; Xie *et al.*, 1995). *Hrs* are composed of repeated sequences encompassing both direct repeats and imperfect pal-

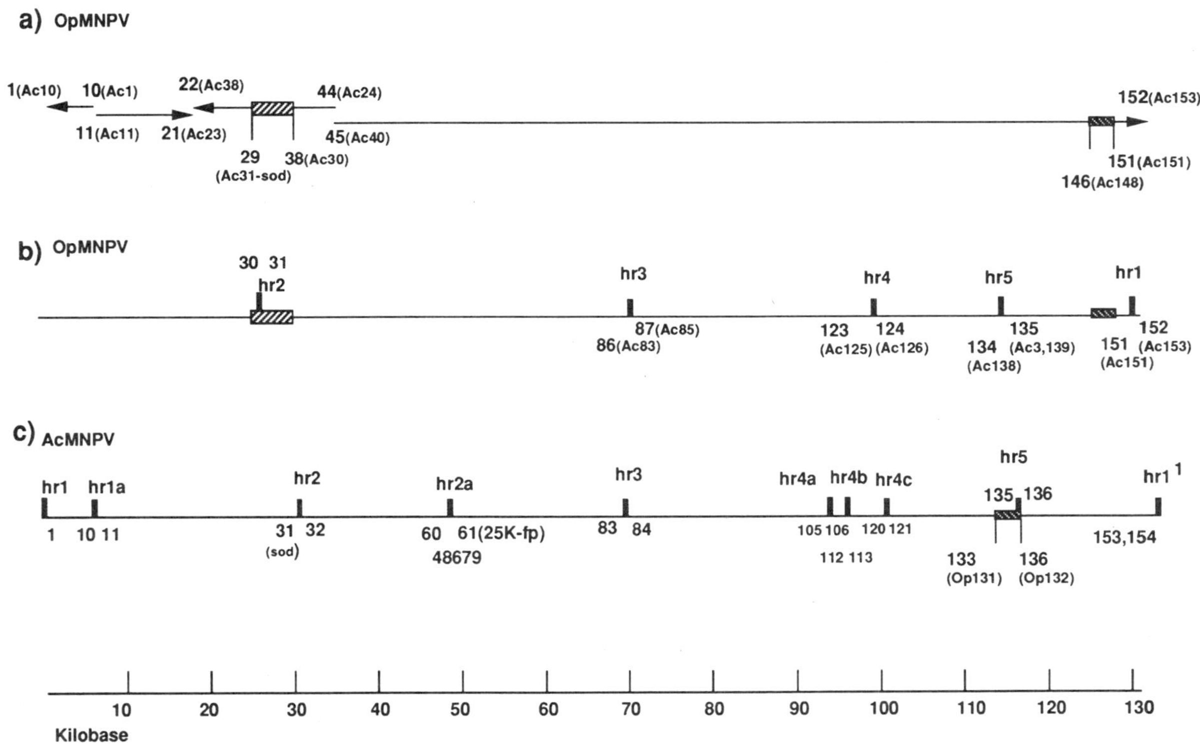

FIGURE 1. Comparison of the major differences in genome organization between OpMNPV and AcMNPV. (a) Map of OpMNPV showing the orientation of homologues of AcMNPV genes. The numbers show the OpMNPV ORFs followed by the AcMNPV (Ac) homologues in parentheses. The arrows indicate the orientation of several regions of OpMNPV relative to the homologous regions in the AcMNPV genome. (b) Context of *hr* sequences in the OpMNPV genome. OpMNPV ORFs that flank the repeated regions are shown. Selected AcMNPV ORFs are designated as above. (c) Context of *hr*s in AcMNPV. The OpMNPV ORF numbers for the homologues bordering the insert between AcMNPV ORFs 133 and 136 are shown in parentheses. The *hr*2a sequence at nucleotide position 48679 is from J. Kuzio (personal communication). AcMNPV ORF numbers used in this figure are from Ayres *et al.* (1994). The crosshatched boxes show the location of major inserts not present in the heterologous virus.

indromic sequences and have closely related counterparts elsewhere in the genome. The AcMNPV *hr*s are the most well-characterized and are located in nine regions of the genome (Fig. 1c). Each *hr* contains one to nine 30-bp imperfect palindromes within a directly repeated sequence (Ayres *et al.*, 1994; J. Kuzio, personal communication). In both AcMNPV and OpMNPV, *hr*s can act as enhancers of RNA polymerase II-mediated transcription (Guarino and Summers, 1986; Theilmann and Stewart, 1992) and also behave as origins of DNA replication in transient replication assays (Ahrens *et al.*, 1995; Kool *et al.*, 1995; Pearson *et al.*, 1992).

Five *hr*s (*hr* 1–5) were identified in the OpMNPV genome that varied from 2 to 10 repeats (Fig. 1b). Some of the regions show very high sequence similarity. For example, *hr*4 is 93% identical to *hr*1 over a 500-bp region. OpMNPV *hr*s contain 30-bp imperfect palindromes embedded in a 66-bp repeat. The consensus palindrome sequence (Fig. 2a) shows 24 out of 30 palindrome matches. This compares with 26 out of 30 matches for the AcMNPV palindrome consensus (Fig. 2b). The sequences located in the direct repeat that form the palindrome flanking sequences are similar in size (21 nucleotides upstream and 15 nucleotides downstream) to the corresponding AcMNPV sequences (22 nucleotides upstream and 20 nucleotides downstream) (Kool *et al.*, 1995). However, the lengths of the sequences separating the direct repeats are highly variable in AcMNPV (0–131 nucleotide) but are not present in the OpMNPV sequences. The OpMNPV palindrome consensus sequence was 57% identical to the AcMNPV consensus sequence (Fig. 2c). Sequences related to AcMNPV and OpMNPV *hr*s have also been reported from *Choristoneura fumiferana* MNPV (CfMNPV) (Xie *et al.*, 1995) and *Anticarsia gemmatalis* MNPV (AgMNPV) (Garcia-Maruniak *et al.*, 1996). The LdMNPV genome has homologous regions that are highly repetitive and contain small palindromic sequences, but do not appear to be closely related to the *hr*s discussed above (Pearson and Rohrmann, 1995).

In addition to the conservation of *hr* sequences in a number of viruses, the genomic context of some of the *hr*s appears to be conserved. The most clear example of conservation of context is *hr*3, which is located between ORFs 83 and 84 in AcMNPV (Fig. 1b,c). Homologues of AcMNPV ORF 83 also flank the left of *hr*3 in OpMNPV (OpMNPV ORF86). The ORFs on the right margin differ between the viruses, but the differences are due to the deletion of ORFs. AcMNPV *hr*3 is flanked on the right by ORF84. However, OpMNPV lacks an ORF84 homologue, and therefore *hr*3 is flanked by OpMNPV ORF87 (a homologue of AcMNPV ORF85). The conservation of location of other *hr*s is not as clear as *hr*3, but the position of several other *hr*s is close but not identical between the two viruses. Major rearrangements, insertions, and deletions relative to AcMNPV were detected near all the OpMNPV *hr*s. Because of their sequence similarity, it is likely that OpMNPV and AcMNPV *hr*s descended from a common viral ancestor that had a set of *hr*s present at several locations in its genome. The close related-

a

```
GCTTTTCGAGAGCGTTCGCACCCGAAAAGC   OpMNPV
  C
******** * ***   *** * ********  24/30
```

b

```
GCTTTACTATTAGAATTCTACTTGTAACGC AcMNPV
                      G      AA
******* * ********** * *******  26/30
```

c

```
GCTTTTCGAGAGCGTTCGC-ACCCGAAAAGC   OpMNPV
  C
::::: : :    :    : ::  : ::::: 17/30
GCTTTACTATTA-GAATTCTACTTGTAACGC AcMNPV
                       G     AA
```

FIGURE 2. Comparison of OpMNPV and AcMNPV *hr* palindrome consensus sequences. (a) The OpMNPV *hr* palindrome consensus sequence. The asterisks below the sequence are nucleotides that contribute to the palindrome. (b) The AcMNPV *hr* consensus sequence. (c) A comparison of the OpMNPV and AcMNPV *hr* consensus sequences. The colons indicate identity; the dash indicates a gap inserted to facilitate the alignment. The consensus sequence is based on a nucleotide being present with a frequency of 70% or more at a given position. If no single nucleotide had a 70% frequency, additional nucleotides are indicated that together are present at 70% or more at these positions.

ness of the *hr* sequence within each viral genome suggests that the sequences undergo coevolution, possibly caused by their interaction with a viral protein (e.g., IE-1) that is also involved in *hr* binding (Choi and Guarino, 1995; Leisy *et al.*, 1995; Rodems and Friesen, 1995). In addition, the *hr*–IE-1 interaction may be modulated by the interaction with host factors. The relative conservation of *hr* positions may be necessitated by their role as transcriptional enhancers or replication origins.

VIII. CONCLUDING REMARKS

The entire sequences of three baculovirus genomes have revealed a number of major differences in this diverse viral family. One of the most fascinating differences is that individual baculoviruses possess different sets of genes; only some genes are found in common. Thus, OpMNPV, for example, has a set of three genes involved in dNTP metabolism that are lacking in AcMNPV and BmNPV. The nature of these three genes suggests that OpMNPV may employ a different strategy for provisioning DNA precursors than the other two viruses. AcMNPV, in contrast, possesses *hcf-1*, a host cell-specific factor that allows AcMNPV to replicate more efficiently in certain insect species; OpMNPV and BmNPV, with different host ranges, lack *hcf-1*. The presence of different genes in different baculovirus genomes suggests that baculovirus genomes are collectively a rich reservoir of genes that, if shared among different genomes through recombination, could provide diversity and plasticity to baculovirus speciation.

Another feature of baculovirus genomes that has been revealed by the sequences of complete genomes is the presence of groups of related genes within the same genome. These include a conotoxinlike gene family, a family of inhibitor of apoptosis (*iap*) genes, and a phosophotyrosine phosphatase gene family. It is likely that these genes were acquired from host species, duplicated in the baculovirus genome, and then evolved to have different functions. A major challenge for future investigation is to understand the function of these and other baculovirus genes. Those genes that appear to be nonessential under laboratory conditions are likely to have a function in the context of field conditions—in populations of different viruses and in populations of different insects; such functional analyses are likely to present new challenges to molecular biologists as well as ecologists.

Homologous regions (*hr*s) appear to be a novel feature of baculovirus genomes. The relatedness in both location and nucleotide sequence of the *hr*s) suggests that they may perform vital functions in the baculovirus life cycle. Although they have been implicated as both transcriptional enhancers and viral replication origins, their origin function appears to be highly specific and they are not replicated by the heterologous virus (Ahrens *et al.*, 1995; Pearson *et al.*, 1993). This suggests that *hr*s may have a high degree of specificity in their interactions with viral replication factors. The apparent conservation of the genome context of *hr* sequences could also suggest that either genome spacing or a relationship with specific ORFs is important for their function. Although the complete sequences of three baculovirus genomes has contributed to our understanding of baculovirus diversity, other baculoviruses have much larger genomes. Sequencing additional baculovirus genomes may enable us to answer questions about baculovirus evolution and their ability to exploit particular hosts.

REFERENCES

Abdul-Jabbar, M., and Synder, L., 1984, Genetic and physiological studies of an *Escherichia coli* locus that restricts polynucleotide kinase and RNA ligase mutants of bacteriophage T4, *J. Virol.* **51:**522.

Ahrens, C. H., Pearson, M. N., and Rohrmann, G. F., 1995, Identification and characterization of a second putative origin of DNA replication in a baculovirus of *Orgyia pseudotsugata*, *Virology* **207:**572.

Ahrens, C. H., Russell, R., Funk, C. J., Evans, J. T., Harwood, S. H., and Rohrmann, G. F., 1997, The sequence of the *Orgyia pseudotsugata* multinucleocapsid nuclear polyhedrosis virus genome, *Virology* **229:**381.

Amitsur, M., Levitz, R., and Kaufmann, G., 1987, Bacteriophage T4 anticodon nuclease, polynucleotide kinase and RNA ligase reprocess the host lysine tRNA, *EMBO J.* **6:**2499.

Amitsur, M., Morad, I., Chapman-Shimshoni, D., and Kaufmann, G., 1992, Hds restriction modification proteins partake in latent anticodon nuclease, *EMBO J.* **8:**2411.

Ayers, M. D., Howard, S. C., Kuzio, J., Lopez-Ferber, M., and Possee, R. D., 1994, The complete DNA sequence of *Autographa californica* nuclear polyhedrosis virus, *Virology* **202:**586.

Beames, B., and Summers, M. D., 1989, Location and nucleotide sequence of the 25 k protein missing from baculovirus few polyhedra (FP) mutants, *Virology* **168:**344.

Becker, D., and Knebel-Mörsdorf, D., 1993, Sequence and temporal appearance of the early transcribed baculovirus gene HE65, *J. Virol.* **67:**5867.

Birnbaum, M. J., Clem, R. J., and Miller, L. K., 1994, An apoptosis-inhibiting gene from a nuclear polyhedrosis virus encoding a polypeptide with Cys/His sequence motifs, *J. Virol.* **68:**2521.

Braunagel, S. C., Daniel, K. D., Reilly, L. M., Guarino, L. A., Hong, T., and Summers, M. D., 1992, Sequence, genomic organisation of the *Eco*RI-A fragment of *Autographa californica* nuclear polyhedrosis virus, and identification of a viral–protein resembling the outer capsid protein VP8 of rotavirus, *Virology* **191:**1003.

Braunagel, S. C., Elton, D. M., Ma, H., and Summers, M. D., 1996a, Identification and analysis of an *Autographa californica* nuclear polyhedrosis virus structural protein of the occlusion-derived virus envelope—ODV-E56, *Virology* **217:**97.

Braunagel, S. C., He, H., Ramamurthy, P., and Summers, M. D., 1996b, Transcription, translation, and cellular localization of three *Autographa californica* nuclear polyhedrosis virus structural proteins: ODV-E18, ODV-E35, and ODV-EC27, *Virology* **222:**100.

Breindl, V., and Jirovec, O., 1935, Polyeder und Polyedervirus im Lichte der Feulgenschen Nucleareaktion, *Vest. Ceskoslov. Zool. Spol., Praha* **3:**9.

Bud, H. M., and Kelly, D. C., 1977, The DNA contained by nuclear polyhedrosis viruses isolated from four *Spodoptera* spp. (Lepidoptera, Noctuidae): Genome size and configuration assessed by electron microscopy, *J. Gen. Virol.* **37:**135.

Burgess, S., 1977, Molecular weights of lepidopteran baculovirus DNAs: Derivation by electron microscopy, *J. Gen. Virol.* **37:**501.

Carson, D. D., Summers, M. D., and Guarino, L. A., 1991, Molecular analysis of a baculovirus regulatory gene, *Virology* **182:**279.

Carstens, E. B., Lu, A. L., and Chan, H. L. B., 1993, Sequence, transcriptional mapping, and overexpression of p47, a baculovirus gene regulating late gene expression, *J. Virol.* **67:**2513.

Child, S. J., Palumbo, G. J., Buller, R. M., and Hruby, D. E., 1990, Insertional activation of the large subunit of ribonucleotide reductase is associated with reduced virulence *in vivo*, *Virology* **174:**625.

Chisholm, G. E., and Henner, D. J., 1988, Multiple early transcripts and splicing of the *Autographa californica* nuclear polyhedrosis virus IE-1 gene, *J. Virol.* **62:**3193.

Choi, J., and Guarino, L. A., 1995, The baculovirus transactivator IE1 binds to viral enhancer elements in the absence of insect cell factors, *J. Virol.* **69:**4548.

Clem, R. J., Fechheimer, M., and Miller, L. K., 1991, Prevention of apoptosis by a baculovirus gene during infection of insect cells, *Science* **254:**1388.

Clem, R. J., Robson, M., and Miller, L. K., 1994, Influence of infection route on the infectivity of baculovirus mutants lacking the apoptosis-inhibiting gene p35 and the adjacent gene p94, *J. Virol.* **68:**6759.

Cochran, M. A., and Faulkner, P., 1983, Location of homologous DNA sequences interspersed at five regions in the baculovirus AcMNPV genome, *J. Virol.* **45:**961.

Corsaro, B. G., and Fraser, M. J., 1987, Characterization of genotypic and phenotypic variation in plaque-purified strains of HzSNPV Elkar isolate, *Intervirology* **28:**185.

Crawford, M., and Miller, L. K., 1988, Characterization of an early gene accelerating expression of late genes of the baculovirus *Autographa californica* nuclear polyhedrosis virus, *J. Virol.* **62:**2773.

Crook, N. E., Clem, R. J., and Miller, L. K., 1993, An apoptosis-inhibiting baculovirus gene with a zinc finger-like motif, *J. Virol.* **67:**2168.

Depew, R. E., and Cozzarelli, N. R., 1974, Genetics and physiology of bacteriophage T4 3′-phosphatase: Evidence for the involvement of the enzyme in T4 DNA metabolism, *J. Virol.* **13:**888.

Durantel, D., Croizier, L., Ayres, M. D., Croizier, G., Possee, R. D., and Lopez-Ferber, M., Natural variation and functional analysis of a predicted baculovirus gene product similar to T4 bacteriophage RNA ligase and polynucleotide kinase, *J. Gen. Virol.*, submitted.

Eldridge, R. L., Li, Y., and Miller, L. K., 1992, Characterization of a baculovirus gene encoding a small conotoxinlike polypeptide, *J. Virol.* **66:**6563.

Fraser, M. J., Cary, L., Boonvishudhi, K., and Wang, H. H., 1995, Assay for movement of lepidopteran transposon IFP2 in insect cells using a baculovirus genome as a target DNA, *Virology* **211:**397.

Friesen, P. D., 1993, Invertebrate transposable elements in the baculovirus genome: Characterization and significance, in: *Parasites and Pathogens of Insects* (N. E. Beckage, S. N. Thompson, and B. A. Federici, eds.), pp. 147–178, Academic Press, San Diego.

Friesen, P. D., and Miller, L. K., 1987, Divergent transcription of early 35- and 94-kilodalton protein genes encoded by the *Hin*dIII-K genome fragment of the baculovirus *Autographa californica* nuclear polyhedrosis virus, *J. Virol.* **61:**2264.

Garcia-Maruniak, A., Pavan, O. H. O., and Maruniak, J. E., 1996, A variable region of *Anticarsia gemmatalis* nuclear polyhedrosis virus contains tandemly repeated DNA sequences, *Virus Res.* **41:**123.

Gearing, K., and Possee, R. D., 1990, Functional analysis of a 603 nucleotide open reading frame in the EcoRI-I fragment of *Autographa californica* nuclear polyhedrosis virus, *J. Gen. Virol.* **71:**251.

Guarino, L. A., 1990, Identification of a viral gene encoding a ubiquitin-protein, *Proc. Natl. Acad. Sci. USA* **87:**409.

Guarino, L. A., and Smith, M. W., 1990, Nucleotide sequence and characterization of the 39k gene region of *Autographa californica* nuclear polyhedrosis virus, *Virology* **179:**1.

Guarino, L. A., and Summers, M. D., 1986, Interspersed homologous DNA of *Autographa californica* nuclear polyhedrosis virus enhances delayed-early gene expression, *J. Virol.* **60:**215.

Guarino, L. A., and Summers, M. D., 1987, Nucleotide sequence and temporal expression of a baculovirus regulatory gene, *J. Virol.* **61:**2091.

Harrap, K. A., and Payne, C. C., 1979, The structural properties and identification of insect viruses, *Adv. Virus Res.* **25:**273.

Hawtin, R. E., Arnold, K., Ayres, M. D., Zanotto, P. M de A., Howard, S. C., Gooday, G. A., Chappell, L. H., Kitts, P. A., King, L. A., and Possee, R. D., 1995, Identification and preliminary characterization of a chitinase gene in the *Autographa californica* nuclear polyhedrosis virus genome, *Virology* **212:**673.

Hawtin, R. E., Zarkowska, T., Arnold, K. A., Thomas, C. A., Gooday, G. W., King, L. A., Kuzio, J. A., and Possee, R. D., 1996, Liquefaction of *Autographa californica* nucleopolyhedrovirus-infected insects is dependent of the integrity of virus-encoded chitinase and cathepsin genes, submitted.

Heldens, J. G. M., VanStrien, E. A., Felmann, A. M., Kulscar, P., Munoz, D., Leisy, D. J., Zuidema, D., Goldbach, R. W., and Vlak, J. M., 1996, *Spodoptera exigua* nucleopolyhedrosvirus deletion mutants generated in cell culture lack virulence *in vivo*, *J. Gen. Virol.* **77:**3127–3134.

Hooft van Iddekinge, B. J. L., Smith, G. E., and Summers, M. D., 1983, Nucleotide sequence of the polyhedrin gene of *Autographa californica* nuclear polyhedrosis virus, *Virology* **131:**561.

Hong, T., Braunagel, S. C., and Summers, M. D., 1994, Transcription, translation, and cellular localization of PDV-E66: A structural protein of the PDV envelope of *Autographa californica* polyhedrosis virus, *Virology* **204:**210.

Jewell, J. E., and Miller, L. K., 1980, DNA sequence homology relationships among 6 Lepidopteran nuclear polyhedrosis viruses, *J. Gen. Virol.* **48:**161.

Jones, J. D. G., Grady, K. L., Suslow, T. V., and Bedbrook, J. R., 1986, Isolation and characterization of genes encoding two chitinase enzymes from *Serratia marcescens, EMBO J.*, **5:** 467.

Kaufmann, G., David, M., Borasio, G. D., Teichmann, A., Paz, A., and Amitsur, M., 1986, Phage and host genetic determinants of the specific anticodon loop cleavages in bacteriophage T4-infected *Escherichia coli* CTr5X, *J. Mol. Biol.* **188:**15.

Kim, D., and Weaver, R. F., 1993, Transcription mapping and functional analysis of the protein tyrosine/serine phosphatase, PTPase gene of the *Autographa californica* nuclear polyhedrosis virus, *Virology* **195:**587.

Knebel-Mörsdorf, M. D., Kremer, A., and Jahnel, F., 1993, Baculovirus gene ME53, which contains a putative zinc finger motif, is one of the major early-transcribed genes, *J. Virol.* **67:**753.

Kogan, P. H., and Blissard, G. W., 1994, A baculovirus gp64 early promoter is activated by host transcription factor binding to CACGTG and GATA elements, *J. Virol.* **68:**813.

Kool, M., and Vlak, J. M., 1993, The structural and functional organization of the *Autographa californica* nuclear polyhedrosis virus genome, *Arch. Virol.* **103:**1.

Kool, M., Ahrens, C., Goldbach, R. W., Rohrmann, G. F., and Vlak, J. M., 1994, Identification of genes involved in DNA replication of the *Autographa californica* baculovirus, *Proc. Natl. Acad. Sci. USA* **91:**11212.

Kool, M., Ahrens, C. H., Vlak, J. M., and Rohrmann, G. F., 1995, Replication of baculovirus DNA, *J. Gen. Virol.* **76:**2103.

Kovacs, G. R., Guarino, L. A., Graham, B. L., and Summers, M. D., 1991, Identification of spliced baculovirus RNAs expressed late in infection, *Virology* **185:** 633.

Krappa, R., and Knebel-Mörsdorf, D., 1991, Identification of the very early transcribed baculovirus gene PE-38, *J. Virol.* **65:**805.

Kumar, S., and Miller, L. K., 1987, Effects of serial passage of *Autographa californica* nuclear polyhedrosis virus in cell culture, *Virus Res.* **7:**335

Kuzio, J., Rohel, D. Z., Curry, C. J., Krebs, A., Carstens, E. B., and Faulkner, P., 1984, Nucleotide sequence of the p10 polypeptide gene of *Autographa californica* nuclear polyhedrosis virus, *Virology* **139:**414.

Kuzio, J., Jaques, R., and Faulkner, P., 1989, Identification of p74, a gene essential for virulence of baculovirus occlusion bodies, *Virology* **173:**759.

Lee, H. H., and Miller, L. K., 1978, Isolation of genotypic variants of *Autographa californica* nuclear polyhedrosis virus, *J. Virol.* **27:**754.

Leisy, D. J., Rasmussen, C., Kim, H.-T., and Rohrmann, G. F., 1995, The *Autographa californica* nuclear polyhedrosis virus homologous region 1a: Identical sequences are essential for DNA replication activity and transcriptional enhancer function, *Virology* **208:**742.

Lerner, D. L., Wagaman, P. C., Phillips, T. R., Prospero-Garcia, O., Henriksen, S. J., Fox, H. S., Bloom, F. E., and Elder, J. H., 1995, Increased mutation frequency of feline immunodeficiency virus lacking functional deoxyuridine-triphosphatase, *Proc. Natl. Acad. Sci. USA* **92:**7480.

Li, Y., and Miller, L. K., 1995a, Expression and functional analysis of a baculovirus gene encoding a truncated protein kinase homolog, *Virology* **206:**314.

Li, Y., and Miller, L. K., 1995b, Properties of a baculovirus mutant defective in the protein phosphatase gene, *J. Virol.* **69:**4533.

Li, Y., Passarelli, A. L., and Miller, L. K., 1993, Identification, sequence, and transcriptional mapping of *lef-3*, a baculovirus gene involved in late and very late gene expression, *J. Virol.* **67:**5260.

Liu, A., Qin, J., Rankin, C., Hardin, S. E., and Weaver, R. F., 1986, Nucleotide sequence of a portion of the *Autographa californica* nuclear polyhedrosis virus genome containing the *Eco*RI site-rich region (*hr5*) and an open reading frame just 5′ of the p10 gene, *J. Gen. Virol.* **67:**2565.

Lu, A., and Carstens, E. B., 1991, Nucleotide sequence of a gene essential for viral DNA replication in the baculovirus *Autographa californica* nuclear polyhedrosis virus, *Virology* **181:**336.

Lu, A., and Carstens, E. B., 1992, Nucleotide sequence and transcriptional analysis of the p80 gene of *Autographa californica* nuclear polyhedrosis virus: A homologue of the *Orgyia pseudotsugata* nuclear polyhedrosis virus capsid-associated gene, *Virology* **190:**201.

Lu, A., and Miller, L. K., 1994, Identification of three late expression factor genes within the 33.8- to 43.4-map-unit region of *Autographa californica* nuclear polyhedrosis virus, *J. Virol.* **68:**6710.

Lu, A., and Miller, L. K., 1995a, The roles of eighteen baculovirus late expression factor genes in transcription and DNA replication, *J. Virol.* **69:**975.

Lu, A., and Miller, L. K., 1995b, Differential requirements for baculovirus late expression factor genes in two cell lines, *J. Virol.* **69:**6265.

Maeda, S., 1994, Genbank Submission L33180.

Maruniak, J. E., Brown, S. E., and Knudson, D. L., 1984, Physical maps of SfMNPV baculovirus DNA and its genomic variants, *Virology* **136:**221.

Mathews, C. K., and van Holde, K. E., 1995, *Biochemistry*, Benjamin/Cummings, Menlo Park, CA.

McLachlin, J. R., and Miller, L. K., 1994, Identification and characterization of vlf-1, a baculovirus gene involved in very late gene expression, *J. Virol.* **68:**7746.

Midgeley, C. A., and Murray, N. E.,. 1985, T4 polynucleotide kinase: cloning of the gene (pseT) and amplification of its product, *EMBO J.* **4:**2695.

Miller, L. K., 1986, The genetics of baculoviruses, in: *The Biology of Baculoviruses* (R. R. Granados and B. A. Federici, eds.), Vol. I, *Biological Properties and Molecular Biology*, pp. 217–238, CRC Press, Boca Raton, FL.

Miller, L. K., and Dawes, K. P., 1978a, Restriction endonuclease analysis for the identification of baculovirus pesticides, *Appl. Environ. Microbiol.* **35:**411.

Miller, L. K., and Dawes, K. P., 1978b, Restriction endonuclease analysis to distinguish two closely related nuclear polyhedrosis viruses: *Autographa californica* MNPV and *Trichoplusia ni* MNPV, *Appl. Environ. Microbiol.* **35:**1206.

Miller, D. W., and Miller, L. K., 1982, A virus mutant with an insertion of a copia-like transposable element, *Nature* **299:**562.

Monsma, S. A., Oomens, A. G. P., and Blissard, G. P., 1996, The GP64 envelope fusion protein is an essential baculovirus protein required for cell-to-cell transmission of infection, *J. Virol.* **70:**4607.

Morad, I., Chapman-Shimshoni, D., Amitsur, M., and Kaufmann, G., 1993, Functional expression and properties of the $tRNA^{lys}$-specific core anticodon nuclease encoded by *Escherichia coli prrC*, *J. Biol. Chem.* **268:**26842.

Morris, T. D., Todd, J. W., Fisher, B., and Miller, L. K., 1994, Identification of *lef-7:* A baculovirus gene affecting late gene expression, *Virology* **200:**360.

Morse, M. A., Marriot, A. C., and Nuttall, P. A., 1992, The glycoprotein of Thogoto virus (a tick-borne orthomyxo-like virus) is related to the baculovirus glycoprotein GP64, *Virology* **186:**640.

Murphy, F. A., Fauquet, C. M., Bishop, D. H. L., Ghabrial, S. A., Jarvis, A. W., Martelli, G. P., Mayo, M. A., and Summers, M. D. (eds.), 1995, *Virus Taxonomy. Classification and Nomenclature of Viruses*, sixth report of the international committee on taxonomy of viruses, Springer-Verlag, New York.

Oellig, C., Happ, B., Muller, T., and Doerfler, W., 1987, Overlapping sets of viral RNAs reflect the array of polypeptides in the *Eco*RI J and N fragments (map positions 81.2 to 85.0) of the *Autographa californica* nuclear polyhedrosis virus genome, *J. Virol.* **61**:3048.

O'Reilly, D. R., and Miller, L. K., 1989, A baculovirus blocks insect molting by producing ecdysteroid UDP-glucosyltransferase, *Science* **245**:1110.

O'Reilly, D. R., and Miller, L. K., 1990, Regulation of expression of a baculovirus ecdysteroid UDP-glucosyl transferase gene, *J. Virol.* **64**:1321.

O'Reilly, D. R., Crawford, A. M., and Miller L. K., 1989, Viral proliferating cell nuclear antigen, *Nature* **337**:606.

Passarelli, A. L., and Miller, L. K., 1993a, Three baculovirus genes involved in late and very late gene expression: ie-1, ie-n, and *lef*-2, *J. Virol.* **67**:2149.

Passarelli, A. L., and Miller, L. K., 1993b, Identification and characterization of *lef-1*, a baculovirus gene involved in late and very late gene expression, *J. Virol.* **67**:3481.

Passarelli, A. L., and Miller, L. K., 1993c, Identification of genes encoding late expression factors located between 56.0 and 65.4 map units of the *Autographa californica* nuclear polyhedrosis virus genome, *Virology* **197**:704.

Passarelli, A. L., and Miller, L. K., 1994a, Identification and transcriptional regulation of the baculovirus *lef-6* gene, *J. Virol.* **68**:4458.

Passarelli, A. L., and Miller, L. K., 1994b, *In vivo* and *in vitro* analyses of recombinant baculoviruses lacking a functional *cg30* gene, *J. Virol.* **68**:1186.

Passarelli, A. L., Todd, J. W., and Miller, L. K., 1994, A baculovirus gene involved in late gene expression encodes a large polypeptide with a conserved motif of RNA polymerases, *J. Virol.* **68**:4673.

Pearson, M. N., and Rohrmann, G. F., 1995, *Lymantria dispar* nuclear polyhedrosis virus homologous regions: Characterization of their ability to function as replication origins, *J. Virol.* **6**:213.

Pearson, M. N., Bjornson, R. M., Pearson, G. D., and Rohrmann, G. F., 1992, The *Autographa californica* baculovirus genome: Evidence for multiple replication origins, *Science* **257**:1382.

Pearson, M. N., Bjornson, R. M., Ahrens, C., and Rohrmann, G. F., 1993, Identification and characterization of a putative origin of DNA replication in the genome of a baculovirus pathogenic for *Orgyia pseudotsugata*, *Virology* **197**:715.

Pelham, H. R., 1990, The retention signal for soluble proteins of the endoplasmic reticulum, *Trends Biochem. Sci.* **15**:483.

Pellock, B. J., Lu, A., Meagher, R. B., Weise, M. J., and Miller, L. K., 1996, Sequence, function, and phylogenetic analysis of an ascovirus DNA polymerase gene, *Virology* **216**:146.

Penner, M., Morad, I., Snyder, L., and Kaufmann, G., 1995, Phage T4-coded Stp: Double-edged effector of coupled DNA and tRNA-restriction systems, *J. Mol. Biol.* **249**:857.

Pennock, G. D., Shoemaker, C., and Miller, L. K., 1984, Strong and regulated expression of *Escherichia coli* β-galactosidase in insect cells with a baculovirus vector, *Mol. Cell. Biol.* **4**:399.

Phizicky, E. M., Consaul, S. A., Nehrke, K. W., and Abelson, J., 1992, Yeast tRNA ligase mutants are nonviable and accumulate tRNA splicing intermediates, *J. Biol. Chem.* **267**:4577.

Pick, L., and Hurwitz, J., 1986, Purification of wheat germ RNA ligase. I. Characterization of a ligase associated 5′-hydroxyl polynucleotide kinase activity, *J. Biol. Chem.* **261**: 6684.

Pick, L., Furneaux, H., and Hurwitz, J., 1986, Purification of wheat germ RNA ligase. II. Mechanisms of action of wheat germ RNA ligase, *J. Biol. Chem.* **261**:6694.

Possee, R. D., Sun, T.-P., Howard, S. C., Ayres, M. D., Hill-Perkins, M., and Gearing, K. L., 1991, Nucleotide sequence of the *Autographa californica* nuclear polyhedrosis 9.4 kbp EcorRI-I and -R (polyhedrin gene) region, *Virology* **185**:229.

Rawlings, N. D., Pearl, L. H., and Buttle, D. J., 1992, The baculovirus *Autographa californica* nuclear polyhedrosis virus genome includes a papain-like sequence, *Biol. Chem. Hoppe-Seyler* **373:**1211.

Reilly, L. M., and Guarino, L. A., 1994, The pk 1 gene of AcMNPV encodes a protein kinase, *J. Gen. Virol.* **75:**2999.

Rodems, S. M., and Friesen, P. D., 1995, Transcriptional enhancer activity of *hr5* requires dual-palindrome half sites that mediate binding of a dimeric form of the baculovirus trans-regulator IE1, *J. Virol.* **69:**5368.

Rohrmann, G. F., Carnegie, J. W., Martignoni, M. E., and Beaudreau, G. S., 1977, Characterization of the genome of the nucleopolyhedrosis bundle virus pathogenic for *Orgyia pseudotsugata, Virology* **80:**421.

Rohrmann, G. F., Martignoni, M. E., and Beaudreau, G. S., 1978, Quantification of two viruses in technical preparations of *Orgyia pseudotsugata* baculovirus by means of buoyant density centrifugation of viral deoxyribonucleic acid, *Appl. Environ. Microbiol.* **35:**690.

Runnels, J. M., Soltis, D., Hey, T., and Synder, L., 1982, Genetic and physiological studies of the role of the RNA ligase of bactgeriophage T4, *J. Mol. Biol.* **154:**273.

Schachtel, G. A., Bucher, P., Mocarski, E. S., Blaisdell, B. E., and Karlin, S., 1991, Evidence for selective evolution in codon usage in conserved amino acid segments of human alphaherpesvirus proteins, *J. Mol. Evol.* **33:**483.

Shippam, C., Wu, X., Stewart, S., and Theilmann, D. A., 1997, Characterization of a unique OpMNPV specific early gene not required for viral infection in tissue culture, *Virology* **227:**447.

Silber, R., Malathi, V. G., and Hurwitz, J., 1972, Purification and properties of bacteriophage T4-induced RNA ligase, *Proc. Natl. Acad. Sci. USA* **69:**3009.

Sirotkin, K., Cooley, W., Runnels, J., and Synder, L., 1978, A role in true late gene expression for the T4 bacteriophage 5′-polynucleotide kinase 3′-phosphatase, *J. Mol Biol.* **123:**221.

Smith, G. E., and Summers, M. D., 1978, Analysis of baculovirus genomes with restriction endonucleases, *Virology* **89:**517.

Snopek, T. J., Wood, W. B., Conley, M. P., Chen, P., and Cozzarelli, N. R., 1977, Bacteriophage T4 RNA ligase is gene *63* product, the protein that promotes tail fiber attachment to the baseplate, *Proc. Natl. Acad. Sci. USA* **74:**3355.

Summers, M., and Anderson, D., 1972a, Characterization of deoxyribonucleic acid isolated from the granulosis viruses of the cabbage looper, *Trichoplusia ni* and the fall armyworm, *Spodoptera frugiperda, Virology* **50:**459.

Summers, M., and Anderson, D., 1972b, Granulosis virus deoxyribonucleic acid: A closed double-stranded molecule, *J. Virol.* **9:**710.

Theilmann, D. A., and Stewart, S., 1992, Tandemly repeated sequence at the 3′ end of the IE-2 gene of the baculovirus *Orgyia pseudotsugata* multicapsid nuclear polyhedrosis virus is an enhancer element, *Virology* **187:**97.

Thiem, S. M., and Miller, L. K., 1989a, Identification, sequence, and transcriptional mapping of the major capsid protein gene of the baculovirus *Autographa californica* nuclear polyhedrosis virus, *J. Virol.* **63:**2008.

Thiem, S. M., and Miller, L. K., 1989b, A baculovirus gene with a novel transcription pattern encodes a polypeptide with a zinc finger and a leucine zipper, *J. Virol.* **63:**4489.

Tilakaratne, N., Hardin, S. E., and Weaver, R. F., 1991, Nucleotide sequence and transcript mapping of the HindIII F region of the *Autographa californica* nuclear polyhedrosis virus genome, *J. Gen. Virol.* **72:**285.

Todd, J. W., Passarelli, A. L., and Miller, L. K., 1995, Eighteen baculovirus genes including *lef-11, p35, 39k* and *p47* support late gene expression, *J. Virol.* **69:**968.

Tomalski, M. D., Wu, J., and Miller, L. K., 1988, The location, sequence, transcription, and regulation of a baculovirus DNA polymerase gene, *Virology* **167:**591.

Tomalski, M. D., Eldridge, R., and Miller, L. K., 1991, A baculovirus homolog of a Cu/Zn superoxide dismutase gene, *Virology* **184:**149.

Tsujibo, H., Orikoshi, H., Tanno, H., Fujimoto, K., Miyamoto, K., Imada, C., Okami, Y., and Inamori, Y., 1993, Cloning-sequence and expression of a chitinase gene from a marine bacterium, *Alteromonas* sp. strain O7, *J. Bacteriol.* **175:**176.

Turelli, P., Petursson, G., Guiguen, F., Mornex, J.-F., Vigne, R., and Querat, G., 1996, Replication properties of dUTPase-deficient mutants of caprine and ovine lentiviruses, *J. Virol.* **70:**1213.

Tyndal, C., Meister, J., and Bickle, T. A., 1994, The *Escherichia coli prr* region encodes a functional type IC DNA restriction system closely integrated with an anticodon nuclease gene, *J. Mol. Biol.* **237:**266.

Uberbacher, E. C., and Mural, R. J., 1991, Locating protein-coding regions in human DNA sequences by a multiple sensor-neural network approach, *Proc. Natl. Acad. Sci. USA* **88:**11261.

Uhlin, U., and Eklund, H., 1994, Structure of ribonucleotide reductase protein R1, *Nature* **370:**533.

Vialard, J. E., and Richardson, C. D., 1993, The 1,629-nucleotide open reading frame located downstream of the *Autographa californica* nuclear polyhedrosis virus polyhedrin gene encodes a nucleocapsid-associated phosphoprotein, *J. Virol.* **67:**5859.

Wagaman, P. C., Hasselkus-Light, C. S., Henson, M., Lerner, D. L., Phillips, T. R., and Elder, J. H., 1993, Molecular cloning and characterization of deoxyuridine triphosphatase from feline immunodeficiency virus (FIV), *Virology* **196:**451.

Weitzmann, M., Possee, R. D., and King, L. A., 1992, Genetic analysis of two strains of *Panolis flammea* multiple nucleocapsid nuclear polyhedrosis virus, *J. Gen. Virol.* **73:**1881.

Whitford, M., and Faulkner, P., 1992, Nucleotide sequence and transcriptional analysis of a gene encoding gp41, a structural glycoprotein of the baculovirus *Autographa californica* nuclear polyhedrosis virus, *J. Virol.* **66:**4763.

Whitford, M., Stewart, S., Kuzio, J., and Faulkner, P., 1989, Identification and sequence analysis of a gene encoding gp67, an abundant envelope glycoprotein of the baculovirus *Autographa californica* nuclear polyhedrosis virus, *J. Virol.* **63:**1393.

Whitt, M. A., and Manning, J. S., 1988, A phosphorylated 34-kda protein and a subpopulation of polyhedrin are thiol linked to the carbohydrate layer surrounding a baculovirus occlusion body, *Virology* **163:**33.

Wilson, M. E., Mainprize, T. H., Friesen, P. D., and Miller, L. K., 1987, Location, transcription, and sequence of a baculovirus gene encoding a small arginine-rich polypeptide, *J. Virol.* **61:**661.

Wu, J., and Miller, L. K., 1989, Sequence, transcription and translation of a late gene of the *Autographa californica* nuclear polyhedrosis virus encoding a 34.8 k polypeptide, *J. Gen. Virol.* **70:**2449.

Wu, X., Stewart, S., and Theilmann, D. A., 1993, Characterization of an early gene conding for a highly basic 8.9kDa protein from the baculovirus OpMNPV, *J. Gen. Virol.* **74:**1591.

Wyatt, G. R., 1953, The nucleic acids of some insect viruses, *J. Gen. Physiol.* **36:**201.

Xie, W.-D., Arif, B., Dobos, P., and Krell, P. J., 1995, Identification and analysis of a putative origin of DNA replication in the *Choristoneura fumiferana* multinucleocapsid nuclear polyhedrosis virus genome, *Virology* **209:**409.

Xu, Q., Phizicky, E. M. Greer, C. L., and Abelson, J. N., 1990a, Purification of yeast transfer RNA ligase, *Methods Enzymol.* **181:**463.

Xu, Q., Teplow, D., Lee, T. D., and Abelson, J. N., 1990b, Domain structure in yeast tRNA ligase, *Biochemistry* **29:**6132.

Yuen, L., Dionne, J., Arif, B., and Richardson, C., 1990, Identification and sequencing of the spheroidin gene of *Choistoneura biennis* entomopoxvirus, *Virology* **175:**427.

CHAPTER 6

Regulation of Baculovirus Early Gene Expression

PAUL D. FRIESEN

I. INTRODUCTION

Baculovirus multiplication culminates in high-level production of two temporally and morphologically distinct forms of infectious progeny (budded virus and occluded virus) in a process unique among animal viruses. Regulation of the 100 or more open reading frames required to accomplish productive infection is highly complex and involves sequential and coordinated expression of early, late, and very late genes. In the cascade of viral regulatory events, successive stages of virus replication are dependent on proper expression of genes within the preceding stage. Thus, critical to baculovirus replicative success is the appropriate expression and regulation of early genes. The products of early viral genes function to both accelerate replicative events and to prepare the host cell for virus multiplication, which represents an enormous tax on cellular biosynthetic capacity. Specific early genes are also essential for virus-mediated regulation of the host, including the control of larval molting and evasion of host antiviral responses such as apoptosis (see Chapters 10 and 11, this volume). Thus, early baculovirus genes collectively contribute to host range determination.

Current evidence indicates that the timing and levels of early (and late) baculovirus gene expression is regulated at the level of transcription. Due to the complexity of regulation, the molecular mechanisms by which these

---

PAUL D. FRIESEN • Institute for Molecular Virology and Department of Biochemistry, Graduate School and College of Agricultural and Life Sciences, University of Wisconsin, Madison, Wisconsin 53706.

The Baculoviruses, edited by Lois K. Miller. Plenum Press, New York, 1997.

large DNA viruses execute their transcriptional program and discriminate between early and late genes are of considerable interest. Such mechanisms include a novel switch from early dependence on host RNA polymerase II to that of a new virus-induced RNA polymerase for late transcription (see Chapter 8, this volume). Key to the understanding of baculovirus early gene regulation has been the identification of *cis*-acting DNA elements within the viral genome and the *trans*-acting protein factors capable of potentiating viral transcription through interaction with these DNA elements. Considerable progress has been made toward defining the *cis*-acting regulatory elements involved in early viral gene expression. Moreover, host factors and baculovirus-encoded transregulators (IE0, IE1, IE2, and PE38) involved in early transcription have been identified and characterized. This chapter reviews the structural organization of early viral promoters and transcriptional enhancers, the function of early viral transregulators, and the molecular mechanisms that contribute to proper early baculovirus gene expression.

II. BACULOVIRUS EARLY GENES: ORGANIZATION AND TRANSCRIPTION PATTERNS

Baculovirus genes expressed early during infection are transcribed from both DNA strands and distributed throughout the circular DNA genome. In the case of *Autographa californica* nuclear polyhedrosis virus (AcMNPV), *Orgyia pseudotsugata* NPV (OpMNPV), and *Bombyx mori* NPV (BmNPV), RNA transcripts of many early genes have been mapped in detail, thereby providing useful information with respect to baculovirus gene structure and organization. In general, there is no apparent physical association or grouping of genes according to temporal expression. Likewise, transcriptional enhancer elements involved in early gene regulation are dispersed throughout the viral genome, including the homologous regions (*hr*s) of AcMNPV (Fig. 1A). Early gene transcription involves the synthesis of RNAs that are capped at their 5′ end (Jun-Chuan and Weaver, 1982) and polyadenylated at their 3′ end in response to consensus poly(A) signals at the 3′ end of most genes. Although many early baculovirus genes are transcribed into a single RNA species, early transcription is often characterized by the synthesis of multiple overlapping RNAs that form individual groups or nests with a common 3′ terminus but different 5′ ends (Friesen and Miller, 1985; Lübbert and Doerfler, 1984a). The functional significance of this pattern is still unclear. The AcMNPV *p35* and *ie-1* loci (Fig. 1B,C) represent two well-studied early transcriptional units that synthesize overlapping and nonoverlapping RNAs; both are described in Section III.

Transcription of many early genes initiates immediately after virus inoculation, since specific RNA transcripts are detectable within the first 15 to 60 min (Krappa and Knebel-Mörsdorf, 1991; Theilmann and Stewart, 1991; Huh and Weaver, 1990a; Nissen and Friesen, 1989; Chisholm and

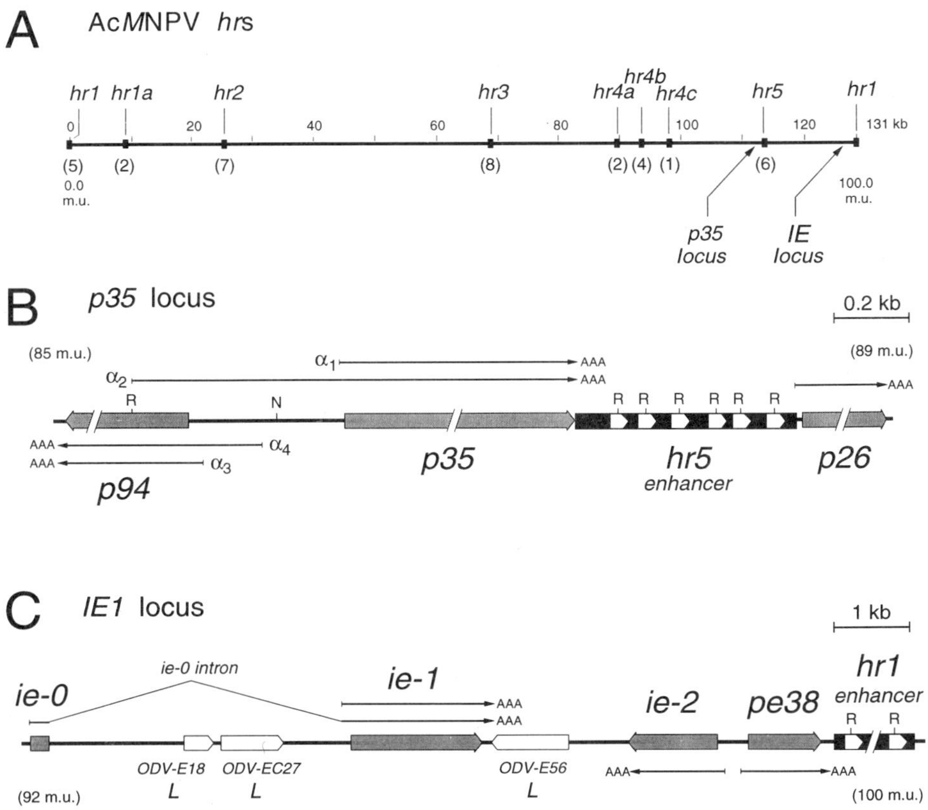

FIGURE 1. Genomic organization of AcMNPV *hr* elements. (A) AcMNPV genome map. Homologous regions (*hrs*) are indicated on the linear map of the AcMNPV genome (Ayres *et al.*, 1994). The number of repetitive palindromic 28-*mer*s present in each *hr* element is indicated in parentheses below the viral genome. (B) Transcriptional organization of the *p35* locus. The *p35* gene (shaded arrow) is flanked on the left by the *p94* gene and on the right by the *hr*5 enhancer (black box) and the *p26* gene. The early *p35* α_1 mRNA and the early *p94* α_3 and α_4 mRNAs initiate from within the *p94–p35* intergenic region. The *hr*5 enhancer (black box) contains six 28-*mer* palindromes (open arrows). (C) Transcriptional organization of the *ie-1* locus. The *ie-1* gene is flanked on the left by the *ie-0* intron that contains late (L) genes *ODV-E18* and *ODV-EC27* and on the right by late gene *ODV-E56*. The immediate early genes *ie-2* and *pe38* are adjacent to the *hr*1 enhancer. Abbreviations: N, *Nru*I; R, *Eco*RI; m.u., map units.

Henner, 1988; Guarino and Summers, 1987). Thus, it is likely that transcription begins as soon as the viral DNA is uncoated in the nucleus. During this early time, the viral DNA genome may adopt a nucleosomelike structure (Wilson and Miller, 1986). Transcription of strictly defined early genes usually peaks between 6 and 12 hr after infection and declines thereafter when late viral transcription is vigorous. The mechanism(s) by which early viral transcription is shut down, like that of host RNA transcription (Ooi and Miller, 1988), is unknown. The finding that the reduction in steady-state levels of some early RNAs can be blocked by metabolic inhibitors, including cycloheximide (inhibitor of protein synthesis) and aphidicolin (inhibitor of viral DNA synthesis), suggests that the turnoff is directly or indirectly due to late gene expression or viral DNA replication (Pullen and Friesen, 1995b; Huh and Weaver, 1990a; Friesen and Miller, 1987; Rice and Miller, 1986). In the case of some early genes, transcription continues late into infection. Late transcription of these early/late genes, including *p35, gp64 efp,* 39K (*pp31*), and *ie-1,* is mediated by late promoter elements that overlap the early promoter (see Section III.D and Chapter 8, this volume).

By definition, a baculovirus early gene is transcribed prior to the initiation of virus DNA replication, which begins 6 to 9 hr after infection (see Chapter 7, this volume). Thus, early transcription is independent of viral DNA replication and late gene expression. Several lines of evidence indicate that early baculovirus promoters are responsive to the host RNA polymerase II and associated host transcription factors. First, naked viral DNA (devoid of virion components) is infectious, indicating that the host transcription machinery is sufficient to initiate productive infection. Second, certain early viral genes are readily and accurately transcribed by *in vitro* reactions composed of nuclear extracts from uninfected host cells (Pullen and Friesen, 1995a; Glocker *et al.,* 1992; Blissard *et al.,* 1992; Hoopes and Rohrmann, 1991), indicating that early promoters respond appropriately to the host RNA polymerase. Third, early viral transcription but not late viral transcription is blocked by the RNA polymerase II inhibitor α-amanitin (Huh and Weaver, 1990b; Grula *et al.,* 1981). Last, early promoters have a structural organization that is similar to that of RNA polymerase II-responsive genes but strikingly different from that of late viral genes (Pullen and Friesen, 1995a; Lu and Carstens, 1993; Guarino and Smith, 1992; Blissard and Rohrmann, 1991; Dickson and Friesen, 1991; Theilmann and Stewart, 1991; Nissen and Friesen, 1989). This distinction between early and late promoters is sufficiently clear that it is often possible to classify the temporal specificity of an unknown baculovirus gene solely on the basis of the nucleotide sequence of its promoter.

III. EARLY BACULOVIRUS PROMOTERS

Current evidence suggests that the promoters for early baculovirus genes are organized to expedite immediate and regulated levels of transcription by mechanisms that exploit host RNA polymerase II and associated host transcription factors. The structure of early promoters therefore mimics that of promoters normally responsive to RNA polymerase II. Moreover, to enhance RNA polymerase II-mediated transcription, the virus encodes early factors that stimulate transcription in a promoter-specific manner. To accomplish regulated expression, early viral promoters are composed of core transcription elements that cooperate with auxiliary *cis*-acting elements located either nearby or within distal transcriptional enhancers (Fig. 1A).

These regulatory elements are responsive to sequence-specific transcription factors.

A. Core Promoter Elements

The core (basal) promoter region of an RNA polymerase II-responsive baculovirus promoter includes sequences surrounding the +1 RNA start site (Fig. 2A). The core functions to provide a basal level of accurate transcriptional initiation that is further augmented or activated by distal *cis*-acting elements. Basal promoters can be categorized as TATA-containing ($TATA^+$), initiator-containing (INR^+), or composite ($TATA^+$, INR^+) promoters (reviewed by Novina and Roy, 1996). Although some baculovirus early promoters remain unclassified, most are represented by these three classes.

1. TATA Elements

The TATA element consists of an A/T-rich motif (consensus TATAA) located 25 to 31 nucleotides upstream from the +1 RNA start site of many RNA polymerase II-responsive promoters. As indicated by the dramatic reduction in promoter activity upon its deletion or replacement, the TATA motif is the principal regulatory element of many baculovirus early promoters (Pullen and Friesen, 1995a; Blissard *et al.*, 1992; Guarino and Smith, 1992; Blissard and Rohrmann, 1991; Dickson and Friesen, 1991). It regulates the rate of transcription initiation and defines the start site for early viral transcription, especially in the absence of an INR motif (Blissard *et al.*, 1992; Guarino and Smith, 1992; Dickson and Friesen, 1991).

In eukaryotes, the TATA element is the recognition sequence and binding site for TATA binding protein (TBP), a factor that mediates assembly of the TFIID transcriptional complex in the rate-limiting step during transcriptional initiation by RNA polymerase II (reviewed by Phillips, 1993). The single copy TBP gene from the permissive lepidopteran host *Spodoptera frugiperda* (SfTBP) has been cloned and sequenced (Rasmussen and Rohrmann, 1994). Although its functionality remains to be demonstrated, the SfTBP protein (307 amino acid residues) has striking sequence similarity to TBP of *Drosophila melanogaster*, especially within the C-terminal 180 residues (93% identity), but differs in its N-terminal domain. Since TBP is a common target of animal virus transactivators during stimulation of viral gene expression, it is expected that baculovirus-mediated enhancement of early promoter activity directly or indirectly involves the host lepidopteran TBP. However, viral factors that interact with the host TBP remain to be identified.

2. Initiator (INR) Motifs

Current evidence suggests that numerous baculovirus early promoters contain a transcriptional INR composed of the heptanucleotide sequence

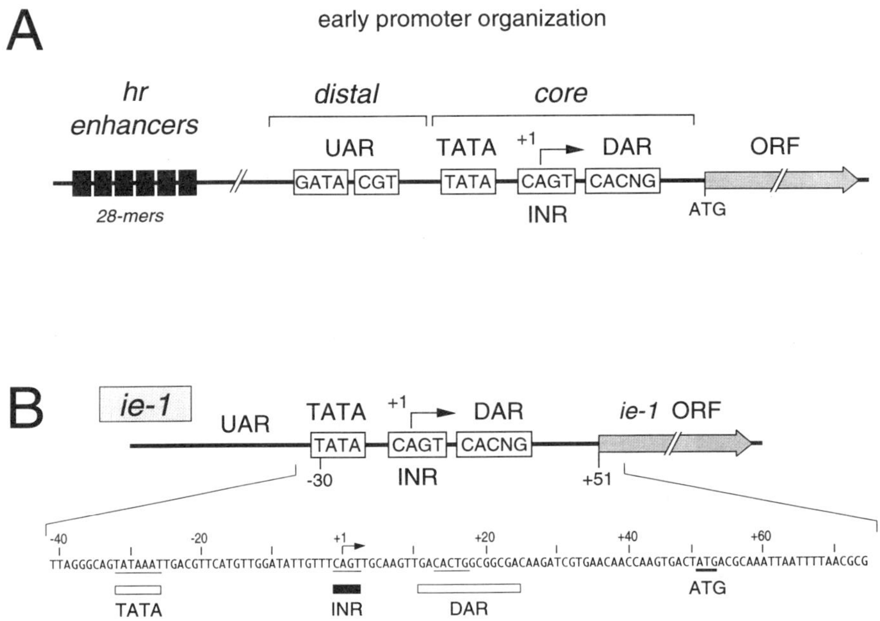

FIGURE 2. Structural organization of early baculovirus promoters. (A) *Cis*-acting early promoter elements. Typical core promoter elements include the TATA element (TATA), CAGT initiator (INR), and DAR motif (CACNG), all located upstream from the open reading frame (ORF) of each gene. Transcription initiates from the CA dinucleotide (position +1) within the CAGT INR, if present. Distal elements within the UAR are composed of specific regulatory sites, including GATA, CGT, and GC motifs. The 28-*mer*-containing *hr* enhancers of AcMNPV function in a position- and orientation-independent manner. The organization and nucleotide sequence of core promoter elements and distal motifs are included for the (B) AcMNPV *ie-1*, (C) AcMNPV *p35*, and (D) OpMNPV *gp64 efp* genes. The early RNA start site (position +1) for each promoter is indicated by an arrow.

ATCA(G/T)T(C/T). These motifs closely match INR motifs of arthropods and to a lesser extent the INR motifs of vertebrates (Hultmark *et al.*, 1986; Cherbas and Cherbas, 1993). By definition, an INR overlaps the RNA start site, contributes to basal promoter activity, and determines the position of the RNA start site in the absence of a TATA element (reviewed by Novina and Roy, 1996). Most conserved among the INR-like motifs located at or near the RNA start site of early baculovirus genes is the tetranucleotide CAGT (Fig. 2A). The CAGT motif functions as an INR within the viral genome as first demonstrated by inactivating the TATA element of the early transregulator gene *ie-1* (Pullen and Friesen, 1995b). In plasmid transfection assays, the CAGT motif also contributes to the overall strength of the TATA-containing promoters of early genes *ie-2, gp64 efp,* and 39K (*pp31*) (Kogan *et al.*, 1995; Blissard *et al.*, 1992; Guarino and Smith, 1992; Carson *et al.*, 1991a). This finding is consistent with CAGT-mediated INR activity. As demonstrated by site-directed mutagenesis, the CA dinulceotide is most

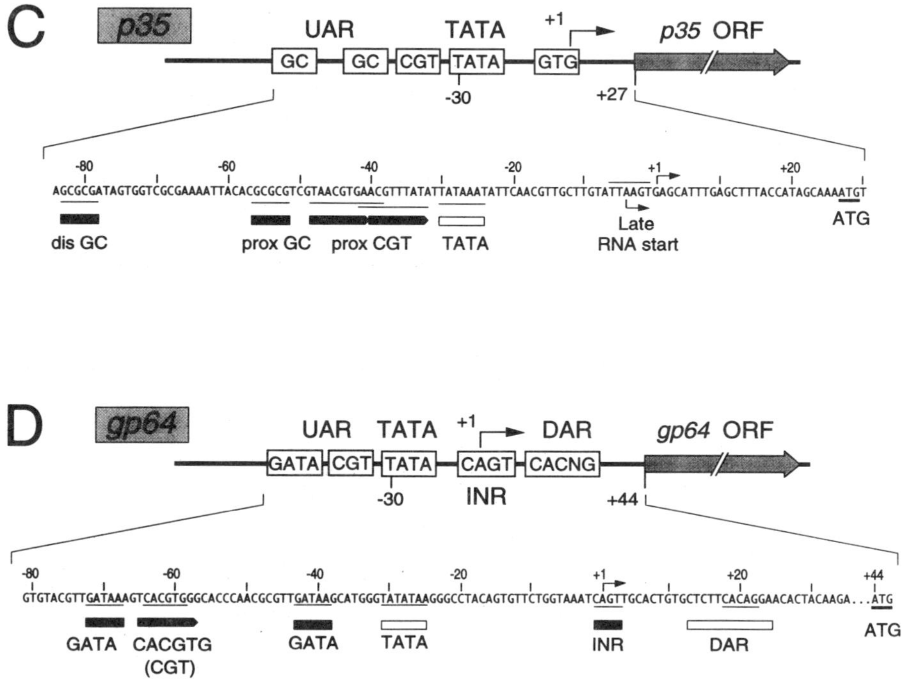

FIGURE 2. *(Continued).*

influential in maintaining regulated levels of CAGT transcriptional initiation, but a consensus CAGT is not required (Kogan *et al.*, 1995; Pullen and Friesen, 1995b).

In vertebrates, the INR binds specific recognition factors that, analogous to TBP, expedite the assembly of the preinitiation TFIID transcription complex at the promoter (reviewed by Burley and Roeder, 1996; Zawel and Reinberg, 1995). Many of the baculovirus CAGT-containing early promoters also possess an upstream TATA element, and thus represent composite (TATA$^+$, INR$^+$) promoters. Thus, both core promoter elements may cooperate to either stabilize the host transcription machinery or enhance recruitment of required factors. It is expected that such interactions accelerate transcription initiation and insure adequate expression of key early genes at the time when DNA template concentrations are at their lowest during infection. This viral strategy that recruits universal host transcription factors may also facilitate early gene expression in a wider variety of tissue types and thereby contribute to viral replication success.

3. Downstream Activating Region Elements

Mutational analyses of the composite *ie-1* and *gp64 efp* promoters have identified a transcriptional element within the 5′ untranslated region, referred to as a downstream activation region (DAR) motif (Kogan *et al.*, 1995; Pullen and Friesen, 1995a). This core regulatory element (Fig. 2A) contributes to basal promoter activity, but has little if any effect on RNA transcript stability. The *ie-1* and *gp64 efp* DARs contain the sequence (A/T)CACNG, which upon deletion eliminates DAR-mediated stimulation of basal promoter activity (Kogan *et al.*, 1995; Pullen and Friesen, 1995a). Due to the relative proximity of the DAR to the CAGT INR motif and its influence on TATA-independent transcription, it is likely that the DAR functions to stabilize protein interactions, including TFIID, at the INR. Host- or viral-encoded factors that interact with DAR sequences remain to be identified. Nonetheless, the presence of similar sequences within the noncoding leader region of other early baculovirus genes suggests that the DAR is a general regulatory motif.

B. Distal Upstream Activating Region Elements

Early baculovirus promoters often possess an upstream activation region (UAR) consisting of one or more *cis*-acting DNA elements that potentiate transcription from the basal promoter (Fig. 2A). Representing a distal regulatory motif that can extend 100 or more nucleotides upstream from the TATA element, the UAR affects the level of transcription but not the position of the RNA start site. Thus, the organization of UAR-containing promoters resembles that of RNA polymerase II promoters of the host. The UAR is functionally separable from the basal promoter as demonstrated by direct substitution of UARs from different promoters (Pullen and Friesen, 1995a) or by UAR deletions (Lu and Carstens, 1993; Krappa *et al.*, 1992; Theilmann and Stewart, 1992a; Theilmann and Stewart, 1991; Blissard and Rohrmann, 1991; Dickson and Friesen, 1991). Likewise, a UAR can stimulate INR-dependent transcription as demonstrated for the *p35* and *gp64* early promoters (Kogan *et al.*, 1995; Pullen and Friesen, 1995b). Consistent with the separability of UAR and basal promoter, the *p35* UAR functionally can be replaced by a baculovirus transcription enhancer in the context of the viral genome (Rodems and Friesen, 1993). Different UARs can exert opposing effects on basal promoter activation as demonstrated for the *p35* and *ie-1* promoters (Pullen and Friesen, 1995a) (Section III.D). Moreover, UAR-mediated promoter stimulation can differ dramatically, depending on whether assayed in uninfected or infected cells by using plasmid transfections or by insertion into the genome of recombinant viruses. The activity of certain UARs can also vary in a host-specific manner (Blissard and Rohrmann, 1991; Theilmann and Stewart, 1991), consistent with the role of host transcription factors in potentiating basal promoter activity through the UAR. Although direct evidence is lacking, it is expected that virus-encoded transcriptional transregulators also affect UAR activity by direct DNA recognition or indirectly through stabilization of host factor interaction.

The analysis of multiple early promoters suggests that individual UAR regulatory motifs are recognized by DNA sequence-specific transcription factors. These factors likely function to stimulate the rate of transcriptional initiation by the RNA polymerase II-complex assembled at the basal promoter. Several distinct baculovirus UAR motifs have been identified. The CGT motif with consensus sequence A(A/T)CGT(G/T) is present within the UAR of the *p35, p94,* 39K (pp31), *p143, gp64,* and *ets* genes, among others. In the context of the AcMNPV genome, the CGT motif alone provides approximately a fivefold stimulation of *p35* basal promoter activity (Dickson and Friesen, 1991). As assayed by plasmid transfections, the CGT motif of the 39K UAR provides a similar stimulation (Guarino and Smith, 1992). The CGT motif bears a striking resemblance to the CACGTG motif first identified in the OpMNPV *gp64* UAR. The CACGTG motif binds an unidentified host factor in as sequence-specific manner as shown by electrophoretic mobility shift assays and is a core motif recognized by a family of eukaryotic transcription factors classified as helix–loop–helix/leucine zipper proteins (Kogan and Blissard, 1994). The GC motif, represented by a GC-rich sequence identical to the E2F binding site in adenovirus promoters, also contributes to UAR activation of the *p35* and 39K promoters (Guarino and Smith, 1992; Dickson and Friesen, 1991). In the case of the *p35* promoter, it is likely that the proximal GC and CGT motifs cooperatively bind interacting factors, since disruption of either motif or insertions that affect the spacing between both motifs eliminate UAR activation of the basal promoter (Rodems, 1995). Another UAR motif referred to as the GATA element was identified in the early *gp64* and *pe38* promoters. This motif resembles the GATA motif of vertebrates and binds a nuclear factor (designated SfNF-1) in host insect cells (Kogan and Blissard, 1994; Krappa *et al.*, 1992). Although mutation of the GATA motif of the *gp64* UAR disrupted factor binding and reduced promoter activity, loss of the same motif within the *pe38* UAR had no effect. The regulatory role of the GATA motif within the context of the viral genome is unknown. It is likely that additional, unrelated baculovirus UAR motifs remain to be identified.

C. Unconventional Promoter Motifs

Several important early baculovirus genes are expressed from promoters that lack recognizable TATA or INR elements. Much less is known about the *cis*-acting elements mediating transcription of such genes. In the case of *dnapol,* which encodes a DNA polymerase activity required for transient plasmid replication (reviewed by Kool *et al.*, 1995), the early promoter lacks both a TATA and CAGT motif. During AcMNPV infection, *dnapol* early transcription is initiated from multiple sites, including the sequence CGTGC (Ohresser *et al.*, 1995; Tomalski *et al.*, 1988) that is positioned at the RNA start site of other genes, including *p143 helicase* and *p47.* Although the *dnapol* promoter is inactive when transfected into uninfected cells, it is highly responsive to the IE1 transregulator (Ohresser *et al.*, 1995). This finding suggests that such unconventional promoters are more dependent on virus-encoded transactivators during infection. Consistent with this possibility, the TATA-containing promoter of *p143* with a CGTGC motif at its RNA start site is inactive in uninfected cells, but is highly responsive to *cis*-linkage with the homologous region *hr*5 transcriptional enhancer and viral transregulators IE1, IE2, and PE38 (Lu and Carstens, 1993).

E. Early Promoter Examples

Most of our knowledge of the organization and regulation of early baculovirus genes has been provided by detailed studies of a limited number of distinct promoters. Even fewer viral promoters have been studied in the context of the viral genome during infection when appropriate host and viral factors are present at physiologically relevant levels. Nonetheless, it is likely that the organization of promoters examined to date are representative of many of the early baculovirus promoters. Key features of the promoter for the *ie-1*, *p35*, and *gp64 efp* genes are described.

1. *ie-1* Promoter

The *ie-1* gene encodes the principal early transregulator protein IE1, which plays a critical role in accelerating baculovirus replication (Section V.A). Its expression is regulated by a typical composite promoter (Fig. 2B) containing TATA and INR elements, both of which are essential for normal *ie-1* expression (Pullen and Friesen, 1995a). Transcription of AcMNPV *ie-1* initiates from the CA dinucleotide of the CAGT-containing INR within the first 15 to 30 min after virus entry (Chisholm and Henner, 1988; Guarino and Summers, 1987). RNAs initiated from this site accumulate late into infection through the activity of an unconventional late promoter motif (Pullen and Friesen, 1995b). The function of IE1 produced late in infection is unknown.

As determined by plasmid transfection assays and *in vitro* transcription reactions using nuclear extracts from uninfected cells, the *ie-1* promoter and associated upstream sequences are active in the absence of other virus-encoded factors (Pullen and Friesen, 1995a; Kovacs *et al.*, 1991a; Theilmann and Stewart, 1991). This responsiveness in uninfected cells has been exploited for the development of plasmid vectors in which the *ie-1* promoter directs expression of foreign genes in stably transfected lepidopteran cells (Section VI). The *cis*-acting regulatory motifs responsible for *ie-1* promoter activity in uninfected cells lie within the UAR. In the case of AcMNPV *ie-1*, sequences between -546 and -34 confer a 1000-fold stimulation of the basal *ie-1* promoter (Pullen and Friesen, 1995a); the UAR motifs involved have not been mapped. The *hr*5 transcriptional enhancer (Section IV.A), but not the *p35* UAR, can functionally replace the *ie-1* UAR in plasmid transfections of uninfected cells. The UAR of the closely related OpMNPV *ie-1* gene also contributes to *ie-1* promoter activity in a cell line-specific manner (Theilmann and Stewart, 1991), suggesting that the *ie-1* UAR contains regulatory motifs that are species specific. The *ie-1* promoter is also stimulated as much as tenfold by its own gene product (IE1), raising the possibility that *ie-1* is autoregulatory during infection (Pullen and Friesen, 1995a; Kovacs *et al.*, 1991; Theilmann and Stewart, 1991).

Surprisingly, the *ie-1* UAR is dispensable for proper regulation of early and late *ie-1* expression in the context of the viral genome (Pullen and Friesen, 1995a). The *ie-1* core promoter, including TATA, INR (CAGT), and DAR motifs (Fig. 2B), is sufficient for proper levels of accurately initiated transcription during infection. Substitution of the *ie-1* UAR with the highly active *p35* UAR stimulated *ie-1* basal transcription approximately 11-fold, and thus indicated that the *ie-1* core promoter was responsive to a heterologous UAR (Pullen and Friesen, 1995a). The lack of *ie-1* UAR-mediated stimulation during infection could be due to localized alterations in viral chromatin or UAR interaction with negative regulators. The finding that the *ie-1* UAR, as well as other early baculovirus promoters, can serve as origins of virus DNA replication in plasmid transfection assays raises the interesting possibility that interaction of DNA replication factors may repress *ie-1* UAR-mediated transcriptional stimulation (Wu and Carstens, 1996); however, it is unknown whether the UARs function as origins of viral DNA replication during infection. These findings suggest that the *ie-1* UAR may be most active immediately after the viral genome is uncoated within the nucleus in order to accelerate *ie-1* expression at the time during infection when virus regulatory factors are limiting.

2. *p35* Promoter

One of the most thoroughly studied combination early/late baculovirus promoters is that for *p35*. The *p35* gene product (P35) is a 35-kDa suppressor of apoptosis that functions by inhibiting a virus-induced CED-3/ICE-like protease (caspase) involved in premature apoptotic death of infected insect cells (LaCount and Friesen, 1997; Bertin *et al.*, 1996; Clem *et al.*, 1991). As such, *p35* provides a dramatic selective advantage to AcMNPV in apoptosis-sensitive cells and contributes to host range determination (Clem and Miller, 1993; Lerch and Friesen, 1993). Since early *p35* expression is involved in efficient suppression of apoptosis (Hershberger *et al.*, 1994), *p35* provides an example of an early gene that has a significant influence on viral replicative success.

p35 is flanked by the early *p94* gene and the *hr*5 transcriptional enhancer (Fig. 1B). It is transcribed within the first hour after virus inoculation to produce the polyadenylated, messenger-active RNA α_1 (Friesen and Miller, 1985, 1987; Huh and Weaver, 1990a). Longer, overlapping RNAs of unknown function are synthesized both early and late after infection. The *p35* and *p94* promoters are located within the 210-base pair (bp) intergenic region from which the *p35* α_1 and *p94* α_3 and α_4 RNAs initiate (Fig. 1B). This promoter region is sufficient for proper regulated early and late transcription of reporter genes when inserted at alternative locations within the AcMNPV genome (Dickson and Friesen, 1991). Within the *p35* promoter (Fig. 2C), the TATA element is the single-most important motif due to its contribution to promoter strength and role in determining the RNA start site (Dickson and Friesen, 1991). The *p35* promoter lacks a CAGT motif. Moreover, there is no apparent requirement for an INR, as suggested by nucleotide insertions between the TATA element and the native RNA start site that always directed transcription approximately 30 bp downstream from the TATA motif without significant change in the rate of initiation.

The *p35* UAR (extending from position −90 to −30) confers a significant 10- to 20-fold stimulation of the basal promoter. The most influential UAR regulatory elements include the GC and CGT motifs (Fig. 2C). Deletion of either motif or insertions that alter the spacing between both motifs eliminated stimulation (Rodems, 1995). These findings suggested that both motifs are recognized by cooperative transcription factors; these host or viral stimulatory factors remain to be identified. In transfection assays, the *p35* promoter is dramatically responsive to the AcMNPV transregulator IE1. Cotransfection with *ie-1* causes more than a 1000-fold stimulation of *p35* promoter activity (Rodems and Friesen, 1993; Nissen and Friesen, 1989). Thus, it is likely that early in infection, IE1 contributes to *p35* regulation. Potential *cis*-acting IE1 response elements have not been identified within the *p35* UAR.

Located immediately downstream from *p35* (Fig. 1B), the transcription enhancer *hr*5 (Section IV.A) also contributes to early *p35* promoter activity. When transfected into uninfected cells, the *p35* promoter plus its UAR is inactive. However, *cis*-linkage to *hr*5 stimulates promoter activity (Nissen and Friesen, 1989), suggesting that *hr*5 enhancement can be mediated in part by host cell factors alone. When *cis*-linked to *hr*5, IE1 further augments *p35* promoter activity to levels approximately 20-fold higher than that in the absence of *hr*5. Direct evidence that *hr*5 affects early *p35* transcription during infection was obtained by deleting the enhancer from the AcMNPV genome (Rodems and Friesen, 1993). Steady-state levels of *p35* transcription were reduced approximately 50%. This modest reduction may be influenced in part by the redundancy of *hr* elements within the genome.

The *p35* promoter also contains a consensus late promoter motif (TTAAG) at position −4 relative to the early +1 RNA start (Fig. 2C). Multiple early baculovirus genes contain late promoter elements that overlap early promoter sequences (Pullen and Friesen, 1995a; Guarino and Smith, 1992; Blissard and Rohrmann, 1991; Nissen and Friesen, 1989). Presumably, such an arrangement prolongs expression of genes required both early and late in infection. Indeed, the *p35* late promoter motif is activated during disappearance of the early *p35* transcripts (Nissen and Friesen, 1989). The overlap between the early and late promoter motifs raises the interesting possibility that viral late transcription factors contribute to the rapid decline in the early transcription due to promoter occlusion.

3. *gp64 efp* Promoter

The *gp64* gene encodes the major envelope glycoprotein of the budded virion, an approximately 67-kDa fusion protein (EFP) required for cell-to-cell transmission of budded virus (Monsma *et al.*, 1996). Expression of *gp64* is regulated by a combination of early and late promoters. The early composite ($TATA^+$, INR^+) promoter (Fig. 2D) has been well studied for OpMNPV by plasmid transfections and *in vitro* transcription assays (Kogan *et al.*, 1995; Kogan and Blissard, 1994; Blissard and Rohrmann, 1991). The TATA element exerts the predominant role in regulating the level of early *gp64* transcription. However, the basal *gp64* promoter also has a TATA-independent activity that is revealed upon elimination of the TATA element (position −29 relative to the +1 RNA start site). Upon TATA inactivation, accurate initiation occurs within the CAGT INR, albeit at much reduced levels (Kogan *et al.*, 1995). Besides the INR, the TATA-independent activity is stimulated by the DAR (positions +14 and +24), which contains a consensus CACNG motif (Fig. 2D). Although the functional significance of the TATA-independent activity is unknown, it may represent a general form of functional redundancy present within other early promoters (Kogan *et al.*, 1995). Such redundancy may insure proper expression of essential viral genes in diverse tissues or insect hosts. It is interesting in this regard that the general organization of promoters for *gp64* and *ie-1*, both essential viral genes, are so similar.

Although the *gp64* promoter is stimulated by the IE1 transregulator, it is also active in uninfected host cells, suggesting that, like the *ie-1* promoter, it is responsive to host transcription factors. Such factors likely recognize distinct motifs within the *gp64* UAR, which stimulates basal promoter activity (Kogan and Blissard, 1994). Two distinct regulatory motifs, GATA and CACGTG, have been identified in the *gp64* UAR (Fig. 2D). In plasmid transfection assays, the stimulatory effect of the two motifs is additive, since elimination of either motif reduced promoter activity by approximately 50%. Sequence-specific binding of host factors was demonstrated for both motifs by electrophoretic mobility shift assays (Kogan and Blissard, 1994). Moreover, mutagenesis of each motif eliminated host factor binding and reduced *gp64* promoter activity, demonstrating a correlation between factor interaction and transcriptional stimulation. The role of GATA and CACGTG motifs for *gp64* expression has not been determined in the context of the viral genome. However, the resemblance of the CACGTG motif to the functional CGT motif within the *p35* UAR (Dickson and Friesen, 1991) suggests that this motif influences promoter activity during infection.

IV. TRANSCRIPTIONAL ENHANCERS

A. Homologous Regions of AcMNPV

In addition to the *cis*-acting UAR elements of early promoters, the baculovirus genome contains multiple copies of repetitive and highly conserved sequences that function as transcriptional enhancers and possible origins of viral DNA replication. In the AcMNPV genome, these elements were first identified on the basis of their sequence similarity (Cochran and Faulkner, 1983) and later by their extraordinary capacity to enhance transcription of viral and heterologous promoters in a position- and orientation-independent manner (Guarino and Summers, 1986b). Sequence analysis of the AcMNPV genome (Ayres *et al.*, 1994) has revealed the presence of eight such homologous regions (*hr*s) designated *hr*1, *hr*1a, *hr*2, *hr*3, *hr*4a, *hr*4b, *hr*4c, *hr*5 (Fig. 1A). The *hr*s range in size from 30 to 800 bp and are distinguished by the number of palindromic repeats within each. Much of our understanding of transcriptional enhancement is derived from studies of the *hr*s of AcMNPV.

1. AcMNPV *hr* Structure and Function

Each of the *hr*s has the capacity to stimulate transcription of *cis*-linked early viral promoters in plasmid transfection assays (Lu and Carstens, 1993; Nissen and Friesen, 1989; Carson *et al.*, 1988; Guarino *et al.*, 1986; Guarino and Summers, 1986b). Enhancement is accomplished by increased transcription of the linked gene by RNA polymerase II, most likely through increased transcriptional initiation (Guarino and Summers, 1986b). The *hr* elements have a complex modular organization characterized by the presence of multiple 60- to 70-bp repeats, each of which contains a 28-bp imperfect palindrome bisected by an *Eco*RI site. Located downstream from the *p35* gene (Fig. 1B), the 484-bp *hr*5 element has an organization typical of the AcMNPV *hr*s. It possesses six direct repeats, each with a 28- to 30-bp palindrome (28-*mer*) flanked on either side by smaller approximately 20-bp direct repeats (Fig. 3). The minimal sequence element required for orientation- and position-independent enhancement of promoter activity by *hr*5 is the 28-*mer* (Rodems and Friesen, 1993). Neither divergent palindromic half-site, $28hs^L$ or $28hs^R$ (Fig. 3), is sufficient for enhancer activity (Leisy *et al.*, 1995; Guarino and Dong, 1994; Rodems and Friesen, 1993). Consistent with a modular organization of the *hr*s, repetition of the 28-*mer* increases promoter enhancement, although to a level less than that provided by the intact *hr*5. The 28-*mer* is also the minimal sequence required for plasmid DNA replication in transfection assays (Leisy *et al.*, 1995; Pearson *et al.*, 1992). The level of transient DNA replication is proportional to the number of 28-*mer* palindromes. These findings suggest that sequence requirements for *hr*-mediated transcriptional enhancement and activity as a DNA replication

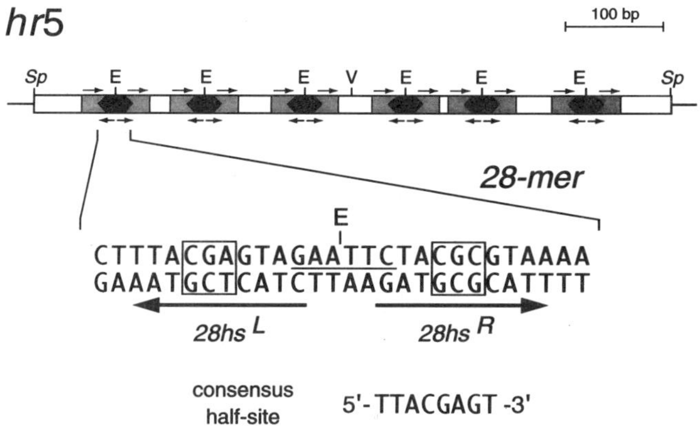

FIGURE 3. Structural organization of homologous region *hr*5. The 484-bp *hr*5 element contains six 60-bp repeats (shaded boxes), each of which includes an imperfect 28-*mer* palindrome (shaded divergent arrows) bisected by an *Eco*RI site. Shorter direct repeats (smaller arrows) are located on either side of the palindrome. The leftmost 28-*mer* (shown below) consists of two inverted half-sites ($28hs^L$ and $28hs^R$) separated by the *Eco*RI site. Nucleotides within each half-site required for 28-*mer* transcriptional enhancer activity are indicated within the boxes. The nucleotide sequence of the consensus 28-*mer* half-site is shown (Kool *et al.*, 1995). Abbreviations: R, *Eco*RI; Sp, *Ssp*I, V, *Eco*RV.

origin are similar, if not identical. Thus, the 28-*mer* may recruit factors that promote both early transcription and DNA replication. As origins of DNA replication (Pearson *et al.*, 1992), the *hr*s also stimulate late gene transcription in transfection assays by increasing reporter plasmid copy number (Lu and Miller, 1995), suggesting that they indirectly contribute to viral late gene expression through the same mechanism.

2. Protein Interactions with the *hr*s

Cis-linkage of *hr*5 to certain early viral promoters (including *p35*, 39K, *ie-2*, and *pl43*) stimulates transcription upon plasmid transfections of uninfected cells. Thus, host factors can mediate *hr* activity in the absence of virus-encoded factors. Nonetheless, *hr*-mediated enhancement is dramatically augmented by the viral transregulator IE1. This potent transactivator (see Section V.A) binds directly or indirectly to *hr* sequences as demonstrated by electrophoretic mobility shift assays (Choi and Guarino, 1995a; Leisy *et al.*, 1995; Rodems and Friesen, 1995; Kovacs *et al.*, 1992; Guarino and Dong, 1991). Nuclear extracts prepared from AcMNPV-infected cells also exhibit an *hr*-binding activity indistinguishable from that of *in vitro* synthesized IE1 (Rodems and Friesen, 1995). As demonstrated by multiple approaches, IE1 binds to the *hr* through interaction with the 28-*mer* (Leisy *et al.*, 1995; Rodems and Friesen, 1995; Guarino and Dong, 1994). Moreover, as predicted from the palindromic structure of the 28-*mer*, IE1 binds as a dimer (Rodems and Friesen, 1995). Optimal IE1 binding and *hr* enhancer activity requires properly spaced 28-*mer* half-sites. Interestingly, the conserved *Eco*RI site is not essential as long as half-site spacing is preserved. Collectively, these data indicate that IE1/28-*mer* interaction is required for IE1-mediated *hr* enhancer activity (Section V.A). Formal demonstration of direct interaction between IE1 and *hr* will require DNA binding by purified IE1.

3. *hr*-Mediated Transcriptional Enhancement during Infection

The unique repetition and distribution of the *hr*s within the AcMNPV genome (Fig. 1A) suggests that these elements play a critical role during virus replication. It has been shown that *hr*5 stimulates promoter-specific transcription within the AcMNPV genome during infection (Rodems and Friesen, 1993). Both the early *p35* and *ie-1* promoters were stimulated when inserted with a linked copy of *hr*5 into the genome of AcMNPV recombinants. In this case, the 28-*mers* was the minimal element required for promoter stimulation. Nonetheless, enhancement by full-length *hr*5 was greater than the sum of its 28-*mer*, suggesting that other *hr* sequences augment enhancer activity. During infection, *hr*5 had no effect on a late viral promoter or a heterologous promoter upon proximal linkage (Rodems and Friesen, 1993). Thus, direct *hr* enhancement may be limited to RNA polymerase II-responsive early promoters. Deletion of *hr*5 from its native position in the AcMNPV genome downstream from *p35* also reduced the steady-state levels of *p35* RNA by 50%. However, *hr*5 deletion had no obvious effect on virus replication in cultured cells (Rodems and Friesen, 1993). This finding is consistent with the predicted functional redundancy of the *hr*s within the viral genome for DNA replication (Pearson *et al.*, 1992).

B. Homologous Region Enhancers of OpMNPV

Homologous regions are dispersed throughout the genome of other baculoviruses (see Chapter 7, this volume). In the OpMNPV genome, a homologous region (0.8 kb) located downstream from the *ie-2* gene functions as a transcriptional enhancer in plasmid transfections (Theilmann and Stewart, 1992b). This OpMNPV enhancer (OpE) consists of 12 tandemly repeated 66-bp elements that only slightly resemble the AcMNPV *hr* repeats. OpE has the capacity to stimulate expression of the early OpMNPV *ie-2* and *pe38* promoters in a position- and orientation-independent manner (Wu *et al.*, 1993; Theilmann and Stewart, 1992b). Although, OpE-mediated enhancement was augmented by IE1, overall levels of stimulation were significantly lower than that conferred by AcMNPV *hr*s. Nonetheless, these findings suggest that enhancer activity by homologous regions are a general feature of baculovirus early gene regulation.

V. BACULOVIRUS-ENCODED EARLY TRANSCRIPTIONAL REGULATORS

A common strategy among nuclear-replicating DNA viruses of animals is to express potent transregulators that function by stimulating early viral gene expression during that period prior to DNA replication when viral DNA template concentrations are lowest. Such factors accelerate replication by insuring that the necessary components for viral DNA synthesis and viral takeover of the host biosynthetic machinery are present at the proper time and level. To date, four baculovirus transregulators have been identified by their capacity to stimulate early viral gene expression: IE0, IE1, IE2, and PE38. The most potent and best studied of these transregulators is IE1.

A. Immediate Early Transregulator IE1

1. IE1 Properties

Encoded by the immediate early gene *ie-1,* IE1 is conserved among the baculoviruses, a feature consistent with its central role in virus replication. Due to its capacity to *trans*-activate multiple early genes at high levels in plasmid transfection assays, IE1 is thought to be the principal transregulator of early baculovirus expression. Nonetheless, the exact role(s) of this multifunctional protein during virus multiplication remains to be determined.

As expected for an early transregulator, *ie-1* (Fig. 1C) is transcribed immediately after infection and its approximately 67-kDa gene product IE1 appears soon thereafter. The steady-state level of IE1 increases during infection (Choi and Guarino, 1995b; Ohresser *et al.,* 1994; Theilmann and Stewart, 1993), most likely the result of late *ie-1* promoter activity (Pullen and Friesen, 1995a). As a mediator of transcription, IE1 must localize to the nucleus of the infected cell. OpMNPV IE1 has also been detected in preparations of extracellular budded virus (Theilmann and Stewart, 1993), raising the possibility that IE1 delivered to the nucleus by viral nucleocapsids may accelerate early gene expression upon uncoating. As an apparent requirement for its activity as a transregulator, IE1 oligomerizes upon synthesis and binds DNA as a dimer (Rodems and Friesen, 1995). Indirect evidence suggests that IE1 is phosphorylated late in infection (Choi and Guarino, 1995a,b); however, the function of this posttranslational modification is unknown.

IE1 was originally identified by its capacity to *trans*-activate the AcMNPV early 39K (*pp31*) promoter in *ie-1* plasmid transfection assays (Guarino and Summers, 1986a, 1987). It is now known that many early baculovirus promoters, as well as heterologous promoters, can be affected to some extent by IE1 in these assays. IE1 potently stimulates transcription of promoters for the 39K, *p35, gp64, p143, dnapol, pe38, lef-1, lef-2,* and *lef-3*

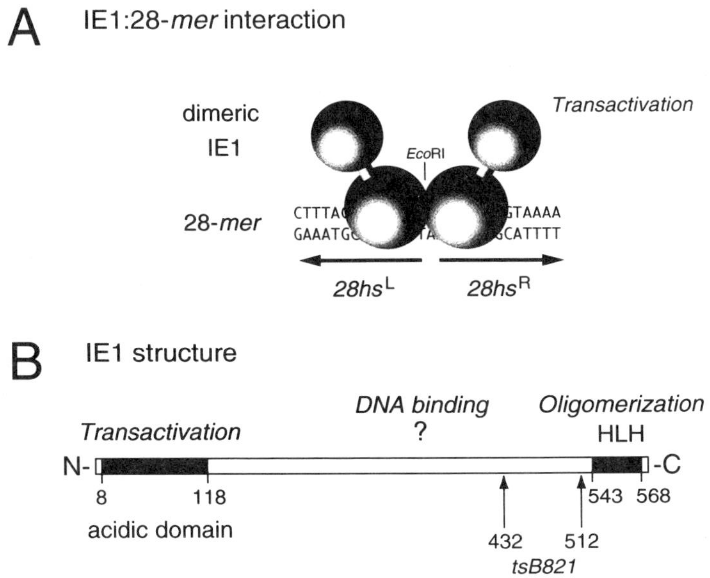

FIGURE 4. IE1 structure and function. (A) Simple model for IE1 interaction with a 28-*mer*. Dimeric IE1 binds to the 28-*mer* by simultaneous recognition of both inverted half-sites ($28hs^L$ and $28hr^R$). Oligomerization and DNA binding are mediated by C-terminal IE1 domains, whereas *trans*-activation is mediated by N-terminal domains. (B) IE1 functional domains. N-terminal IE1 residues 8 to 118, consisting of acidic stretches of residues, are required and sufficient for DNA-dependent IE1 *trans*-activation. IE1 oligomerization is mediated by a helix–loop–helix (HLH)-like domain that includes residues 543 to 568. Located to the C-terminal side of the *trans*-activation domain, the DNA binding domain(s) remains to be identified. Residues 432 and 512 are associated with the temperature-sensitive lesion within IE1 of AcMNPV mutant tsB821.

genes, as well as the *ie-1* promoter itself (Pullen and Friesen, 1995a; Ribeiro *et al.*, 1994; Lu and Carstens, 1993; Passarelli and Miller, 1993; Guarino and Smith, 1992; Blissard and Rohrmann, 1991; Kovacs *et al.*, 1991a; Theilmann and Stewart, 1991; Nissen and Friesen, 1989). No clearly recognizable IE1-response element has been identified within these viral promoters. In transfection assays, IE1 dramatically augments expression of genes that are *cis*-linked to the *hr* enhancers, as first demonstrated for the 39K promoter (Guarino *et al.*, 1986; Guarino and Summers, 1986b). The AcMNPV *p35* promoter is particularly responsive, since it exhibits a 3000-fold stimulation in the presence of IE1 and *hr*5 (Nissen and Friesen, 1989).

IE1 binding to the *hr* occurs through interaction with the 28-*mer* palindromes (Fig. 3) as shown by site-directed mutagenesis and competition with 28-*mer*-containing DNA probes (Choi and Guarino, 1995b; Leisy *et al.*, 1995; Rodems and Friesen, 1995; Guarino and Dong, 1991). IE1 binds as a dimer (Fig. 4A) by a mechanism that involves simultaneous recognition of both 28-*mer* half-sites (Rodems and Friesen, 1995). The finding that nucleotide replacements within the center of the 28-*mer* half-sites (consensus sequence 5′-TTACGAGT-3′) reduced IE1 binding and simultaneously reduced or eliminated position- and orientation-independent stimulation by the 28-*mer* suggests that IE1 *trans*-activation is the direct result of *hr* interaction (Rodems and Friesen, 1995). Although IE1 has the capacity to bind to a single half-site, IE1 stimulation of *hr* enhancer activity requires proper interaction with both half-sites (Leisy *et al.*, 1995; Guarino and Dong, 1994; Rodems and Friesen, 1993). This finding indicated that IE1 binding alone is not sufficient for *hr* activation and suggests that an undefined molecular event(s) is required subsequent to protein binding. Such events might include conformational changes in IE1 itself, an induced change in the structure of the *hr* that promotes transcriptional initiation in a position- and orientation-independent manner, or relief from transcriptional repression caused by chromatin structure. Alternatively, IE1 might stabilize host factor interactions with the *hr*, thereby stimulating transcriptional activity.

Current evidence suggests that IE1 *trans*-activation involves *hr* enhancer-dependent and DNA-independent mechanisms. In an enhancer-dependent model (Fig. 5A), the binding of IE1 to the 28-*mer* leads to transcriptional stimulation through direct or indirect protein contact with TFIID-associated factors of the preinitiation complex that could include TATA binding protein-associated factors (TAFs) or the basal transcription complex. Alternatively, IE1 could alter DNA or chromatin conformation, thereby stimulating transcriptional initiation at the affected promoter. *Cis*-linkage of *hr*5 to a promoter reduces the IE1 concentration required to obtain comparable levels of promoter activation in the absence of *hr*5 (Choi and Guarino, 1995b). This finding, combined with the lower IE1 levels early in infection, suggests that IE1 *trans*-activation of early viral genes is primarily mediated through the *hr*s within the viral genome. In a DNA-independent model (Fig. 5B), IE1 interacts directly or indirectly with the transcriptional initiation complex without binding to DNA in a sequence-specific manner. This model could explain the high degree of IE1-responsiveness of certain promoters in the absence of recognizable IE1 binding sites. Not ruled out is the possibility that IE1 interacts nonspecifically with transfected reporter plasmids causing promoter stimulation (Choi and Guarino, 1995b).

Although the physiological relevance is unknown, IE1 also has the capacity to repress transcription as demonstrated by decreased *ie-0* and *ie-2* promoter activity in transfection assays (Carson *et al.*, 1991b; Kovacs *et al.*, 1991a). IE1-mediated repression of *ie-2* may be due to the presence of a 28-*mer* half-site (consensus 5′-TTACGAGT-3′) located between the TATA and CAGT motifs that could bind IE1 such that it inhibits rather than simulates transcription (Leisy *et al.*, 1996). Consistent with this model is the finding that IE1 mutations that disrupt DNA binding failed to repress *ie-0* transcription, whereas IE1 mutations defective for *trans*-activation but capable of DNA binding repressed as efficiently as wild-type IE1 (Kovacs *et al.*, 1992). It is noteworthy that IE1 can stimulate (rather than repress) *ie-0*

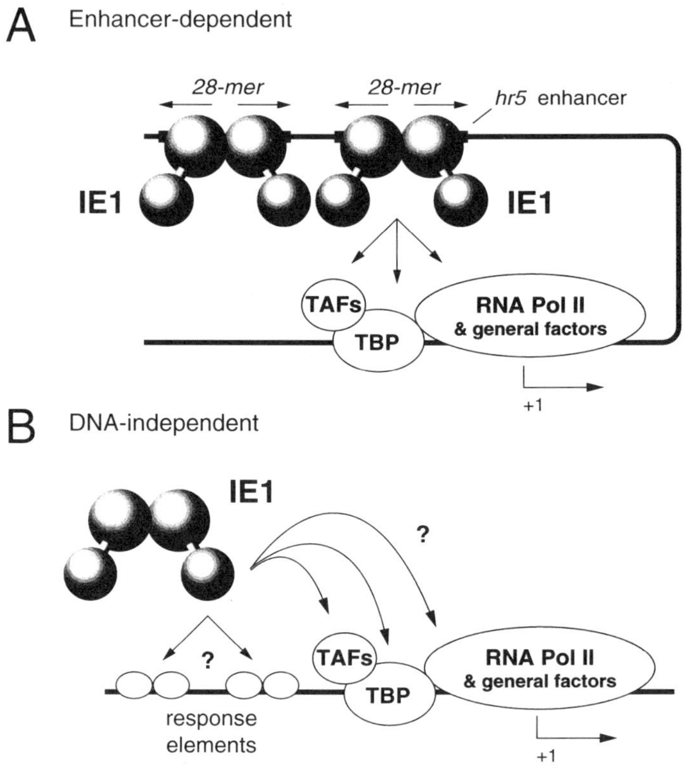

FIGURE 5. Models of IE1 *trans*-activation. (A) Enhancer (*hr*)-dependent *trans*-activation. Binding of dimeric IE1 to the palindromic 28-*mer* repeats within a distal *hr* element leads to IE1-mediated activation of the transcription initiation machinery at the affected promoter through mechanisms that facilitate assembly of the TFIID initiation complex or activate host-encoded TATA binding protein associated factors (TAFs). Alternatively, IE1 binding to the *hr* could alter DNA conformation that is transmitted to the promoter in a position-independent manner. (B) DNA-independent *trans*-activation. In the absence of DNA binding, IE1 causes direct stimulation of the transcriptional initiation complex at the promoter. IE1 could function by interacting with transcription factors bound to distal UAR response elements, TAFs, or TFIID directly.

promoter activity when *cis*-linked to *hr*5, suggesting that IE1 repression is context dependent (Riberio *et al.*, 1994).

2. IE1 Structure

The 582-residue IE1 protein from AcMNPV has a modular structure consisting of separable *trans*-activation and DNA binding domains (Fig. 4). Progressive deletions from either end of IE1 indicated that residues required for *trans*-activation and DNA (*hr*) binding are located within the N- and C-termini, respectively (Kovacs *et al.*, 1992). Moreover, N-terminal IE1 deletions that eliminated *trans*-activation retained DNA binding activity and indicated that these functions are separable, a property typical of transactivators. By using constructed GAL4-IE1 fusions in combination with an early viral promoter containing *cis*-linked GAL4 DNA binding sites, it was determined that IE1 residues 8 to 118 are sufficient for DNA-dependent *trans*-activation (Rodems *et al.*, 1997). This N-terminal domain consists of stretches of acidic residues (Fig. 4B) comprising the *trans*-activation domain of other viral transactivators (Kovacs *et al.*, 1992; Guarino and Summers, 1987). At its C-terminal end, IE1 contains a domain required for oligomerization and *hr* binding (Fig. 4B). A four–three hydrophobic repeat (residues 543 to 568) within a predicted helix–loop–helix domain (Kovacs *et al.*, 1992) is required for IE1 dimerization (Rodems *et al.*, 1997). However, the IE1 residues directly involved in 28-*mer* binding remain to be determined. Interestingly, the highest degree of similarity between AcMNPV and OpMNPV IE1 lies within the C-terminus, including the oligomerization domain (Theilmann and Stewart, 1991; Guarino and Summers, 1987). IE1 nuclear localization signals remain to be identified.

Dominant negative mutations of IE1 should prove useful in probing IE1 molecular mechanisms and identifying interacting factors. In transfection assays, C-terminal IE1 deletions that retained the N-terminal *trans*-activation domain dominantly interfered with the capacity of wild-type IE1 to stimulate *hr*-dependent promoter activity (Kovacs *et al.*, 1992). This finding suggested that the IE1 *trans*-activation domain(s) interacts with cellular factors required for *hr*-mediated enhancer activity. Likewise, dominant interfering mutations of IE1 that retain the C-terminal oligomerization domain have been identified (Pullen *et al.*, 1996). The interfering activity of these IE1 mutations suggested that IE1–IE1 interaction is required for *trans*-activation in a process that can be blocked by heterodimerization with *trans*-activation-defective IE1.

3. Roles of IE1 during Infection

Despite the requirement of IE1 in *trans*-activation and DNA replication assays involving plasmid transfections, information on the exact function of IE1 during baculovirus infection is limited. Characterization of AcMNPV mutant tsB821, carrying a temperature-sensitive mutation in *ie-1*, indicated that IE1 is required for proper timing of early gene expression and viral DNA replication (Ribeiro *et al.*, 1994; Miller *et al.*, 1983). The temperature-sensitive lesion is attributed to substitutions $Ala^{432} \rightarrow Val$ and $Asp^{512} \rightarrow Asn$ (Fig. 4B) that reduced but did not eliminate IE1-mediated *trans*-activation of multiple early viral promoters (*cis*-linked to *hr*5) at the nonpermissive temperature (Ribeiro *et al.*, 1994). The reduction in IE1 *trans*-activation was correlated with reduced *hr*5 binding activity at the nonpermissive temperature (Choi and Guarino, 1995c). Although the temperature-sensitive mutation altered conserved IE1 residues, the function of the domain affected is unknown.

Whereas tsB821 exhibits normal replication at the permissive temperature (23°C), viral DNA replication and budded virus production are delayed for approximately 12 hr at the nonpermissive temperature (33°). Viral late gene expression is also reduced (Choi and Guarino, 1995c; Ribeiro *et al.*, 1994). Surprisingly, tsB821-inoculated cells undergo a variety of responses including full infection, no infection, or apoptotic death in a process that may be cell cycle dependent (Ribeiro *et al.*, 1994). Thus, due to this varied response and the likelihood that the temperature-sensitive mutation is partially active, it is difficult to assign an infection-specific function to IE1.

Upon plasmid transfection of apoptosis-sensitive SF21 cells, *ie-1* has the capacity to induce apoptosis in a significant fraction of the cell population (Prikhod'ko and Miller, 1996). The IE1-induced apoptosis is inhibitable by baculovirus apoptotic suppressors *p35* and *iap* (see Chapter 10, this volume). Although the mechanism(s) by which transfected IE1 triggers apoptosis is unknown, possibilities include *trans*-activation of cellular death genes or effects on DNA replication and the cell cycle. It is unknown whether IE1 induces apoptosis in the context of baculovirus infection (Prikhod'ko and Miller, 1996). Nonetheless, these findings suggest that IE1 affects host cell homeostasis during infection.

B. Immediate Early Transregulator IE0

Encoded by the gene *ie-0* (Fig. 1C), IE0 is an immediate early transregulator with properties similar but distinguishable from those of IE1. The 636-residue IE0 is identical to IE1 (582 residues) except for 54 additional residues at the IE0 N-terminus (Chisholm and Henner, 1988). The extended N-terminus of IE0 is derived from an RNA splicing event in which a small exon located ~4 kb upstream from *ie-1* is fused with the *ie-1* open reading frame, producing the early *ie-0* mRNA (Fig. 1C). In general, RNA splicing occurs only rarely during baculovirus gene expression (Kovacs *et al.*, 1991b; Lübbert and Doerfler, 1984b). *ie-0* RNAs are detected immediately after infection, peak early (between 3 to 5 hr), and disappear later (Pullen and Friesen, 1995a; Kovacs *et al.*, 1991a; Chisholm and Henner, 1988). Thus, the temporal expression of *ie-0* is consistent with early function. As determined by transfection assays, IE0 is a *trans*-activator capable of stimulating transcription from the *ie-1* and 39K promoters, but not from its own promoter (Kovacs *et al.*, 1991a). Unlike IE1, IE0-mediated *trans*-activation of the 39K promoter requires *cis*-linkage to the *hr*5 enhancer. It is unclear whether this *hr* requirement extends to other IE0-responsive early promoters. Consistent with the capacity of IE0 to stimulate *hr*-linked promoters, *in vitro* synthesized IE0 exhibits *hr* binding activity as judged by normal interaction with 28-*mer*-containing DNA probes in electrophoretic mobility shift assays (Rodems *et al.*, 1996). The role of IE0 in regulating viral replication during infection is unknown.

C. Immediate Early Transregulator IE2

ie-2 was also identified during a screen for AcMNPV genes (Fig. 1C) capable of *trans*-activation of the 39K promoter in plasmid transfection assays (Carson *et al.*, 1988). Its gene product, IE2 (formerly designated IE-N), is a 47-kDa nuclear-associated protein that stimulates transcription of several early baculovirus promoters, including its own. However, the level of stimulation (from two- to threefold) is significantly lower than that conferred by IE1. In transfection assays, IE2 stimulates plasmid DNA replication probably through the indirect *trans*-activation of genes required for replication (Kool *et al.*, 1995). Current evidence indicates that IE2 affects the rate of transcription through the function of an acidic domain (residues 198 to 206) at its C-terminus (Yoo and Guarino, 1994a,b). IE2 contains a number of additional motifs typically found in transcription regulators (Carson *et al.*, 1988), including a leucine zipper and a RING finger. Since there is no evidence yet for DNA binding by IE2, it has been suggested that IE2 stimulates transcription indirectly through interaction with auxiliary transcription factors (Yoo and Guarino, 1994b). IE2 is conserved between baculoviruses, including OpMNPV (Theilmann and Stewart, 1992a). During AcMNPV and OpMNPV infection, IE2 is synthesized early and disappears later, suggestive of early regulatory function (Krappa *et al.*, 1995; Stewart and Theilmann, 1993). IE2 and the transregulator PE38 colocalize to punctate nuclear structures in transfected cells (Krappa *et al.*, 1995). However, the exact role of IE2 during infection remains to be determined.

D. Transactivator PE38

The immediate early *pe38* (or *p34*) gene (Fig. 1C) encodes the 38-kDa nuclear protein PE38 (Wu *et al.*, 1993; Krappa and Knebel-Mörsdorf, 1991). Sequence analysis predicts that PE38 contains an N-terminal RING finger and a C-terminal leucine zipper motif, typical of transcriptional regulators. Transcription of *pe38* is initiated from the divergent *pe38* and *ie-2* promoter region of AcMNPV (Fig. 1C) immediately after infection, producing RNA transcripts that peak 12 hr after infection and disappear thereafter. Consistent with an early replicative function, PE38 is selectively synthesized early in infection (Krappa *et al.*, 1995, Wu *et al.*, 1993). When produced in uninfected cells by plasmid transfection, PE38 accumulates in punctate nuclear structures in a pattern similar to that of IE2. On the basis of stimulation of early promoter activity in transfection assays, PE38 is a viral transregulator (Lu and Carstens, 1993; Passarelli and Miller, 1993; Wu *et al.*, 1993). PE38 stimulated expression from the *p143* helicase promoter (~fourfold), but had no effect on the 39K (pp31) promoter (Lu and Carstens, 1993); in contrast, both promoters are highly responsive to IE1 *trans*-activation. Although additional PE38-responsive genes need identification, these findings suggest that PE38 exhibits a restricted *trans*-activation range compared to IE1. Potential PE38 response elements within the *p143* promoter have not been identified, nor is the function of PE38 during infection known.

VI. APPLICATIONS: EARLY BACULOVIRUS PROMOTERS FOR FOREIGN GENE EXPRESSION

As a direct result of the molecular characterization of early viral promoters, baculovirus promoter-based vectors have been developed for the expression of foreign genes in uninfected insect cells. In particular, the AcMNPV *ie-1* and OpMNPV *gp64 efp* promoters have been exploited for constitutive expression in stably transfected lepidopteran cells (Monsma *et al.*, 1996; Cartier *et al.*, 1994; Jarvis *et al.*, 1990). Plasmid expression vectors have been designed in which the *ie-1* promoter plus the associated *ie-1* UAR directs high-level expression of inserted foreign genes in response to host transcription factors. Improved *ie-1* expression vectors have been designed (Fig. 6A) in which the full-length *ie-1* promoter plus its associated DAR is linked to the *hr*5 enhancer (Cartier *et al.*, 1994). These *hr*5–*ie-1* promoter-based expression vectors provide high-level, constitutive expression not

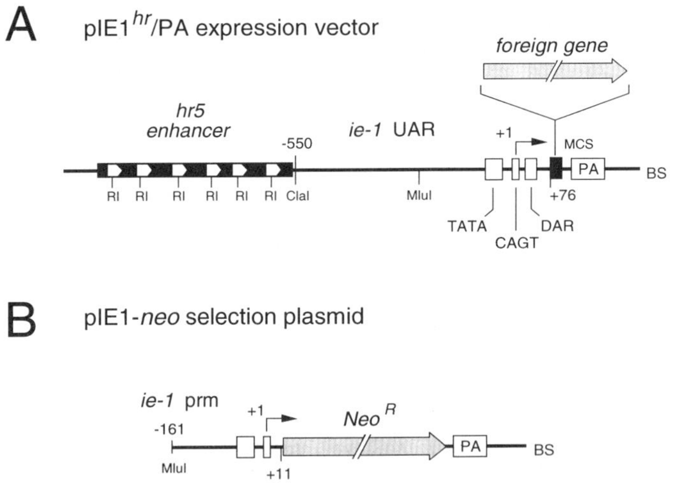

FIGURE 6. Immediate early expression vectors for insect cells. (A) Expression vector pIE1^{hr}/PA. Foreign gene expression is directed by the *ie-1* promoter consisting of sequences from −550 to +76 that include the core promoter elements TATA, INR, and DAR, along with the *ie-1* upstream activating region (UAR). Foreign genes are inserted at the multiple cloning site (MCS) located upstream from a polyadenylation signal (PA). The *hr*5 enhancer is positioned upstream from the *ie-1* promoter. (B) pIE1-*neo*^{R} plasmid. The truncated *ie-1* promoter (−161 to +11) directs expression of the neomycin resistance gene in transfected cells. Both plasmids are Bluescript (BS)-based vectors (Cartier *et al.*, 1994).

only in lepidopteran cells, but also in cultured dipteran cells, including those from *Aedes albopictus* mosquitoes and *Drosophila melanogaster* (Shotkoski *et al.*, 1996; Harper, 1995). The versatility of these expression vectors in diverse insects suggests that the *ie-1* promoter is responsive to highly conserved invertebrate transcription factors. Stable transfection of cultured lepidopteran cells is accomplished by transfection of the *ie-1* promoter-containing vector along with a neomycin-resistance vector (Fig. 6B) also under control of the *ie-1* promoter for selection of G418-resistant cells (Jarvis *et al.*, 1990). Although the physical status of the vector DNA (e.g., integrated or episomal) is unknown in stably transfected cells, the *ie-1* promoter remains responsive to viral *trans*-activation, since expression is boosted upon baculovirus infection (Cartier *et al.*, 1994; Jarvis, 1993).

Constitutive expression of foreign or viral genes in stably transfected invertebrate cells has provided new opportunities to investigate both host cell and viral replicative events. Included to date are studies on glycosylation and secretory pathways in insect cells (Jarvis *et al.*, 1990), mechanisms of apoptotic regulation (Cartier *et al.*, 1994), the physiology of ion channel action in invertebrate cells (Shotkoski *et al.*, 1996), and the generation of helper cell lines for the propagation of novel baculovirus mutants (Monsma *et al.*, 1996).

VII. SUMMARY

Proper expression of baculovirus early genes is critical to the replication success of these unusually prolific DNA viruses. The baculoviruses have therefore evolved mechanisms that exploit the host cell transcriptional machinery to expedite immediate and regulated levels of early gene expression. Early transcription is accomplished in part by the use of promoters that closely resemble RNA polymerase II-responsive promoters of the host. As such, molecular studies of these baculovirus early promoters continue to provide important insight into the regulation of transcription and gene expression of invertebrates. In a strategy that accelerates early gene expression and prepares the host cell for viral replication, the baculoviruses employ potent viral transregulators that potentiate transcription through mechanisms that remain to be determined. These viruses also make use of a novel arrangement of repetitive DNA elements (*hr*s) that function as efficient early transcriptional enhancers and possible origins of DNA replication. Further investigation of the molecular interaction between the *hr*s and virus-encoded transregulators should provide increased understanding of essential replication events. Moreover, such studies will likely uncover novel mechanisms for eukaryotic transcriptional enhancement and virus–host interactions. Last, strategically designed early baculovirus promoters for foreign gene expression in insects have been a direct outcome of baculovirus transcriptional studies. These virus-based vectors should continue to have numerous applications, including, for example, use in development of genetic transformation systems for nondrosophilid insects.

ACKNOWLEDGMENTS. P. D. F. wishes to thank the members of his laboratory, in particular Steve Rodems, Steve Pullen, Mark Nissen, Julie Dickson, Pam Hershberger, and Vicki Olson, who have contributed to studies on early baculovirus gene expression. Steve Pullen and Steve Rodems also contributed to several figures herein. The author acknowledges support from the National Institutes of Allergy and Infectious Diseases through Public Health Service grant AI25557 and USDA Federal Hatch support through the Wisconsin Agricultural Experimental Station (WIS03367).

REFERENCES

Ayres, M. D., Howard, S. C., Kuzio, J., Lopez, F. M., and Possee, R. D., 1994, The complete DNA sequence of *Autographa californica* nuclear polyhedrosis virus, *Virology* **202:**586–605.

Bertin, J., Mendrysa, S. M., LaCount, D. J., Guar, S., Krebs, J. F., Armstrong, R. C., Tomaselli, K. J., and Friesen, P. D., 1996, Apoptotic suppression by baculovirus P35 involves cleavage by and inhibition of a virus-induced CED-3/ICE-like protease, *J. Virol.* **70:**6251–6259.

Blissard, G. W., and Rohrmann, G. F., 1991, Baculovirus gp64 gene expression: Analysis of sequences modulating early transcription and transactivation by IE1, *J. Virol.* **65:**5820–5827.

Blissard, G. W., Kogan, P. H., Wei, R., and Rohrmann, G. F., 1992, A synthetic early promoter from a baculovirus: Roles of the TATA box and conserved start site of CAGT sequence in basal levels of transcription, *Virology* **190:**783–793.

Burley, S. K., and Roeder, R. G., 1996, Biochemistry and structural biology of transcription factor IID (TFIID), *Annu. Rev. Biochem.* **65:**769–799.

Carson, D. D., Guarino, L. A., and Summers, M. D., 1988, Functional mapping of an AcNPV immediately early gene which augments expression of the ie-1 trans-activated 39K gene, *Virology* **162:**444–451.

Carson, D. D., Summers, M. D., and Guarino, L. A., 1991a, Molecular analysis of a baculovirus regulatory gene, *Virology* **182:**279–286.

Carson, D. D., Summers, M. D., and Guarino, L. A., 1991b, Transient expression of the *Autographa californica* nuclear polyhedrosis virus immediate-early gene, IE-N, is regulated by three viral elements, *J. Virol.* **65:**945–951.

Cartier, J. L., Hershberger, P. A., and Friesen, P. D., 1994, Suppression of apoptosis in insect cells stably transfected with baculovirus *p35*: Dominant interference by N-terminal sequences *p35*$^{1-76}$, *J. Virol.* **68:**7728–7737.

Cherbas, L., and Cherbas, P., 1993, The arthropod initiator: The capsite consensus plays an important role in transcription, *Insect Biochem. Molec. Biol.* **23:**81–90.

Chisholm, G. E., and Henner, D. J., 1988, Multiple early transcripts and splicing of the *Autographa californica* nuclear polyhedrosis virus *ie-1* gene, *J. Virol.* **62:**3193–3200.

Choi, J., and Guarino, L. A., 1995a, The baculovirus transactivator IE1 binds to viral enhancer elements in the absence of insect cell factors, *J. Virol.* **69:**4548–4551.

Choi, J., and Guarino, L. A., 1995b, Expression of the IE1 transactivator of *Autographa californica* nuclear polyhedrosis virus during viral infection, *Virology* **209:**99–107.

Choi, J., and Guarino, L. A., 1995c, A temperature-sensitive IE1 protein of *Autographa californica* nuclear polyhedrosis virus has altered transactivation and DNA binding activities, *Virology* **209:**90–98.

Clem, R. J., and Miller, L. K., 1993, Apoptosis reduces both the *in vitro* replication and the *in vivo* infectivity of a baculovirus, *J. Virol.* **67:**3730–3738.

Clem, R. J., Fechheimer, M., and Miller, L. K., 1991, Prevention of apoptosis by a baculovirus gene during infection of insect cells, *Science* **254:**1388–1390.

Cochran, M. A., and Faulkner, P., 1983, Location of homologous DNA sequences interspersed at five regions in the baculovirus AcMNPV genome, *J. Virol.* **45:**961–970.

Dickson, J. A., and Friesen, P. D., 1991, Identification of upstream promoter elements mediating early transcription from the 35,000-molecular-weight protein gene of *Autographa californica* nuclear polyhedrosis virus, *J. Virol.* **65:**4006–4016.

Friesen, P. D., and Miller, L. K., 1985, Temporal regulation of baculovirus RNA: Overlapping early and late transcripts. *J. Virol.* **54:**392–400.

Friesen, P. D., and Miller, L. K., 1987, Divergent transcription of early 35- and 94-kilodalton protein genes encoded by the *Hin*dIII K genome fragment of the baculovirus *Autographa californica* nuclear polyhedrosis virus, *J. Virol.* **61:**2264–2272.

Glocker, B., Hoopes, R. J., and Rohrmann, G. F., 1992, *In vitro* transactivation of baculovirus early genes by nuclear extracts from *Autographa californica* nuclear polyhedrosis virus-infected *Spodoptera frugiperda* cells, *J. Virol.* **66:**3476–3484.

Grula, M. A., Buller, P. L., and Weaver, R. F., 1981, α-Amanitin-resistant viral RNA synthesis in nuclei isolated from nuclear polyhedrosis virus-infected *Heliothis zea* larvae and *Spodoptera frugiperda* cells, *J. Virol.* **38:**916–921.

Guarino, L. A., and Dong, W., 1991, Expression of an enhancer-binding protein in insect cells transfected with the *Autographa californica* nuclear polyhedrosis virus IE1 gene, *J. Virol.* **65:**3676–3680.

Guarino, L. A., and Dong, W., 1994, Functional dissection of the *Autographa californica* nuclear polyhedrosis virus enhancer element hr5, *Virology* **200:**328–335.

Guarino, L. A., and Smith, M., 1992, Regulation of delayed-early gene transcription by dual TATA boxes, *J. Virol.* **66:**3733–3739.

Guarino, L. A., and Summers, M. D., 1986a, Functional mapping of a *trans*-activating gene required for expression of a baculovirus delayed-early gene, *J. Virol.* **57:**563–571.

Guarino, L. A., and Summers, M. D., 1986b, Interspersed homologous DNA of *Autographa californica* nuclear polyhedrosis virus enhances delayed-early gene expression, *J. Virol.* **60:**215–233.

Guarino, L. A., and Summers, M. D., 1987, Nucleotide sequence and temporal expression of a baculovirus regulatory gene, *J. Virol.* **61:**2091–2099.

Guarino, L. A., Gonzalez, M. A., and Summers, M. D., 1986, Complete sequence and enhancer function of the homologous DNA regions of *Autographa californica* nuclear polyhedrosis virus, *J. Virol.* **60:**224–229.

Harper, T., 1995, *Characterization of the Proteins Encoded by Nodavival RNA3*, Ph.D. thesis, University of Wisconsin–Madison.

Hershberger, P. A., LaCount, D. J., and Friesen, P. D., 1994, The apoptotic suppressor P35 is required early during baculovirus replication and is targeted to the cytosol of infected cells, *J. Virol.* **68:**3467–3477.

Hoopes, R. J., and Rohrmann, G. F., 1991, *In vitro* transcription of baculovirus immediate early genes: Accurate mRNA initiation by nuclear extracts from both insect and human cells, *Proc. Natl. Acad. Sci. USA* **88:**4513–4517.

Huh, N. E., and Weaver, R. F., 1990a, Categorizing some early and late transcripts directed by the *Autographa californica* nuclear polyhedrosis virus, *J. Gen. Virol.* **71:**2195–2200.

Huh, N. E., and Weaver, R. F., 1990b, Identifying the RNA polymerases that synthesize specific transcripts of the *Autographa californica* nuclear polyhedrosis virus, *J. Gen. Virol.* **71:**195–202.

Hultmark, D., Klemenz, R., and Gehring, W. J., 1986, Translational and transcriptional control elements in the untranslated leader of the heat-shock gene hsp22, *Cell* **44:**429–438.

Jarvis, D. L., 1993, Effects of baculovirus infection on IE1-mediated foreign gene expression in stably transformed insect cells, *J. Virol.* **67:**2583–2591.

Jarvis, D. L., Fleming, J. G. W., Kovacs, G. R., Summers, M. D., and Guarino, L. A., 1990, Use of early baculovirus promoters for continuous expression and efficient processing of foreign gene products in stably transformed lepidopteran cells, *Bio/Technology* **8:**950–955.

Jun-Chuan, Q., and Weaver, R. F., 1982, Capping of viral RNA in cultured *Spodoptera frugiper-*

da cells infected with *Autographa californica* nuclear polyhedrosis virus, *J. Virol.* **43:**234–240.

Kogan, P. H., and Blissard, G. W., 1994, A baculovirus gp64 early promoter is activated by host transcription factor binding to CACGTG and GATA elements, *J. Virol.* **68:**813–822.

Kogan, P. H., Chen, X., and Blissard, G. W., 1995, Overlapping TATA-dependent and TATA-independent early promoter activities in the baculovirus gp64 envelope fusion protein gene, *J. Virol.* **69:**1452–1461.

Kool, M., Ahrens, C. H., Vlak, J. M., and Rohrmann, G. F., 1995, Replication of baculovirus DNA, *J. Gen. Virol.* **76:**2103–2118.

Kovacs, G. R., Guarino, L. A., and Summers, M. D., 1991a, Novel regulatory properties of the Ie1 and IE0 transactivators encoded by the baculovirus *Autographa californica* multicapsid nuclear polyhedrosis virus, *J. Virol.* **65:**5281–5288.

Kovacs, G. R., Guarino, L. A., Graham, B. L., and Summers, M.D., 1991b, Identification of spliced baculovirus RNAs expressed late in infection, *Virology* **185:**633–643.

Kovacs, G. R., Choi, J., Guarino, L. A., and Summers, M. D., 1992, Functional dissection of the *Autographa californica* nuclear polyhedrosis virus immediate-early 1 transcriptional regulatory protein, *J. Virol.* **66:**7429–7437.

Krappa, R., and Knebel-Mörsdorf, D., 1991, Identification of the very early transcribed baculovirus gene PE-38, *J. Virol.* **65:**805–812.

Krappa, R., Behn, K. A., Jahnel, F., Doerfler, W., and Knebel-Mörsdorf, D., 1992, Differential factor binding at the promoter of early baculovirus gene PE38 during viral infection: GATA motif is recognized by an insect protein, *J. Virol.* **66:**3494–3503.

Krappa, R., Roncarati, R., and Knebel-Mörsdorf, D., 1995, Expression of PE38 and IE2; viral members of the C_3HC_4 finger family, during baculovirus infection: PE38 and IE2 localize to distinct nuclear regions. *J. Virol.* **69:**5287–5293.

LaCount, D. J., and Friesen, P. D., 1997, The role of early and late replication events in the induction of apoptosis by baculoviruses, *J. Virol.* **71:**1530–1537.

Leisy, D. J., Rasmussen, C., Kim, H.-T., and Rohrmann, G. F., 1995, The *Autographa californica* nuclear polyhedrosis virus homologous region 1a: Identical sequences are essential for DNA replication activity and transcriptional enhancer function, *Virology* **208:**742–752.

Leisy, D. J., Rasmussen, C., and Rohrmann, G. F., 1996, American Society for Virology, Abstracts.

Lerch, R. A., and Friesen, P. D., 1993, The 35-kilodalton protein gene (*p35*) of *Autographa californica* nuclear polyhedrosis virus and the neomycin resistance gene provide dominant selection of recombinant baculoviruses, *Nucleic Acids Res.* **21:**1753–1760 (Erratum, **21:**2962).

Lu, A., and Carstens, E. B., 1993, Immediate-early baculovirus genes transactivate the p143 gene promoter of *Autographa californica* nuclear polyhedrosis virus, *Virology* **195:**710–718.

Lu, A., and Miller, L. K., 1995, The roles of eighteen baculovirus late expression factor gene in transcription and DNA replication, *J. Virol.* **69:**975–982.

Lübbert, H., and Doerfler, W., 1984a, Transcription of overlapping sets of RNAs from the genome of *Autographa californica* nuclear polyhedrosis virus: A novel method for mapping RNAs, *J. Virol.* **52:**255–265.

Lübbert, H., and Doerfler, W., 1984b, Mapping of early and late transcripts encoded by the *Autographa californica* nuclear polyhedrosis virus: Is viral RNA spliced? *J. Virol.* **50:**497–506.

Miller, L. K., Trimarchi, R., Browne, D., and Pennock, G. D. A., 1983, A temperature sensitive mutant of the baculovirus *Autographa californica* nuclear polyhedrosis virus defective in early function required for further gene expression, *Virology* **126:**376–380.

Monsma, S. A., Oomens, A. G. P., and Blissard, G. W., 1996, The GP64 envelope fusion protein is an essential baculovirus protein required for cell-to-cell transmission of infection, *J. Virol.* **70:**4607–4616.

Nissen, M. S., and Friesen, P. D., 1989, Molecular analysis of the transcriptional regulatory region of an early baculovirus gene, *J. Virol.* **63:**493–503.
Novina, C. D., and Roy, A. L., 1996, Core promoters and transcriptional control, *Trends Gent.* **12:**351–355.
Ohresser, M., Morin, N., Cerutti, M., and Delsert, C., 1994, Temporal regulation of a complex and unconventional promoter by viral products, *J. Virol.* **68:**2589–2597.
Ooi, B. G., and Miller, L. K., 1988, Regulation of host RNA levels during baculovirus infection, *Virology* **166:**515–523.
Passarelli, A. L., and Miller, L. K., 1993, Three baculovirus genes involved in late and very late gene expression: *ie-1, ie-n,* and *lef-2, J. Virol.* **67:**2149–2158.
Pearson, M., Bjornson, R., Pearson, G., and Rohrmann, G., 1992, The *Autographa californica* baculovirus genome: Evidence for multiple replication origins, *Science* **257:**1382–1384.
Phillips, S. E. V., 1993, Saddling up that TATA box, *Curr. Biol.* **3:**112–114.
Prikhod'ko, E. A., and Miller, L. K., 1996, Induction of apoptosis by baculovirus transactivator IE1, *J. Virol.* **70:**7116–7124.
Pullen, S. S., 1996, Regulation of expression and functional analysis of the baculovirus transregulator IE1, Ph.D. thesis, University of Wisconsin–Madison.
Pullen, S. S., and Friesen, P. D., 1995a, Early transcription of the ie-1 transregulator gene of *Autographa californica* nuclear polyhedrosis virus is regulated by DNA sequences within its 5′ noncoding leader region, *J. Virol.* **69:**156–165.
Pullen, S. S., and Friesen, P. D., 1995b, The CAGT motif functions as an initiator element during early transcription of the baculovirus transregulator *ie-1, J. Virol.* **69:**3575–3583.
Rasmussen, C., and Rohrmann, G. F., 1994, Characterization of the *Spodoptera frugiperda* TATA-binding protein: Nucleotide sequence and response to baculovirus infection, *Insect Biochem. Mol. Biol.* **24:**699–708.
Ribeiro, B. M., Hutchinson, K., and Miller, L. K., 1994, A mutant baculovirus with a temperature-sensitive Ie-1 transregulatory protein, *J. Virol.* **68:**1075–1084.
Rice, W. C., and Miller, L. K., 1986, Baculovirus transcription in the presence of inhibitors and in nonimpressive *Drosophila* cells, *Virus Res.* **6:**155–172.
Rodems, S. M., and Friesen, P. D., 1993, The hr5 transcriptional enhancer stimulates early expression from the *Autographa californica* nuclear polyhedrosis virus genome but is not required for virus replication, *J. Virol.* **67:**5776–5785.
Rodems, S. M., and Friesen, P. D., 1995, Transcriptional enhancer activity of *hr*5 requires dual palindrome half-sites that mediate binding of a dimeric form of the baculovirus transregulator IE1, *J. Virol.* **69:**5368–5375.
Rodems, S. M., Pullen, S. S., and Friesen, P. D., 1997, submitted.
Shotkoski, F., Zhang, H. G., Jackson, M. B., and ffrench-Constant, R. H., 1996, Stable expression of insect GABA receptors in insect cell lines. Promoters for efficient expression of *Drosophila* and mosquito Rdl GABA receptors in stably transformed mosquito cells, *FEBS Lett.* **380:**257–262.
Theilmann, D. A., and Stewart, S., 1991, Identification and characterization of the IE-1 gene of *Orgyia pseudotsugata* multicapsid nuclear polyhedrosis virus, *Virology* **180:**492–508.
Theilmann, D. A., and Stewart, S., 1992a, Molecular analysis of the *trans*-activating IE-2 gene of *Orgyia pseudotsugata* multicapsid nuclear polyhedrosis virus, *Virology* **187:**84–96.
Theilmann, D. A., and Stewart, S., 1992b, Tandemly repeated sequence at the 3′ end of the IE-2 gene of the baculovirus *Orgyia pseudotsugata* multicapsid nuclear polyhedrosis virus is an enhancer element, *Virology* **187:**97–106.
Theilmann, D. A., and Stewart, S., 1993, Analysis of the *Orgyia pseudotsugata* multicapsid nuclear polyhedrosis virus trans-activators IE-1 and IE-2 using monoclonal antibodies, *J. Gen. Virol.* **74:**1819–1826.
Tomalski, M. D., Wu, J. G., and Miller, L. K., 1988, The location, sequence, transcription, and regulation of a baculovirus DNA polymerase gene, *Virology* **167:**591–600.

Wilson, M. E., and Miller, L. K., 1986, Changes in the nucleoprotein complexes of a baculovirus DNA during infection, *Virology* **151**:315–328.

Wu, X., Stewart, S., and Theilmann, D. A., 1993, Alternative transcriptional initiation as a novel mechanism for regulating expression of a baculovirus *trans* activator, *J. Virol.* **67**:5833–5842.

Wu, Y., and Carstens, E. B., 1996, Initiation of baculovirus replication: Early promoter regions can function as infection-dependent replicating sequences in a plasmid-based replication assay, *J. Virol.* **70**:6967–6972.

Yoo, S., and Guarino, L. A., 1994a, Functional dissection of the ie2 gene product of the baculovirus *Autographa californica* nuclear polyhedrosis virus, *Virology* **202**:164–172.

Yoo, S., and Guarino, L. A., 1994b, The *Autographa californica* nuclear polyhedrosis virus *ie-2* gene encodes a transcriptional regulator, *Virology* **202**:746–753.

Zawel, L., and Reinberg, D., 1995, Common themes in assembly and function of eukaryotic transcription complexes, *Annu. Rev. Biochem.* **64**:533–561.

CHAPTER 7

Baculovirus DNA Replication

ALBERT LU, PETER J. KRELL, JUST M. VLAK, AND GEORGE F. ROHRMANN

I. INTRODUCTION

Baculovirus replication initiates a cascade of gene expression that ultimately results in the production of progeny virus. This cascade is regulated at different points during the replication cycle: early gene expression is regulated by the interaction of *cis*-acting elements, viral *trans*-acting factors, and host factors (see Chapter 6, this volume), and late/very late gene expression is regulated by a combination of viral DNA replication, *cis*-acting elements as well as early and late viral factors (see Chapter 8, this volume).

Most of the initial characterization of baculovirus DNA replication has been done using *Autographa californica* nuclear polyhedrosis virus (AcMNPV). In *Spodoptera frugiperda* cells infected with AcMNPV, DNA replication is detected by 6 hr postinfection (p.i.) and continues until about 18 hr p.i., after which the level of DNA replication declines (Tjia *et al.*, 1979). Purified viral DNA is infectious when transfected into uninfected insect cells (Burand *et al.*, 1980; Carstens *et al.*, 1980; Potter and Miller, 1980), demonstrating that proteins associated with mature virions are not essential for viral DNA replication. In addition, viral DNA synthesis was

ALBERT LU • DuPont Agricultural Products, DuPont Stine-Haskell Research Center, Newark, Delaware 19714-0030. PETER J. KRELL • Department of Microbiology, University of Guelph, Guelph, Ontario N1G 2W1 Canada. JUST M. VLAK • Department of Virology, Wageningen Agricultural University, 6709 PD Wageningen, The Netherlands. GEORGE F. ROHRMANN • Department of Microbiology, Oregon State University, Corvallis, Oregon 97331-7301.

The Baculoviruses, edited by Lois K. Miller. Plenum Press, New York, 1997.

shown to be dependent on prior viral protein synthesis (Erlandson *et al.*, 1985; Gordon and Carstens, 1984; Wang and Kelly, 1983; Kelly, 1982; Kelly and Lescott, 1976), suggesting that baculovirus gene products are needed for DNA replication. Early studies on baculovirus-infected cells demonstrated the presence of a novel DNA polymerase activity that is distinct from host cell DNA polymerases (Mikhailov *et al.*, 1986; Wang and Kelly, 1983; Miller *et al.*, 1981). The identification of temperature-sensitive AcMNPV mutants defective in DNA replication also suggested that baculoviruses encode their own replicative proteins (Brown *et al.*, 1979; Lee and Miller, 1978).

Cis- and *trans*-acting elements associated with DNA replication have been identified in the genomes of several baculoviruses using a number of different strategies. Putative *cis*-acting elements that may be involved in the initiation of DNA synthesis have been identified by the characterization of defective viral genomes generated by serial passage of the virus in tissue culture (Kool *et al.*, 1991, 1993a; Lee and Krell, 1992) and from transient replication studies that rely on the ability of plasmids containing these elements to be amplified when transfected into baculovirus-infected cells (Ahrens *et al.*, 1995a; Leisy *et al.*, 1995; Pearson and Rohrmann, 1995; Xie *et al.*, 1995; Lee and Krell, 1994; Leisy and Rohrmann, 1993; Pearson *et al.*, 1992, 1993). Viral genes involved in DNA replication have been identified by the isolation and characterization of conditional–lethal mutants defective in viral DNA replication (Ribeiro *et al.*, 1994; Lu and Carstens, 1991; Gordon and Carstens, 1984; Brown *et al.*, 1979; Lee and Miller, 1978) and by transient replication assays in which segments of baculovirus genomes are tested for their ability to support replication of cotransfected origin-containing plasmid DNA (Ahrens and Rohrmann, 1995a,b; Ahrens *et al.*, 1995b; Lu and Miller, 1995a; Kool *et al.*, 1994a,c).

This work has led to the identification and characterization of a number of putative origin sequences as well as a set of baculovirus genes that are likely to be involved in the replication of baculovirus DNA. Some of the functions of these gene products can be inferred by homology to other replicative proteins, while the function of the others are currently unknown. In this chapter, we review our knowledge of those structural elements and viral gene products of nucleopolyhedroviruses that have been found to be involved in the replication of baculovirus DNA.

II. *CIS*-ACTING ELEMENTS

A. Defective Genomes and DNA Replication

Insight into the identification of possible replication origins was facilitated by the analysis of defective genomes of AcMNPV that arise from serial undiluted passage of the virus in cell culture (Lee and Krell, 1992, 1994; Kool *et al.*, 1993a). These defective viruses are propagated along with helper wild-type virus and gradually evolve into heterogeneous populations composed of virions that lack major segments of their genomes and instead contain tandemly repeated viral sequences that behave as replication origins. Defective viruses derived from AcMNPV contain large deletions, particularly within the 1.7 to 45.0 map units (m.u.) region of the genome (Kool *et al.*, 1993a). Analysis of these defective genomes identified four distinct regions that appear to be amplified upon extended passage. Two of these regions, corresponding to 50.1 to 53.2 and 87.2 to 88.9 m.u. of AcMNPV, contain homologous regions (*hrs*) (Kool *et al.*, 1993a) (see Section II.B). Another region encompassing the 85.0 to 87.2 m.u. region does not contain *hr* sequences (Lee and Krell, 1992). Both *hr* and non-*hr* sequences can support transient plasmid replication (Kool *et al.*, 1994b), suggesting that two different classes of replication origins are present in the AcMNPV genome. The occurrence of these particular sequences in defective genomes combined with the results of transient replication assays strongly suggest that they serve as bonafide initiation sites for AcMNPV DNA replication.

B. Homologous Regions as Replication Origins

All well-characterized baculovirus genomes contain a set of closely related sequences known as homologous regions, which are interspersed throughout the genome (see Chapter 5, this volume). Homologous regions from several baculoviruses share a number of common sequence features: (1) a core sequence consisting of an imperfect palindrome flanked by direct repeats, and (2) multiple copies of this core sequence separated by variable lengths of DNA. In AcMNPV, *hr*s consist of one to eight copies of a repeated sequence composed of 30-base pair (bp) palindromes flanked by 20-bp direct repeats and separated by about 80- to 120-bp of DNA (Ayres *et al.*, 1994; Guarino *et al.*, 1986). These imperfect palindromes can form cruciform structures *in vitro*; however, it is not known whether such structures play a direct role in *hr* function (Rasmussen *et al.*, 1996). While *hr*s share significant intragenomic sequence homology, their relatedness between viruses can be variable (Pearson and Rohrmann, 1995; Majima *et al.*, 1993; Thielmann and Stewart, 1992; Guarino *et al.*, 1986; Arif and Doerfler, 1984; Kuzio and Faulkner, 1984; Cochran and Faulkner, 1983). Homologous regions from AcMNPV and *Orgyia pseudotsugata* MNPV (OpMNPV) can function as *cis*-acting enhancers of IE-1-mediated early gene expression (see Chapter 6, this volume), and specific sequences within the AcMNPV *hr*s have been shown to interact with IE-1 (Rasmussen *et al.*, 1996; Choi and Guarino, 1995; Leisy *et al.*, 1995; Rodems and Friesen, 1995; Guarino and Dong, 1991, 1994). Interspersed homologous sequences also have been mapped to the genome of the *Trichoplusia ni* granulosis virus (TnGV); however, in contrast to *hr*s, palindromic or repetitive sequences are absent in these regions (Hashimoto *et al.*, 1996).

The *hr* sequences have been shown to confer upon a plasmid the ability to replicate in insect cells in an infection-dependent manner using a *Dpn*I-based assay system (Ahrens *et al.*, 1995a; Pearson and Rohrmann, 1995; Xie *et al.*, 1995; Lee and Krell, 1994; Kool *et al.*, 1993a,b; Leisy and Rohrmann, 1993; Pearson *et al.*, 1992, 1993). This system discriminates between replicated and nonreplicated plasmid DNAs based on their sensitivity to *Dpn*I (Peden *et al.*, 1980). Plasmid DNAs prepared in Dam$^+$ bacteria are methylated and retain their methylation in eukaryotic cells in the absence of plasmid replication and are consequently sensitive to *Dpn*I digestion. In contrast, plasmids containing a replication origin are replicated when introduced into baculovirus-infected cells by transfection, and the daughter strands, which are not methylated, are resistant to *Dpn*I cleavage. This transient plasmid amplification assay has formed the basis for most of the functional analyses of putative baculovirus origin sequences.

These analyses have demonstrated that a single palindrome from an AcMNPV *hr* can support limited plasmid DNA replication (Leisy *et al.*, 1995; Pearson *et al.*, 1992), although the relative efficiency of replication of a particular *hr*-containing plasmid increases as the number of palindromes present in that *hr* increases (Leisy *et al.*, 1995; Pearson *et al.*, 1992). Plasmids containing a half of the palindrome or palindromes modified by disruption to the core *Eco*RI site are severely compromised in their ability to replicate in infected cells (Wu and Carstens, 1996; Leisy *et al.*, 1995). Disruption of the *Eco*RI site by the addition or deletion of 4 bp did not affect the ability of IE-1 to bind to the palindrome, suggesting that multiple viral and/or host factors interact at the origin to initiate DNA replication (Leisy *et al.*, 1995) (see Section V) or that IE-1 binding with the mutant sequence is not a functional interaction.

Elements flanking *hr* sequences have been shown to be necessary for optimal infection-dependent plasmid replication (Leisy *et al.*, 1995; Pearson and Rohrmann, 1995; Xie *et al.*, 1995). Two of these elements have been characterized in *Lymantria dispar* MNPV (LdMNPV) and *Choristoneura fumiferana* MNPV (CfMNPV) and are composed of multiple repeats that are relatively A/T-rich (Pearson and Rohrmann, 1995; Xie *et al.*, 1995) and can enhance replication independent of their orientation relative to the *hr* sequence (Pearson and Rohrmann, 1995). In the case of LdMNPV, the presence of the A/T-rich domain juxtaposed adjacent to an *hr* sequence stimulates plasmid replication approximately 70-fold relative to *hr* alone (Pearson and Rohrmann, 1995). The presence of A/T-rich regions adjacent to replication origins has been well documented (DePamphilis, 1993). Interaction of the *hr* sequence with an origin-binding protein may result in the melting of DNA within these A/T-rich regions, thereby facilitating entry of the replication complex.

Although *hr*s from different baculoviruses generally share similar structural features, the ability of *hr*s from different baculoviruses to function as replication origins in heterologous systems is variable and may, in part, reflect the host specificity of baculovirus replication. Plasmids containing an AcMNPV *hr* replicate poorly when transfected into OpMNPV-infected Ld652-Y cells (Ahrens *et al.*, 1995a; Pearson *et al.*, 1993); however, such plasmids show substantial replication in CfMNPV-infected Cf-124-T cells

(Xie *et al.*, 1995). Conversely, replication origins derived from OpMNPV, CfMNPV, or Spodoptera exigua MNPV (SeMNPV) do not replicate in AcMNPV-infected Sf-9 cells (Ahrens *et al.*, 1995a; Xie *et al.*, 1995; Pearson *et al.*, 1993; Muñoz and P. J. Krell, personal communication), suggesting that AcMNPV *hr*s may be more promiscuous than their counterparts in OpMNPV and CfMNPV.

The identification and conservation of multiple copies of *hr* elements in those baculoviruses that have been characterized suggest that they play an important role in virus replication. Currently, there is no direct evidence that *hr*s function as origins of replication in the context of a virus infection, particularly since other non-*hr* origins have been described (see Section II.C). However, an *in vivo* role for *hr*5 as an enhancer of early gene expression in AcMNPV has been documented (Rodems and Friesen, 1993). The presence of multiple *hr*s throughout the genome may have evolved to provide redundancy in the initiation of viral DNA replication to ensure that DNA replication (and early gene expression) occurs efficiently in the event that a single *hr* is deleted. This theory is supported by the results of studies demonstrating that deletion of a single *hr* element from the genome of AcMNPV and *Bombyx mori* NPV (BmNPV) does not affect the phenotype of the virus *in vitro* or *in vivo* (Majima *et al.*, 1993; Rodems and Friesen, 1993).

C. Non-*hr* Origins of Replication

Non-*hr* origins have been reported in two baculoviruses (Kool *et al.*, 1994b; Pearson *et al.*, 1993). These regions contain unique palindromic and repetitive sequences that are not found in *hr* sequences and are relatively complex in organization, often involving multiple domains that collectively are necessary for maximal replication efficiency.

In AcMNPV, sequences within the *Hin*dIII-K region (84.9 to 87.3 m.u.) that are enriched and tandemly repeated in defective AcMNPV genomes (Lee and Krell, 1994) also support replication of plasmids in transient replication assays (Kool *et al.*, 1994b). In one report, plasmids containing this non-*hr* origin of replication (*ori-K*) replicated as efficiently as *hr*5-containing plasmids (Kool *et al.*, 1994b). However, in another study, *Hin*dIII-K replicated much less efficiently than *hr*-containing plasmids (Leisy and Rohrmann, 1993). Deletion analysis of the *Hin*dIII-K region indicated that sequences needed for optimal replication efficiency are contained within a relatively large fragment between 84.9 to 85.9 m.u. within the p94 gene. Of the five domains that constitute this non-*hr*, only domains III and IV appeared to be essential for *ori* activity. Domain III is A/T rich and contains several imperfect palindromes that are distinct from those found in the *hr*s of AcMNPV. The function of this region as an *ori* sequence *in vivo* is unknown, but its conservation in defective AcMNPV genomes (Lee and Krell, 1994) and in the genome of another closely related baculovirus that is missing the p94 gene (Kool *et al.*, 1994b) suggests that it may play an important role in baculovirus replication.

A non-*hr* origin of replication has also been identified in the genome of OpMNPV (Pearson *et al.*, 1993). Sequences located within the 7.0 to 11.3 m.u. region (*Hin*dIII-N) confer upon plasmids the ability to replicate when transfected into OpMNPV-infected Ld652-Y cells. The replication efficiency of plasmids containing this non-*hr* sequence is significantly less than plasmids containing OpMNPV *hr* sequences (Ahrens *et al.*, 1995a). The organization of this non-*hr* appears to be relatively complex, since deletion of any portion of *Hin*dIII-N was found to reduce the replication efficiency, suggesting that sequences affecting *ori* activity are located throughout this region. Sequence analysis identified a wide variety of direct and inverted repeats, palindromic sequences, and transcription factor binding sites, none of which have been correlated with *ori* activity (Pearson *et al.*, 1993).

Additional non-*hr* sequences have been identified, but they have not been further characterized using baculovirus infection-dependent replication assays (Lee *et al.*, 1992; van Iddekinge, 1986; Blinov *et al.*, 1984).

III. RELATIONSHIP BETWEEN TRANSCRIPTION ELEMENTS AND ORIGINS OF DNA REPLICATION

The intimate relationship between RNA transcription and initiation of DNA replication has been repeatedly demonstrated in eukaryotic systems (reviewed in DePamphilis, 1996; Held and Heintz, 1992). In many instances, transcriptional elements are also components of eukaryotic origins of DNA replication, and in several cases, such as in SV40 and polyomaviruses, these elements function as enhancers of gene expression (reviewed in DePamphilis, 1988). Although baculovirus *hr* sequences appear to function as both enhancers of early gene expression and as origins of replication, at least in the transient replication assay system, whether *hr*s function as replication origins *in vivo* remains to be resolved.

This issue is further complicated by the identification of non-*hr* sequences that also function as origins in infection-dependent replication assays and by the observation that plasmids containing sequences derived from the promoter regions of early baculovirus genes replicate in this assay system (Wu and Carstens, 1996). In many of these cases, plasmid replication is equivalent to that observed with plasmids containing *hr* sequences (Wu and Carstens, 1996). This situation is analogous to the situation in adenovirus and papillomaviruses where factors, NF-I and NF-III or E2, respectively, bind to transcriptional elements located within the *ori*-region (Pruijn *et al.*, 1987; Rosenfeld *et al.*, 1987; Spalholz *et al.*, 1987; Wides *et al.*, 1987). The baculovirus transcriptional activator protein, IE-1, is known to interact with specific sequences within the *hr*s and to also transactivate reporter genes under the transcriptional control of early baculovirus promoters (see Chapter 6, this volume, and Section V). Therefore, it is possible that putative replication origins correlate with sites of IE-1 binding or the binding that results from the interaction of IE-1 and a specific host factor. Currently, it is not known whether sequences in the non-*hr* regions such as the AcMNPV *ori-K* or the OpMNPV *Hin*dIII-N fragment also interact with their respective IE-1 proteins or whether differences exist in the transacting factor requirements between *hr*s and non-*hr*s.

This correlation between plasmid replication and the presence of an early AcMNPV promoter suggests that *hr*s and early promoter elements direct infection-dependent replication by recruiting viral–host transcription factors that may modify the nucleoprotein structure of the DNA and subsequently allow the formation of an initiation complex on the DNA. The finding that IE-1 is included in the set of baculovirus genes essential for transient plasmid replication supports this view (see Section V). Furthermore, it has been shown that a set of plasmids containing early baculovirus genes involved in DNA replication also coreplicated with plasmids containing *hr* sequences when cotransfected into insect cells (Kool *et al.*, 1994a,b; Lu and Miller, 1995a). The identification of several types of replication origins indicates that, *in vivo*, specific structural constraints such as DNA–nucleosome interactions and DNA topology may preclude initiation at most sequences.

IV. THE STRUCTURE OF REPLICATED DNA

Hr-containing plasmid DNA transiently replicated in infected cells appear to be in a high molecular weight form, indicating that under these conditions replication does not lead to the production of an exact replica of the input circular plasmid DNA. Analysis of replicated plasmid DNA by partial digestion with a restriction enzyme that cuts the input plasmid at a single site resulted in DNA fragments that were multiples of unit length, suggesting that the DNA was present as a concatemer containing a number of copies of the plasmid (Leisy and Rohrmann, 1993; Xie *et al.*, 1995). Such a structure could be indicative of rolling circle-type replication. In addition, defective genomes also appeared to consist of concatemers of sequences from the *Hin*dIII-K region (Lee and Krell, 1992, 1994). The identification of viral replication intermediates and the mechanism by which a concatemeric structure could be resolved into unit length, circular genome segments has not been reported.

V. GENES INVOLVED IN BACULOVIRUS DNA REPLICATION

A variety of strategies are employed for the replication of viral genomes and these are dependent on the expression of a combination of viral or host genes, which vary depending on the viral system (DePamphilis, 1996). To identify baculovirus-encoded genes that influence baculovirus-specific DNA replication, an approach similar to that employed for the identification of the herpes simplex virus 1 (HSV-1) replication genes (Challberg, 1986) was used. A set of overlapping cosmid clones encompassing the complete genome were transfected into uninfected insect cells along with a reporter plasmid containing a putative replication origin. A *Dpn*I assay was used to identify a minimal set of cosmids that replicated the plasmid DNA. Subclones from these cosmids were then used to isolate and identify the genes within the essential cosmids that were involved in DNA replication. These included five essential (*p143, ie-1, lef-1, lef-2,* and *lef-3*) and five stimulatory genes (*dnapol, p35, ie-2, lef-7,* and *pe38*) from AcMNPV (Lu and Miller, 1995a; Kool *et al.*, 1994,c) and a related set of genes from OpMNPV (Ahrens and Rohrmann, 1995a,b; Ahrens *et al.*, 1995b).

A. DNA POLYMERASE (*dnapol*) AND PROLIFERATING CELL NUCLEAR ANTIGEN

Early investigations reported the presence of a new DNA polymerase activity induced in baculovirus-infected cells (Kelly, 1981, 1982; Miller *et al.*, 1981). Subsequently, a 3′→5′ exonuclease activity specific for single-stranded DNA was shown to be associated with BmNPV DNA polymerase (Mikhailov *et al.*, 1986). In other DNA polymerases, this activity is associated with proofreading newly synthesized DNA and hydrolyzing mismatched nucleotides at the primer terminus.

The location and sequence has been determined for genes from six different baculoviruses that encode predicted proteins of about 115 kDa and that contain motifs conserved among a number of DNA polymerases (Ahrens and Rohrmann, 1996; Chaeychomsri *et al.*, 1995; Liu and Carstens, 1995; Cowan *et al.*, 1994; Bjornson *et al.*, 1992; Tomalski *et al.*, 1988). Although one study suggested that baculovirus DNA polymerase was stimulatory for DNA replication in the transient assay (Lu and Miller, 1995a), other reports indicated that it was essential (Ahrens and Rohrmann, 1995a; Kool *et al.*, 1994a,c; Pearson *et al.*, 1993). The fact that under certain conditions the baculovirus DNA polymerase homologue is not required for transient DNA replication suggests that the set of essential replication genes may be able to act in concert with a host DNA polymerase to replicate DNA at a suboptimal level. The compatibility of the baculovirus replication factors with other DNA polymerases is also reflected in substitution experiments using the enzyme from diverse viruses. Both the AcMNPV and OpMNPV DNA polymerases are able to substitute for each other in the transient replication assay (Ahrens and Rohrmann, 1996). Similarly, an ascovirus DNA polymerase was able to substitute for the AcMNPV DNA polymerase in this assay (Pellock *et al.*, 1996).

A protein called proliferating cell nuclear antigen (PCNA) is an important factor that facilitates the processivity of some classes of DNA polymerases. Although the AcMNPV genome contains a PCNA homologue with 42% amino acid identity to rat PCNA (O'Reilly *et al.*, 1989), it is neither essential nor stimulatory in the transient replication assay (Kool *et al.*, 1994a). However, in a mutant virus in which the PCNA gene was interrupted, the pattern of late gene expression that is dependent on DNA replication was delayed (Crawford and Miller, 1988). Although PCNA is highly conserved between many organisms, it is not present in the BmNPV genome (S. Maeda, personal communications) and the PCNA homologue in OpMNPV has only 31% amino acid sequence identity with the predicted AcMNPV PCNA. Therefore, the role of this gene in baculovirus DNA replication remains to be elucidated. For more discussion on PCNA, see Chapter 10 (this volume).

B. Helicase

A baculovirus gene with limited sequence similarity to helicases was originally identified by sequencing an open reading frame (ORF) containing a temperature-sensitive mutation that resulted in defective DNA synthesis (Lu and Carstens, 1991). This gene (*p143*) encodes a predicted protein of 143 kDa that contains a number of motifs characteristic of helicases including NTP-binding and DNA–RNA unwinding motifs. *p143* has been shown to possess a nonspecific double-stranded DNA binding activity (Laufs *et al.*, 1997), supporting its function as a helicase; however, it has yet to be demonstrated to possess helicase activities. In addition, sequences in the baculovirus helicase that are involved in specifying host range have been identified (Croizier *et al.*, 1994; Maeda *et al.*, 1993) (see Chapter 9, this volume). BmNPV and AcMNPV infect *B. mori* (Bm) and *S. frugiperda* (Sf21) cells, respectively, but cannot replicate in the heterologous cell line. However, AcMNPV recombinants with a fragment of the BmNPV helicase gene replacing the corresponding AcMNPV sequence are capable of replicating in both Bm and Sf21 cells. This mutant region was delimited to a 79-bp region (with only four altered codons) (Croizier *et al.*, 1994). AcMNPV recombinants containing this portion of the BmNPV helicase gene also replicated in *B. mori* larvae (Croizier *et al.*, 1994), which are not normally permissive for AcMNPV. It has recently been reported that, although the AcMNPV helicase mutants replicate efficiently in both Sf9 and BmN cells when a high multiplicity of infection (MOI) inoculum is used, low MOI infections resulted in much lower yields of extracellular virus in Sf9 cells (Kamita and Maeda, 1996). In such low MOI infections, DNA replication and late gene expression appears to be inhibited and can only reach normal levels at higher MOI. Therefore, single copies of the mutant helicase gene do not support productive replication in Sf9 cells. This appears to be due to a reduction in the activity of the putative helicase. The ability of the helicase mutant to replicate in BmN cells appears to be due to a reduced cytotoxicity of the helicase gene product in these cells. This reduced cytotoxicity may also be due to a reduction in the activity of the putative DNA helicase.

C. IE-1

Immediate early gene 1 (*ie-1*) is the only baculovirus gene for which splicing has been reported (Chisholm and Henner, 1988). The unspliced form encodes a predicted protein of 67 kDa (Guarino and Summers, 1987), whereas splicing results in 54 additional amino acids at the N-terminus. Plasmids expressing unspliced *ie-1* are essential for transient baculovirus DNA replication (Ahrens and Rohrmann, 1995b; Lu and Miller, 1995a; Kool *et al.*, 1994a).

ie-1 activates a variety of baculovirus early gene promoter–reporter constructs when they are cotransfected into uninfected insect cells (Lu and Carstens, 1993; Rodems and Friesen, 1993; Blissard and Rohrmann, 1991; Guarino and Summers, 1986). This activation is greatly enhanced when the constructs are linked to *hr* sequences. *Hr*-containing DNA has been shown to bind IE-1 in gel shift assays (Choi and Guarino, 1995; Leisy *et al.*, 1995; Rodems and Friesen, 1995). Deletion analysis of AcMNPV *ie-1* indicated the presence of separate transcriptional activation and DNA-binding domains (Kovacs *et al.*, 1992).

The requirement for *ie-1* in baculovirus DNA replication may result from its function in activating the expression of early genes, some of which are required for viral DNA replication; however, a direct role in origin binding and initiation of the early steps leading to the assembly of a replication complex is also possible. For additional information on IE-1, see Chapter 6 (this volume).

D. LEF-1

The *lef-1* gene has been identified in AcMNPV (Passarelli and Miller, 1993a), OpMNPV (Ahrens and Rohrmann, 1995a), *Buzura suppressaria* NPV (BsSNPV) (Z. H. Hu and J. M. Vlak, unpublished data) and in two spruce budworm baculoviruses, CfMNPV, and CfDEF (another strain of defective CfMNPV) (Barrett *et al.*, 1996). The amino acid sequence identity between the LEF-1 proteins ranged from 55 to 79%, and in each of these baculoviruses, *lef-1* is located adjacent to the *egt* gene. *lef-1* was initially recognized as an early gene important for late and very late gene expression (Passarelli and Miller, 1993a). It was later shown that *lef-1* was essential for transient DNA replication for both AcMNPV and OpMNPV (Ahrens and Rohrmann, 1995a; Lu and Miller, 1995a; Kool *et al.*, 1994a).

Alignment of LEF-1 sequences from AcMNPV, OpMNPV, CfMNPV, and CfDEF identified four conserved domains with amino acid sequence identities from 69 to 100% (Barrett *et al.*, 1996). Three of these domains are also conserved in the DNA primase genes of several organisms, suggesting that LEF-1 may be a baculovirus primase.

E. LEF-2

lef-2 is essential for baculovirus late gene expression in transient expression assays (Passarelli and Miller, 1993b) and was also found to be essential for DNA replication (Ahrens and Rohrmann, 1995b; Lu and Miller, 1995a; Kool *et al.*, 1994a). In *S. frugiperda* cells infected with an AcMNPV mutant with an altered *lef-2* gene, late gene expression was compromised, but DNA replication appeared to be unaffected (Merrington *et al.*, 1996). This may indicate that *lef-2* has two independent roles: one in late gene expression and a second in DNA replication. In addition, yeast two-hybrid and glutathione S transferase interaction assays both indicate that LEF-2 interacts with LEF-1, suggesting that these proteins may form a hetero-oligomeric complex involved in replication (Evans *et al.*, 1997).

F. LEF-3

AcMNPV *lef-3*, an essential gene for DNA replication in transient assays, is predicted to be a protein of 385 amino acids (44.5 kDa). Biochemical evidence suggests that the AcMNPV LEF-3 is a single-strand DNA binding (SSB) protein (Hang *et al.*, 1995). Extracts from AcMNPV-infected cell nuclei were fractionated by binding to single-stranded DNA agarose columns. A protein was found to bind to these columns with high affinity and eluted at salt concentrations greater than 0.9 M. This SSB was similar in size to that predicted for LEF-3. Moreover, it was shown by immunological analysis to be equivalent to the expected *lef-3* gene product. The purified SSB protein has a preference for single-stranded DNA and demonstrated nonspecificity and cooperativity of binding on DNA.

G. P35

The AcMNPV *p35* gene was found to be greatly stimulatory or essential for plasmid DNA replication in transient assays, depending on the assay conditions (Lu and Miller, 1995a; Kool *et al.*, 1994a). The known function of *p35* as an apoptotic suppressor of AcMNPV-infected *S. frugiperda* cells (Clem *et al.*, 1991) (see Chapter 10, this volume) suggests that the role of *p35* in these assays is to prevent apoptosis in transfected cells triggered by either plasmid DNA replication or a product of one or more replication genes. The ability of a separate class of inhibitor of apoptosis (*iap*) genes to substitute for *p35* in these assays (Lu and Miller, 1995a) supports this view. In addition, treatment of cells with aphidicolin blocks apoptosis induced by infection of cells with a *p35* mutant of AcMNPV, suggesting that DNA replication or expression of a late gene product(s) is responsible for induction of apoptosis (Clem and Miller, 1994).

One of the replication genes, *ie-1*, triggers apoptosis when transfected into uninfected *S. frugiperda* cells (Prikhod'ko and Miller, 1996). Apoptosis induced by transient expression of *ie-1* can be blocked by *p35* or a functional *iap*. This effect of IE-1 was suggested to be related to the potential of high levels of *ie-1* expression in transfected cells to stimulate cellular DNA synthesis that is not linked to the cell cycle and consequently induce an apoptotic response (Prikhod'ko and Miller, 1996). Therefore, *p35* may be stimulatory in the replication assay because it suppresses death of transfected cells caused by their response to *ie-1* or a combination of *ie-1* expression and plasmid DNA replication.

H. IE-2 and PE-38

Two other genes, *ie-2* and *pe-38*, which stimulate DNA replication, encode transactivators of early gene transcription (Lu and Carstens, 1993; Carson et al., 1988). In particular, PE-38 contributes to the activation of expression of the baculovirus helicase homologue, whereas IE-2 stimulates *pe-38* (Lu and Carstens, 1993) and *ie-1* expression (Yoo and Guarino, 1994). The stimulatory role of *ie-2* and *pe-38* therefore may involve their activation of essential replication genes. The contribution of *ie-2* to transient plasmid replication has been found to be cell-line dependent (Lu and Miller, 1995b). In contrast to the situation in Sf-21 cells, *ie-2* does not have a role in plasmid replication in TN-368 cells, a cell line derived from *Trichoplusia ni*, suggesting that IE-2 may have species-specific effects on viral DNA replication (see Chapter 9, this volume).

I. LEF-7

The *lef-7* gene encodes a protein of 226 amino acids (26.7 kDa) that stimulates CAT expression in transient expression assays using a reporter plasmid containing *cat* under the control of either the p39-capsid protein or polyhedrin promoters (Morris *et al.*, 1994). Subsequently, it was determined that *lef-7* stimulated DNA replication in transient replication assays. LEF-7 has 21% sequence similarity to the HSV-1 UL 29 gene (encoding a SSB) and contains two single-stranded DNA binding motifs (Lu and Miller, 1995a). A cell-line-specific effect on plasmid replication has been demonstrated for *lef-7* (Lu and Miller, 1995b) (see Chapter 9, this volume). Lef-7 provides a stimulatory role in plasmid replication in Sf-21 but not in TN-368 cells, suggesting that, like *ie-2, hcf-1,* and *p35, lef-7* may be a host range determinant.

J. HCF-1

Host cell-specific factor 1 (*hcf-1*) encodes a relatively cysteine-rich polypeptide that is differentially involved in transient plasmid replication in a cell-line-specific manner (Lu and Miller, 1995b). The *hcf-1* was required for transient replication in *T. ni* but not *S. frugiperda* cells. Subsequently, it has been demonstrated that AcMNPV DNA replication in *T. ni* cells is compromised if *hcf-1* is disrupted or deleted from the genome (Lu and Miller, 1996). A homologue of *hcf-1* has not been identified in the genome of BmNPV (S. Maeda, personal communication) or OpMNPV (Ahrens *et al.*, 1997). A potential role for *hcf-1* in baculovirus host range and virulence is discussed in Chapter 9 (this volume).

K. Host Factors

Baculovirus DNA replication may depend on host factors in addition to the viral factors described above. The observation in the transient DNA replication assay that specific baculovirus–cell line combinations are unable to replicate plasmids containing sequences that behave as replication origins in heterologous virus–cell systems could indicate incompatibility at either the level of viral genes or host factors. The observation that four altered codons in the helicase gene of AcMNPV allows it to replicate in *B. mori* cells, an otherwise nonpermissive cell line (Croizier *et al.*, 1994; Maeda *et al.*, 1993), suggests that there may be an interaction between the viral helicase and a cellular protein, and this interaction is cell line and virus species specific.

Recently, a 38-kDa host nuclear protein has been implicated in specific binding to AcMNPV *hr*1 (Habib *et al.*, 1996). This interaction requires the *hr* palindrome as well as the short flanking sequences on either side. Binding is not observed if nuclear extracts are dephosphorylated. The binding of this protein may be important for the enhancer function of *hr*1. However, no data are available to establish a direct role for this protein in AcMNPV infection.

VI. A MODEL FOR THE INITIATION OF BACULOVIRUS DNA REPLICATION

The identification of six viral genes essential for origin-specific DNA replication in transient DNA replication assays is similar to the number of genes found to be required for DNA replication in the herpesvirus family. Although the exact functions of these baculovirus genes await further biochemical analyses, we can predict, based on analogy to other better-characterized systems such as the herpesvirus system, that the baculovirus replisome includes a DNA polymerase, a helicase, a primase, a primase-associated protein, and SSBs involved in origin recognition and stabilization of single-stranded regions of the replication fork.

The presence of conserved amino acid motifs found in other replicative proteins strongly suggest that DNAPOL, P143, and LEF-1 function within the replisome complex as a DNA polymerase, a helicase, and a primase, respectively. Since LEF-3 cooperatively binds to single-stranded DNA, its role in DNA replication may be to bind to single-stranded DNA formed at the replication fork by the unwinding of parental duplex DNA by P143. In herpesviruses, the helicase is a component of a multisubunit complex composed of helicase (UL5), primase (UL52), and DNA binding (UL-9) activities (Bruckner *et al.*, 1991; Crute *et al.*, 1989). If a similar complex is found in baculovirus-infected cells, then P143 may interact with LEF-1, possibly through its leucine zipper motif (Lu and Carstens, 1991). The interaction of LEF-1 and LEF-2 in the yeast two-hybrid system (Evans *et al.*, 1997) suggests that LEF-2 might function as a primase-associated protein, analogous to UL9 in the herpesvirus system.

At this time we do not know whether the initiation of baculovirus DNA replication requires an additional origin-binding protein; however, if there is such a protein, IE-1 is a promising candidate, given its requirement in transient assays and its ability to bind to *hr*s. If IE-1 does function in this capacity, then it is likely that the DNA unwinding activity normally associated with origin-binding proteins such as UL8 (Bruckner *et al.*, 1991) is supplied by P143, since it is the only protein in the set of essential baculovirus proteins to possess a conserved ATP binding motif. A role for P143 in the initiation of DNA replication is consistent with the phenotype of conditional lethal mutants of AcMNPV containing a mutation within the p143 gene (Gordon and Carstens, 1984).

Thus, a possible scenario for the contribution of these essential genes to baculovirus DNA replication would be that the initiation of replication would be mediated by the binding of IE-1 to the *hr*s, leading to localized melting of duplex DNA that would allow the assembly of a complex of P143, LEF-1, and LEF-2 at the origin. This complex would subsequently also prime lagging strand DNA for the polymerase, while simultaneously unwinding DNA at the replication fork. The single-stranded regions resulting from unwinding of the DNA would be stabilized by LEF-3. The contribution of accessory factors such as P35, LEF-7, IE-2, and PE-38 would then be to maximize DNA replication in a species-specific manner, presumably in response to the absence of specific host factors or the presence of host factors that would otherwise interfere with the viral replication process. The clear involvement of P35 in blocking cellular apoptosis highlights the fact that the contribution of these genes may not be directly related to the replication process per se.

It is unclear to what extent cellular factors participate in the synthesis of baculovirus DNA. One obvious gene that must be supplied by the host cell that is not present in the set of replicative baculovirus genes is a topoisomerase, which would be responsible for relieving the torsional stress caused by unwinding of duplex DNA by P143. Furthermore, evidence exists that a host DNA polymerase can substitute for the baculovirus DNA polymerase in transient replication assays (Lu and Miller, 1995a; G. F.

Rohrmann, personal communication), suggesting that the baculovirus replisome is able to interact with cellular replicative enzymes. In herpesviruses, it has been suggested that initiation of DNA replication is mediated by UL9 interacting with a cellular DNA polymerase rather than the viral polymerase (Huberman, 1995). The differential requirement for HCF-1 and LEF-7 in plasmid DNA replication in transient assays performed in Sf-21 and TN-368 cells (Lu and Miller, 1995b) suggests that cellular factors do play a significant role in baculovirus DNA replication.

VII. FUTURE RESEARCH

Baculoviruses appear to be different from other virus systems because of the variety and number of sequences within their genomes that have the capacity to initiate replication in a transient assay. By analogy with other better-characterized eukaryotic systems, DNA replication probably can initiate at multiple origins that are distributed throughout the genome. The precise site of initiation, however, may be of little consequence to the final production of daughter genomes if the majority of baculovirus DNA is amplified via a rolling circle mode of replication. Under these conditions, a population of viral genomes would utilize a variety of origins with differing levels of initiation efficiency, but each initiation event would lead to a similar number of replicated genomes.

In addition to understanding how baculovirus DNA replication is initiated, delineating the mechanism for producing mature unit length genomes will be a major challenge. Infection-dependent replication of origin-containing plasmid DNA appears to result in the production of concatemers of plasmid DNA of large but indeterminate size. There is no evidence for resolution of these structures to unit length plasmids. Therefore, the replication of this DNA may be incomplete and suggests that other viral sequences are involved in resolution to unit length genomes or that there is a size recognition mechanism that processes DNA to a certain size. In the latter case, the plasmid concatemers may have been processed to fall within the recognized size.

Finally, the facility with which DNA undergoes recombination in baculovirus-infected cells, in conjunction with the presence of homologous regions in all baculovirus genomes that have been examined in detail, could indicate that a much more complex replication mechanism is employed. It could result in multiple repeated reinitiation within replicating DNA molecules, leading to the production of complex branched structures similar to those described for HSV-1 (Severini *et al.*, 1996). Resolution of such structures could involve interaction of yet-unidentified enzymes or recombination involving *hr* sequences.

In contrast to the mechanisms involved in replication initiation and resolution, the mechanism of genome sequence elongation is likely to be similar to other systems. Substantial progress has been made in identifying and clarifying the roles of different replication factors. Understanding their precise roles in replication initiation and chain elongation would be greatly facilitated by the development of an *in vitro* replication system. The identification of baculovirus-encoded genes involved in DNA replication in transient replication assays will facilitate developing such a system. We look forward to the development of a model for the interaction and roles of each of these factors in the near future.

ACKNOWLEDGMENTS. We thank George Kamita for helpful suggestions. This review was supported in part by NSF grant MCB9630769 to G. F. R. and is technical report no. 11148 from Oregon State University Agricultural Experiment Station.

REFERENCES

Ahrens, C. H., and Rohrmann, G. F., 1995a, Identification of essential *trans*-acting regions required for DNA replication of the *Orgyia pseudotsugata* multinucleocapsid nuclear polyhedrosis virus: lef-1 is an essential replication gene, *Virology* **207:**417.

Ahrens, C. H., and Rohrmann, G. F., 1995b, Replication of *Orgyia pseudotsugata* baculovirus DNA: ie-1 and lef-2 are essential and ie-2, p34 and Op-iap are stimulatory genes, *Virology* **212:**650.

Ahrens, C. H., and Rohrmann, G. F., 1996, The DNA polymerase and helicase genes of a baculovirus of *Orgyia pseudotsugata, J. Gen. Virol.* **77:**825.

Ahrens, C. H., Pearson, M. N., and Rohrmann, G. F., 1995a, Identification and characterization of a second putative origin of DNA replication in a baculovirus of *Orgyia pseudotsugata, Virology* **207:**572.

Ahrens, C. H., Carlson, C., and Rohrmann, G. F., 1995b, Identification, sequence and transcriptional analysis of lef-3, gene essential for *Orgyia pseudotsugata* baculovirus DNA replication, *Virology* **210:**417.

Ahrens, C. H., Russell, R. L. Q., Funk, C. J., Evans, J. T., Harwood, S. H., and Rohrmann, G. F., 1997, The sequence of the *Orgyia pseudotsugata* multinucleocapsid nuclear Polyhedrosis genome, *Virology* **229:**381–399.

Arif, B., and Doerfler, W., 1984, Identification and localization of reiterated sequences in the *Choristoneura fumiferana* MNPV genome, *EMBO J.* **3:**525.

Ayres, M. D., Howard, S. C., Kuzio, J., Lopez-Ferber, M. and Possee, R. D., 1994, The complete DNA sequence of *Autographa californica* nuclear polyhedrosis virus, *Virology* **202:**586.

Barrett, J. W., Lauzon, H. A. M., Mercuri, P. S., Krell, P. J., Sohi, S. S., and Arif, B. M., 1996, The putative LEF-1 proteins from two distinct *Choristoneura fumiferana* multiple nucleopolyhedroviruses share domain homology to eukaryotic primases, *Virus Genes* **13:**229.

Bjornson, R. M., Glocker, B., and Rohrmann, G. F., 1992, Characterization of the nucleotide sequence of the *Lymantria dispar* nuclear polyhedrosis virus DNA polymerase gene region. *J. Gen. Virol.* **73:**3177.

Blinov, V. M., Gutorov, V. V., Holodilov, N. G., Iljichev, A. A., Karginov, V. A., Mikrjukov, N. N., Mordvinov, B. A., Nikonov, I. V., Petrov, N. A., Urmanov, I. H., and Vasilenko, S. K., 1984, Nucleotide sequence of the *Galleria mellonella* nuclear polyhedrosis virus origin of DNA replication, *FEBS Lett.* **167:**254.

Blissard, G. W., and Rohrmann, G. F., 1991, Baculovirus gp64 gene expression: Analysis of sequences modulating immediate early transcription and transactivation by IE1, *J. Virol.* **65:**5820.

Bruckner, R. C., Crute, J. J., Dodson, M. S., and Lehman, I. R., 1991, The herpes simplex virus 1 origin binding protein: A DNA helicase, *J. Biol. Chem.* **266**:2669.

Brown, M., Crawford, A. M., and Faulkner, P. F., 1979, Genetic analysis of a baculovirus *Autographa californica* nuclear polyhedrosis virus: Isolation of temperature-sensitive mutants and assortment into complementation groups, *J. Virol.* **31**:190.

Burand, J. P., Summers, M. D., and Smith, G. E., 1980, Transfection with baculovirus DNA, *Virology* **101**:286.

Carson, D. D., Guarino, L. A., and Summers, M. D., 1988, Functional mapping of an AcNPV immediate early gene which augments expression of the IE-1 *trans*-activated 39k gene, *Virology* **162**:444.

Carstens, E. B., Tjia, S. T., and Doerfler, W., 1980, Infectious DNA from *Autographa californica* nuclear polyhedrosis virus, *Virology* **101**:311.

Chaeychomsri, S., Ikeda, M., and Kobayashi, M., 1995, Nucleotide sequence and transcriptional analysis of the DNA polymerase gene of *Bombyx mori* nuclear polyhedrosis virus, *Virology* **206**:435.

Challberg, M. D., 1986, A method for identifying the viral genes required for herpesvirus DNA replication, *Proc. Natl. Acad. Sci. USA* **83**:9094.

Chisholm, G. E., and Henner, D. J., 1988, Multiple early transcripts and splicing of the *Autographa californica* nuclear polyhedrosis virus IE-1 gene, *J. Virol.* **62**:3193.

Choi, J., and Guarino, L. A., 1995, The baculovirus transactivator IE1 binds to viral enhancer elements in the absence of insect cell factors, *J. Virol.* **69**:4548.

Clem, R. J., and Miller, K. K., 1994, Control of programmed cell death by the baculovirus genes p35 and iap, *Mol. Cell. Biol.* **14**:5212.

Clem, R. J., Fecheimer, M., and Miller, L. K., 1991, Prevention of apoptosis by a baculovirus gene during infection of insect cells, *Science* **254**:1388.

Cochran, M. A., and Faulkner, P., 1983, Location of homologous DNA sequences interspersed at five regions in the baculovirus AcMNPV genome, *J. Virol.* **45**:961.

Cowan, P., Bulach, D., Goodge, K., Robertson, A., and Tribe, D. E., 1994, Nucleotide sequence of the polyhedrin gene region of *Helicoverpa zea* single nucleocapsid nuclear polyhedrosis virus: Placement of the virus in lepidopteran nuclear polyhedrosis virus group II, *J. Gen. Virol.* **75**:3211.

Crawford, A. M., and Miller, L. K., 1988, Characterization of an early gene accelerating expression of late genes of the baculovirus *Autographa californica* nuclear polyhedrosis virus, *J. Virol.* **62**:2773.

Croizier, G., Croizier, L., Argaud, O., and Poudevigne, D., 1994, Extension of *Autographa californica* nuclear polyhedrosis virus host range by interspecific replacement of a short DNA sequence in the p143 helicase gene, *Proc. Natl. Acad. Sci. USA* **91**:48.

Crute, J. J., Tsurumi, T., Zhu, L., Weller, S. K., Olivo, P. D., Challberg, M. D., Mocarski, E. S., and Lehman, I. R., 1989, Herpes simplex virus 1 helicase-primase: A complex of three herpes-encoded gene products, *Proc. Natl. Acad. Sci. USA* **86**:2186.

DePamphilis, M. L., 1988, Transcriptional elements as components of eukaryotic origins of DNA replication, *Cell* **52**:635.

DePamphilis, M. L., 1993, Eukaryotic DNA replication: Anatomy of an origin, *Annu. Rev. Biochem.* **62**:29.

DePamphilis, M. L., (ed.), 1996, *DNA Replication in Eukaryotic Cells*, Plainview, NY, Cold Spring Harbor Laboratory Press.

Erlandson, M. A., Gordon, J., and Carstens, E. B., 1985, Size and map locations of early transcription products on the *Autographa californica* nuclear polyhedrosis virus genome, *Virology* **142**:12.

Evans, J. T., Leisy, D. J., and Rohrmann, G. F., 1997, Characterization of the interaction between the baculovirus replication factors, LEF-1 and LEF-2, *J. Virol.* **71**:3114.

Gordon, J. D., and Carstens, E. B., 1984, Phenotypic characterization and physical mapping of a temperature-sensitive mutant of *Autographa californica* nuclear polyhedrosis virus defective in DNA synthesis, *Virology* **138**:69.

Guarino, L. A., and Dong, W., 1991, Expression of an enhancer-binding protein in insect cells transfected with the *Autographa californica* nuclear polyhedrosis virus IE1 gene, *J. Virol.* **65**:3676.

Guarino, L. A., and Dong, W., 1994, Functional dissection of the *Autographa californica* nuclear polyhedrosis enhancer element hr5, *Virology* **200**:328.

Guarino, L. A., and Summers, M. D., 1986, Functional mapping of a *trans*-activating gene required for expression of a baculovirus delayed-early gene, *J. Virol.* **57**:563.

Guarino, L. A., and Summers, M. D., 1987, Nucleotide sequence and temporal expression of a baculovirus regulatory gene, *J. Virol.* **61**:2091.

Guarino, L. A., Gonzalez, M. A., and Summers, M. D., 1986, Complete sequence and enhancer function of the homologous DNA regions of *Autographa californica* nuclear polyhedrosis virus, *J. Virol.* **60**:224.

Habib, S., Pandey, S., Chatterji, U., Burma, S., Ahmad, R., Jain, A., and Hasnain, S. E., 1996, Bifunctionality of the AcMNPV homologous region sequence (hr1): Enhancer and ori functions have different sequence requirements, *DNA Cell Biol.* **15**:737.

Hang, X., Dong, W., and Guarino, L. A., 1995, The lef-3 gene of *Autographa californica* nuclear polyhedrosis virus encodes a single-stranded DNA-binding protein, *J. Virol.* **69**:3924.

Hashimoto, Y., Hayashi, K., Okuno, Y., Hayakawa, T., Saimoto, Granados, R. R., and Matsumoto, T., 1996, Physical mapping and identification of interspersed homologous sequences in the *Trichoplusia ni* granulosis virus genome, *J. Gen. Virol.* **77**:555.

Held, P. G., and Heintz, N. H., 1992, Eukaryotic replication origins, *Biochem. Biophys. Acta* **1130**:235.

Huberman, J. A., 1995, Prokaryotic and eukaryotic replicons, *Cell* **82**:535.

Kamita, S. G., and Maeda, S., 1996, Abortive infection of the baculovirus *Autographa californica* nuclear polyhedrosis virus in Sf-9 cells after mutation of the putative helicase gene, *J. Virol.* **70**:6244.

Kelly, D. C., 1981, Baculovirus replication: Stimulation of thymidine kinase and DNA polymerase activities in *Spodoptera frugiperda* cells infected with *Trichoplusia ni* nuclear polyhedrosis virus. *J. Gen. Virol.* **52**:313.

Kelly, D. C., 1982, Baculovirus replication. *J. Gen. Virol.* **63**:1.

Kelly, D. C., and Lescott, T., 1976, The effects of inhibitors of nucleic acid and protein synthesis on the replication of *Spodoptera frugiperda* nuclear polyhedrosis virus in cultured *Spodoptera frugiperda* cells, *Intervirology* **7**:360.

Kool, M., Voncken, J. W., van Lier, F. L. J., Tramper, J., and Vlak, J. M., 1991, Detection and analysis of *Autographa californica* nuclear polyhedrosis virus mutants with defective interfering properties, *Virology* **183**:739.

Kool, M., van den Berg, P. M. M. M., Tramper, J., Goldbach, R. W., and Vlak, J. M., 1993a, Location of two putative origins of DNA replication of *Autographa californica* nuclear polyhedrosis virus. *Virology* **192**:94.

Kool, M., Voeten, J. T. M., Goldbach, R. W., Tramper, J., and Vlak, J. M., 1993b, Identification of seven putative origins of *Autographa californica* MNPV DNA replication, *J. Gen. Virol.* **74**:2661.

Kool, M., Ahrens, C., Goldbach, R. W., Rohrmann, G. F., and Vlak, J. M., 1994a, Identification of genes involved in DNA replication of the *Autographa californica* baculovirus, *Proc. Natl. Acad. Sci. USA* **91**:11212.

Kool, M., Goldbach, R. W., and Vlak, J. M., 1994b, A putative non-hr origin of DNA replication of the *Hin*dIII-K fragment of *Autographa californica* nuclear polyhedrosis virus, *J. Gen. Virol.* **75**:3345.

Kool, M., Voeten, J. T. M., and Vlak, J. M., 1994c, Functional mapping of regions of the *Autographa californica* nuclear polyhedrosis viral genome required for DNA replication, *Virology* **198**:680.

Kovacs, G. R., Choi, J., Guarino, L. A., and Summers, M. D., 1992, Functional dissection of the *Autographa californica* nuclear polyhedrosis virus immediate-early 1 transcriptional regulatory protein, *J. Virol.* **66:**7429.

Kuzio, J., and Faulkner, P., 1984, Region of repeated DNA in the genome of *Choristoneura fumiferana* nuclear polyhedrosis virus, *Virology* **139:**185.

Laufs, S., Lu, A., Arrell, K., and Carstens, E. B., 1997, *Autographa californica* nuclear polyhedrosis virus is a DNA binding protein, *Virology,* **228:**98.

Lee, H. H., and Miller, L. K., 1978, Isolation of genotypic variants of *Autographa californica* nuclear polyhedrosis virus, *J. Virol.* **27:**754.

Lee, H. Y., and Krell, P. J., 1992, Generation and analysis of defective genomes of *Autographa californica* nuclear polyhedrosis virus, *J. Virol.* **66:**4339.

Lee, H. Y., and Krell, P. J., 1994, Reiterated DNA fragments in defective genomes of *Autographa californica* nuclear polyhedrosis virus are competent for AcMNPV-dependent DNA replication, *Virology* **202:**418.

Lee, H. Y., Arif, B., Dobos, P., and Krell, P., 1992, Identification of bent DNA and ARS fragments in the genome of *Choristoneura fumiferana* nuclear polyhedrosis virus, *Virus Res.* **24:**249.

Leisy, D. J., and Rohrmann, G. F., 1993, Characterization of the replication of plasmids containing hr sequences in baculovirus infected *Spodoptera frugiperda* cells, *Virology* **196:**722.

Leisy, D. J., Rasmussen, C., Him, H.-T., and Rohrmann, G. F., 1995, The *Autographa californica* nuclear polyhedrosis virus homologous region 1a: Identical sequences are essential for DNA replication activity and transcriptional enhancer function, *Virology* **208:**742.

Liu, J. J., and Carstens, E. B., 1995, Identification, localization, transcription, and sequence analysis of the *Choristoneura fumiferana* nuclear polyhedrosis virus DNA polymerase gene, *Virology* **209:**538.

Lu, A., and Carstens, E. B., 1991, Nucleotide sequence of a gene essential for viral DNA replication in the baculovirus *Autographa californica* nuclear polyhedrosis virus, *Virology* **181:**336.

Lu, A., and Carstens, E. B., 1993, Immediate-early baculovirus genes transactivate the p143 gene promoter of *Autographa californica* nuclear polyhedrosis virus, *Virology* **195:**710.

Lu, A., and Miller, L. K., 1995a, The roles of eighteen baculovirus late expression factor genes in transcription and DNA replication, *J. Virol.* **69:**975.

Lu, A., and Miller, L. K., 1995b, Differential requirements for baculovirus late expression factor genes in two cell lines, *J. Virol.* **69:**6265.

Lu, A., and Miller, 1996, Species-specific effects of the hef-1 gene on baculovirus virulence, *J. Virol.* **70:**5123.

Maeda, S., Kamita, S. G., and Kondo, A., 1993, Host range expansion of *Autographa californica* nuclear polyhedrosis virus (NPV) following recombination of a 0.6-kilobase-pair DNA fragment originating from *Bombyx mori* NPV, *J. Virol.* **67:**6234.

Majima, K., Kobara, R., and Maeda, S., 1993, Divergence and evolution of homologous regions of *Bombyx mori* nuclear polyhedrosis virus, *J. Virol.* **67:**7513.

Merrington, C.L., Kitts, P. A., King, L. A., and Possee, R. D., 1996, An *Autographa californica* nucleopolyhedrovirus lef-2 mutant: Consequences for DNA replication and late gene expression, *Virology* **217:**338.

Mikhailov, V. S., Marlyev, K. A., Ataeva, J. O., Kullyev, P. K., and Atrazhev, A. M., 1986, Characterization of 3′ to 5′ exonuclease associated with DNA polymerase of silkworm nuclear polyhedrosis virus, *Nucleic Acids Res.* **14:**3841.

Miller, L. K., Jewell, J. E., and Browne, D., 1981, Baculovirus induction of a DNA polymerase, *J. Virol.* **40:**305.

Morris, T. D., Todd, J. W., Fisher, B., and Miller, L. K., 1994, Identification of lef-7: A baculovirus gene affecting late gene expression, *Virology* **200:**360.

O'Reilly, D. R., Crawford, A. M., and Miller, L. K., 1989, Viral proliferating cell nuclear antigen, *Nature* **337:**606.

Passarelli, A. L., and Miller, L. K., 1993a, Identification and characterization of lef-1, a baculovirus gene involved in late and very late gene expression, *J. Virol.* **67:**3481.

Passarelli, A. L., and Miller, L. K., 1993b, Three baculovirus genes involved in late and very late gene expression: ie-1, ie-n, and lef-2, *J. Virol.* **67**:2149.

Pearson, M. N., and Rohrmann, G. F., 1995, *Lymantria dispar* nuclear polyhedrosis virus homologous regions: Characterization of their ability to function as replication origins, *J. Virol.* **69**:213.

Pearson, M. N., Bjornson, R. M., Pearson, G. D., and Rohrmann, G. F., 1992, The *Autographa californica* baculovirus genome: Evidence for multiple replication origins, *Science* **257**:1382.

Pearson, M. N., Bjornson, R. M., Ahrens, C., and Rohrmann, G. F., 1993, Identification and characterization of a putative origin of DNA replication in the genome of a baculovirus pathogenic for *Orgyia pseudotsugata*, *Virology* **197**:715.

Peden, K. W. C., Pipas, J. M., Pearson-White, S., and Nathans, D., 1980, Isolation of mutants of an animal virus in bacteria, *Science* **209**:1392.

Pellock, B. J., Lu, A., Meagher, R. B., Weise, M. J., and Miller, L. K., 1996, Sequence, function, and phylogenetic analysis of an ascovirus DNA polymerase gene, *Virology* **216**:146.

Potter, K. N., and Miller, L. K., 1980, Transfection of two invertebrate cell lines with DNA of *Autographa californica* nuclear polyhedrosis virus, *J. Invertebr. Pathol.* **36**:431.

Prikhod'ko, E. A., and Miller, L. K., 1996, Induction of apoptosis by baculovirus transactivator IE1, *J. Virol.* **70**:7116.

Pruijn, G. J., van Driel, W., van Miltenburg, R. T., and van der Vliet, P. C., 1987, Promoter and enhancer elements containing a conserved motif are recognized by nuclear factor III, a protein stimulating adenovirus DNA replication, *EMBO J.* **6**:3771.

Rasmussen, C., Leisy, D. J., Ho, P. S., and Rohrmann, G. F., 1996, Structure–function analysis of the *Autographa californica* multinucleocapsid nuclear polyhedrosis virus homologous region palindromes, *Virology* **224**:235.

Ribeiro, B. M., Hutchinson, K., and Miller, L. K., 1994, A mutant baculovirus with a temperature-sensitive IE-1 transregulatory protein, *J. Virol.* **68**:1075.

Rodems, S. M., and Friesen, P. D., 1993, The hr5 transcriptional enhancer stimulates early expression from the *Autographa californica* nuclear polyhedrosis virus genome but is not required for virus replication, *J. Virol.* **67**:5776.

Rodems, S. M., and Friesen, P. D., 1995, Transcriptional enhancer activity of hr5 requires dual-palindrome half sites that mediate binding of a dimeric form of the baculovirus transregulator IE1, *J. Virol.* **69**:5368.

Rosenfeld, P. J., O'Neill, E. A., Wides, R. J., and Kelly, T. J., 1987, Sequence-specific interactions between cellular DNA-binding proteins and the adenovirus origin of DNA replication, *Mol. Cell. Biol.* **7**:875.

Severini, A., Scraba, D. G., and Tyrell, D. L. J., 1996, Branched structures in the intracellular DNA of herpes simplex virus type I, *J. Virol.* **70**:3169.

Spalholz, B. A., Lambert, P. F., Yee, C. L., and Howley, P. M., 1987, Bovine papillomavirus transcriptional regulation: Localization of the E2-responsive elements of the long control region, *J. Virol.* **61**:2128.

Theilmann, D. A., and Stewart, S., 1992, Tandemly repeated sequence at the 3′ end of the IE-2 gene of the baculovirus *Orgyia pseudotsugata* multicapsid nuclear polyhedrosis virus is an enhancer element, *Virology* **187**:97.

Tjia, S. T., Carstens, E. B., and Doerfler, W., 1979, Infection of *Spodoptera frugiperda* cells with *Autographa californica* nuclear polyhedrosis virus. II. The viral DNA and the kinetics of its replication, *Virology* **99**:391.

Tomalski, M. D., Wu, J., and Miller, L. K., 1988, The location, sequence, transcription and regulation of a baculovirus DNA polymerase gene, *Virology* **167**:591.

van Iddekinge, H. B. J. L., Meyer, D., Zabel, P., and Vlak, J. M., 1986, Detection of sequences in *Autographa californica* nuclear polyhedrosis virus DNA, that act as autonomous replicating sequences (ARS) in yeast, *Arch. Virol.* **88**:279.

Wang, X., and Kelly, D. C., 1983, Baculovirus replication: Purification and identification of the *Trichoplusia ni* nuclear polyhedrosis virus-induced DNA polymerase, *J. Gen. Virol.* **64**:2229.

Wides, R. J., Challberg, M. D., Rawlins, D. R., and Kelly, T. J., 1987, Adenovirus origin of DNA replication: Sequence requirements for replication *in vitro, Mol. Cell. Biol.* **7**:864.

Wu, Y., and Carstens, E. B., 1996, Initiation of baculovirus DNA replication: Early promoter regions can function as infection-dependent replicating sequences in a plasmid-based replication assay, *J. Virol.* **70**:6967.

Xie, W.-D., Arif, B., Dobos, P., and Krell, P. J., 1995, Identification and analysis of a putative origin of DNA replication in the *Choristoneura fumiferana* multinucleocapsid nuclear polyhedrosis virus genome, *Virology* **209**:409.

Yoo, S., and Guarino, L. A., 1994, The *Autographa californica* nuclear polyhedrosis virus ie-2 gene encodes a transcriptional regulator, *Virology* **202**:746.

CHAPTER 8

Regulation of Baculovirus Late and Very Late Gene Expression

ALBERT LU AND LOIS K. MILLER

I. INTRODUCTION

The late stage of baculovirus infection is defined as those events that occur following the initiation of viral DNA replication and is usually subdivided into late and very late phases. These phases coincide with the production of the budded and occluded forms of the virus, respectively (see Chapters 2 and 4, this volume). Budded viruses (BVs) are produced by the packaging of newly synthesized viral DNA into nucleocapsids that leave the nucleus and bud through the cell's plasma membrane. Occluded viruses (OVs) are produced later when the nucleocapsids become enveloped within the nucleus; the resulting virions are then embedded in a crystalline protein matrix, thereby forming nuclear bodies known as polyhedra or granules, depending on their morphological appearance, which, in turn, specifies their genus (nucleopolyhedrosis virus or granulosis virus, respectively).

Both the late and very late phases are characterized by the nature of the genes actively transcribed during that phase. During the late phase, genes encoding structural proteins of the nucleocapsid, such as *vp39* and *p6.9*, the major capsid protein and basic core protein, respectively (see Chapter 2, this

ALBERT LU • Dupont Agricultural Products, DuPont Stine-Haskell Research Center, Newark, Delaware 19714-0030. LOIS K. MILLER • Departments of Entomology and Genetics, University of Georgia, Athens, Georgia 30602-2603.

The Baculoviruses, edited by Lois K. Miller. Plenum Press, New York, 1997.

volume), are abundantly transcribed. Genes encoding occlusion-related polypeptides such as polyhedrin or granulin, the major crystalline proteins of polyhedra or granules, are transcribed primarily during the very late phase. Another abundantly expressed very late gene is *p10*, which encodes a polypeptide that affects nuclear disintegration in the final phases of cell death.

Most of the information on the regulation of baculovirus late and very late gene expression derives from work on *Autographa californica* nuclear polyhedrosis virus (AcMPNV) replication in cell lines derived from the fall armyworm *Spodoptera frugiperda* (e.g., SF-21 cells). Late AcMNPV genes are transcribed primarily between 6 and 24 hr postinfection (p.i.), whereas very late genes are transcribed in a rapid burst beginning around 18 hr p.i. and continuing through 72 hr p.i. Transcription of both late and very late genes depends on viral DNA replication; a block or delay in viral DNA replication results in a parallel effect on late gene expression. The dramatic increase in viral gene transcription during the late and very late phases of infection contrasts with the decline in the steady-state levels of host mRNAs.

In this chapter, we review current information relating to the mechanism of baculovirus late and very late gene regulation. Baculoviruses induce a novel multisubunit RNA polymerase that is responsible for the transcription of late and very late genes. This polymerase or associated factors requires the sequence TAAG as an essential promoter element, and transcription is initiated within this sequence. *In vitro* transcription assays have been developed that accurately initiate transcription from late and very late gene promoters. A number of viral genes have been identified as being involved in late gene transcription, and some of these genes appear to be RNA polymerase subunits. The distinction between late and very late promoters involves an additional promoter element found between the TAAG site and the translational initiation codon. Recognition or utilization of this promoter element includes additional viral gene products including very late factor-1 (VLF-1), which has homology to integrase/resolvases.

II. A NOVEL VIRUS-INDUCED RNA POLYMERASE

During virus infection, a novel virus-induced RNA polymerase is synthesized that is responsible for the transition from early to late and very late gene expression (Beniya *et al.*, 1996; Yang *et al.*, 1991; Huh and Weaver, 1990a,b; Fuchs *et al.*, 1983; Grula and Weaver, 1981; Grula *et al.*, 1981). The presence of the novel RNA polymerase can be detected with nuclear runon assays using nuclei isolated at various times during infection of cells incubated in the presence or absence of α-amanitin (Huh and Weaver, 1990a,b; Grula *et al.*, 1981); synthesis of viral RNA becomes increasingly resistant to inhibition by α-amanitin, an inhibitor of host RNA polymerase II, as the infection progresses. In SF-21 and SF-9 cells, early AcMNPV transcripts are transcribed until 6 to 8 hr p.i. by an RNA polymerase that is sensitive to α-amanitin (Hoopes and Rohrmann, 1991; Huh and Weaver, 1990a,b; Fuchs *et al.*, 1983). From this point, the degree to which viral transcription is sensitive to α-amanitin decreases until about 18 hr p.i., at which time viral transcription is primarily resistant to α-amanitin (Huh and Weaver, 1990a,b; Fuchs *et al.*, 1983). A chromatographically distinct α-amanitin-resistant RNA polymerase activity is observed in AcMNPV-infected SF-21 cells (Beniya *et al.*, 1996; Xu *et al.*, 1995; Yang *et al.*, 1991; Fuchs *et al.*, 1983; Grula *et al.*, 1981).

Glycerol gradient analysis of the enriched RNA polymerase suggests that it is a large, multisubunit enzyme; nine polypeptides are observed in the peak gradient fractions that are associated with α-amanitin-resistant RNA polymerase activity (Beniya *et al.*, 1996). The enzyme is able to initiate transcription specifically from late baculovirus promoters in an *in vitro* transcription system (see Section IV). The sizes of the nine polypeptides, 155, 117, 102, 85, 40, 37, 35, 34, and 31 kDa (Beniya *et al.*, 1996), are consistent with the predicted sizes of several transcription-related viral proteins identified through transient expression analyses (see Section V). A different chromatographic procedure has allowed the separation of two different enzyme activities, both of which recognize late and very late baculovirus promoters but differ in their ability to initiate transcription from the very late promoter (Xu *et al.*, 1995).

Thus, transcription of baculovirus genes involves a novel RNA polymerase that is probably either partially or entirely encoded by the viral genome, and this enzyme is probably further modified, possibly by subunit addition, to recognize very late promoters. This view is further supported by promoter analyses (Section III) and by identification of viral genes involved in late and very late gene expression (see Section V and VI).

III. LATE AND VERY LATE PROMOTER STRUCTURES

TAAG is an essential element of most, if not all, late and very late baculovirus promoters. Transcriptional mapping studies have defined the transcriptional initiation site of a wide variety of late and very late transcripts to be located within a TAAG motif (Eldridge *et al.*, 1992; Lu and Carstens, 1992; Whitford and Faulkner, 1992; Possee *et al.*, 1991; Tomalski *et al.*, 1991; Guarino, 1990; Guarino and Smith, 1990; Nissen and Friesen, 1989; Thiem and Miller, 1989; Wu and Miller, 1989; Wilson *et al.*, 1987; Guarino and Summers, 1986; Friesen *et al.*, 1986; Howard *et al.*, 1986; Kuzio *et al.*, 1984). Late promoters have been found to initiate from (A/G/T)TAAG, but none have been found to initiate from CTAAG to date. TAAG is found less frequently in the AcMNPV genome than expected for a random sequence (Morris and Miller, 1994). Mutational analyses of a late and a very late promoter confirm that TAAG is the most crucial promoter element for late and very late promoters (Morris and Miller, 1994; Ooi *et al.*, 1989).

A comprehensive study of the effects of mutations on the function of the late *vp39* promoter (Morris and Miller, 1994; Thiem and Miller, 1990) has provided considerable insight into the sequence requirements of a late promoter. The *vp39* promoter is a complex late promoter containing three late transcriptional start sites within TAAGs located at -57, -105, and -321 relative to the translational start site (Thiem and Miller, 1989). A series of deletion mutations starting 5′ of the most distal TAAG sequence and extending toward the translational start site revealed that each of the three TAAGs functions independently and that each TAAG remains functional until the TAAG itself is deleted (Thiem and Miller, 1989). Furthermore, if the proximal *vp39* TAAG-containing region is joined distally to sequences containing the very late *polh* promoter, both promoters are active and the *vp39* TAAG retains its late temporal regulation, while the *polh* promoter retains features of its very late regulation.

Linker-scan mutational analysis of the 99-base pair (bp) region immediately upstream of the translational start site of *vp39,* including the proximal TAAG motif at -57, showed that the TAAG motif is the primary promoter element. Linker mutations affecting the TAAG motif abolish promoter activity (Morris and Miller, 1994), while mutations affecting the 8-bp region immediately upstream or the 6-bp region immediately downstream of the TAAG motif reduce promoter activity by 90 and 60%, respectively. Mutations introduced outside of this 18-bp sequence encompassing the TAAG motif do not significantly affect expression from the *vp39* promoter, indicating that the promoter is predominantly, if not exclusively, located within the 18-bp region. A synthetic promoter consisting of only the 18 bp encompassing the proximal TAAG directs temporally accurate late gene expression, although the level of expression is approximately one fourth that of the entire (464 bp) promoter (Morris and Miller, 1994). Thus, a very small DNA sequence (<18 bp) can act as a late promoter; the TAAG motif is the most critical element of this late promoter; and the context of the TAAG modulates the activity of the TAAG. Although other late promoters have not been studied as extensively as the *vp39* promoter, these basic conclusions also hold for late promoters embedded within the more complex dual early/late promoters of p35 (Dickson and Friesen, 1991; Nissen and Friesen, 1989) and 39K (Guarino and Smith, 1992).

Mutational analyses of two very late gene promoters, *polh* and *p10,* indicate that the TAAG sequence also plays a critical role in very late promoter activity (Qin *et al.*, 1989; Weyer and Possee, 1989; Ooi *et al.*, 1989; Rankin *et al.*, 1988; Possee and Howard, 1987). The promoters of the AcMNPV *p10* and *polh* genes have very limited sequence homology but do share a similar sequence, ATAAGT/AATT, at the transcriptional start site (Weyer and Possee, 1989). A more general version of this sequence, (T/A)ATAAGNA(T/A/C) T(T/A)T, is conserved in the promoters of all known polyhedrin, granulin, and *p10* genes (Rohrmann, 1986). Progressive deletion mutations into the *polh* or *p10* promoters from the 5′ to 3′ direction show that sequences upstream of TAAG are relatively unimportant for *p10* or *polh* promoter function in transient expression assays. Linker-scan mutations within the region 12 to 22 bp upstream of the *polh* TAAG motif increased expression from this promoter by approximately 50% when tested in the context of the viral genome (Ooi *et al.*, 1989; Rankin *et al.*, 1988). This effect may be due to the elimination of a weak repressor binding site or may simply influence the context of the TAAG element (e.g., a higher A/T content). These mutations increased expression during the very late phase, not during the late phase; if a repressor does bind to this region, it is not responsible for repressing expression during the late phase nor is it responsible for the bulk of expression during the very late phase. Mutations affecting the TAAG site, however, completely abolish reporter gene expression. Linker mutations altering the TAAG sequence of the *polh* promoter in the context of the viral genome result in dramatic (>1000-fold) decreases in *polh* mRNA levels and transcription rates (Ooi *et al.*, 1989).

A minimum promoter comprised of the 18 bp surrounding the *polh* TAAG sequence behaves only as a weak late promoter (Morris and Miller, 1994); thus, the context of the TAAG sequence alone does not dictate very late gene regulation. This 18-bp sequence from the *polh* promoter is considerably weaker as a promoter than the equivalent 18-bp sequence from the *vp39* promoter. Thus, *polh* may not compete well for the TAAG-initiating viral RNA polymerase activity during the late phase; the context of the TAAG is clearly important for the level of transcriptional initiation from the TAAG motif during the late phase of infection (Morris and Miller, 1994). Furthermore, unlike the late promoters, additional sequences beyond those immediately surrounding the TAAG are involved in the determination of the temporal regulation of this promoter (Morris and Miller, 1994).

An additional element downstream of the TAAG site is required for the "burst" of expression observed for very late promoters. Deletions extending 3′ to 5′ from the translational start codon toward the TAAG motif suggested that an additional element within the region between TAAG and the translational initiation codon was important for expression from the *p10* and *polh* promoters (Qin *et al.*, 1989; Weyer and Possee, 1989; Matsuura *et al.*, 1987; Possee and Howard, 1987). A study of linker-scan mutations in the context of the viral genome ascertained that the element within the 5′ untranslated leader region was a promoter element. Mutations within the 50 bp between the TAAG and the *polh* translational initiation codon reduce expression between 2- and 20-fold specifically during the very late phase. Determination of the effects of these mutations on RNA initiation site, RNA stability, and RNA initiation rates show that this element within the region specifying the untranslated leader region regulates the rate of transcriptional initiation from the very late TAAG (Ooi *et al.*, 1989; Rankin *et al.*, 1988).

The region between the TAAG and the *polh* ATG initiation codon is required for the burst of activity from the promoter during the very late phase of infection (Ooi *et al.*, 1989), and the presence of this additional "burst" element is the major difference observed between late and very late promoters. The mutational data collectively indicate that a repressor is not involved in the down-regulation of the *polh* promoter during the late phase nor in the down-regulation of late promoters during the very late phase unless the repressor binds directly to the sequence including and immediately downstream of TAAG sequence (Ooi *et al.*, 1989; Morris and Miller, 1994). Therefore, it is likely that activation of the *polh* (and *p10*) promoters primarily involves the specific binding of a viral factor(s) to a sequence(s) within the burst sequence. Alternatively, this burst element may specifically respond to alterations in its context (e.g., supercoiling). Placement of the *polh* burst element immediately downstream of the minimal 18-bp *vp39* promoter sequence, however, did not convert this promoter to a very late promoter (Morris and Miller, 1994). Thus, the presence of a burst element alone is not sufficient for the high expression levels at very late times p.i. The exact position of the burst element relative to TAAG or the combination of the TAAG context and the burst element may be important for very late promoter activity.

Although similarities exist at the TAAG sites of the *polh* and *p10* promoters, these two promoters differ substantially in the sequence of the burst element (Rohrmann, 1986) and differ subtly in their regulation (Roelvink *et al.*, 1992; Min and Bishop, 1991). Expression of a reporter gene from the *p10* promoter precedes *polh*-promoted expression by several hours (Roelvink *et al.*, 1992). It is possible that the *p10* promoter competes more successfully than the *polh* promoter for limited very late factor(s). Promoter competition experiments support this view. Viruses in which either the *p10* or *polh* promoter and their respective coding sequences have been deleted from the genome were constructed to determine whether deletion of one promoter effectively increased the availability of shared factors for transactivation of the other (Chaabihi *et al.*, 1993). Whereas the *p10* promoter/gene deletion virus expressed higher levels of *polh* mRNA, no effect on *p10* expression was observed in viruses carrying a *polh* promoter/gene deletion (Chaabihi *et al.*, 1993). This result suggests that the *p10* promoter may compete more successfully than the *polh* promoter for common factors. An alternate explanation is that additional promoter-specific factors may differentially influence *polh* and *p10* expression. More recently, a mutation in a very late expression factor gene, *vlf-1,* was identified that affects both *polh* and *p10* expression, but *p10* expression is affected to a lesser degree than *polh* (McLachlin and Miller, 1994) (see Section VI).

Gel retardation assays using extracts from uninfected and infected cells indicate that one or more host factors can interact with the *polh* and *p10* promoters (Jain and Hasnain, 1996; Burma *et al.*, 1994; Etkin *et al.*, 1994). A 30-kDa phosphorylated protein called the polyhedrin promoter binding protein (PPBP) binds to a hexanucleotide motif, AAATAAA; this sequence is found immediately 5′ of the *polh* TAAG motif and also within the burst element of the *polh* promoter (Burma *et al.*, 1994). Sequences downstream of AATAAA appear to contribute to PPBP binding, since binding activity is lost when the TAAG motif is replaced by an unrelated sequence. It is not clear what role PPBP has, if any, in *polh* promoter regulation. The fact that the AATAAA site immediately upstream of TAAG can be extensively mutated without significant impact on expression from the *polh* promoter (Ooi *et al.*, 1989) suggests that PPBP does not play a major role in *polh* regulation *in vivo.* A 30-kDa host protein is also reported to bind to the *p10* promoter, and binding of this protein to the *p10* promoter is also competed by the AATAAA hexanucleotide (Jain and Hasnain, 1996). Again, evidence that this protein, which may be identical to PPBP, influences *p10* expression *in vivo* is lacking.

Another host factor of about 200 kDa interacts with sequences located between -72 and -86 bp relative to the *polh* translational start site (Etkin *et al.*, 1994). This sequence, located 21 to 34 bp upstream of the TAAG site, includes two GATA motifs. Theoretically, this factor might be responsible for the subtle repressive activity of this region of the promoter; mutations affecting this region up-regulate *polh* promoter activity by 50% very late in infection (Ooi *et al.*, 1989). Since a promoter containing only the first 64 bp of the *polh* promoter and lacking the region bound by this factor still exhibits strong, very late promoter activity (Thiem and Miller, 1990), this factor is probably not a major player in the regulation of very late expression from the *polh* promoter. No direct correlation between the binding of this factor and an effect on *polh* transcriptional initiation has been established to date.

IV. *IN VITRO* TRANSCRIPTION ASSAYS

In vitro transcription assays hold promise as a means of studying the biochemical function of the viral factors thought to be involved in late and very late gene transcription, as well as providing a possible means of identifying additional factors, especially host factors, that may be involved. Two *in vitro* transcription systems that support late and very late gene transcription have been described (Xu *et al.*, 1995; Glocker *et al.*, 1993).

Extracts prepared from AcMNPV-infected cells during the late to very late phase support transcription initiating accurately from the TAAG motifs of both the late *vp39* and very late *polh* promoters (Beniya *et al.*, 1996; Glocker *et al.*, 1993). The maintenance of low levels of Mg^{2+} (1 to 2 mM) is crucial to detecting significant levels of transcription from these promoters (Glocker *et al.*, 1993). *In vitro* transcription from the late and very late promoters is insensitive to tagetitoxin, an inhibitor of RNA polymerase III, and to α-amanitin. This is consistent with the view that this transcription is mediated by a viral RNA polymerase rather than a host RNA polymerase. Transcription can be monitored by primer extension assays that distinguish TAAG-initiated transcripts from transcripts initiated randomly or from early promoters.

An *in vitro* transcription assay based on cytidine-free (C-free) cassettes as reporters for 39K and *polh* promoter activity has also been described (Xu *et al.*, 1995). Utilization of C-free cassettes circumvents some of the problems with background from nonspecific transcriptional initiation. Because an important promoter element is located downstream of the transcriptional initiation point of very late promoters, a systematic study of the effects of replacing C residues in this region was conducted. The C-free *polh* cassette constructed appears to be regulated similarly to the wild-type promoter in transient expression assays. Cell extracts prepared at 12 or 36 hr p.i. support accurately initiated, α-amanitin-resistant transcription from templates containing the 39K and *polh* promoter/C-free cassettes (Xu *et al.*, 1995). Furthermore, cell extracts prepared at 36 hr p.i. support about fivefold higher levels of transcription from the *polh* promoter than from the 39K late promoter than observed with 12 hr p.i. extracts. This observation suggests that additional virus-specific factor(s) are present in the 36 hr p.i. extract that can distinguish between late and very late promoters. Fractionation of the 36 hr p.i. nuclear extracts by chromatography resulted in the identification of two fractions—the 0.3 and 0.5 M KC1 phosphocellulose fractions—that had distinguishable transcriptional activity on the late and very late reporter plasmids. The 0.3 M fraction provides 10-fold higher levels of transcripts from the late 39K promoter than from the *polh* promoter whereas the 0.5 M fraction exhibits a twofold higher level of *polh* promoter-directed transcription than 39K-promoted transcription (Xu *et al.*, 1995). These results suggest that it should be possible to further define the differential regulation of late and very late transcription biochemically.

There are two notable features about these *in vitro* transcription systems that may provide insight into the nature of late gene transcription. First, replication of the DNA template was not required in either system, although it appears to be required *in vivo* (Rice and Miller, 1986; Gordon and Carsten, 1984). It is possible that transcription from naked or nicked DNA templates does not require DNA replication or that the apparent dependence of late gene expression on DNA replication *in vivo* is not mimicked in these *in vitro* assays (see Section VII). Second, initiation of transcription from late and very late promoters does not require a preincubation step (Xu *et al.*, 1995). This implies that the polymerase does not require the formation of a multifactor preinitiation complex to form on the template before transcriptional initiation can occur. This correlates well with the fact that the viral-induced polymerase can be purified as a multisubunit enzyme that supports late gene transcription without additional factor addition (Beniya *et al.*, 1996).

V. VIRAL GENES REGULATING LATE GENE EXPRESSION

AcMNPV genes involved in regulating baculovirus late gene expression were identified following the development of a transient expression assay in which a reporter gene under the control of the late *vp39* promoter is transactivated by a combination of viral genes following transfection of the reporter plasmid (Passarelli and Miller, 1993a,b). The original reporter plasmid used in this assay, pCAPCAT (Thiem and Miller, 1990), contains the chloramphenicol acetyltransferase (*cat*) gene under *vp39* promoter control, as well as an AcMNPV homologous repeat (*hr*) region. The presence of an *hr* sequence is necessary for AcMNPV-activated plasmid DNA replication in transfected SF cells (Kool *et al.*, 1993; Leisy and Rohrmann, 1993; Pearson *et al.*, 1992) and is thought to serve primarily as an origin of plasmid replication in this assay, although it may serve other functions as well (see Chapters 6 and 7, this volume). When pCAPCAT is cotransfected into SF-21 cells in the presence of 12 clones that collectively represent the entire AcMNPV genome (Fig. 1), the level of CAT activity observed in cell lysates prepared at 48 hr posttransfection (p.t.) are approximately 1000-fold higher than those observed in the absence of viral sequences (Passarelli and Miller, 1993a). Removal of certain clones from this genomic library results in substantial (>20-fold) decreases in transactivation levels. Substitution of transactivating clones with smaller subclones provided a rapid and accurate method for mapping and identifying the genes responsible for the transactivation.

Eighteen AcMNPV genes, *ie-1, ie-2, lefs 1–11, dnapol, p143, p47, p35,* and 39K, are required for optimal transactivation of expression from the late *vp39* and *p6.9* promoters and the very late *polh* and *p10* promoters (Todd *et al.*, 1995, 1996; Lu and Miller, 1994; Morris *et al.*, 1994; Passarelli *et al.*, 1994; Li *et al.*, 1993; Passarelli and Miller, 1993a–c, 1994) (Table 1). These genes are referred to as late gene expression factor (LEF) genes. Only two of these genes, *ie-1* and *ie-2,* also transactivate transient early gene expression. Many of these *lef*s have been transcriptionally analyzed, and they are expressed during the early stage of infection (Morris *et al.*, 1994; Passarelli and Miller, 1993b, 1994; Carstens *et al.*, 1993; Li *et al.*, 1993; Lu and Carstens, 1992; Carson *et al.*, 1991; Guarino and Smith, 1990; Chisholm and Henner, 1988; Tomalski *et al.*, 1988; Friesen and Miller, 1987; Guarino and Summers, 1987). Two additional genes, ORF 16 and 17, of AcMNPV (Ayres *et al.*, 1994) were reported to transactivate expression from a putative late promoter in transient expression assays (Guarino and Summers, 1988); however, deletion of one of these genes (ORF16/DA-26) has no apparent effect on viral replication in cell culture or in insects (O'Reilly *et al.*, 1990). Thus, the involvement of these two genes in late expression remains to be confirmed using a well-defined late promoter.

A *lef* library consisting of 18 plasmids, each carrying one of the *lef*s, is sufficient to support expression from reporter plasmids under late and very

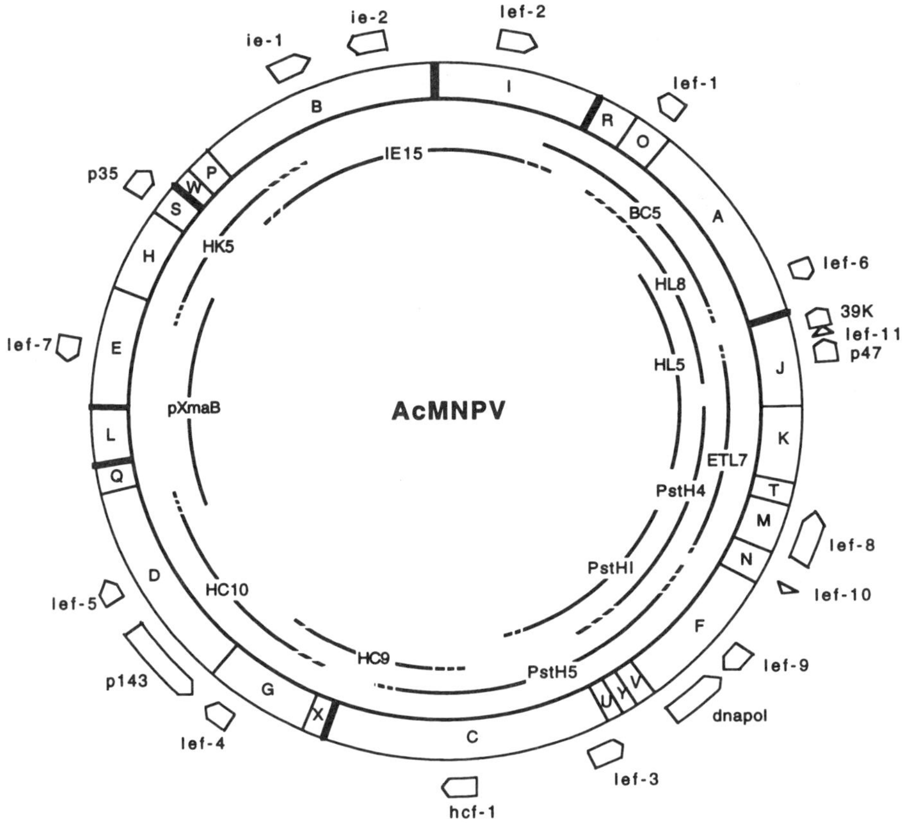

FIGURE 1. Circular representation of the *Eco*RI restriction map of AcMNPV showing the location and identity of *lefs* (outside) and the 12 clones comprising the AcMNPV clone library (inside). Bold lines represent the location of homologous repeat (*hr*) regions within the genome. Broken lines at the ends of the library clones indicate the putative positions of the ends of the clones.

late promoter control (Todd *et al.*, 1995, 1996; Lu and Miller, 1995a). Omission of any of the *lef*s dramatically reduces expression 50- to 100-fold from the *vp39* promoter. Removal of any one of the 18 *lef* plasmids from the *lef* library results in a significant reduction in the steady-state levels of *cat* RNA (Lu and Miller, 1995a); thus, none of the LEFs appear to act at the level of translational regulation.

The roles of the 18 *lef*s have been further defined using a transient plasmid DNA replication assay (Lu and Miller, 1995a). Nine of the *lef*s, *ie-1, ie-2, lef*s *1–3, lef-7, p143, dnapol,* and *p35,* influence the steady-state levels of plasmid DNA in the transfected cells, and thus are involved directly or indirectly in the replication or stability of the reporter plasmid (Lu and Miller, 1995a; Kool *et al.*, 1994; see Chapter 7, this volume). Thus, the

TABLE I. AcMNPV Late Expression Factor Genes

Gene	Map location (m.u.)[a]	Size (kDa)	Function[b]/homology
lef-2	2.3–2.8	23.9	DNA replication/transcription[c]
lef-1	8.5–7.8	30.8	DNA replication
lef-6	17.5–17.9	20.4	Transcription/viral RNA polymerase motif
39K/pp31	21.8–22.5	31.3	Transcription
lef-11	22.7–22.4	13.1	Transcription
p47	24.9–24.0	47.5	Transcription[c]
lef-8	32.2–30.3	101.8	RNA polymerase motif
lef-10	33.7–33.9	8.7	Transcription
lef-9	36.7–37.9	59.3	RNA polymerase motif
dnapol	41.3–39.1	114.3	DNA polymerase
lef-3	44.0–43.1	44.5	DNA replication/SSBP
hcf-1	44.9–45.5	34.4	DNA replication[d]
lef-4	57.2–58.2	53.9	Transcription[c]
p143	63.0–60.3	143.2	Helicase motifs[c]
lef-5	64.2–64.8	31.0	Transcription
lef-7	78.6–78.1	26.6	DNA replication[d]/SSBP motif
p35	87.0–87.7	34.8	Inhibitor of apoptosis
ie-1	95.0–96.3	66.9	*hr* Binding/transactivator[c]
ie-2/ie-n	98.7–97.8	47.0	Transactivator[d]

[a]Direction indicated as well.
[b]Based on transient expression and DNA replication assays.
[c]Phenotype also detectable by analysis of mutants.
[d]Host- or cell-specific effects observed.

transient expression assay appears to reflect the dependence of baculovirus late gene expression on DNA replication, although DNA replication also effects template copy number.

Although the role of some of these replication-associated *lef*s such as *dnapol, lef-3* and *p143* are obviously related to viral DNA replication (see Chapter 7, this volume), some of the *lef*s are probably not directly involved in DNA replication, while others may have roles in both DNA replication and gene expression. IE-1 and IE-2 are known to activate expression from early promoters and their primary role may be to transactivate other *lef*s including the early replicative *lef*s. IE-1 is known to transactivate *lef-3, lef-7, p143, dnapol, p35,* and *ie-2* promoters in transient expression assays (Morris *et al.*, 1994; Ohresser *et al.*, 1994; Lu and Carstens, 1993; Nissen and Friesen, 1989; Guarino and Summers, 1986). However, IE-1 recognizes and binds to *hr* regions (Choi and Guarino, 1995; Rodems and Friesen, 1995; Guarino and Dong, 1991) and may therefore also be directly involved in DNA replication (see Chapters 6 and 7, this volume). Temperature shift studies using a temperature-sensitive mutant of IE-1 have suggested that IE-1 is not involved directly in late gene expression but may be involved in very late gene expression (Choi and Guarino, 1995).

LEF-2 may also be involved in DNA replication, late gene expression, and very late gene expression; VLD-1, a virus carrying a mutant allele of *lef-2,* is delayed in late gene expression and defective in very late gene expression (Merrington *et al.,* 1996). Although a delay in DNA replication has not been observed for this mutant, a delay in the initial round(s) of infection can be difficult to detect experimentally. The interpretation that this gene is involved in very late gene expression should be viewed with caution, since many incidental factors can affect very late gene expression and alterations in earlier stages of baculovirus infection may curtail very late gene expression in a nonspecific manner.

The antiapoptotic gene *p35* acts as a replicative *lef* in transient plasmid replication assays (Lu and Miller, 1995a; Kool *et al.,* 1994) and affects the steady-state levels of reporter mRNA in the late transient expression assays (Lu and Miller, 1995a). The effect of *p35* in these assays is related to its antiapoptotic activity, since active members of another class of baculovirus apoptotic inhibitors, the inhibitor of apoptosis (*iap*) genes, can fully substitute for *p35* in these assays (Lu and Miller, 1995a). Furthermore, the effect of omitting *p35* from the *lef* library becomes more pronounced the later the transfected cells are harvested, suggesting that *p35* prolongs plasmid stability and/or gene expression (Todd *et al.,* 1996). A slight effect of *p35* on transient early gene expression has also been reported (Gong and Guarino, 1994). Since P35 forms stable complexes with members of the ced-3/ICE-like cysteine protease family that trigger apoptosis and inhibit them in a stoichiometric fashion (Bump *et al.,* 1995; see Chapter 10, this volume), the effect of *p35* in these transient expression assays is most likely related to preventing cell death.

Species specificity and possibly tissue specificity were observed in the requirement for three replicative *lef*s. *Ie-2* and *lef-7* exert little or no effect on plasmid DNA replication or transient late or very late gene expression in TN-368 cells derived from *Trichoplusia ni* (Lu and Miller, 1995b). An additional replicative *lef,* host cell-specific factor-one (*hcf-1*), is required for *vp39-* and *polh*-promoter-directed *cat* expression in the TN-368 cell line, but *hcf-1* cannot substitute for either of these genes in SF-21 cells. *Hcf-1* is not required for late gene expression in SF-21 cells and appears to play a host-specific role during viral infection (Lu and Miller, 1995b; see Chapter 9, this volume).

At least three of the transcription-specific *lef*s are likely to be components of the virus-encoded RNA polymerase (Lu and Miller, 1994; Passarelli *et al.,* 1994). Two of these LEFs, LEF-8 and LEF-9 contain amino acid sequence motifs that are conserved in prokaryotic and eukaryotic RNA polymerases (Lu and Miller, 1994; Passarelli *et al.,* 1994), while LEF-6 shares some sequence similarity to the 19-kDa subunit of the vaccinia virus DNA-dependent RNA polymerase (Passarelli and Miller, 1994). LEF-8 is predicted to encode a 102-kDa polypeptide with a conserved motif, GXKX$_4$HGQ/NKG, which is found in the B or B′ subunit of DNA-directed RNA polymerases from a diverse group of organisms including bacteria, yeast, plants, invertebrates, and vertebrates (Passarelli *et al.*, 1994). In every case, the location of this motif is positionally conserved (at the carboxy-terminus) and is thought to be an essential component at the catalytic site of the polymerase (Schultz *et al.*, 1993). LEF-8 may correspond to the 102-kDa subunit of the viral-specific RNA polymerase reported by Beniya *et al.* (1996). LEF-9 is 59 kDa and contains a short sequence motif similar to a conserved motif identified in the largest subunit of prokaryotic and eukaryotic DNA-directed RNA polymerases (Lu and Miller, 1994). The predicted polypeptide sizes of LEF-6 and LEF-9 do not correlate well with the sizes of the polypeptides reported to co-purify with viral RNA polymerase activity (Beniya *et al.*, 1996). The remaining *lef*s, *p47, lef-4, lef-5,* 39K, *lef-10,* and *lef-11,* appear to have no significant homology to genes in existing sequence databases.

The roles of *p47* and *lef-4* in late gene transcription have been established by independent methods; these genes were identified by two conditional lethal temperature-sensitive mutants of AcMNPV, ts317 and ts538, respectively (Carstens *et al.*, 1993, 1994; Partington *et al.*, 1990). At the nonpermissive temperature these mutants synthesize viral DNA, but they are defective in late and very late gene expression. The temperature-sensitive defect is exerted at the level of transcription or RNA stability, since late *p6.9* and *vp39* transcripts as well as *p10* and *polh* transcripts are severely reduced or absent from mutant-infected cells (Carstens *et al.*, 1994; Partington *et al.*, 1990). Both mutants are defective in extracellular virus production. *p47* is transcribed early in infection as part of a multicistronic message that also contains an uncharacterized ORF (*p43*), *lef-11,* and 39K (Carstens *et al.*, 1993). This 4.0-kb transcript appears to be the only transcript encoding P47. *lef-4* is located immediately upstream and in the opposite direction to *vp39* (Passarelli and Miller, 1993c); the promoters of these two genes apparently overlap.

The 39K gene (also known as pp31) encodes a phosphoprotein that is produced both early and late in infection and is associated with the virogenic stroma in the nucleus (Guarino *et al.*, 1992). This structure is a network of electron-dense filaments that is believed to be the site of viral DNA replication and transcription and also clearly serves as the site for the morphogenesis of nucleocapsids (Fraser, 1986; Granados and Lawler, 1981; see Chapter 4, this volume). The localization of 39K to the virogenic stroma is consistent with its role in late and very late gene transcription.

VI. VIRAL GENE(S) REGULATING VERY LATE GENE EXPRESSION

Characterization of tsB837, a temperature-sensitive mutant of AcMNPV that is defective in occlusion body production (Lee and Miller, 1979), led to the identification of a gene, very late expression factor-1 gene (*vlf-1*), which appears to regulate very late gene transcription (McLachlin and Miller, 1994).

The tsB837 mutant produces BVs at the nonpermissive temperature but is defective in the production of polyhedra; plaques produced by this virus are similar in size and number to wild-type plaques but lack polyhedra (Lee and Miller, 1979). At the nonpermissive temperature, synthesis of late proteins proceeds normally, but polyhedrin and P10 levels are reduced in tsB837-infected cells (McLachlin and Miller, 1994). The failure of tsB837 to accumulate very late proteins is due to reduced steady-state levels of the very late RNAs. Thus, tsB837 is defective in either the transcription or stabilization of very late genes. The reduction in the levels of *p10* RNA is not as pronounced as that observed for *polh* RNA (McLachlin and Miller, 1994). This difference may be due to a differential effect of the temperature-sensitive mutation on VLF-1 interaction with these promoters or may reflect their differential regulation.

The tsB837 mutation was mapped to a single base pair change in *vlf-1,* which is predicted to direct the synthesis of a 44-kDa polypeptide, VLF-1, with sequence similarity to a family of prokaryotic and yeast integrase/resolvases (McLachlin and Miller, 1994). Proteins belonging to this family, which includes *cre* of phage P1, Tn554 and Tn4430 transposons, FLP of *S. cerevisiae,* and lambda phage integrase (*int*), all possess two distinct regions of homology. This family of genes has general sequence similarity in two different regions, A and B, which are separated by approximately 50 to 100 residues. Only one residue (Arg) is strictly conserved in region A, while region B contains several conserved amino acid residues in the form of $HXXRHX_9GX_{11-12}HX_{8-9}Y$. VLF-1 has all of these sequence features except for the first histidine of region B. The equivalent histidine has been site-specifically mutagenized in the yeast FLP; the mutant retains DNA binding activity but is defective in strand synapsis, with the alignment of two different double-stranded DNA strands during recombination (Parsons *et al.,* 1988). Thus, VLF-1 might have DNA binding activity but probably lacks integrase activity. Region B of VLF-1 also bears some similarity to a motif found in a superfamily of RNA helicases that falls outside of the helicase domains (McLachlin and Miller, 1994). The significance of these motifs in VLF-1 is unknown; there is no evidence that baculoviruses integrate into the host genome or that they rearrange viral sequences during very late gene expression.

VLF-1 preferentially stimulates very late gene expression in transient expression assays (Todd *et al.,* 1996). Replacement of clones containing *vlf-1* from the genomic clone library with *lef*s found in this region results in transient expression from both late and very late promoters. However, addition of a plasmid containing *vlf-1* further increases expression from the very late *polh* and *p10* promoters approximately five- to tenfold but has little or no effect on the late *vp39* and *p6.9* promoters (Todd *et al.,* 1996). Thus, the specific effect of VLF-1 on very late gene expression can also be observed in transient assays. The transcriptional organization of the *vlf-1* region is complex; at least two early and four late transcripts traverse this region (McLachlin and Miller, 1994).

The approximately five- to tenfold activation that *vlf-1* exerts on *polh* expression in transient assays is lower than expected for such a robust promoter, suggesting either that additional genes are involved in very late gene expression or that the transient expression system is not fully mimicking *in vivo* infection conditions. The fact that *polh* is being expressed in the absence of *vlf-1* suggests that the expression that is observed from the *polh* promoter in this system may be due primarily to the weak activity of *polh* as a late promoter. Prolonged late expression would effectively increase the background level of *polh* transcription and reduce the relative level of activation that can be observed for *vlf-1* in normal infections.

A systematic search of the AcMNPV genome using transient expression assays revealed no additional genes that differentially affected only very late gene expression (Todd *et al.*, 1996). This suggests that VLF-1 may be the major transactivator of very late gene expression. However, the comparatively weak activity of this promoter in this assay leaves room for additional explanations of these results. Assuming VLF-1 has DNA binding activity like other members of the integrase family, a possible hypothesis for VLF-1 function is that VLF-1 interacts with the burst element of very late promoters and thereby directs the viral-specific RNA polymerase to these promoters.

As noted previously (Section V), some of the *lef*s may play a role in more than one aspect of very late gene expression, including DNA replication, *lef* expression, late gene expression, and very late gene expression. The transient expression assay is unable to distinguish which of the *lef*s are directly required for very late gene expression in its current form. As noted in Section V, a role for both IE-1 and LEF-2 in very late gene expression has been proposed based on the behavior of viruses with mutations in these genes. Unlike the *vlf-1* mutant tsB837, these mutants exhibit additional defects, including a delay in late gene expression and a reduction in the number of progeny virus produced, so that the mutational effect is not specific for very late gene expression. More work is necessary to determine whether these genes affect very late gene expression directly or indirectly.

VII. ADDITIONAL ASPECTS OF LATE AND VERY LATE GENE REGULATION

A. The Role of DNA Replication in Viral Gene Transcription

A dependence of late transcription on DNA replication is observed in normal infections. This dependence has been demonstrated during infection in multiple ways. Treatment of infected cells with aphidicolin, an inhibitor of both cellular and viral DNA polymerases (Miller *et al.*, 1981), completely inhibits late gene expression and blocks the infection at the transition point between early and late gene expression (Rice and Miller, 1986). The dependence is also observed with viral mutants. A null mutant of the proliferating cell nuclear antigen (PCNA) homologue is slightly delayed in both DNA replication and late gene expression (O'Reilly *et al.*, 1989; Crawford and Miller, 1988). As expected, PCNA was not required in transient DNA replication or late gene expression assays (Kool *et al.*, 1994; Passarelli *et al.*, 1994); a subtle difference in temporal regulation would not be expected to be observable in these types of assays. A temperature-sensitive mutation in the helicase-related polypeptide p143, ts8, results in defective DNA replication and a block in late gene transcription at the nonpermissive temperature (Erlandson *et al.*, 1985; Gordon and Carstens, 1984). A dual role for p143 in both replication and late gene expression could be an alternate explanation for this effect. It is unlikely that the dependence of late transcription on DNA replication is simply due to template copy number. In the case of aphidicolin inhibition, late gene transcription is totally blocked, but early gene transcription is easily observed from the same number of templates.

Nuclear run-on assays using cell nuclei isolated from AcMNPV-infected cells treated with aphidicolin also indicate that α-amanitin-resistant transcription depends on DNA replication during infection (Huh and Weaver, 1990b). Direct addition of aphidicolin to the nuclear run-on assays, however, does not affect transcription, indicating that preinitiated transcription complexes are unaffected by DNA replication.

Current *in vitro* transcription assays do not reflect a dependence of late gene expression on DNA replication. At least two explanations are possible for this observation. If the dependence of baculovirus late transcription depends on nicks in the DNA template introduced by viral DNA replication as is observed for the dependence of T4 bacteriophage late gene expression on phage DNA replication, then nicks present in templates used in the *in vitro* transcription system allow late transcription in the absence of DNA replication. Alternatively, if DNA replication is required to displace proteins on the viral DNA templates during infection in order to initiate late gene expression, transcription could proceed on naked DNA templates used in the *in vitro* assays. It is clear that continuous DNA replication is not required in these *in vitro* transcription systems (Xu *et al.*, 1995; Glocker *et al.*, 1993).

Whether very late gene transcription is coupled to DNA replication is not known. Although very late gene expression depends on DNA replication during infection, that dependence may be related to a requirement for late gene expression to activate very late gene expression.

B. The Regulation of Host Gene Expression

During the late phase of infection, the steady-state levels of host cellular RNAs decline (Ooi and Miller, 1988) and host protein synthesis ceases (Carstens *et al.*, 1979). The mechanism governing the decline in host RNAs is not known; the decline appears to be controlled by one or more late viral genes based on (1) the sensitivity of host RNA decline to aphidicolin, which blocks late but not early gene expression, and (2) the resistance of the decline to cycloheximide, which inhibits early gene expression (Ooi and Miller, 1988). Although some α-amanitin-sensitive viral gene transcription appears to continue during the late phase of infection (Huh and Weaver, 1990a), it is not clear that host RNA synthesis continues during the late phase. A lack of host RNA synthesis would contribute to the decline in the steady-state levels of these mRNAs. Unlike the viral DNA that appears to adopt a unique nucleoprotein structure during the late phase of infection, the host DNA remains in a chromatinlike structure (Wilson and Miller, 1986), and this may account for differential viral and host α-amanitin sensitive RNA synthesis. The viral and host DNAs are also probably located in different compartments of the infected nucleus, with active transcription most likely occurring in the virogenic stroma; however, this remains to be demonstrated experimentally.

It is also not clear whether there is a specific translational mechanism for down-regulating expression from early and host RNAs versus late or very late RNAs; the shutoff of host proteins seems to be more dramatic than what can be accounted for by the decline in host RNA levels. Theoretically, a distinction could be made between RNA polymerase II-mediated and viral RNA polymerase-mediated transcripts, since the latter invariably have a TAAG (or AAG) sequence at their 5′ end. The question of a specific mechanism for the shutoff of host protein synthesis remains to be addressed experimentally.

C. The Role of Translational Regulation

There is no firm evidence for translational regulation from any mRNAs during the course of wild-type baculovirus infection. Some evidence suggests that sequences immediately surrounding the translational initiation codon of *polh* influence translational initiation from this message (Luckow and Summers, 1988; Kang *et al.*, 1987), but much of the effect of this sequence may be explained by general features of eukaryotic translation initiation control (Kozak, 1991). Many late transcripts have mini-open reading frames (ORFs) or are dicistronic or polycistronic and regulation of expression of downstream ORFs may occur in such transcripts.

D. The Regulation of Flanking Gene Expression

Different temporal classes of baculovirus transcripts are dispersed throughout the genome, and it is likely that transcription of one gene may influence the regulation of flanking gene(s). One study demonstrated that transcription from the *polh* promoter down-regulates the steady-state levels of late transcripts initiated downstream of *polh* and synthesized in the opposite direction as *polh* RNAs (Ooi and Miller, 1990). The affected transcripts are those initiated from the late ORF9 (also known as the 1629 ORF) promoter. Two size classes of transcripts are produced from the ORF9 promoter: 1.7-kb and 3.2-kb RNAs. The larger RNAs traverse not only ORF9, but also traverse both the *polh* gene (ORF8) in the antisense direction and ORF7 (or ORF603) in the sense direction. The 1.7-kb RNAs traverse ORF9 as well as overlapping the 3′ end of *polh* RNAs in the antisense direction. Unlike many late RNAs, the steady-state levels of both the 1.7- and 3.2-kb RNAs rapidly decline upon the initiation of very late gene expression. The decline in the levels of these late RNAs during the very late phase is blocked by a mutation in the TAAG sequence of the *polh* promoter, indictating that the initiation of *polh* transcription is responsible for the rapid decline of the 1.7- and 3.2-kb late RNA levels. It is not known whether this is due to the disruption of transcription from the ORF9 promoter by RNA polymerase that initiated transcription from the *polh* TAAG and are extending through ORF9 in an antisense direction. An alternative possibility is that *polh* RNAs hybridize with the 1.7- and 3.2-kb RNAs and thereby activate a double-stranded RNA-directed turnover mechanism that eliminates the hybrids, causing the observed decline in the 1.7- and 3.2-kb RNAs.

Another study provided additional evidence that upstream and downstream flanking regions can affect late gene expression (Gross and Rohrman, 1993). The polyhedron envelope protein (PEP) gene (ORF 131/calyx/pp34) is the third ORF of five tandem ORFs that are transcribed late or very late in infection from their respective promoters, resulting in a family of overlapping RNAs with coterminal 3′ ends. Elimination of transcription from upstream ORFs by TAAG mutations or by deletion of the two upstream ORFs altogether increased PEP gene expression between 17 and 35%, suggesting a potential role for promoter occlusion in the regulation of baculovirus late gene expression. Furthermore, deletions of sequences 3′ to the PEP gene were found to reduce PEP gene promoter activity, suggesting a role for 3′ flanking sequences in regulating late gene expression.

There will probably be other examples of *cis*-acting transcriptional regulation of flanking genes, since baculovirus genes are compactly arranged and early, late, and very late genes are dispersed throughout the genome.

VIII. SUMMARY

The body of work reviewed here suggests that baculovirus late and very late gene expression is regulated primarily at the level of transcription. Transient expression assays have identified a large number of LEF genes that are involved in late and very late gene expression. A subset of these genes encode DNA replication factors indicative of a relationship between DNA replication and late gene expression. It is likely that at least some of the transcriptional LEFs are subunits of the viral RNA polymerase, based on the presence of conserved RNA polymerase motifs and the multisubunit structured of the partially purified polymerase.

Both late and very late genes probably share a common transcriptional apparatus, since all of the LEFs affect both late and very late gene expression and both promoter classes contain an essential TAAG promoter that serves as the transcriptional initiation site. However, very late genes exhibit an additional level of control that is dependent on expression of at least one late gene product, VLF-1, and at least one additional promoter element, the burst element. VLF-1 may regulate expression by interaction either directly or indirectly with this burst element located downstream of the transcriptional initiation site.

The exact functions of each of these LEFs in late and very late gene expression remain to be explored, but should be facilitated by the development of *lef*-based *in vitro* transcription systems. Knowledge of the functions of the LEFs and how they interact with each other during infection will provide insight into the novel regulatory processes that affect baculovirus host range and expression of heterologous genes in the baculovirus expression vector system.

REFERENCES

Ayres, M. D., Howard, S. C., Kuzio, J., Lopez-Ferber, M., and Possee, R. D., 1994, The complete sequence of *Autographa californica* nuclear polyhedrosis virus, *Virology* **202:**586.

Beniya, H., Funk, C. J., Rohrmann, G. F., and Weaver, R. F., 1996, Purification of a virus-induced RNA polymerase from *Autographa californica* nuclear polyhedrosis virus-infected *Spodoptera frugiperda* cells that accurately initiates late and very late transcription *in vitro, Virology* **216:**12.

Bump, N. J., Hackett, M., Hugunin, M., Seshagiri, S., Brady, K., Chen, P., Ferenz, C., Franklin, S., Ghayur, T., Li, P., Mankovich, J., Shi, L., Greenburg, A. H., Miller, L. K., and Wong, W. W., 1995, Inhibition of ICE family proteases by baculovirus antiapoptotic protein p35, *Science* **269:**1865.

Burma, S., Mukherjee, B., Jain, A., Habib, S., and Hasnain, S., 1994, An unusual 30-kDa protein binding to the polyhedrin promoter of *Autographa californica* nuclear polyhedrosis virus, *J. Biol. Chem.* **269:**2750.

Carson, D. D., Summers, M. D., and Guarino, L. A., 1991, Molecular analysis of a baculovirus regulatory gene, *Virology* **182:**279.

Carstens, E. B., Tjia, S. T., and Doerfler, W., 1979, Infection of *Spodoptera frugiperda* cells with *Autographa californica* nuclear polyhedrosis virus. I. Synthesis of intracellular proteins after virus infection, *Virology* **99:**386.

Carstens, E. B., Lu, A. L., and Chan, H., 1993, Sequence, transcriptional mapping, and overexpression of *p47,* a baculovirus gene regulating late gene expression, *J. Virol.* **67:**2513.

Carstens, E. B., Chan, H., Yu, H., Williams, G. V., and Casselman, R., 1994, Genetic analyses of temperature-sensitive mutations in baculovirus late expression factors, *Virology* **204:**323.

Chaabihi, H., Ogliastro, M.-H., Martin, M., Giraud, C., Devauchelle, G., and Cerutti, M., 1993, Competition between baculovirus polyhedrin and *p10* gene expression during infection of insect cells, *J. Virol.* **67:**2664.

Chisholm, G. E., and Henner, D. J., 1988, Multiple early transcripts and splicing of the *Autographa californica* nuclear polyhedrosis virus IE-1 gene, *J. Virol.* **62:**3193.

Choi, J., and Guarino, L. A., 1995, A temperature-sensitive IE1 protein of *Autographa californica* nuclear polyhedrosis virus has altered transactivation and DNA binding activities, *Virology* **209:**90.

Crawford, A. M., and Miller, L. K., 1988, Characterization of an early gene accelerating expression of late genes of the baculovirus *Autographa californica* nuclear polyhedrosis virus, *J. Virol.* **62:**2773.

Dickson, J. A., and Friesen, P. D., 1991, Identification of upstream promoter elements mediating early transcription from the 35,000 molecular weight protein gene of *Autographa californica* nuclear polyhedrosis virus, *J. Virol.* **65:**4006.

Eldridge, R., Li, Y., and Miller, L. K., 1992, Characterization of a baculovirus gene encoding a small conotoxinlike polypeptide, *J. Virol.* **66:**6563.

Erlandson, M. A., Gordon, J. D., and Carstens, E. B., 1985, Size and map locations of early transcription products on the *Autographa californica* nuclear polyhedrosis virus genome, *Virology* **142:**12.

Etkin, E., Carp-Weiss, L., and Levi, B.-Z., 1994, *Spodoptera frugiperda* Sf-9 cells nuclear factor binds to a specific sequence on the baculovirus polyhedrin promoter, *Virus Res.* **31:**343.

Fraser, M. J., 1986, Ultrastructural observations of virion maturation in *Autographa californica* nuclear polyhedrosis virus infected *Spodoptera frugiperda* cells, *J. Ultrastruct. Mol. Struct. Res.* **95:**189.

Friesen, P. D., and Miller, L. K., 1987, Divergent transcription of early 35- and 94-kilodalton protein genes encoded by the *Hin*dIII-K genome fragment of the baculovirus *Autographa californica* nuclear polyhedrosis virus, *J. Virol.* **61:**2264.

Friesen, P. D., Rice, W. C., Miller, D. W., and Miller, L. K., 1986, Bidirectional transcription from a solo long terminal repeat of the retrotransposon TED: Symmetrical RNA start sites, *Mol. Cell. Biol.* **6:**1599.

Fuchs, L. Y., Woods, M. S., and Weaver, R. F., 1983, Viral transcription during *Autographa californica* nuclear polyhedrosis virus infection: A novel RNA polymerase induced in infected *Spodoptera frugiperda* cells, *J. Virol.* **48:**641.

Glocker, B., Hoopes, Jr., R. R., Hodges, L., and Rohrmann, G. F., 1993, *In vitro* transcription from baculovirus late gene promoters: Accurate mRNA initiation by nuclear extracts prepared from infected *Spodoptera fruigiperda* cells, *J. Virol.* **67:**3771.

Gong, M., and Guarino, L. A., 1994, Expression of the 39K promoter of *Autographa californica* nuclear polyhedrosis virus is increased by the apoptotic suppressor P35, *Virology* **204:**38.

Gordon, J. D., and Carstens, E. B., 1984, Phenotypic characterization and physical mapping of a temperature sensitive mutant of *Autographa californica* nuclear polyhedrosis virus defective in DNA synthesis, *Virology* **138:**69.

Granados, R. R., and Lawler, L. A., 1981, *In vivo* pathway of *Autographa californica* baculovirus invasion and infection, *Virology* **108:**297.

Gross, C. H., and Rohrmann, G. F., 1993, Analysis of the role of 5′ promoter elements and 3′ flanking sequences on the expression of a baculovirus polyhedron envelope protein gene, *Virology* **192:**273.

Grula, M. A., and Weaver, R. F., 1981, An improved method for isolation of *Heliothis zea* DNA-dependent RNA polymerase: Separation and characterization of a form III RNA polymerase activity, *Insect Biochem.* **11:**149.

Grula, M. A., Buller, P. L., and Weaver, R. F., 1981, alpha-Amanitin resistant viral RNA synthesis in nuclei isolated from nuclear polyhedrosis virus-infected *Heliothis zea* larvae and *Spodoptera frugiperda* cells, *J. Virol.* **38:**916.

Guarino, L. A., 1990, Identification of a viral gene encoding a ubiquitin-like protein, *Proc. Natl. Acad. Sci. USA* **87:**409.

Guarino, L. A., and Dong, W., 1991, Expression of an enhancer-binding protein in insect cells transfected with the *Autographa californica* nuclear polyhedrosis virus IE1 gene, *J. Virol.* **65:**3676.

Guarino, L. A., and Smith, M. W., 1990, Nucleotide sequence and characterization of the 39K region *Autographa californica* nuclear polyhedrosis virus, *Virology* **179:**1.

Guarino, L. A., and Smith, M. W., 1992, Regulation of delayed-early gene transcription by dual TATA boxes, *J. Virol.* **66:**3733.

Guarino, L. A., and Summers, M. D., 1986, Functional mapping of a *trans*-activating gene required for expression of a baculovirus delayed-early gene, *J. Virol.* **57**:563.

Guarino, L. A., and Summers, M. D., 1987, Nucleotide sequence of a baculovirus regulatory gene, *J. Virol.* **61**:2091.

Guarino, L. A., and Summers, M. D., 1988, Functional mapping of *Autographa californica* nuclear polyhedrosis virus genes required for late gene expression, *J. Virol.* **62**:463.

Guarino, L. A., Dong, W., Xu, B., Broussard, D. R., Davis, R. W., and Jarvis, D. L., 1992, Baculovirus phosphoprotein pp31 is associated with virogenic stroma, *J. Virol.* **66**:7113.

Hoopes, R. R., and Rohrmann, G. F., 1991, *In vitro* transcription of baculovirus immediate early genes: Accurate mRNA initiation by nuclear extracts from both insect and human cells, *Proc. Natl. Acad. Sci. USA* **88**:4513.

Howard, S. C., Ayres, M. D., and Possee, R. D., 1986, Mapping the 5′ and 3′ ends of *Autographa californica* nuclear polyhedrosis virus polyhedrin mRNA, *Virus Res.* **5**:109.

Huh, N. E., and Weaver, R. F., 1990a, Categorizing some early and late transcripts directed by the *Autographa californica* nuclear polyhedrosis virus, *J. Gen Virol.* **70**:2195.

Huh, N. E., and Weaver, R. F., 1990b, Identifying the RNA polymerases that synthesize specific transcripts of the *Autographa californica* nuclear polyhedrosis virus, *J. Gen. Virol.* **71**:195.

Jain, A., and Hasnain, S. E., 1996, A 30-kDa host protein bind to two very-late baculovirus promoters, *Eur. J. Biochem.* **239**:384.

Kang, C. Y., Bishop, D. H. L., Seo, J. S., Marsuura, Y., and Choe, M., 1987, Secretion of particles of hepatitis B surface antigen from insect cells using a baculovirus vector, *J. Gen. Virol.* **68**:2607.

Kool, M., van den Berg, M. M. M., Tramper, J., Goldbach, R. W., and Vlak, J. M., 1993, Location of two putative origins of DNA replication of *Autographa californica* nuclear polyhedrosis virus, *Virology* **192**:94.

Kool, M., Ahrens, C. H., Goldbach, R. W., and Rohrmann, G. F., 1994, Identification of genes involved in DNA replication of the *Autographa californica* baculovirus, *Proc. Natl. Acad. Sci. USA* **91**:11212.

Kozak, M., 1991, Structural feature in eukaryotic mRNAs that modulate the initiation of translation, *J. Biol. Chem.* **266**:19867.

Kuzio, J., Rohel, D. Z., Curry, C. J., Krebs, A., Carstens, E. B., and Faulkner, P. F., 1984, Nucleotide sequence of the p10 polypeptide gene of *Autographa californica* nuclear polyhedrosis virus, *Virology* **139**:414.

Lee, H. H., and Miller, L. K., 1979, Isolation, complementation, and initial characterization of temperature sensitive mutants of the baculovirus *Autographa californica* nuclear polyhedrosis virus, *J. Virol.* **31**:240.

Leisy, D. J., and Rohrmann, G. F., 1993, Characterization of the replication of plasmids containing hr sequences in baculovirus-infected *Spodoptera frugiperda* cells, *Virology* **196**:722.

Li, Y., Passarelli, A. L., and Miller, L. K., 1993, Identification, sequence, and transcriptional mapping of *lef-3,* a baculovirus gene involved in late and very late gene expression, *J. Virol.* **67**:5260.

Lu, A., and Carstens, E. B., 1992, Transcriptional analysis of the EcoRI-D region of the baculovirus *Autographa californica* nuclear polyhedrosis virus identifies an early 4.0-kilobase RNA encoding the essential p143 gene, *J. Virol.* **66**:655.

Lu, A., and Carstens, E. B., 1993, Immediate-early baculovirus genes transactivate the *p143* gene promoter of *Autographa californica* nuclear polyhedrosis virus, *Virology* **195**:710.

Lu, A., and Miller, L. K., 1994, Identification of three late expression factor genes within the 33.8- to 43.3-map-unit region of *Autographa californica* nuclear polyhedrosis virus, *J. Virol.* **68**:6710.

Lu, A., and Miller, L. K., 1995a, The roles of eighteen baculovirus late expression factor genes in transcription and DNA replication, *J. Virol.* **69**:975.

Lu, A., and Miller, L. K., 1995b, Differential requirements for baculovirus late expression factor genes in two cell lines, *J. Virol.* **69**:6265.

Luckow, V. A., and Summers, M. D., 1988, Signals important for high-level expression of foreign genes in *Autographa californica* nuclear polyhedrosis virus expression vectors, *Virology* **167**:56.

Matsuura, Y., Possee, R. D., Overton, H. A., and Bishop, D. H. L., 1987, Baculovirus expression vectors: The requirements for high level expression of proteins including glycoproteins, *J. Gen. Virol.* **68**:1233.

McLachlin, J. R., and Miller, L. K., 1994, Identification and characterization of *vlf-1*, a baculovirus gene involved in very late gene expression, *J. Virol.* **68**:7746.

Merrington, C. L., Kitts, P. A., King, L. A., and Possee, R. D., 1996, An *Autographa californica* nucleopolyhedrovirus lef-2 mutant: Consequences for DNA replication and very late gene expression, *Virology* **217**:338.

Miller, L. K., Jewell, J. E., and Browne, D., 1981, Baculovirus induction of a DNA polymerase, *J. Virol.* **40**:305.

Min, M.-K., and Bishop, D. H. L., 1991, Transcriptional analyses of baculovirus polyhedrin and foreign gene expression relative to baculovirus p10 mRNA levels. *J. Gen. Virol.* **72**:2551.

Morris, T. D., and Miller, L. K., 1994, Mutational analysis of a baculovirus major late promoter. *Gene* **140**:147.

Morris, T. D., Todd, J. W., Fisher, B., and Miller, L. K., 1994, Identification of *lef-7*, a baculovirus gene affecting late gene expression. *Virology* **200**:360.

Nissen, M. S., and Friesen, P. D., 1989, Molecular analysis of the transcriptional regulatory region of an early baculovirus gene. *J. Virol.* **63**:493.

Ohresser, M. N., Morin, N., Cerutti, M., and Delsert, C., 1994, Temporal regulation of a complex and unconventional promoter by viral products. *J. Virol.* **68**:2589.

Ooi, B. G., and Miller, L. K., 1988, Regulation of host RNA levels during baculovirus infection. *Virology* **166**:515.

Ooi, B. G., and Miller, L. K., 1990, Transcription of the baculovirus polyhedrin gene reduces the levels of an antisense transcript initiated downstream. *J. Virol.* **64**:3126.

Ooi, B. G., Rankin, C., and Miller, L. K., 1989, Downstream sequence augment transcription from the essential initiation site of a baculovirus polyhedrin gene, *J. Mol. Biol.* **210**:721.

O'Reilly, D. R., Crawford, A. M., and Miller, L. K., 1989, Viral proliferating cell nuclear antigen, *Nature* **337**:606.

O'Reilly, D. R., Passarelli, A. L., Goldman, I. F., and Miller, L. K., 1990, Characterization of the DA26 gene in a hypervariable region of the *Autographa californica* nuclear polyhedrosis virus genome, *J. Gen. Virol.* **71**:1029.

Parsons, R. L., Prasad, P. V., Harshey, R. M., and Jayaram, M., 1988, Step-arrest mutants of FLP recombinase: Implications for the catalytic mechanism of DNA recombination, *Mol. Cell. Biol.* **8**:3303.

Partington, S., Yu, H., Lu, A., and Carstens, E. B., 1990, Isolation of temperature sensitive mutants of *Autographa californica* nuclear polyhedrosis virus: Phenotype of baculovirus mutants defective in very late gene expression, *Virology* **175**:91.

Passarelli, A. L., and Miller, L. K., 1993a, Three baculovirus genes involved in late and very late gene expression. *ie-1, ie-n*, and *lef-2*, *J. Virol.* **67**:2149.

Passarelli, A. L., and Miller, L. K., 1993b, Identification and characterization of *lef-1*, a baculovirus gene involved in late and very late gene expression, *J. Virol.* **67**:3481.

Passarelli, A. L., and Miller, L. K., 1993c, Identification of genes encoding late expression factors located between 56.0 and 65.4 map units of the *Autographa californica* nuclear polyhedrosis virus genome, *Virology* **197**:704.

Passarelli, A. L., and Miller, L. K., 1994, Identification and transcriptional regulation of the baculovirus *lef-6* gene, *J. Virol.* **68**:4458.

Passarelli, A. L., Todd, J. W., and Miller, L. K., 1994, A baculovirus gene involved in late gene expression predicts a large polypeptide with a conserved motif of RNA polymerases, *J. Virol.* **68**:4673.

Pearson, M. N., Bjornson, R. M., Pearson, G., and Rohrmann, G. F., 1992, The *Autographa californica* baculovirus genome: Evidence for multiple replication origins, *Science* **257**:1382.

Possee, R. D., and Howard, S. C., 1987, Analysis of the polyhedrin gene promoter of the *Autographa californica* nuclear polyhedrosis virus, *Nucleic Acids Res.* **15**:10233.

Possee, R. D., Sun, T.-P., Howard, S. C., Ayres, M. D., Hill-Perkin, M., and Gearing, K. L., 1991, Nucleotide sequence of the *Autographa californica* nuclear polyhedrosis virus 9.4 kbp EcoRI-I and -R (polyhedrin gene) region, *Virology* **185**:229.

Qin, J., Liu, A., and Weaver, R. F., 1989, Studies on the control region of the *p10* gene of the *Autographa californica* nuclear polyhedrosis virus, *J. Gen. Virol.* **70**:1273.

Rankin, C., Ooi, B. G., and Miller, L. K., 1988, Eight base pairs encompassing the transcriptional start point are the major determinant for baculovirus polyhedrin gene expression, *Gene* **70**:39.

Rice, W. C., and Miller, L. K., 1986, Baculovirus transcription in the presence of inhibitors and in nonpermissive *Drosophila* cells, *Virus Res.* **6**:155.

Rodems, S. M., and Friesen, P. D., 1995, Transcriptional enhancer activity of hr5 requires dual-palindrome half sites that mediate binding of a dimeric form of the baculovirus trans-regulator IE1, *J. Virol.* **69**:5368.

Roelvink, P. W., van Meer, M. M. M., de Kort, C. A. D., Possee, R. D., Hammock, B. D., and Vlak, J. M., 1992, Dissimilar expression of *Autographa californica* multiple nucleocapsid nuclear polyhedrosis virus polyhedrin and *p10* genes, *J. Gen. Virol.* **73**:1481.

Rohrmann, G. F., 1986, Polyhedrin structure, *J. Gen. Virol.* **67**:1499.

Schultz, P., Celia, H., Riva, M., Sentenac, A., and Oudet, P., 1993, Three-dimensional model of yeast RNA polymerase I determined by electron microscopy of two-dimensional crystals, *EMBO J.* **12**:26017.

Thiem, S. M., and Miller, L. K., 1989, Identification, sequence, and transcriptional mapping of the major capsid protein gene of the baculovirus *Autographa californica* nuclear polyhedrosis virus, *J. Virol.* **63**:2008.

Thiem, S. M., and Miller, L. K., 1990, Differential gene expression mediated by late, very late, and hybrid baculovirus promoters, *Gene* **91**:87.

Todd, J. W., Passarelli, A. L., and Miller, L. K., 1995, Eighteen baculovirus genes, including *lef-11, p35,* 39K, and *p47,* support late gene expression, *J. Virol.* **69**:968.

Todd, J. W., Passarelli, A. L., Lu, A., and Miller, L. K., 1996, Factors regulating baculovirus late and very late gene expression in transient-expression assays, *J. Virol.* **70**:2307.

Tomalski, M. D., Wu, J., and Miller, L. K., 1988, The location, sequence, transcription and regulation of a baculovirus DNA polymerase gene, *Virology* **167**:591.

Tomalski, M. D., Eldridge, R., and Miller, L. K., 1991, A baculovirus homolog of a Cu/Zn superoxide dismutase gene, *Virology* **184**:149.

Weyer, U., and Possee, R. D., 1989, Analysis of the promoter of the *Autographa californica* nuclear polyhedrosis virus *p10* gene, *J. Gen. Virol.* **70**:203.

Whitford, M., and Faulkner, P. F., 1992, Nucleotide sequence and transcriptional analysis of a gene encoding gp41, a structural glycoprotein of the baculovirus *Autographa californica* nuclear polyhedrosis virus, *J. Virol.* **66**:4763.

Wilson, M. E., and Miller, L. K., 1986, Changes in the nucleoprotein complexes of a baculovirus DNA during infection, *Virology* **151**:315.

Wilson, M. E., Mainprize, T. H., Friesen, P. D., and Miller, L. K., 1987, Location, transcription and sequence of a baculovirus gene encoding a small arginine-rich polypeptide, *J. Virol.* **61**:661.

Wu, J., and Miller, L. K., 1989, Sequence, transcription and translation of a late gene of the *Autographa californica* nuclear polyhedrosis virus encoding a 34.8 k polypeptide, *J. Gen. Virol.* **70**:2449.

Xu, B., Yoo, S., and Guarino, L. A., 1995, Differential transcription of baculovirus late and very late promoters: Fractionation of nuclear extracts by phosphocellulose chromoatography, *J. Virol.* **69**:2912.

Yang, C. L., Stetler, D. A., and Weaver, R. F., 1991, Structural comparison of the *Autographa californica* nuclear polyhedrosis virus-induced RNA polymerase and the three nuclear RNA polymerases from the host, *Spodoptera frugiperda, Virus Res.* **20**:251.

CHAPTER 9

The Molecular Basis of Baculovirus Host Range

LOIS K. MILLER AND ALBERT LU

I. INTRODUCTION

Because baculoviruses are currently important in a broad spectrum of biotechnologies ranging from biological pest control to gene expression vectors, an understanding of the molecular basis of the restriction of the replication of individual members of this virus family to a relatively small number of species is important. Understanding the nature of such host range restrictions will allow more accurate predictions regarding the potential effects of modifying the viral genomes and the impact of such modifications on the ability of the viruses to infect beneficial or nontarget organisms.

There are several levels at which restrictions to baculovirus replication are likely to be encountered. One level is tissue specificity; different baculoviruses exhibit distinctive tissue tropism and this tropism may vary depending on the host species (see Chapter 3, this volume). Another level is species specificity, where replication of a given virus in a particular species may be severely, if not completely, impaired. Although some baculoviruses appear to replicate in only a very limited number of insect species, others are known to replicate in numerous species, although their relative ability to infect and replicate in the different species may vary greatly. A third level of restriction is the limitation of baculovirus infection to hosts within the phylum Arthropoda.

LOIS K. MILLER • Departments of Entomology and Genetics, University of Georgia, Athens, Georgia 30602-2603 ALBERT LU • DuPont Agricultural Products, DuPont Stine-Haskell Research Center, Newark, Delaware 19714-0030.

The Baculoviruses, edited by Lois K. Miller. Plenum Press, New York, 1997.

The fact that baculoviruses have been positively identified only in arthropods but are found in crustaceans and insects (and possibly arachnids, since mites have viruses previously classified as nonoccluded baculoviruses) suggests that their evolution as a distinct virus family may have begun during the Silurian period of the Paleozoic era between 400 to 450 million years ago. Coevolution with the rapidly diverging arthropod phylum in the ensuing years would result in unique features responsible for their current restriction to this phylum. Determining what these very ancient features are is likely to be central to understanding the truly unique qualities of the baculovirus family. Interaction with members of other virus families (e.g., acquisition of other genes from other arthropod viruses such as entomopoxviruses) during their evolution would be expected to affect properties such as tissue tropism, but would probably not fundamentally change their phylogenetic or species restrictions. However, determining the nature of the tissue- and species-specific differences is crucial to understanding how to predict or control the host-specificity and behavior of individual members of the family.

Identifying the nature of the blocks and the factors governing them will be challenging. Studies on the nature of the blocks has been pursued at a variety of levels for the past 25 years as part of the assessment of the safety of baculoviruses as biopesticides. The technology employed in these studies has been refined as new technologies have developed. In the past 5 years, several factors have been discovered that affect the host range of baculoviruses, and research to understand the molecular basis of their action is ongoing. It is likely that more factors will be found and that an understanding of their mode of action will require a far deeper understanding of the general mechanism by which baculoviruses invade their host organisms and usurp the molecular machinery of their host cells. Although studying virus behavior in cell culture cannot completely substitute for analysis of virus behavior at the organismal level, it is an essential step in understanding the most fundamental of the restrictions on virus replication, and therefore is the most logical starting place in defining the molecular basis for baculovirus host specificity.

In this chapter, we summarize the information available concerning the molecular basis of the restriction of baculovirus replication to specific hosts or cell types. Most of the work has been pursued using budded virus in cell cultures, and in many cases the distinction between cell line and tissue or species specificity has not been made; the properties of a cell line may or may not be representative of all the tissue of the species from which they have been derived. In general, it appears that baculoviruses are able to enter a phylogenetically broad range of insect cells, but appear to be unable to reach the nucleus of most vertebrate cell types, a notable exception being mammalian hepatocytes. The expression of baculovirus genes in nonpermissive cells appears to be blocked at different points in different semi- and nonpermissive cells. In insect cells, efficient expression of viral-borne genes under host or early viral promoter control is observed, but late gene expression is severely reduced. Five different baculovirus genes have been shown to influence the ability of AcMNPV to replicate in certain cell lines; the antiapoptotic inhibitors *p35* and *iap,* the helicase homologue *p143,* host cell factor *hcf-1,* and host range factor *hrf-1* genes. Two of these genes have also been found to exert species-specific effect.

II. RESTRICTIONS AT THE LEVEL OF VIRUS ENTRY INTO CELLS

A. The Influence of the Occlusion Matrix

The presence of the crystalline protein matrix of occluded virions presents one barrier to the ability of baculoviruses to infect nontarget organisms. During a normal infection of an insect, occlusion bodies are ingested as contaminants of the food or water supply, and the alkaline (pH>10.0) midgut of the insect host facilitates the dissolution of the matrix, releasing the virions in the midgut where they initiate the primary infection (see Chapter 3, this volume). Organisms that lack such alkaline conditions in their digestive tract or other potential points of entry (e.g., respiratory tract) should not be susceptible to infection by the occluded form; studies on birds, mammals, and wasps have found that occlusion bodies pass through the alimentary canal with no obvious effect on the organisms (Morel and Fouillaud, 1992; Lautenschlager and Podgwaite, 1979; Entwistle *et al,* 1978). If occlusion bodies are the primary transmissible form of the virus in the environment, then organisms lacking an alkaline entry point would normally not even encounter an "active" form of the virus. However, in the event that the organism was infected by an active form of the virus (e.g., an alkaline-liberated virus, a preoccluded virus, or a budded virus), then these "active" forms of the virus would probably be able to enter surface cells such as those of the digestive or respiratory tract and possibly express some early genes. Because of restrictions at the molecular level within cells of nonpermissive organisms, most viruses would probably not replicate within these cells but a virus might theoretically act as a gene delivery vector in the cells it enters (see Section II.B).

Thus, although the host range limitation presented by the occlusion body matrix is limited, it does have practical implications in areas such as the use of the viruses as biopesticides. The occluded form is usually the form of the virus applied in the field and the occlusion matrix provides an additional layer of safety in preventing virus entry into cells of nontarget organisms. Clearly, however, the primary factors governing the host range of a virus are its ability to enter and to reach the appropriate compartment of the host cell, execute the timely and efficient expression of its genes, and produce gene products that can interact effectively with host biosynthetic and signal transduction pathways.

B. Entry of Budded or Occlusion-Derived Viruses into Insect Cells

Although the host range and tissue tropism of many mammalian viruses is governed to a significant extent by the ability of the virus to interact with a specific type of receptor on the cell surface, this has not been found to be a limiting feature of baculovirus host range to date. The fact that budded virus (BV) and occlusion-derived viruses (ODV) enter cells by different means (Chapter 4, this volume) complicates this issue somewhat. ODVs apparently enter midgut epithelial cells by direct fusion of the viral envelope with the plasma membrane of the cell, whereas BVs normally enter cells by endocytosis (Volkman *et al,* 1986; Volkman and Goldsmith, 1985). Fusion of the BV envelope with the endosomal membrane is driven by low pH and is mediated by the gp64 envelope fusion protein, which is not found in ODV (Monsma *et al.*, 1996; Blissard and Wenz, 1992; see Chapter 2, this volume).

Since ODVs initiate the primary infection of the insect, their ability to enter midgut cells is crucial and would be expected to affect the host range of the baculovirus. However, very little information is available on the ability of ODV to enter midgut cells. One study found that there are saturable attachment sites for ODV on the brush border membranes of midgut cells, suggesting the presence of receptor sites (Horton and Burand, 1993). However, the density of these sites appears to be high and suggests that the moiety (moieties) serving as the "attachment site" is (are) either highly abundant or relatively nonspecific. The ability of ODV to enter midgut cells of a range of different species has not been conducted.

In considering host range limitations of baculoviruses, however, the focus of attention should not be exclusively on ODV, since BVs are responsible for the systemic infection of the organism, and if BVs cannot enter the cells of the species in question, the infection of a few midgut cells by ODV would probably have little or no effect on the organism (Monsma *et al.*, 1996; McNitt *et al.*, 1995) (see Section III.A). Thus, the behavior of BVs is also of central importance in host range considerations. Numerous studies have been done on *Autographa californica* nuclear polyhedrosis virus (AcMNPV) BV entry into nonpermissive insect and mammalian cells. Although AcMNPV is unable to produce progeny virus (i.e., it does not productively infect) in cell lines derived from a number of phylogenetically diverse insect species, the budded form of the virus is nevertheless able to enter and deliver its DNA to the nucleus of these cells as determined by the expression of genes under host or early viral promoter control (see Section III.A). A different picture emerges from similar studies for most mammalian cell types in which the DNA does not reach the nucleus in an expressible form, although hepatocytes appear to be an exception (see Section IV). The data suggest that there is no specific receptor protein for AcMNPV entry unless the receptor is highly conserved. Alternatively, the virus may recognize a relatively common residue (e.g., a sugar moiety of a membrane glycoprotein) found on the surface of insect cells and hepatocytes.

III. THE BEHAVIOR OF AcMNPV IN SEMI- AND NONPERMISSIVE INSECT CELLS

A. Evidence for Entry into a Wide Range of Insect Cells

Infection of semi- or nonpermissive lepidopteran and dipteran cell lines with recombinant AcMNPV containing one or more reporter genes under the regulatory control of an insect promoter (H) or an early (E), late (L), or very late (VL) baculovirus promoter resulted in expression from host and early viral promoters, but expression from late and very late promoters was variable, depending on the cell line (Morris and Miller, 1992, 1993; Carbonell *et al.*, 1985). In one study (Carbonell *et al.*, 1985), *Drosophila melanogaster* (DL-1) cells and *Aedes aegypti* (mosquito) midgut cells were infected with an AcMNPV recombinant containing two reporter genes, one under the control of the Rous sarcoma virus-long terminal repeat (RSV-LTR) and the other under AcMNPV *polh* promoter control; substantial levels of expression were observed from the RSV-LTR but little or no expression was observed from the *polh* promoter.

Similar results were obtained for a variety of semi- or nonpermissive lepidopteran and dipteran cell lines infected with recombinant viruses containing a reporter gene under control of either H, E, L, or VL promoters (Morris and Miller, 1992); most of these semi- and nonpermissive insect cells supported substantial levels of expression from H and E promoters but decreased levels of expression from the L and VL promoters. Furthermore, *in vivo* expression of a paralytic mite toxin was detected in wasps *(Polistes metricus)* fed *Spodoptera frugiperda* larvae infected with AcMNPV recombinants expressing the toxin gene under the control of the *Drosophila hsp70* gene promoter (McNitt *et al.*, 1995) but not with recombinants expressing the gene under *polh* promoter control. In this series of experiments, the wasps ingested larvae containing BV, OV, and preoccluded virions and it is not clear which form of the virus was responsible for the apparent expression of the toxin gene in the wasps; but since OV are deposited intact in the nest by the wasp larvae as a component of their meconium [the first and final excrement in larvae of this species (see Morel and Fouillaud, 1992)], the expression observed is probably due to either BV or preoccluded virus.

The ability of AcMNPV to express genes under H or E promoter control in these different cells lines indicates that AcMNPV can adsorb, penetrate, uncoat, and present its DNA in an expressible form in the nucleus of a diverse range of insect cells. Thus, the block to productive infection of semi-and nonpermissive insect cells is subsequent to the expression of early genes.

B. Nature of the Blocks to Virus Replication in Selected Insect Cells

1. AcMNPV Gene Expression in Semi- and Nonpermissive Cells

Comparison of the relative activities of virus-borne H, E, L, and VL gene promoters in permissive and several semi- and nonpermissive insect cell lines provides insight into the nature of the blocks to virus replication in these selected cell lines (Morris and Miller, 1992, 1993). In permissive cells (i.e., SF-21 cells) there is a successive increase in E to VL promoter activities as infection progresses. In contrast, the ability of semi- and nonpermissive cells to support expression from each successive temporal promoter class tends to decline as the infection progresses (Morris and Miller, 1992). However, many of these cell lines do support significant levels of L gene expression, but the activity of the L promoter is substantially reduced compared to its activity in permissive cells (Morris and Miller, 1992, 1993). The *hsp70* promoter performed well in most cell lines compared to its performance in the fully permissive SF-21 cell line and it outperformed the viral promoters tested in four nonpermissive cells, including *D. melanogaster* cells (Morris and Miller, 1992).

Consistent with the observation that some late gene expression is observed in these cell lines, evidence of viral DNA replication was found in cell lines derived from *Bombyx mori* (BmN-4), *Choristoneura fumiferana* (CF-1), *Mamestra brassicae* (MaBr-3), *Lymantria dispar* (Ld652Y), and *D. melanogaster* (Dm), although no viral DNA replication was observed in the *Helicoverpa zea* (Hz1b3) cell line (Morris and Miller, 1993). A delay in viral DNA replication was detected in cell lines derived from *C. fumiferana* (Liu and Carstens, 1993; Morris and Miller, 1993). Thus, it appears that the block to productive infection in most of these cell lines occurs during or after viral DNA replication. However, a block may occur at an earlier point that does not affect viral DNA replication but is required for subsequent aspects of the infection process. The viral DNA replication data revealed no direct correlation between the level of E or L promoter activity and the level of viral DNA replication, although it is not clear that the levels of DNA replication were quantitatively assessed these experiments. The absence of detectable levels of AcMNPV DNA replication in Hz1b3 cells suggests that the progress of AcMNPV replication in this cell line is blocked at a stage during or preceding the initiation of DNA synthesis.

Fluorescence-activated cell sorting (FACS) analysis of the level of reporter gene expression per individual cell provides a picture of how gene expression is distributed within the population of cells and assesses whether the expression observed by assaying reporter gene product activity is reflective of the population in general or is restricted to a small proportion of the population. FACS analysis of selected insect cell lines infected with AcMNPV recombinants carrying a reporter gene under E, L, or VL promoters shows that the cell populations behave in a relatively uniform manner; there are not simply a few cells in the population that support high levels of expression (Morris and Miller, 1993). In the fully permissive SF-21 cell line, many (but not all) cells show detectable levels of expression from an E promoter and virtually all (ca. 100%) of the cells support detectable and high levels of expression from the L and VL promoters (Morris and Miller, 1993); the levels of expression observed in SF-21 cells are described as "productive" levels of expression for each promoter class. (It is noteworthy that expression from the early ETL promoter is the most variable, even in a fully permissive cell line such as SF-21; levels of expression from this promoter are undetectable in over 50% of the cells, but >98% of the entire cell population subsequently support high and relatively uniform levels of L and VL expression. The reason for this cellular variation in early gene expression is not known.) In some nonpermissive cell lines such as CF-1, a similar proportion of the cells support expression from the E promoter, but compared to SF-21 cells, only a very small proportion of CF-1 cells support "productive" levels of L expression and <5% support productive levels of VL expression (Morris and Miller, 1993).

Similar results were reported for AcMNPV infection of the Cf124T cell line (Liu and Carstens, 1993). Thus, the inability of *C. fumiferana* cells to support AcMNPV BV and OV production is not due to a problem with virus entry, but is due to a subsequent problem that affects the progressive of infection beyond the DNA replication step. AcMNPV-induced apoptosis in a midgut cell line from *C. fumiferana*, CF-203, has been reported (Palli *et al.*, 1996). Infection of this cell line with CfMNPV prior to AcMNPV infection results in a fully permissive AcMNPV infection, suggesting that in this particular cell line, CfMNPV provides a transacting factor(s) that inhibits AcMNPV-induced apoptosis. Other nonpermissive cell lines, such as Hz1b3 and Dm cells, exhibit more severe defects, but again there is remarkably little variation in the response of the cell population as a whole. In all cases, there is a larger proportion of cells that support "productive" levels of early gene expression and increasingly smaller proportions of cells that support productive levels of L and VL expression (Morris and Miller, 1993).

Thus, the restriction to productive infection in most insect cell lines is at a postentry level and is different for each cell line tested. It is likely that multiple cell-line-specific factors influence the ability of AcMNPV to establish a productive infection. Some of these factors are discussed in Section III.B and III.C.

2. The Block to AcMNPV Replication in Ld652Y Cells

The AcMNPV infection of Ld652Y cells is characterized by exceptionally low levels of viral protein synthesis and a total shutoff of all host and viral protein synthesis by 16 to 20 hr postinfection (p.i.) (Guzo *et al.*, 1992; Morris and Miller, 1992). AcMNPV-specific transcription is detected from all temporal classes at levels almost equivalent to that found in permissive cells (Guzo *et al.*, 1992; Morris and Miller, 1992), but very few virus-specific proteins are observed during the course of infection (McClintock *et al.*, 1986) and reporter protein levels are exceptionally low (Morris and Miller, 1992, 1993). Since viral DNA is replicated and L and VL genes are transcribed in these cells, at least low levels of early proteins involved in DNA replication and late transcription must be present in AcMNPV-infected Ld652Y cells (Morris and Miller, 1993). Viral RNAs appear to be transported efficiently from the nucleus to the cytoplasm, and RNA isolated from infected cells can be successfully translated *in vitro* (Guzo *et al.*, 1992), indicating that the block to productive infection of AcMNPV-infected Ld652Y cells occurs at the level of translation. The block to AcMNPV infection of Ld652Y cells can be overcome by superinfection of Ld652Y cells with LdMNPV (McClintock and Dougherty, 1987), indicating that LdMNPV must express at least one *trans*-acting factor that can effectively extend the host range of AcMNPV to Ld652Y cells.

The LdMNPV gene that encodes this factor was mapped by cotransfecting AcMNPV genomic DNA with individual cosmid clones from an overlapping LdMNPV library and screening for the production of BVs and the presence of OVs in Ld652Y cells (Thiem *et al.*, 1996). Those clones that enabled AcMNPV to productively infect Ld652Y cells were subcloned to identify the specific gene involved. Using this strategy, a novel gene, host range factor 1 (*hrf-1*), was identified (Thiem *et al.*, 1996). The *hrf-1* encodes an acidic polypeptide (pI of 4.61) of 25.7 kDa, with no striking sequence homology to genes or proteins within the nucleotide or protein databases, although the amino acid composition of HRF-1 suggests some similarity to GADD proteins. Interestingly, one GADD gene is able to restore protein synthesis in cells infected with an HSV mutant that exhibited a defect in protein synthesis (He *et al.*, 1996). Recombinants of AcMNPV expressing *hrf-1* productively infect Ld652Y cells, resulting in the expression of L and VL viral proteins and the production of both BV and OV (Thiem *et al.*, 1996). These recombinants can infect *Lymantria dispar* larvae, indicating that *hrf-1* is a host range gene at the species level, not simply a cell-line- or tissue-specific factor (S. Thiem, personal communication).

3. The Block to AcMNPV Infection in *B. mori* Cells

AcMNPV and BmNPV have nonoverlapping host range characteristics despite the relatively high DNA homology between the two viruses. SF-21 and BmN cells are nonpermissive to BmNPV and AcMNPV infection, respectively (Summers *et al.*, 1978). Some recombinants generated by coinfection of AcMNPV-OT2 and BmNPV-T3 variants (Kondo and Maeda, 1991) or by cotransfection of AcMNPV-OT2 DNA with portions of the BmNPV genome in SF-21 cells (Mori *et al.*, 1992) have been found to possess an expanded host range that allows AcMNPV-based recombinants to replicate in the normally nonpermissive BmN cell line. Most of these recombinant genomes were generated by multiple recombination events in several different regions of the genome (Mori *et al.*, 1992; Kondo and Maeda, 1991).

A series of backcross infections of one of these expanded host range viruses (Kondo and Maeda, 1991) with AcMNPV-OT2 in SF-21 cells resulted in an expanded host range recombinant virus, eh2-AcNPV, that differed from AcMNPV-OT2 by a single detectable *Hin*dIII fragment (Maeda *et al.*, 1993). Subsequent analysis of the effects of cotransfecting AcMNPV DNA with individual eh-AcNPV *Hin*dIII fragments in BmN cells identified a 572-bp fragment derived from the BmNPV-T3 *p143* gene that was responsible for the ability of eh-AcMNPV to infect BmN cells (Maeda *et al.*, 1993). The fragment corresponds to amino acids 413 to 602 of the BmNPV DNA helicase homologue (*p143*) and differs from the corresponding region of the cognate AcMNPV *p143* protein by 14 out of 109 amino acids. AcMNPV recombinants containing this 572-bp region from BmNPV are able to infect both BmN cells and SF-9 cells productively (Maeda *et al.*, 1993).

Additional studies have demonstrated that allelic replacement of only a 79-bp region corresponding to amino acids 550 to 576 of *p143* is enough to extend the host range of AcMNPV (1.2) to Bm5 cells (Croizier *et al.*, 1994). This 79-bp region of BmNPV-SC7 differs from the homologous region in the AcMNPV *p143* gene by only four amino acids. Sequence analysis of this region in these host range extended viruses passaged in *B. mori* larvae indicated that only three of these substitutions are necessary for this host range extension (Croizier *et al.*, 1994). Collectively, these independent findings indicate that the region between amino acids 550 to 576 of *p143* can influence the host range of a baculovirus. Recently, a single amino acid change from Ser to Asp within this region of *p143* was demonstrated to be sufficient to expand the host range of AcMNPV (Kamita and Maeda, 1996a).

The role *p143* plays in host range determination in *B. mori* cells is unknown, but co-infection of BmN cells with AcMNPV-OT2 and BmNPV-T3 results in the premature cessation of host and viral protein synthesis by 12 hr p.i., even though viral transcription appears to be normal (Kamita and Maeda, 1993). This block in protein synthesis does not occur when BmN cells are co-infected with BmNPV and eh2-AcNPV, suggesting that expression of AcMNPV *p143* is cytotoxic in BmN cells and results in the shutoff of the host cell translational machinery, perhaps by triggering a cellular antiviral defense mechanism. The cytotoxic effect of AcMNPV *p143* in BmN cells, however, has been shown to be attenuated for AcMNPV-L1 (Kamita and Maeda, 1993), the AcMNPV strain used in the host range studies of

Morris and Miller (1992, 1993). Low multiplicity infection (less than 1) of Sf-9 cells with eh2-AcNPV results in an abortive infection, suggesting that these amino acid changes within *p143* have an effect on the ability of AcMNPV to replicate in Sf-9 cells (Kamita and Maeda, 1996b). The infectivity of eh2-AcNPV in *S. frugiperda* larvae was reduced relative to wild-type AcNPV (Kamita and Maeda, 1996b).

It is noteworthy that the *p143* gene of LdMNPV is unable to replace the AcMNPV *p143* gene in transient DNA replication assays in SF-21 cells, although the LdMNPV DNA polymerase gene is capable of substituting for the AcMNPV DNA polymerase gene in these assays (Ahrens and Rohrmann, 1996). This result underscores the importance of *p143* in influencing baculovirus host range.

C. The Role of Other *lef*s in AcMNPV Host Range

1. The Role of *hcf-1* in Host Range and/or Tissue Specificity

Eighteen AcMNPV genes known as late expression factor genes (*lef*s) are sufficient to support expression of a reporter gene under late viral promoter control in SF-21 cells in transient-expression assays (See Chapter 8, this volume). This same complement of genes, known as the *lef* library, however, are not able to support expression from this late promoter in another cell line permissive for AcMNPV infection, TN-368, derived from *Trichoplusia ni* (Lu and Miller, 1995a). Addition of another gene known as host cell-specific factor-1, *hcf-1* (ORF70) (Ayres *et al.*, 1994), to the *lef* library allows efficient late reporter gene expression in transfected TN-368 cells (Lu and Miller, 1995a). The *hcf-1* is required for expression of reporter genes under late and very late but not early promoter control in TN-368 cells. Thus, *hcf-1* can be considered a host-cell-specific *lef*. The *hcf-1* gene is predicted to encode a putative 34-kDa cysteine-rich polypeptide that may possess zinc binding domains but bears no clear relationship to other known proteins in existing databases. The existence of homologues of *hcf-1* in other baculoviruses remains to be determined; a search of the genome sequence of BmNPV did not identify an *hcf-1* homologue.

The *hcf-1* has both species-specific and tissue-specific effects on AcMNPV host range as determined by the analysis of *hcf-1* null mutants in insects as well as in cell culture (Lu and Miller, 1996). Null *hcf-1* mutants replicate normally in SF-21 cells with no discernible differences relative to wild-type AcMNPV in the temporal expression of viral proteins during infection, kinetics of viral DNA replication, or the production of BVs. Furthermore, disruption or deletion of *hcf-1* in the AcMNPV genome did not affect the infectivity (LD_{50}) or the virulence (LT_{50}) of the virus in *S. frugiperda* larvae (Lu and Miller, 1996). In contrast, *hcf-1*$^{-}$ mutants exhibited a mutant phenotype in cell lines derived from *T. ni* as well as *T. ni* larvae. In TN-368 cells, a cell line derived from adult ovarian tissues, *hcf-1* mutants were defective in viral DNA replication and late gene expression, and both host and viral protein synthesis were completely shut off by 18 hr p.i. BV production in TN-368 infected cells was reduced at least 100-fold relative to wild-type AcMNPV (Lu and Miller, 1996). The *hcf-1*$^{-}$ mutant-infected BTI-TN5B1-4 (also known as Hi-5) cells, a cell line derived from *T. ni* eggs, exhibited a less severe phenotype in which infection appeared to be only delayed by about 24 hr. The *T. ni* larvae infected with these mutants either orally or by hemocoelic injection displayed reduced virulence (a significantly longer LT_{50}) relative to wild type. In addition, there was a 50-fold decrease in the infectivity of *hcf-1* mutants by hemocoelic injection of BV. No difference in infectivity was observed when *T. ni* larvae were orally infected with *hcf-1* mutant OVs (Lu and Miller, 1996).

Thus, *hcf-1* exerts both species-specific and tissue-specific effects on the replicative ability of AcMNPV in insects. Since *hcf-1* null mutants exhibited no difference in oral infectivities between the two species, *hcf-1* may not be a host range determinant *per se,* if AcMNPV is transmitted exclusively by oral infection. However, *hcf-1* does affect the virulence of AcMNPV in *T. ni* larvae by both oral and hemocoelic infection routes. This would be expected to provide the virus with a significant competitive advantage in nature.

2. The Possible Roles of *ie-2* and *lef-7* in Host Range Restriction

The finding that *hcf-1* was required for transient expression of a reporter gene under late viral promoter control in a cell-line-specific manner suggested that other AcMNPV *lef*s may be differentially required for late gene expression between SF-21 and TN-368 cells. When the effect of omitting individual *lef*s from the *lef* library was monitored, *ie-2, lef-7,* and *p35* were found to exert little or no effect on late reporter gene expression in TN-368 cells (Lu and Miller, 1995a). IE-2 and LEF-7 act as auxiliary factors in AcMNPV origin-directed reporter plasmid DNA replication assays in SF-21 cells (Kool *et al.,* 1994; Lu and Miller, 1995b). IE-2 is known to augment IE-1-mediated transactivation of reporter genes under early promoter control (Carson *et al.,* 1991) and LEF-7 has homology to the herpesvirus UL29 family of single-stranded DNA binding proteins (Lu and Miller, 1995b). The influence of *ie-2* and *lef-7* on AcMNPV host range will need to be determined using mutant viruses in insect cells and larvae.

3. The Role of Antiapoptotic Genes in Baculovirus Host Range

The differential requirement for *p35* in the replication and expression of a reporter gene under late viral promoter control in SF-21 and TN-368 cells (Lu and Miller, 1995a) confirms and extends previous observations that *p35* is nonessential for virus replication in TN-368 cells and *T. ni* larvae, and thus is a host range determinant (Clem and Miller, 1993, 1994; Hershberger *et al.*, 1992; Clem *et al.*, 1991). AcMNPV mutants lacking a functional *p35* gene are severely impaired in their ability to replicate in SF-21 cells or infect *S. frugiperda* larvae, but they replicate normally in TN-368 cells or *T. ni* larvae (see Chapter 10, this volume). Thus, *p35* affects the host range of AcMNPV in *S. frugiperda* larvae. However, *p35* also exerts tissue-specific effects in *T. ni* larvae; a reduction in OV production and a lack of melting are observed in *T. ni* larvae infected with *p35* mutants. Thus, *p35* exerts both tissue and species-specific effects (see Chapter 10, this volume).

A *p35* gene is not found in the genome of *Orgyia pseudotsugata* nuclear polyhedrosis virus (OpMNPV), a closely related baculovirus with a relatively narrow host range compared to AcMNPV (Gombart *et al.*, 1989). Instead, another gene, *iap* (for inhibitor of apoptosis), functions to prevent apoptosis in the absence of a *p35* homologue (Birnbaum *et al.*, 1994). Homologues of *iap* are present in the AcMNPV genome, but these genes are apparently nonfunctional as inhibitors of apoptosis in SF-21 cells (Clem and Miller, 1994). Recombinant AcMNPV that contain *iap* genes from other baculoviruses in place of *p35* are able to replicate normally in SF-21 cells, indicating that *iap* can also be considered to influence the host range of AcMNPV (Birnbaum *et al.*, 1994; Crook *et al.*, 1993). A review of the roles of *p35* and *iap* in apoptosis is presented in Chapter 10 (this volume).

Considering the fact that AcMNPV carries at least one gene that selectively facilitates its replication in *T. ni* larvae and at least one gene that selectively facilitates its replication in *S. frugiperda* larvae suggests that the ability of baculoviruses to infect alternate species may be an important aspect of their natural life history.

IV. THE BEHAVIOR OF BACULOVIRUSES IN VERTEBRATE CELLS

A. Baculovirus Interaction with Vertebrate Cells

Many studies have assessed the ability of baculoviruses to enter, infect, or persist in mammalian cells. One very early study reported overt virus infection following transfection of human amnion cells with DNA of *B. mori* NPV (Himeno *et al.*, 1967), but this has never been confirmed. Another study reported isotopic labeling of AcMNPV virions in a Chinese hamster cell line (McIntosh and Shamy, 1980), but the significance of isotopic labeling under the conditions used (e.g., long isotopic labeling periods) has never been established. With these exceptions, the general consensus from the numerous other studies is that AcMNPV can enter vertebrate cells, but there are no cytopathic effects (with the exception of one report that described mild toxicity that was observed for inactivated virus too), even with very high multiplicities of infection.

Several studies on AcMNPV interaction with vertebrate cells are particularly noteworthy. Volkman and Goldsmith (1983) surveyed 35 cell lines, 23 of human origin and 12 of nonhuman vertebrate origin. Each cell line was incubated at two different temperatures (28 or 37°C), the cells were observed at four different times after inoculation (from 16 to 168 hr), and virus entry was monitored by a peroxidase–antiperoxidase assay using an antibody raised to ODV. BV, ODV, preoccluded virus, or hemolymph-derived virus were used as inocula. Uptake of the ODV form of the virus appeared to be most efficient and human lung carcinoma cells were the most active in virus uptake. Electron microscopy confirmed the presence of nucleocapsids in the cytoplasm and vacuoles of these cell, but no evidence of virus particles was found in the nucleus nor was there evidence of active viral gene expression. Two murine liver cell lines exhibited little or no uptake of the various virus forms in this study. Another study utilizing poikilothermic vertebrate cells to provide a more optimal temperature (27°C) for AcMNPV entry also found evidence of AcMNPV entry into the cytoplasm of cells but no evidence of virus replication (Brusca *et al.*, 1986). Hartig *et al.*, (1989, 1991) designed several viral replication, persistence, transformation, and toxicity assays; all assays were negative for baculovirus inoculations, except mild toxicity of AcMNPV was reported in several mammalian cell cultures, but this was observed for psoralen-inactivated virus as well. Studies by Riemann and Miltenburger (1993) confirmed by electron microscopy that AcMNPV enters mammalian cells but that they are found only in phagocytic vacuoles or the cytoplasm, not in the nucleus. No cytopathic effect was observed nor was there any increase in chromosomal abnormalities.

As soon as techniques were developed to generate recombinants of AcMNPV expressing foreign genes, Carbonell *et al.* (1985) initiated experiments designed to study the expression of viral genes in nontarget cells including mammalian cells. Using the BV form of a recombinant AcMNPV expressing the highly sensitive reporter gene, chloramphenicol acetyl transferase (CAT) gene, under the control of the strong RSV-LTR promoter and the *Escherichia coli lacZ* gene under polyhedrin promoter control, they observed a very low level of CAT activity following continuous exposure of a very high concentration of virus (800 pfu/cell) in mouse L29 cells. Subsequent experiments using protein synthesis inhibitors showed that this CAT activity was entering the cell as a component of the virus particles rather than being synthesized by the L29 cells; association of CAT enzyme with virus particles was confirmed by sucrose gradient analysis (Carbonell and Miller, 1987). Similar results were obtained (Carbonell and Miller, 1987) with the human lung carcinoma cell line of Volkman and Goldsmith (1983). CAT activity in the cells decreased with time and no cytopathic effects were detected. No β-galactosidase activity above background levels was detected in nontarget cells. Carbonell and Miller (1987) concluded that the virus was probably entering the cells but that the viral DNA was not reaching the nucleus in an expressible form.

B. AcMNPV-Directed Gene Expression in Mammalian Hepatocytes

Two recent reports demonstrate that AcMNPV can enter some cells of hepatic origin and express reporter genes under the control of strong vertebrate viral promoters (Boyce and Bucher, 1996; Hofmann *et al.*, 1995). Both groups were motivated to study hepatocytes because these cells have receptors on their cell surface that recognize asialoglycoproteins and target such proteins for turnover in the liver. The cultured insect cell lines that are commonly used for baculovirus expression vector work produce glycoproteins that lack terminal sialic acid residues (Ogonah *et al.*, 1996; Jarvis and Finn, 1995; Davidson and Castellino, 1991; Kuroda *et al.*, 1990). Thus, it was proposed that hepatocytes might recognize asialoglycoproteins on the surface of baculovirus BVs and facilitate their entry into hepatocytes (Boyce and Bucher, 1996; Hofmann *et al.*, 1995). However, the situation is more complex than originally envisioned, since nonhepatic cells expressing the asialoglycoprotein receptor do not show significant baculovirus-vectored gene expression (Hofmann *et al.*, 1995). Another motivation for studying hepatocytes is that the liver secretes many blood proteins and is therefore often the target of human genetic therapy, since many genetic defects entail the inability of the liver to produce active blood proteins.

Both studies found that infection of hepatic cell lines with AcMNPV expressing a sensitive reporter gene under a strong vertebrate viral promoter, the RSV-LTR or the cytomegalovirus (CMV) immediate early gene promoter, resulted in reporter gene expression (Boyce and Bucher, 1996; Hofmann *et al.*, 1995). The most susceptible hepatic cells lines were Huh7, derived from a hepatic carcinoma, and HepG2, derived from a human hepatoma. Most studies were performed on these tumor cell lines, although primary hepatocytes also exhibit some expression. The vast majority of cells derived from nonhepatic sources showed little or no reporter gene expression. No expression from the viral *polh* promoter was detected. High multiplicities of infection (e.g., 10 to 1000 pfu/cell) were used for most studies and the level of expression in Huh7 and HepG2 cells increased with increasing virus dose. One study demonstrated that cycloheximide, an inhibitor of protein synthesis, could block 90% of the expression observed (Boyce and Bucher, 1996), confirming that reporter gene expression was occurring in the mammalian cells. Both studies showed that expression was inhibited by chloroquine, which blocks the acidification of the endosome, suggesting that the virus is entering via an endosomal pathway. There was no evidence of cytopathic effect in the AcMNPV-infected cells, even when the cells were inoculated with an enormous dose of virus (1500 pfu/cell). The lack of cytopathic effect was confirmed by showing that the plating efficiency of the cells was unchanged after 2 days of treatment with the virus (Hofmann *et al.*, 1995).

Based on these results, both groups suggest that AcMNPV virions might be useful as gene delivery vectors for human gene therapy directed at liver cell modification. On the positive side, the viruses exert no observable cytopathic effect on the numerous cell lines tested, even at very high virus doses. Furthermore, they do not replicate in mammalian cells and would therefore be expected to deliver genes with no additional complications. Baculoviruses are also capable of carrying rather large stretches of foreign DNA as well as retrotransposons; retroelements might provide an efficient means of integrating the delivered genes into the host genome. On the negative side, high doses of virus appear to be required to obtain efficient gene delivery to the majority of cells in the population. In addition, tumor cell lines appear to be more efficient in virus uptake than do primary hepatocytes. Baculoviruses, however, are useful in a variety of other ways and it would not be surprising if they also find application in the human gene therapy area.

It is important to note that even though baculoviruses may be able to enter at least some tissues of mammals, this in no way influences the "host range" of the viruses or their safety as insect pest control agents, although the fact that these viruses can enter human cells reinforces the recommendation that attention to safety should be taken when working with baculovirus gene expression vectors. However, most baculovirus vectors carry foreign genes under very late viral promoter control, and it is unlikely that a very late viral promoter will be activated in human cells, even if the virus escapes the many other barriers blocking access to "receptive" cell surfaces. Nevertheless, common sense precautions are important in all laboratory work.

Baculoviruses have been in contact with vertebrates for millions of years, and human have lived in close contact with them since the emergence of the species. There is no evidence that baculoviruses influence vertebrate or human health in any manner. The data that have been provided confirm that baculoviruses cause no adverse affects on mammalian cells, even at doses that are difficult to achieve in the laboratory.

V. SUMMARY

Baculovirus restriction to the phylum Arthropoda coupled with their prevalence in the environment suggest that there are fundamental molecular reasons for the inability of members of this family to successfully invade vertebrates. The occlusion matrix surrounding the virions responsible for the primary infection of insect hosts probably plays little or no role in host restriction, although it does provide an additional barrier to the ability of the virions of the environmentally prevalent form of the virus to even make contact with the cells of most vertebrate hosts. More likely, the mechanisms limiting host range reside in the ability of the viral gene products to interact effectively with appropriate host counterparts within the cell. For the most part, baculovirus host range does not appear to be significantly limited at the point of entry into cells. This is particularly true for AcMNPV entry into insect cells, but it may be an additional factor in their inability to invade mammalian cells, since the virus cannot deliver its DNA in an expressible form to cells derived from most mammalian tissues.

Even when a baculovirus is able to enter a nonpermissive cell and deliver its DNA to the nucleus, additional barriers to infection are found in mammalian cells and in most nontarget insect cells. It is likely that there are substantially more barriers in mammalian cells than there are in insect cells, and it will be important to understand the mechanistic nature of these barriers in both mammalian and insect cells. Data thus far indicate that insect cells display a variety of responses to AcMNPV infection, ranging from virtually no viral gene expression, as in the LD652Y cell line, to substantial levels of early and late gene expression, as in the BmN4 cell line.

It is notable that addition of or changes in a single gene can alter some of the host range properties of a baculovirus in insect hosts; the genes induced appear to counteract one or more host defense mechanisms. The *hrf-1* gene of LdMNPV is an example of how the ability of a baculovirus (AcMNPV) to replicate in an insect cell line (LD652Y) was increased by the addition of a single gene to the AcMNPV genome. In other examples, changing 3 bp of the AcMNPV *p143* gene increases its ability to replicate in BmN4 cells; the presence of *p35* allows the BV form of AcMNPV to replicate in *S. frugiperda* cells and larvae; and the presence of the *hcf-1* gene allows AcMNPV to replicate in TN-368 cells and optimizes its virulence in *T. ni* larvae. It is striking that three of these genes (*p143*, *p35*, and *hcf-1*) have been identified as genes that influence the replication or stability of plasmids carrying a homologous region sequence. However, three genes (*hcf-1*, *hrf-1*, and *p35*) appear to counteract a host defense response. Although *p143* is thought to act as a DNA helicase in the replication assay, the role that the 3 bp affected by extended host range viruses seems to involve the shutoff of host protein synthesis. Replication of AcMNPV DNA does occur in LD652Y cells, but further progression of infection is blocked. Collectively, these results suggest that *p143* also has a role in defending against a host defense mechanism. This role may be indirect; if *p143* is unable to coordinate both DNA replication and gene transcription in a precise manner, the host defense system may win the virus–cell conflict.

In the future, it will also be necessary to distinguish cell-culture-specific effects from those that are meaningful at the organismal level. Replication in cell culture at relatively high multiplicities of infection may bear little reflection on the ability of a virus to successfully cause disease in an individual or in an insect population. It is likely that the "practical" host range of a virus (i.e., the host range that would be of significance in the environment) is controlled by multiple genes. Not only will the presence or absence of such genes be important, but the adaptation of their sequences to a specific host or complex of hosts is also likely to be important. Understanding the nature of the most central host range limitations to baculovirus replication will provide tremendous insight into fundamental mechanisms that dictate the outcome of a virus infection of a host cell.

REFERENCES

Ahrens, C. H. and Rohrmann, G. F., 1996, The DNA polymerase and helicase genes of a baculovirus of *Orgyia pseudotsugata*, *J. Gen. Virol.* 77:825.

Ayres, M. D., Howard, S. C., Kuzio, J., Lopez-Ferber, M., and Possee, R. D., 1994, The complete DNA sequence of *Autographa californica* nuclear polyhedrosis virus, *Virology* **202:**586.

Birnbaum, M. J., Clem, R. J., and Miller, L. K., 1994, An apoptosis-inhibiting gene from a nuclear polyhedrosis virus encoding a polypeptide with cys/his sequence motifs, *J. Virol.* **68:**2521.

Blissard, G. W., and Wenz, J., 1992, Baculovirus gp64 envelope glycoprotein is sufficient to mediate pH-dependent membrane fusion, *J. Virol.* **66:**6829.

Boyce, F. M., and Bucher, N. L. R., 1996, Baculovirus-mediated gene transfer into mammalian cells, *Proc. Natl. Acad. Sci USA* **93:**2348.

Brusca, J., Summers, M. D., Couch, J., and Courtney, L., 1986, *Autographa californica* nuclear polyhedrosis virus efficiently enters but does not replicate in poikilothermic vertebrate cells, *Intervirology* **26:**207.

Carbonell, L. F., and Miller, L. K., 1987, Baculovirus interaction with non-target organisms a virus-borne reporter gene is not expressed in two mammalian cell lines, *Appl Environ Microbiol.* **53:**1412.

Carbonell, L. F., Klowden, M. J., and Miller, L. K., 1985, Baculovirus-mediated expression of bacterial genes in dipteran and mammalian cells, *J. Virol.* **56:**153.

Carson, D. D., Summers, M. D., and Guarino, L. A., 1991, Molecular analysis of a baculovirus regulatory gene, *Virology* **182:**279.

Clem, R. J., and Miller, L. K., 1993, Apoptosis reduces both the *in vitro* replication and the *in vivo* infectivity of a baculovirus, *J. Virol.* **67:**3730.

Clem, R. J., and Miller, L. K., 1994, Control of programmed cell death by the baculovirus genes *p35* and *iap*, *Mol. Cell. Biol.* **14:**5212.

Clem, R. J., Fechheimer, M., and Miller, L. K., 1991, Prevention of apoptosis by a baculovirus gene during infection of insect cells, *Science* **254:**1388.

Croizier, G., Croizier, O. A., and Poudevigne, D., 1994, Extension of *Autographa californica* nuclear polyhedrosis virus host range by interspecific replacement of a short DNA sequence in the p143 helicase gene, *Proc. Natl. Acad. Sci. USA* **91:**48.

Crook, N. E., Clem, R. J., and Miller, L. K., 1993, An apoptosis-inhibiting baculovirus gene with a zinc-finger-like motif, *J. Virol.* **67:**2168.

Davidson, D. J., and Castellino, F. J., 1991, Asparagine-linked oligosaccharide processing in lepidopteran insect cells. Temporal dependence of the nature of the oligosaccharides assembled on asparagine-289 of recombinant human plasminogen produced in baculovirus vector infected *Spodoptera frugiperda* (IPLB-SF-21-AE) cells, *Biochemistry* **30:**6167.

Entwistle, P. F., Adams, P. H. W., and Evans, H. F., 1978, Epizootiology of a nuclear polyhedrosis virus in European spruce sawfly (*Gilpinia hercyniae*): The rate of passage of infective virus through the gut of birds during cage tests, *J. Invertebr. Pathol.* **15:**173.

Gombart, A. F., Blissard, G. W., and Rohrmann, G. F., 1989, Characterization of the genetic organization of the *Hin*d-III M region of the multicapsid nuclear polyhedrosis virus of *Orgyia pseudotsugata* reveals major differences among baculoviruses, *J. Gen. Virol.* **70:**1815.

Guzo, D., Rathburn, H., Guthrie, K., and Dougherty, E., 1992, Viral and host cellular transcription in *Autographa californica* nuclear polyhedrosis virus-infected gypsy month cell lines, *J. Virol.* **66:**2966.

Hartig, P. C., Chapman, M. A., Hatch, G. G., and Kawanishi, C. Y., 1989, Insect virus: Assays for toxic effects and transformation potential in mammalian cells, *Appl. Environ. Microbiol.* **55:**1916.

Hartig, P. C., Cardon, M. C., and Kawanishi, C. Y., 1991, Insect virus assays for viral replication and persistence in mammalian cells, *J. Virol. Methods* **31:**335.

He, B., Chou, J., Liebermann, D. A., Hoffman, B., and Roizman, B., 1996, The carboxyl terminus of the murine MyD116 gene substitutes for the corresponding domain of the gamma(1)34.5 gene of herpes simplex virus to preclude the premature shutoff of total protein synthesis in infected human cells, *J. Virol.* **70:**84.

Hershberger, P. A., Dickson, J. A., and Friesen, P. D., 1992, Site-specific mutagenesis of the 35-kilodalton protein gene encoded by *Autographa californica* nuclear polyhedrosis virus: Cell line-specific effects on virus replication, *J. Virol.* **66:**5525.

Himeno, M., Sakai, F., Onodera, K., Nakai, H., Fukada, T., and Kawade, Y., 1967, Formation of nuclear polyhedral bodies and nuclear polyhedrosis virus of silkworm in mammalian cells infected with viral DNA, *Virology* **33:**507.

Hofmann, C., Sandig, V., Jennins, G., Rudolph, M., Schlag, P., and Strauss M., 1995, Efficient gene transfer into human hepatocytes by baculovirus vectors, *Proc. Natl. Acad. Sci. USA* **92:**10099.

Horton, H. M., and Burand, J. P., 1993, Saturable attachment sites for polyhedron-derived baculovirus on insect cells and evidence for entry via direct membrane fusion, *J. Virol.* **67:**1860.

Jarvis, D. L., and Finn, E. E., 1995, Biochemical analysis of the N-glycosylation pathway in baculovirus-infected lepidopteran insect cell, *Virology* **212:**500.

Kamita, S. G., and Maeda, S., 1993, Inhibition of *Bombyx mori* nuclear polyhedrosis virus (NPV) replication by the putative DNA helicase gene of *Autographa californica* NPV, *J. Virol.* **67:**6239.

Kamita, S. G., and Maeda, S., 1996a, Sequencing of the putative DNA helicase-encoding gene of *Bombyx mori* nuclear polyhedrosis virus and fine-mapping of a region involved in host range expansion, *Gene* (accepted).

Kamita, S. G., and Maeda, S., 1996b, Abortive infection of the baculovirus *Autographa californica* nuclear polyhedrosis virus in Sf-9 cells after mutation of the putative DNA helicase gene, *J. Virol.* **70:**6244.

Kondo, A., and Maeda, S., 1991, Host range expansion by recombination of the baculoviruses *Bombyx mori* nuclear polyhedrosis virus and *Autographa californica* nuclear polyhedrosis virus, *J. Virol.* **65:**3625.

Kool, M., Ahrens, C. H., Goldbach, R. W., and Rohrmann, G. F., 1994, Identification of genes involved in DNA replication of the *Autographa californica* baculovirus, *Proc. Natl. Acad. Sci. USA* **91:**11212.

Kuroda, D., Geyer, H., Geyer, R., Doerfler, W., and Klenk, H.-D., 1990. The oligosaccharides of influenza virus hemagglutinin expressed in insect cells by a baculovirus vector, *Virology* **174:**418.

Lautenschlager, R. A., and Podgwaite, J. D., 1979, Passage rates of nucleopolyhedrosis virus by avian and mammalian predators of the gypsy moth, *Environ. Entomol.* **8:**210.

Liu, J. J., and Carstens, E. B., 1993, Infection of *Spodoptera frugiperda* and *Choristoneura fumiferana* cell lines with the baculovirus *Choristoneura fumiferana* nuclear polyhedrosis virus, *Can. J. Microbiol.* **39:**932.

Lu, A., and Miller, L. K., 1995a, Differential requirements for baculovirus late expression factor genes in two cell lines, *J. Virol.* **69:**6265.

Lu, A., and Miller, L. K., 1995b, The roles of eighteen baculovirus late expression factor genes in transcription and DNA replication, *J. Virol.* **69:**975.

Lu, A., and Miller, L. K., 1996, Species-specific effects of the *hcf-1* gene on baculovirus virulence, *J. Virol.* **70:**5123.

Maeda, S., Kamita, S. G., and Kondo, A., 1993, Host range expansion of *Autographa californica* nuclear polyhedrosis virus (NPV) following recombination of a 0.6-kilobase-pair DNA fragment originating from *Bombyx mori* NPV, *J. Virol.* **67**:6234.

McClintock, J. T., and Dougherty, E. M., 1987, Superinfection of baculovirus-infected gypsy moth cells with the nuclear polyhedrosis viruses of *Autographa californica* and *Lymantria dispar*, *Virus Res.* **7**:351.

McClintock, J. T., Dougherty, E. M., and Weiner, R. M., 1986, Semi-permissive replication of a nuclear polyhedrosis virus of *Autographa californica* in a gypsy moth cell line, *J. Virol.* **57**:197.

McIntosh, A. H., and Shamy, R., 1980, Biological studies of a baculovirus in a mammalian cell line, *Intervirology* **13**:331.

McNitt, L., Espelie, K. E., and Miller, L. K., 1995, Assessing the safety of toxin-producing baculovirus biopesticides to a nontarget predator, the social wasp *Polistes metricus* Say, *Biol. Control* **5**:267.

Monsma, S. A., Oomens, A. G. P., and Blissard, G. W., 1996, The gp64 envelope fusion protein is an essential baculovirus protein required for cell-to-cell transmission of infection, *J. Virol.* **70**:4607.

Morel, G., and Fouillaud, M., 1992, Presence of microorganisms and viral inclusion bodies in the nests of the paper wasp *Polistes hebraeus* Fabricius (Hymenoptera, Vespidae), *J. Invertebr. Pathol.* **60**:210.

Mori, H., Nakazawa, H., Shirai, N., Shibata, N., Sumida, M., and Matsubara, F., 1992, Foreign gene expression by a baculovirus vector with an expanded host range, *J. Gen Virol.* **73**:1877.

Morris, T. D., and Miller, L. K., 1992, Promoter influence on baculovirus-mediated gene expression in permissive and non-permissive insect cell lines, *J. Virol.* **66**:7397.

Morris, T. D., and Miller, L. K., 1993, Characterization of productive and non-productive AcMNPV infection in selected insect cell lines, *Virology* **197**:339.

Ogonah, O. W., Freedman, R. B., Jenkins, N., Patel, K., and Rooney, B. C., 1996, Isolation and characterization of an insect cell line able to perform complex N-linked glycosylation on recombinant proteins, *Bio/Technology* **14**:197.

Palli, S. R., Caputo, G. F., Sohi, S. S., Brownwright, A. J., Ladd, T. R., Cook, B. J., Primavera, M., Arif, B. M., and Retnakaran, A., 1996, CfMNPV blocks AcMNPV-induced apoptosis in a continuous midgut cell line, *Virology* **222**:201.

Reimann, R., and Miltenberger, H. G., 1983, Cytogenetic studies in mammalian cells after treatment with insect pathogenic viruses (Baculoviridae). II *in vitro* studies with mammalian cell lines. *Entomophaga* **28**:33.

Summers, M. D., Volkman, L. E., and Hsieh, C-H., 1978, Immunoperoxidase detection of baculovirus antigens in insect cells, *J. Gen Virol.* **40**:545.

Thiem, S. M., Du, X., Quentin, M. E., and Berner, M. M., 1996, Identification of a baculovirus gene that promotes *Autographa californica* nuclear polyhedrosis virus replication in a nonpermissive insect cell line, *J. Virol.* **70**:2221.

Volkman, L. E., and Goldsmith, P. A., 1983, *In vitro* survey of *Autographa californica* nuclear polyhedrosis virus interaction with nontarget vertebrate host cells, *Appl. Environ. Microbiol.* **45**:1085.

Volkman, L. E., and Goldmith, P. A., 1985, Mechanism of neutalization of budded *Autographa californica* nuclear polyhedrosis virus by a monoclonal antibody: Inhibitation of entry by adsorptive endocytosis, *Virology* **143**:185.

Volkman, L. E., Goldmith, P. A., and Hess, R. T., 1986, Alternate pathway of entry of budded *Autographa californica* nuclear polyhedrosis virus: Fusion at the plasma membrane, *Virology* **148**:288.

CHAPTER 10

Regulation of Programmed Cell Death by Baculoviruses

Rollie J. Clem

I. INTRODUCTION TO PROGRAMMED CELL DEATH AND APOPTOSIS

A. Description of Apoptosis

Programmed cell death is the normal fate of most eukaryotic cells. The term "programmed" refers to the genetic pathways existing within all cells that, when stimulated in the proper context, result in cell suicide. This differs from necrosis, which is cell death caused by gross injury or insult and is beyond the control of the cell (Walker *et al.*, 1988). Apoptosis is a particular form of programmed cell death; the word *apoptosis*, which in Greek means "leaves falling from trees," was originally used to reflect the stereotypical morphological events that often are associated with programmed cell death, especially in vertebrates (Kerr *et al.*, 1972). Rapid progress has been made in the past 5 years in the elucidation of the pathways involved in signaling and carrying out programmed cell death (Fraser and Evan, 1996; Martin and Green, 1995; Steller, 1995; Tewari and Dixit, 1996; Vaux and Strasser, 1996; Williams and Smith, 1993).

The process of apoptosis is characterized by distinct morphological characteristics including surface blebbing, cell shrinkage, chromatin condensation, and nuclear disassembly. In contrast to necrotic cell death, the plasma membrane of an apoptotic cell remains intact for an extended period

ROLLIE J. CLEM • Department of Molecular Microbiology and Immunology, Johns Hopkins School of Hygiene and Public Health, Baltimore, Maryland 21205.
The Baculoviruses, edited by Lois K. Miller. Plenum Press, New York, 1997.

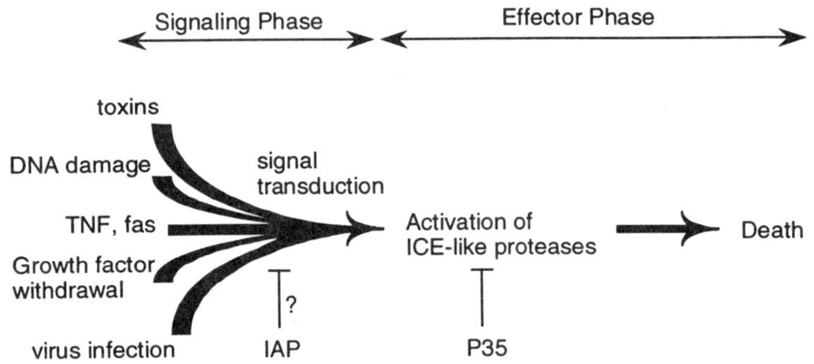

FIGURE 1. A simplified model of the pathways involved in programmed cell death. In the signaling phase, multiple signal transduction pathways can be stimulated by various death stimuli. These multiple pathways are thought to all culminate in the activation of the effector pathway, which involves (among other things) activation of ICE-like proteases. These proteases are thought to be directly responsible for the death of the cell. The baculovirus antiapoptotic proteins P35 and IAP are shown; P35 directly inhibits many of these ICE-like proteases, while IAP is thought to act at an unknown earlier step.

of time. There are changes in the membrane of an apoptotic cell, however, that allow it to be rapidly recognized by other cells and phagocytized in tissues (Fadok *et al.*, 1992). Late in the process of apoptosis, the DNA is digested by an endogenous endonuclease, resulting in internucleosomal cleavage (Wyllie, 1980).

On the molecular level, the process of apoptosis can be divided into signaling and effector phases (Fig. 1). The signaling phase involves detection of the death signal and transduction of the signal to the death machinery. This signal can originate either from outside the cell (through the binding of a ligand to a receptor) or from within the cell (for example, detection of toxins, radiation-induced damage, or virus infection). Given the multitude of death stimuli and the fact that various stimuli can have opposite effects in different cell types, there are probably multiple signal transduction pathways that are involved in communicating the signal. The complexity of these pathways is illustrated by the number of molecules that have been identified that can, when overexpressed, affect apoptosis either positively or negatively (see the reviews cited above). However, it appears that these multiple signaling pathways all converge in a common effector pathway whose job it is to carry out the death program. This effector pathway appears to be highly conserved among different organisms. The principle players in the effector pathway appear to be cysteine proteases related to interleukin-1β converting enzyme (ICE) (Martin and Green, 1995). The ICE-like proteases are synthesized as inactive precursors and are themselves activated by proteolytic cleavage. Furthermore, the ICE-like enzymes appear to cleave each other, and thus there is accumulating evidence for a protease cascade involved in the process of death. If true, then once this cascade is initiated, the death program would not require the synthesis of new proteins.

Presumably, at some point in the cascade the ICE-like proteases must digest other substrates that are actually responsible for the death of the cell. The best example so far is the cleavage of nuclear lamins, which results in disassembly of the nucleus (Takahashi *et al.*, 1996). Other proteins have been shown to be cleaved by ICE or related proteases, but the physiological significance of these events is less clear (Martin and Green, 1995).

B. The Importance of Apoptosis

Apoptosis plays a vital role in many normal situations in both vertebrate and invertebrate organisms (Wyllie *et al.*, 1980). From this perspective, one of the most advantageous characteristics of apoptosis is the maintenance of cellular membrane integrity during apoptotic death, which prevents the leakage of intracellular components into the extracellular milieu and the inflammation that would result. Apoptosis is particularly prominent during development, in processes such as tissue involution and remodeling (Ucker, 1991). Apoptosis is also important in normal cell turnover in nearly all tissues, especially those with high turnover rates such as the skin, the intestinal epithelium, and the blood and lymph systems. This is dramatically illustrated by the fact that defects in apoptosis in these tissues can lead to the development of tumors (Thompson, 1995). Apoptosis is also important for the functioning of the vertebrate immune system, where it is involved in cell killing by cytotoxic T cells and in the selection of memory B cells (Duvall and Wyllie, 1986). At this point, the best-defined role for programmed cell death in invertebrates is during embryonic development and metamorphosis (Lockshin, 1985; Steller and Grether, 1994; Truman and Schwartz, 1984); but by analogy with higher organisms, programmed cell death is undoubtedly also important in a multitude of other invertebrate processes, including tissue homeostasis and defense against microbial pathogens (Clouston and Kerr, 1985; Martz and Howell, 1989).

C. The Influence of Apoptosis on Viral Replication Strategies

In recent years, the study of programmed cell death has influenced nearly every subdiscipline in biology, with virology being no exception. The number of viruses that have been shown to induce and/or block apoptosis continues to grow rapidly (Razvi and Welsh, 1995; Shen and Shenk, 1995). It is now widely recognized that the ability of individual cells to direct their own suicide has had an important influence on the evolution of viruses and their strategies for replication. This is due to the fact that many viruses provoke the suicide response very early in their replication process, and that the premature death of the host cell can have a negative influence on the ability of the virus to replicate or establish persistence. Thus, viruses have had to either evolve mechanisms to prevent cell suicide or find ways to cope with replicating in an apoptotic environment. In many cases the path chosen seems to have been determined by the coding capacity of the viral genome. Thus, many large DNA viruses have acquired genes that are able to directly intercede in the apoptotic pathway and prolong the life of the infected cell until after they have finished replicating, while smaller DNA and RNA viruses often have the ability to replicate efficiently in apoptotic cells. For example, the RNA virus Sindbis virus seems to replicate more efficiently in apoptotic than in nonapoptotic cells (Ubol *et al.*, 1994).

The mechanisms by which viruses induce apoptosis are in most cases not well understood. In some cases, overexpression of a single viral gene product has been shown to induce apoptotic death, but whether these proteins actually trigger apoptosis during viral infection is difficult to prove (Laurent-Crawford *et al.*, 1993; Li *et al.*, 1995; Noteborn *et al.*, 1994; Prikhod'ko and Miller, 1996; Rao *et al.*, 1992; Suarez *et al.*, 1996; Tyler *et al.*, 1995). Disregulation of the cell cycle has been proposed as a potential mechanism of apoptosis induction, with the transforming protein E1A of adenovirus being the best example (Debbas and White, 1993).

The strategies employed by viruses to block apoptosis are many and varied. For example, the adenoviruses appear to have evolved several mechanisms of counteracting host cell apoptosis. The adenovirus early gene E1A induces p53-dependent apoptotic death, which can be counteracted by either of two polypeptides (19K and 55K) encoded by the E1B gene. E1B-19K is a homologue of the *bcl-2* gene (Chiou *et al.*, 1994), the prototype of a cellular gene family whose members are involved in cell death regulation (Farrow and Brown, 1996). E1B-55K binds to and inactivates the cellular p53 protein, another cellular protein important in signaling apoptosis in certain situations (Debbas and White, 1993). Other viral gene products that inactivate p53, such as SV40 T antigen and papillomavirus E6, are also capable of inhibiting p53-dependent apoptosis (Shen and Shenk, 1995). Finally, the E3 gene of adenovirus also encodes proteins that are able to inhibit the killing of infected cells by cytotoxic T lymphocytes or tumor necrosis factor (Wold and Gooding, 1991).

In addition to adenovirus, several other DNA viruses contain homologues of the *bcl-2* family, including BHRF1 of Epstein–Barr virus (Henderson *et al.*, 1993), ORF 16 of herpesvirus saimiri (Nava *et al.*, 1996a), KSbcl-2 of human herpesvirus 8 (Cheng *et al.*, 1996), and LMW5-HL of African swine fever virus (Neilan *et al.*, 1993). Poxviruses also encode genes that can inhibit or delay host cell apoptosis, including the serpins crmA (Spi-2) (Ray *et al.*, 1992) and Spi-1 (Brooks *et al.*, 1995) and the ankyrin repeat-containing protein CHOhr (Ink *et al.*, 1995). The CrmA gene product blocks death through its ability to inhibit ICE proteases (Ray *et al.*, 1992). Cytomegalovirus IE1 and IE2 are transcription factors that are able to inhibit apoptosis induced by an adenovirus mutant lacking E1B (Zhu *et al.*, 1995). Finally, LMP-1 of

Epstein–Barr virus is able to promote the survival of latently infected lymphocytes, possibly through a signal transduction mechanism (Mosialos *et al.*, 1995).

Understanding the mechanisms by which viruses manipulate normal host functions to their own advantage has been of vital importance in furthering our understanding of many cellular processes, including cell cycle regulation, tumorigenesis, RNA transcription and processing, and protein translation. In a similar fashion, the study of baculovirus-induced apoptosis has led to a greater understanding of the genetic pathways of cell death and their similarities between insects and other organisms, as well as provided insight into the role of apoptosis in defense against viral pathogens.

II. INDUCTION OF APOPTOSIS BY BACULOVIRUSES

A. Characteristics of AcMNPV-Induced Apoptosis

The baculovirus AcMNPV was one of the first viruses known to induce and inhibit apoptosis (Clem *et al.*, 1991). A spontaneous mutant of AcMNPV known as the annihilator mutant (vAcAnh) was isolated as an occlusion negative plaque during the construction of an expression vector. SF-21 cells, derived from *Spodoptera frugiperda,* did not produce polyhedra when infected with vAcAnh; but unexpectedly, the virus did not contain the foreign gene of interest. Moreover, vAcAnh-infected cells died much more rapidly than cells infected with wild-type AcMNPV, and the morphology of the dying cells was distinctly different from that of wild-type-infected cells. The first clue that the cells were undergoing apoptosis was an obvious blebbing of the plasma membrane beginning at 9 hr postinfection (p.i.). The onset of blebbing in vAcAnh-infected SF-21 cells is somewhat asynchronous, with most of the cells starting to bleb between 12 and 24 hr p.i. (Clem *et al.*, 1991). Examination of individual cells over time reveals that this blebbing process intensifies over the following 1 to 2 hr, and cytoplasmic and nuclear contents are shed into large membrane-bound vesicles called apoptotic bodies (Fig. 2). These apoptotic bodies continue to exclude vital dyes for a number of hours after they are formed. In addition to the blebbing of cytoplasm and formation of apoptotic bodies, other characteristics typical of vertebrate apoptosis are also exhibited by vAcAnh-infected SF-21 cells, including nuclear condensation and budding of nuclear material into the apoptotic bodies, a retention of intact mitochondria (and exclusion of vital dyes) until late in the apoptotic process, and internucleosomal cleavage of cellular DNA beginning between 6 and 12 hr p.i. (Clem *et al.*, 1991). The SF-9 cell line, a clonal derivative of SF-21, also underwent apoptosis when infected with vAcAnh (R. J. Clem and L. K. Miller, unpublished results). The apoptotic death that these cells undergo appears to be remarkably similar to that of mammalian cells, except that the blebbing of mammalian cells is usually much more rapid.

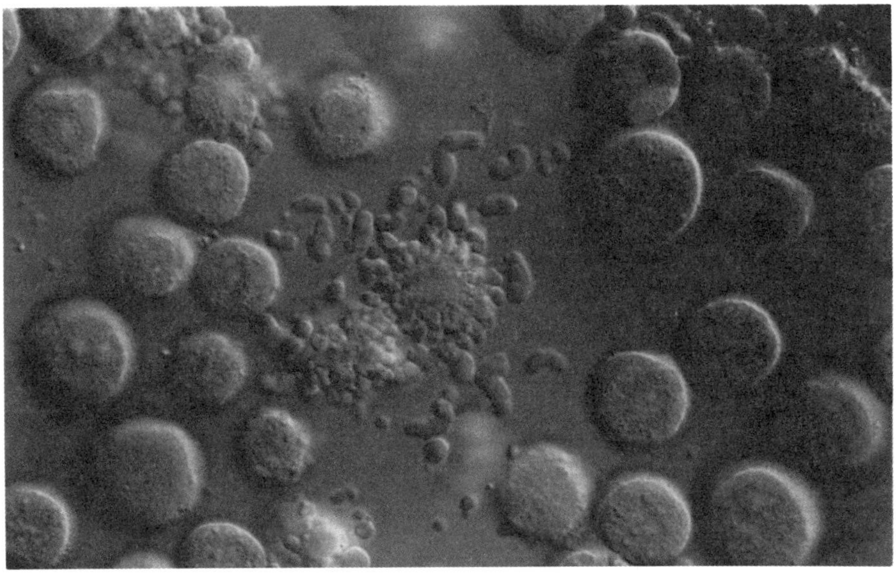

FIGURE 2. Photomicrograph of SF-21 cells infected with the AcMNPV *p35* mutant vAcAnh, 12 hr p.i. The two cells near the center are in the later stages of blebbing; note the formation of large blebs (apoptotic bodies) at the cell surface. By 48 hr p.i., all of the cells will have gone through the blebbing process. Many of the free apoptotic bodies were produced from blebbing cells that have already disintegrated.

It seems that at least the initial stages of apoptosis are triggered by wild-type AcMNPV infection of SF-21 cells, since a transient blebbing occurs that is similar to the initial stages of blebbing induced by vAcAnh (Clem *et al.*, 1991). However, the blebbing seen in wild-type-infected cells is only transient, and it eventually disappears. The cells remain viable for several days before succumbing to a death that appears to be necrotic. The failure of the apoptotic response to progress in wild-type-infected cells suggested that a gene product either encoded or induced by AcMNPV was able to intercede and prevent apoptosis from reaching its conclusion. Marker rescue assays determined that the apoptotic death triggered by vAcAnh could be suppressed by a gene or genes in the EcoRI-S fragment of the AcMNPV genome, and nucleotide sequencing revealed a deletion in the *p35* gene of vAcAnh (Clem *et al.*, 1991). The role of the *p35* gene in blocking apoptosis was confirmed by the replacement of a portion of *p35* in wild-type AcMNPV with the *lac*Z gene. This *p35* mutant virus had the same phenotype as vAcAnh, confirming that the *p35* gene was responsible for blocking the apoptotic response during wild-type infection.

Neither wild-type AcMNPV nor *p35* mutant viruses induced apoptosis or transient blebbing in the TN-368 cell line derived from *Trichoplusia ni* (Clem *et al.*, 1991). The phenotype of *p35* mutant viruses appeared to be completely normal in these cells. This was important as it allowed for the production of working stocks of the mutant viruses, which replicated poorly in SF-21 cells.

Even though *p35* mutant infection triggered apoptosis in SF-21 cells, it was initially difficult to prove that the *p35* gene product (P35) was acting directly on a cellular apoptotic pathway and not by somehow suppressing viral initiation of apoptosis. An important finding was that the RNA synthesis inhibitor actinomycin D could also trigger rapid apoptosis in these cells (Crook *et al.*, 1993). Actinomycin D-induced apoptosis appears identical to *p35* mutant-induced death, and this led to the demonstration that P35 blocks apoptosis that is triggered by a nonviral signal in the absence of other viral genes (Cartier *et al.*, 1994; Clem and Miller, 1994a).

Despite the presence of the *p35* gene, wild-type AcMNPV induced apoptosis in the cell lines SL2 from *Spodoptera littoralis* (Chejanovsky and Gershburg, 1995) and CF-203 from *Choristoneura fumiferana* (Palli *et al.*, 1996). Internucleosomal cleavage of cellular DNA is also observed in these cell lines, and AcMNPV-induced apoptosis appears ultrastructurally similar in SL2 and SF-21 cells. However, the apoptosis observed in CF-203 cells appears to differ somewhat from that of SF-21 cells. While apoptosis triggered by *p35* mutant infection in SF-21 cells is identical in appearance to that triggered by actinomycin D, there are morphological differences between virus-induced death and actinomycin D-induced death in CF-203 cells (Palli *et al.*, 1996). Since CF-203 is the only one of the three cell lines that has been examined during apoptosis by electron microscopy (Palli *et al.*, 1996), it is difficult to draw any solid conclusions at this point concerning the subtleties of the morphological differences between apoptotic death in the three cell lines.

B. Possible Mechanisms of Induction of Apoptosis by AcMNPV

Pretreatment of *p35* mutant-infected SF-21 cells with the DNA synthesis inhibitor aphidicolin blocks virus-induced apoptosis (Clem and Miller, 1994a). This result, along with the timing of apoptosis induction (6 to 9 hr p.i.), suggests that the initiation of apoptosis is somehow related to the transition from the early to late stages of infection. During this crucial time, several important events are occurring simultaneously in the infected cell: (1) viral DNA synthesis is initiated, (2) late gene expression begins, and (3) the synthesis of cellular RNA and protein, along with the expression of early viral genes, is gradually shut down. However, it is difficult to determine which of these processes is responsible for the induction of apoptosis, since they are so closely interconnected. At this point the data support a scenario in which there are multiple apoptotic signals generated during AcMNPV infection, with each being sufficient to trigger a suicide response by the infected cell.

1. Block in RNA Synthesis

It is clear that inhibiting RNA synthesis is sufficient to induce apoptosis in SF-21 cells (Clem and Miller, 1994a). Treatment with actinomycin D, which intercalates into DNA and prevents RNA elongation, 5,6-dichlorobenzimidazole riboside, a nucleoside analogue, or α-amanatin, which binds to and inactivates the large subunit of RNA polymerase II, each induces rapid apoptosis in these cells (Clem and Miller, 1994a). The fact that three drugs, which each inhibits RNA synthesis by a different mechanism, all induce apoptosis with similar kinetics suggests that the induction is due to a specific effect. These results led to the suggestion that the cessation of host mRNA synthesis seen at late times in AcMNPV infection (Ooi and Miller, 1988) could be responsible for the induction of apoptosis (Clem and Miller, 1994a). Although the active process of apoptosis often requires RNA and protein synthesis, inhibition of these processes has been shown to induce apoptotic death in some situations (Raff *et al.*, 1993). Presumably the continued synthesis of a short-lived inhibitory protein is required to prevent the initiation of death in these cells. It is puzzling that the protein synthesis inhibitor cycloheximide does not induce apoptotic death in SF-21 cells, despite the fact that it is potently active in this cell line (Clem and Miller, 1994a). It may be that the synthesis of a specific RNA species is required to inhibit death, or perhaps the process of transcription itself along with its concomitant changes in chromatin structure is somehow involved. It also may simply be that cycloheximide is a more leaky drug than the RNA synthesis inhibitors tested, and only low levels of an inhibitory protein are required for function.

2. Viral DNA Replication

It is equally plausible that an unplanned round of DNA synthesis is what triggers the apoptotic death of AcMNPV-infected SF-21 cells. When the viral genes necessary for DNA synthesis are transiently expressed in SF-21 cells, very little DNA can be recovered unless *p35* is included in the transfections, suggesting that apoptosis is triggered by the expression of the viral DNA synthesis machinery (Lu and Miller, 1995). However, the expression of one of these genes, IE-1, has been shown to be capable of inducing apoptosis by itself (see Section II.B.3) (Prikhod'ko and Miller, 1996), and including *p35* in transient gene expression assays that contain IE-1 also increases reporter gene expression (Gong and Guarino, 1994; Todd *et al.*, 1995, 1996). The ability of another antiapoptotic gene, Cp-*iap* (see Section III.B), to replace *p35* in these transient assays is further evidence that apoptosis is involved (Lu and Miller, 1995; Prikhod'ko and Miller, 1996; Todd *et al.*, 1995). As discussed above, it may be that either stimulus (cessation of RNA synthesis or induction of viral DNA synthesis) is sufficient to induce apoptosis in SF-21 cells, although it seems that transient expression of IE-1 or the DNA synthesis machinery is not nearly as efficient at inducing apoptosis as is inhibition of RNA synthesis (Prikhod'ko and Miller, 1996).

3. Possible Apoptosis-Stimulating Viral Genes

Expression of the early viral transactivator IE-1 is sufficient to induce apoptosis in a subpopulation of transfected cells (Prikhod'ko and Miller, 1996). The observation that only about 25% of cells transfected with IE-1 undergo apoptosis suggests a possible connection with the cell cycle. However, it is uncertain whether expression of IE-1 is the major trigger for apoptosis in infected cells (Prikhod'ko and Miller, 1996), since during infection IE-1 would still be expressed in the presence of aphidicolin, which blocks virus-induced apoptosis (Clem and Miller, 1994a). However, these results could be consistent if IE-1 induces apoptosis by directly stimulating an unscheduled round of cellular DNA synthesis (similar to adenovirus E1A) and aphidicolin acts by interceding at a point downstream of the IE-1 trigger. Whether or not aphidicolin blocks IE-1 induced apoptosis in uninfected cells has not been reported, but it has no effect on actinomycin D-induced death (Clem and Miller, 1994a). Another piece of evidence questioning whether IE-1 is the major apoptotic trigger in infected cells is the observation that the temperature-sensitive IE-1 mutant tsB821 induces apoptosis in SF-21 cells at the nonpermissive temperature (Ribeiro *et al.*, 1994). However, the level of apoptosis stimulated by tsB821 is not high and it is possible that a small amount of functional IE-1 is produced in these cells at the nonpermissive temperature. It is also noteworthy that the apoptosis that does occur in tsB821-infected cells coincides with the initiation of DNA synthesis, which is delayed (Ribeiro *et al.*, 1994).

Another possible player in AcMNPV-induced death is the *p10* gene. Viral mutants lacking a functional *p10* gene fail to lyse nuclei at late times (van Oers *et al.*, 1993; Williams *et al.*, 1989). This function of *p10* appears to be analogous to that of the adenovirus death protein (Tollefson *et al.*, 1996). The E3 region of adenovirus encodes a small (11.6 kDa) glycoprotein [adenovirus death protein (ADP)] that localizes in the nuclear membrane at late times in adenovirus-infected cells. The nuclei of cells infected with adenovirus mutants lacking ADP also do not lyse, similar to that of *p10* mutants of AcMNPV. However, transient expression of *p10* in uninfected SF-21 cells does not appear to have any deleterious consequences (R. Clem and J. M. Hardwick, unpublished results). The death of both adenovirus-infected cells and wild-type AcMNPV-infected SF-21 cells does not appear to be apoptotic in nature, but these deaths are in a sense programmed, since a specific viral gene product is required for the final lysis of the infected cells or nuclei.

C. Induction of Apoptosis by Other Baculoviruses

In addition to AcMNPV, other baculoviruses have also been shown to induce apoptosis in lepidopteran cells. Mutants of the closely related virus *Bombyx mori* NPV (BmNPV) lacking functional *p35* also induce an apoptotic response in BmN cells (Kamita *et al.*, 1993), although only a portion of the infected cells undergoes apoptosis, and the apoptotic response is more delayed than in vAcAnh-infected SF-21 cells. This delay may explain why virus protein synthesis and replication appear normal despite the induction of apoptosis (Kamita *et al.*, 1993).

The insect virus Hz-1, which was formerly classified in the baculovirus family but is now temporarily unclassified, has also been shown to induce apoptosis upon superinfection of SF-9 cells (Lee *et al.*, 1993). Although previously uninfected SF-9 cells died by necrosis when infected with Hz-1, persistently infected SF-9 cell lines underwent extensive apoptosis when superinfected with wild-type Hz-1. The few surviving cells continued to divide and eventually replaced the monolayer. The production of viral progeny in the apoptotic cells was greatly diminished, suggesting that the process of apoptosis may contribute to the establishment of persistent infection (Lee *et al.*, 1993).

III. ANTIAPOPTOTIC GENES OF BACULOVIRUSES

A. The *p35* Gene

1. Identification of the *p35* Gene

The *p35* gene was first identified in 1987 by sequence analysis of the region adjacent to the insertion of the transposable element TED (Friesen and Miller, 1987). Although the function of P35 was unknown until the mapping of the spontaneous mutation in vAcAnh in 1991 (Clem *et al.*, 1991), the transcriptional regulation of *p35* had already been studied extensively by that time (see Chapter 6, this volume). Transcription of *p35* is from a complex promoter containing both early and late start sites (Dickson and Friesen, 1991; Friesen and Miller, 1987). Although expressed at higher levels late in infection, early synthesis of P35 is required to prevent virus-induced apoptosis (Crook *et al.*, 1993; Hershberger *et al.*, 1994). The *p35* gene has no detectable similarity to other known genes, except for the *p35* homologue present in BmNPV, which is 90% identical to AcMNPV *p35* at the amino acid level and also plays a role in preventing apoptosis induced by BmNPV in *B. mori* cells (Kamita *et al.*, 1993).

2. Characteristics of the P35 Polypeptide

The protein encoded by the AcMNPV *p35* gene (P35) is 299 amino acids in length and has a predicted molecular mass of 34.8 kDa. The P35 protein has no recognizable sequence motifs and lacks a signal sequence. It is localized in the cytoplasm in both infected and plasmid transfected cells (Hershberger *et al.*, 1994). P35 is also present in budded virions (Hershberger *et al.*, 1994), although the significance of this is unclear, since the protein must be synthesized at early times to prevent apoptosis (Crook *et al.*, 1993; Hershberger *et al.*, 1994). The only remarkable features of the sequence of the P35 protein are several clusters of charged residues (Fig. 3), including lysine-rich domains at the center of the molecule and at its carboxyl-terminus, and additional highly charged domains in the amino-terminal half of the protein (Bertin *et al.*, 1996; Clem *et al.*, 1996b). Based on hydrophobicity calculations, these highly charged domains are predicted to be exposed on the outer surface of the molecule, and thus may participate in protein–protein interactions.

The P35 polypeptide is remarkably sensitive to mutation, including insertions, deletions, and amino acid substitutions. In the most comprehensive example of P35 mutagenesis to date, 11 out of 18 small (two codon) insertions throughout P35 resulted in nonfunctional proteins (Bertin *et al.*, 1996). Most of the null mutations were in the amino-terminal half of the protein. However, the extreme carboxyl-terminus of P35 is also important for its function, since mutations in this region also abolish antiapoptotic activity (Bertin *et al.*, 1996; Hershberger *et al.*, 1992). In addition, Bertin *et al.* (1996) defined three charged regions in the amino-terminal half of P35 that were designated CHR1, CHR2, and CHR3. Substitution of selected charged residues with alanine in these regions resulted in loss of antiapoptotic function in 3 of 11 cases.

Several lines of evidence suggest that P35 must complex with itself or other proteins for stability and/or function. The amino-terminal portion of the molecule can serve as a dominant negative inhibitor of P35 function (Cartier *et al.*, 1994). When a construct encoding only the amino-terminal 76 amino acids was stably expressed in SF-21 cells, the cells still underwent apoptosis upon wild-type AcMNPV infection, despite the expression of full-length P35 protein. There was little or no accumulation of full-length P35 protein in these infected cells despite the fact that other viral proteins accumulated to normal levels, suggesting that an amino-terminal domain of P35 must interact with either itself or another protein in order to stabilize the entire P35 protein (Cartier *et al.*, 1994). Many of the null mutants produced by Bertin *et al.* (1996) also exhibited decreased steady-state levels of P35 protein, both in apoptotic SF-21 cells and in nonapoptotic TN-368 cells. The overall tertiary structure of P35 was not predicted to be greatly perturbed by these minor mutations, suggesting that protein–protein interactions were being disrupted (Bertin *et al.*, 1996).

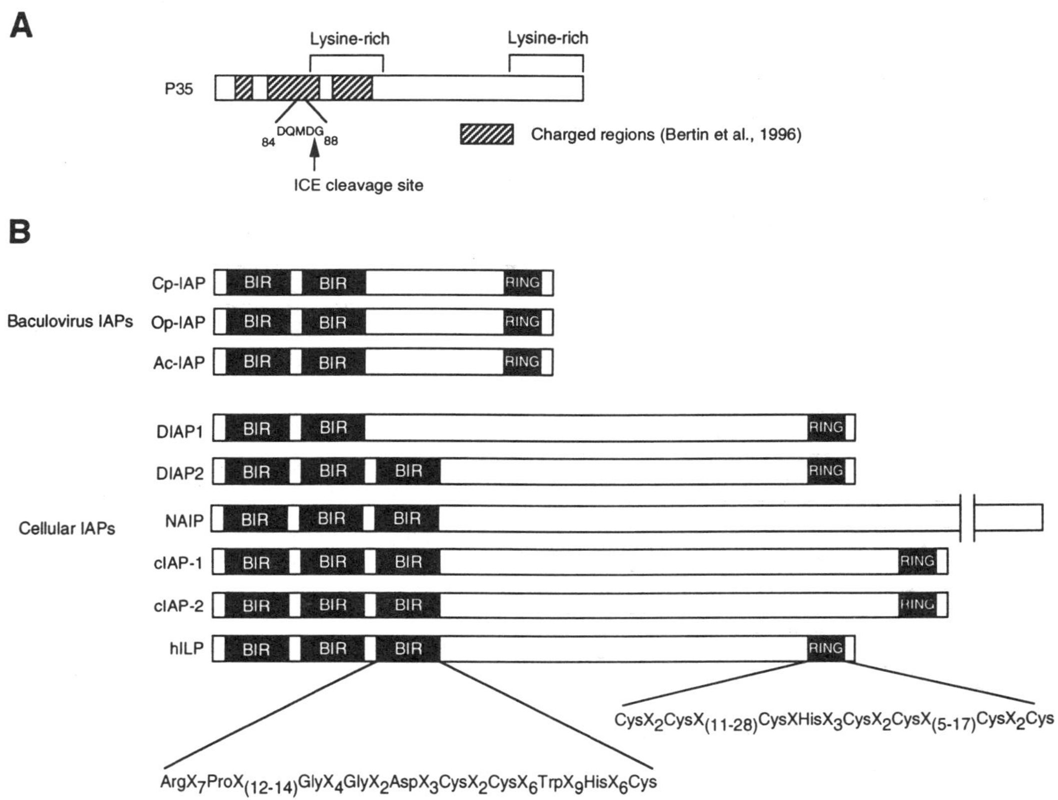

FIGURE 3. (A) Characteristics of the AcMNPV P35 protein. The lysine-rich and charged regions discussed in the text are indicated, as is the amino acid sequence (single letter code) surrounding the ICE cleavage site immediately following position 87. (B) The known homologues of baculovirus IAP. The baculovirus IAPs are smaller (around 30 kDa) and contain only two BIRs, while the cellular homologues have three BIRs (with the exception of DIAP1) and additional sequences in the region between the BIRs and the RING finger, which makes their predicted molecular mass approximately 70 kDa for cIAP-1 and -2, and approximately 55 kDa for hILP, DIAP1, and DIAP2. The NAIP protein does not contain a RING finger, and is approximately 140 kDa. The consensus sequences for the BIRs and the IAP RING finger are shown below, with X indicating any amino acid.

3. P35 Inhibits Programmed Cell Death in a Wide Range of Organisms

The *p35* gene was one of the first antiapoptotic genes discovered. As such, there was considerable interest by others in the cell death field in determining whether P35 could block cell death in other paradigms. To this end, *p35* was expressed in organisms ranging from nematodes to human cells in order to determine whether it could block programmed cell death in these diverse situations. As a result of these experiments, P35 is the most widely acting antiapoptotic gene known. It has been shown to block developmentally programmed cell death in the nematode *Caenorhabditis elegans* (Sugimoto *et al.*, 1994; Xue and Horvitz, 1995) and in *Drosophila melanogaster* (Hay *et al.*, 1994), as well as apoptotic death in mammalian neural cells (Rabizadeh *et al.*, 1993), sympathetic neurons (Martinou *et al.*, 1995), breast carcinoma cells and B lymphocytes (Beidler *et al.*, 1995), and neuroblastoma cells and fibroblasts (Nava *et al.*, in preparation). The death stimuli utilized in these studies also varied widely. The ability of P35 to block many different death stimuli suggests that it acts in the effector pathway, possibly at one of the final execution steps (Fig. 1).

4. P35 Is an Inhibitor of ICE-like Proteases

Given its ability to block cell death in a wide variety of organisms, P35 also must act at a highly conserved step in the death pathway. This is indeed the case, as the P35 polypeptide has been shown to potently inhibit several members of the ICE family of cysteine proteases (Bertin *et al.*, 1996; Bump *et al.*, 1995; Xue and Horvitz, 1995). This family of related enzymes is present in nematodes, insects, and mammals, and they appear to play a central role in carrying out apoptotic death (Martin and Green, 1995). The ICE proteases appear to be highly specific, and they are unusual in that they cleave after aspartate residues. The only other known protease with a requirement for aspartate in the P1 position is the serine protease granzyme B, which is involved in the killing of cells by cytotoxic T lymphocytes (Shi *et al.*, 1992). The ability of P35 to inhibit ICE proteases has been demonstrated *in vitro* using purified proteins as well as in cells. When expressed in COS cells, P35 inhibits the cleavage of pro-interleukin-1β by ICE (Bump *et al.*, 1995). *In vitro*, purified P35 can efficiently block the activity of purified mammalian ICE homologues including ICE itself, ICH-1, ICH-2, and CPP32 (Bertin *et al.*, 1996, Bump *et al.*, 1995), as well as the nematode ICE homologue CED-3 (Bertin *et al.*, 1996; Xue and Horvitz, 1995). As expected, P35 also inhibits an ICE-like activity induced in SF-21 cells by AcMNPV infection (Bertin *et al.*, 1996). However, P35 has no inhibitory activity against granzyme B (Bump *et al.*, 1995). Although both competitive (Xue and Horvitz, 1995) and noncompetitive (Bump *et al.*, 1995) models have been proposed to explain how P35 inhibits ICE, the ability of P35 to inhibit ICE at an equimolar ratio indicates that it is acting as an irreversible inhibitor rather than a competitive substrate (Bertin *et al.*, 1996; Bump *et al.*, 1995).

P35 is cleaved by ICE into two fragments, which then form a stable complex with the enzyme (Bump *et al.*, 1995). The cleavage occurs between residues Asp87 and Gly88 of P35, after an aspartate residue that is the characteristic cleavage site for the ICE-like proteases (Fig. 3). This cleavage is essential for both inhibition and complex formation (Bertin *et al.*, 1996; Bump *et al.*, 1995; Xue and Horvitz, 1995). The cleavage site is within one of the highly charged regions of P35, and thus is predicted to be exposed on the outer surface. Cleavage after residue 87 (the P1 position) is necessary but not sufficient for antideath function; mutation of residue 84 (the P4 position) results in a protein that is still cleaved but is no longer protective in SF-21 cells, providing further evidence that the cleaved protein must remain bound to the active site of ICE in order to function as an inhibitor (Bertin *et al.*, 1996). Interestingly, the P4 mutant retains its ability to inhibit mammalian CPP32 and ICE *in vitro* (Bertin *et al.*, 1996). This tolerance at the P4 position fits with the ability of P35 to inhibit ICE-like enzymes from a variety of organisms.

P35 functions analogously to another viral inhibitor of ICE, the cowpox CrmA protein (Ray *et al.*, 1992). However, P35 and CrmA differ in several important ways. While CrmA is a member of the serpin family of protease inhibitors, P35 has no sequence similarity to serpins. P35 is also unable to inhibit granzyme B, while CrmA does inhibit this enzyme (Bump *et al.*, 1995; Quan *et al.*, 1995). P35 also appears to prevent apoptosis in a broader fashion than does CrmA, since CrmA does not block developmental cell death in *Drosophila* or *C. elegans* (Xue and Horvitz, 1995).

B. The *iap* Gene Family

1. *iap* Genes of Baculoviruses

Once it was clear that AcMNPV contained a gene that could block apoptosis, the vAcAnh mutant was used to search for antiapoptotic genes in other baculoviruses (Birnbaum *et al.*, 1994; Crook *et al.*, 1993). A genetic screen was used that took advantage of the fact that vAcAnh-infected SF-21 cells do not produce occlusion bodies. By cotransfecting vAcAnh DNA and genomic DNA from other baculoviruses into SF-21 cells, antiapoptotic genes were identified by screening for the presence of occlusion bodies in the transfected cells several days after transfection. Since the heterologous viruses did not replicate in SF-21 cells, the presence of occlusion bodies indicated that an antiapoptotic gene was present in the heterologous virus that blocked apoptotic death of the cells, allowing the production of occlusion bodies by vAcAnh. A limited screen of several other baculoviruses resulted in the discovery of a second family of antiapoptotic genes, called *iap* (*i*nhibitor of *ap*optosis). To date, three different baculoviruses are known to contain *iap* genes. The genomic DNA from *Cydia pomonella* granulosis virus (CpGV) (Crook *et al.*, 1993) and *Orgyia pseudotsugata* NPV (OpMNPV (Birnbaum *et al.*, 1994) was able to rescue occlusion body formation when cotransfected into SF-21 cells with vAcAnh DNA, and the genes responsible for rescuing vAcAnh were found to be homologous (Cp-*iap* and Op-*iap*, respectively). A third baculovirus *iap* gene was identified in the AcMNPV genome by sequence homology (Crook *et al.*, 1993), but the AcMNPV *iap* gene does not appear to have antiapoptotic activity (Clem and Miller, 1994a). Although an additional gene in AcMNPV was later named *iap2* (Ayres *et al.*, 1994), this gene has no significant homology to the other known *iap* homologues other than the presence of a RING finger and so does not appear to be a true *iap*-like gene.

In addition to the significance of finding a new family of antiapoptotic genes, the identification of Cp-*iap* and Op-*iap* was also interesting from the point of view of baculovirus evolution. The localization of Cp-*iap* on the CpGV genome was the first successful use of a marker rescue technique that relied on recombination between different baculoviruses. The Cp-*iap* gene was localized by using the marker rescue assay described above and a cosmid library containing the entire CpGV genome in overlapping segments (Crook *et al.*, 1993). Further subcloning of the rescuing cosmid localized the Cp-*iap* gene to the extreme left end of the SalI-B fragment of the genome. In spite of hybridization data suggesting that there is low sequence homology between CpGV and AcMNPV outside the gradulin/polyhedrin genes (Crook *et al.*, 1993), viruses were isolated from the marker rescue assay that retained the ability to block apoptosis. Thus, it was of interest to examine the recombination events that occurred in the marker rescue assay. Recombinant viruses were isolated from cotransfections using vAcAnh and the entire CpGV genome or a plasmid containing only the region immediately surrounding the Cp-*iap* gene. Despite three rounds of plaque purification, the recombinants that arose by cotransfection of the two intact viral genomes consisted of a heterogeneous mixture of virus genotypes, with the Cp-*iap* gene present only in a subpopulation of the virus genotypes. The recombinants that were obtained from cotransfection of vAcAnh DNA and the plasmid containing Cp-*iap*, however, appeared to be genetically homogenous. The inability to obtain stable recombinant viruses when the two virus genomes were used suggests that the likelihood of recombination events occurring between distantly related baculoviruses is low.

The discovery of an *iap* homologue in OpMNPV also had interesting evolutionary connotations, since OpMNPV had been previously shown to lack the genomic region in AcMNPV containing *p35* and *p94*, in spite of a general overall similarity in genome organization between these two viruses (Gombart *et al.*, 1989). The Op-*iap* gene was localized to the *Hin*dIII-K region of OpMNPV by a strategy similar to that used for Cp-*iap* (Birnbaum *et al.*, 1994). This genomic position is approximately the same as the genomic position of the nonfunctional Ac-*iap* in AcMNPV (Birnbaum *et al.*, 1994). The presence of a functional *iap* together with the lack of *p35* and *p94* in OpMNPV infers that either AcMNPV acquired *p35* and *p94* or OpMNPV lost the two genes at some point after the divergence of these two related viruses. In either case, there appears to be an evolutionary association between *p35* and *p94*, suggesting they may have associated functions as well (see Section IV.B.2).

Like *p35*, the baculovirus *iap* genes are also able to inhibit apoptosis in SF-21 cells induced by actinomycin D (Clem and Miller, 1994a), as well as apoptosis in other organisms, including *Drosophila* (Hay *et al.*, 1995) and mammalian cells (Duckett *et al.*, 1996; Uren *et al.*, 1996). However, preliminary results suggest that they do not appear to be as indiscriminate as *p35* in their ability to block cell death. For example, they do not provide as robust protection as does *p35* in Sindbis virus-infected baby hamster kidney cells and murine neuroblastoma cells (Nava *et al.*, in preparation), and Op-*iap* apparently has no antiapoptotic activity in rat neural cells under the same conditions that *p35* efficiently blocks apoptosis (D. Bredesen, personal communication). The apparent inability of the IAPs to function in all situations suggests these proteins may act at a step more upstream than P35 (see Section III.B.4).

2. Sequence Motifs Found in IAP Proteins

The IAP proteins contain two easily identifiable sequence motifs (Fig. 3). The first is a type of zinc finger known as a RING finger, which is found near the carboxyl-termini of the IAPs, and the second is two to three novel imperfect repeated sequences near the amino-termini known as baculovirus IAP repeats (BIRs). The RING finger and BIR motifs both exhibit characteristics of metal-coordinating sequences and each have been shown experimentally to bind zinc (Barlow *et al.*, 1994; Borden *et al.*, 1995; Clem *et al.*, in preparation).

The RING finger motif was first described in 1991 (Freemont *et al.*, 1991), and is now found in approximately 50 proteins, including the AcMNPV proteins IE-2, PE-38, CG30, IAP2, and Ac-IAP. These proteins are involved in a variety of functions, including several that are involved in transcriptional activation and genomic recombination (Clem and Miller, 1994b). In spite of this, sequence-specific DNA binding has not been demonstrated for any RING finger motif, and it has been proposed that they may instead be important in protein–protein interactions (Berg and Shi, 1996). Solution structures of two RING fingers have been determined (Barlow *et al.*, 1994; Borden *et al.*, 1995). These two RING fingers share the same basic core structure, consisting of a "cross-brace" with the first and third pairs of cysteine and histidine residues binding one atom of zinc, and the second and fourth pair binding another zinc atom. While the cores of the two RING fingers were the same, there were significant differences in their overall three-dimensional structure. These variations in tertiary structure suggest that different RING fingers could have quite different functions, which could explain the diversity of activities associated with RING finger-containing proteins. The RING finger motifs found in the IAPs differ from other RING fingers in that they contain three amino acids between the conserved histidine and the fourth conserved cysteine residues (Fig. 3), rather than the two amino acids found in other known RING fingers.

The BIRs are repeated sequences found near the amino termini of IAP proteins (Fig. 3). An individual BIR is approximately 65 amino acids in length and has a conserved CysX_2Cys motif near its center followed by a HisX_6Cys motif near its carboxyl end (where X is any amino acid) (Clem *et al.*, 1996b). There are also six other residues that are absolutely conserved in all known BIRs, as well as a number of other highly conserved residues. The BIRs are found only in a very limited set of proteins, including the IAPs, NAIP (see Section III.B.3), an incomplete open reading frame from Chilo iridescent virus (Crook *et al.*, 1993), and a structural protein called A224L from African swine fever virus (Chacon *et al.*, 1995). The ability of the BIRs to bind zinc (Clem *et al.*, in preparation) is no doubt important in allowing the IAP proteins to assume their proper conformation.

3. Cellular Homologues of Baculovirus *iap*

Recently a number of cellular homologues of baculovirus *iap* have been identified (Fig. 3). The first was a gene called NAIP, which encodes a human protein that is frequently mutated in patients with spinal muscular atrophy, a neurodegenerative disease that affects infants (Roy *et al.*, 1995). The NAIP protein has three BIRs at its amino-terminus but no other detectable homology to the IAPs. At 140 kDa, it is a much larger molecule than the baculovirus IAPs, which are around 30–35 kDa in mass.

In addition to NAIP, three other human proteins have been identified that appear to be even closer relatives of the baculovirus IAPs (Clem *et al.*, in preparation, Duckett *et al.*, 1996; Liston *et al.*, 1996; Rothe *et al.*, 1995a; Uren *et al.*, 1996). These proteins were independently identified by several different groups, either by screening expressed sequence tag databases with the baculovirus *iap* sequences or through their interactions with known proteins (see Section III.B.4). Since they were identified multiple times, each gene has several different names; they are cIAP-1/hIAP-2/MIHB, cIAP-2/hIAP-1/MIHC, and hILP/X-IAP/MIHA. The human IAPs are more similar than NAIP to the baculovirus genes in that they contain a RING finger at their carboxyl-termini (Fig. 3). However, like NAIP, the human IAPs have three rather than two BIRs. They also have an additional 150 or so amino acids between the third BIR and the RING finger not found in the baculovirus IAPs. The hILP protein appears to be localized in the cytoplasm (Duckett *et al.*, 1996). The cIAP-1 and 2 genes are located near the same locus on human chromosome 11, while hILP is located on the X chromosome.

Like their baculoviral counterparts, the human IAPs appear to protect cells against apoptosis when overexpressed. However, their ability to block death may be cell type specific and/or death stimulus specific. According to one report, overexpression of all three human IAPs as well as NAIP can protect cells against apoptotic death (Liston *et al.*, 1996), but so far these results have not been verified for cIAP-1 and/or cIAP-2 in other laboratories (Clem *et al.*, in preparation; Uren *et al.*, 1996). hILP/X-IAP, on the other hand, does seem to have strong antiapoptotic activity in each system where it has been tested so far (Duckett *et al.*, 1996; Liston *et al.*, 1996; Uren *et al.*, 1996).

Two *Drosophila* IAP-homologous proteins have also been described, DIAP1 and DIAP2/DILP/DIHA (Duckett *et al.*, 1996; Hay *et al.*, 1995; Liston *et al.*, 1996; Uren *et al.*, 1996). The DIAP1 protein is the product of the *thread* locus. There are several allelic mutants of *thread*, most of which are lethal (Hay *et al.*, 1995). Overexpression of either DIAP1 or -2 is able to partially prevent cell death in the developing eye, including both developmental cell death and death induced by overexpression of *rpr* or *hid*, two *Drosophila* genes that have been shown to be involved in programmed cell death (Grether *et al.*, 1995; White *et al.*, 1994). Both DIAP1 and -2 are similar in structure to the other known IAPs, with DIAP1 containing two BIRs and DIAP2 containing three BIRs (Hay *et al.*, 1995). Both proteins have a RING finger at their carboxyl termini.

4. How Do the IAPs Block Apoptosis?

Knowledge of the mechanism by which IAPs block cell death is still limited, but several clues exist. One of the most interesting to date is the interaction of cIAP-1 and 2 with components of the tumor necrosis factor (TNF) signaling pathway. TNF is a pleiotropic cytokine that binds to two known receptors, TNF receptors 1 and 2 (TNFR1 and -2) (Tartaglia *et al.*, 1993). Binding of TNF to TNFR1 can induce apoptosis, proliferation, or activation of the antiviral transcription factor NF-kB, depending on the situation. Whether the binding of TNF to TNFR2 can induce apoptosis is less clear, but it does activate NF-kB. A class of proteins that interact with TNFR2 called tumor necrosis factor receptor-associated factors (TRAFs) have been described (Cheng *et al.*, 1995; Hu *et al.*, 1994; Rothe *et al.*, 1994; Song *et al.*, 1995). One of the TRAFs, TRAF2, is involved in NF-kB signaling via TNFR2 (Rothe *et al.*, 1995b). The proteins encoded by cIAP-1 and cIAP-2 interact with the TRAF-1 and -2 proteins both *in vitro* and in cells (Rothe *et al.*, 1995a; Uren *et al.*, 1996). The BIRs are necessary and sufficient for this interaction (Rothe *et al.*, 1995a).

The ability of cIAP-1 and 2 to bind TRAF1 and -2 suggests that the IAPs may block death through interactions with components of the TNF pathway or other similar signal transduction pathways. This is an attractive hypothesis: it would fit with the apparent inability of the IAPs to protect against apoptosis in all situations, since these signal transduction pathways would only be active in particular cell types. However, the importance of the interaction of IAPs with TRAFs in protecting against apoptosis is as yet unclear, since the TRAFs themselves have not been shown to be involved in regulating cell death, but instead appear to regulate NF-kB (Hsu *et al.*, 1996). Also, despite their ability to bind TRAF1 and -2, overexpression of cIAP-1 and -2 does not affect NF-kB signaling (Rothe *et al.*, 1995). Thus, the interaction of cIAP-1 and -2 with TRAF1 and -2 has no functional significance to date.

Interestingly, hILP does not interact with any of the six known TRAFs (C. Duckett, personal communication), despite the fact that hILP appears to have the strongest antiapoptotic function of the three human IAPs (Clem *et al.*, in preparation). This suggests hILP may act via a different mechanism than cIAP-1 and -2, although the possibility remains that hILP interacts with other, as yet unknown, TRAF homologues.

Other clues to the function of the IAPs come from the dissection of their functional motifs. When the BIR or RING finger domains of Cp-IAP and Ac-IAP were exchanged, it was found that hybrid proteins containing either the BIRs or RING finger of Ac-IAP were inactive, suggesting that both domains are important for antiapoptotic function (Clem and Miller, 1994a). In addition, Cp-IAP constructs lacking only the RING finger are also inactive both in SF-21 cells (R. J. Clem and L. K. Miller, unpublished results) and in mammalian cells (Uren *et al.*, 1996). However, the BIRs and ring finger of the cellular IAPs appear to have antiapoptotic and proapoptotic activities, respectively. When a version of DIAP1 lacking the RING finger was expressed in the developing *Drosophila* eye, it still prevented cell death as well as the full-length protein and actually blocked *hid*-dependent death more efficiently than the full-length protein (Hay *et al.*, 1995). A similar version of the BIR domains of human cIAP-1 was able to block some normally occurring cell death in the eye, whereas the full-length cIAP-1 protein was inactive (Hay *et al.*, 1995). In addition, expression of a DIAP1 gene lacking the BIRs appeared to induce extra cell death in the developing eye (Hay *et al.*, 1995). Finally, truncated versions of cIAP-1 and hILP (but not cIAP-2) that lack the RING finger also block Sindbis virus-induced apoptosis, even though full-length cIAP-1 has no antideath activity in this assay (Clem *et al.*, in preparation).

Taken together, these results indicate that the BIRs are crucial for antiapoptotic function of the both viral and cellular IAPs. However, the role of the RING finger appears to be different in the cellular and viral homologues. While the RING finger appears to be required for the function of the baculovirus IAPs, removal of the RING finger actually increases the antiapoptotic activity of at least one of the cellular proteins, suggesting that the viral and cellular IAP homologues may be interacting with different proteins or even acting in different pathways.

IV. EFFECTS OF APOPTOSIS ON BACULOVIRUS REPLICATION

A. Consequences at the Cellular Level

It was obvious when vAcAnh was first isolated that the apoptotic response of the SF-21 cells was having profound effects on viral replication, since it was very difficult to obtain high titer viral stocks using these cells. In addition, titering the virus was difficult since the plaques were very small and occlusion negative. Thus, some of the first experiments done to characterize the vAcAnh mutant were aimed at characterizing the defects in virus replication in SF-21 cells.

1. Effects on Viral Gene Expression and DNA Replication

The examination of protein synthesis profiles derived from SF-21 cells infected with *p35* mutant viruses revealed several major defects in viral protein synthesis (Clem and Miller, 1993; Hershberger *et al.*, 1992). There was a delay in the appearance of early viral proteins, as well as a reduction in their overall accumulation, compared to cells infected with wild-type AcMNPV. In addition, there was no observable synthesis of any late or very late viral proteins in mutant-infected cells. Finally, host protein synthesis was not efficiently shut off by 18 hr p.i. in the mutant-infected cells as it is in wild-type-infected cells. However, both host and viral protein synthesis terminated by 24 hr after infection.

Inspection of the steady-state levels of viral transcripts corroborated the protein synthesis data (Clem and Miller, 1993). The accumulation of transcripts from both an early and a late viral gene was delayed and reduced in SF-21 cells infected with a *p35* mutant virus compared to cells infected with wild-type AcMNPV. Furthermore, transcripts from the very late polyhedrin gene were almost undetectable in mutant-infected cells. Expression of the reporter gene *lac*Z from the very late polyhedrin promoter in *p35* mutant viruses further verified the almost complete lack of very late gene expression in *p35* mutant-infected SF-21 cells (Hershberger *et al.*, 1992).

Viral DNA synthesis was also affected in SF-21 cells infected with *p35* mutant viruses (Hershberger *et al.*, 1992). The decrease in DNA synthesis was approximately equivalent to the decrease in late gene expression (five- to tenfold).

In contrast to the profound effects on viral gene expression seen in SF-21 cells, there were no observable defects in TN-368 cells infected with *p35* mutants (Clem and Miller, 1993; Hershberger *et al.*, 1992). This included both the timing of appearance and levels of accumulation of early, late, and very late proteins and transcripts, and the timing of the shutoff of host protein synthesis.

2. Effects on Virus Production

As one would expect from the defective patterns of gene expression, *p35* mutants do not replicate efficiently in SF-21 cells. The yield of budded virus is greatly reduced, and there is no observable production of occluded virus (Clem and Miller, 1993; Hershberger *et al.*, 1992). The magnitude of the reduction in budded virus yield depends on the amount of input virus (Hershberger *et al.*, 1992). In cells infected with *p35* mutants at a high multiplicity of infection, there is a 50- to 100-fold reduction in budded virus yield compared to *p35*-containing viruses (Clem and Miller, 1993; Hershberger *et al.*, 1992). However, the timing of the first appearance of newly synthesized virus appears to be roughly similar between the *p35* mutants and control viruses (Clem and Miller, 1993).

Viral replication was normal in TN-368 cells, including normal yields of budded and occluded virus (Clem and Miller, 1993; Hershberger *et al.*, 1992). Thus, P35 is not required for efficient viral replication in the TN-368 cell line. The observation that *p35* mutants replicate completely normally in TN-368 cells, coupled with the result that replication of *p35* mutants in SF-21 cells can be restored to normal by the unrelated antiapoptotic genes Cp-*iap* and Op-*iap*, is strong evidence that the defects associated with the lack of P35 in SF-21 cells are due to apoptosis.

The question of why TN-368 cells are resistant to apoptosis induction by *p35* mutant infection is an interesting one. This cell line appears to be generally resistant to the induction of apoptosis, since it is also resistant to actinomycin D (R. J. Clem and L. K. Miller, unpublished results) and expression of IE-1 (Prikhod'ko and Miller, 1996). However, apoptosis can be triggered in TN-368 cells by overexpression of ICE (Bump *et al.*, 1995). These results suggest that TN-368 cells either lack a component of the death signaling pathway or overexpress an antiapoptotic protein. This resistance to apoptosis induction is also seen at the organismal level (see Section IV.B). Interestingly, *T. ni* larvae are also more resistant to the effects of ionizing radiation than are many other lepidopterans, including *S. frugiperda* (Koval, 1996).

B. Consequences at the Organismal Level

One of the major concerns early on in these studies was whether the induction of apoptosis by *p35* mutant viruses was a biologically relevant process or simply an artifact of cell culture and unique to the SF-21 and SF-9 cell lines. Specifically, given the abortive replication of viruses lacking *p35* in SF-21 cells, we were interested in the possibility that apoptosis could function as an antiviral defense in insects (Clouston and Kerr, 1985; Martz and Howell, 1989).

1. Effects on Virus Infectivity and Replication

Both *S. frugiperda* and *T. ni* larvae are highly susceptible to infection with wild-type AcMNPV; injection of as little as 50 to 100 or 1 to 10 plaque-forming units is sufficient to kill 50% of larvae from each species, respectively (Clem and Miller, 1993). However, *S. frugiperda* larvae are extremely resistant to infection by *p35* mutant viruses both by injection of budded virus into the hemocoel and by *per os* infection (Clem and Miller, 1993; Clem *et al.*, 1994). At least 1000-fold more *p35* mutant virus is required than wild-type virus to kill *S. frugiperda* larvae by hemocoelic injection, while the difference in the amount of occluded virus required for *per os* infectivity is approximately 25-fold. The reasons for the larger difference by injection is unknown; it may be related at least in part to differences in the sensitivity of different tissue to apoptosis or to the presence of multiple enveloped virions per AcMNPV occlusion body (Clem *et al.*, 1994). The infectivity of *p35* mutant viruses in *T. ni* larvae is equivalent to that of wild-type AcMNPV by both routes of infection (Clem and Miller, 1993; Clem *et al.*, 1994).

What is the evidence that this defect in infectivity is due to an apoptotic response by the cells of the *S. frugiperda* larvae? Direct evidence, such as an observation of apoptosis in larval tissues and correlation with a lack of infectivity, is lacking at this point. However, there is compelling indirect evidence in support of a role for apoptosis in the reduction in infectivity:

1. Virus mutants lacking *p35* exhibit normal infectivity, both by injection and *per os*, in *T. ni* larvae, the species from which the TN-368 cell line is derived (Clem and Miller, 1993; Clem *et al.*, 1994). Thus, the induction of apoptosis in SF-21 cells but not TN-368 cells correlates with infectivity at the organismal level.
2. The induction of apoptosis in SF-21 cells results in the severe impairment of viral replication, including a block in the production of occluded virus. Similarly, even in *S. frugiperda* larvae that die from *p35* mutant infection, the yield of occluded virus is reduced 900-fold compared to wild-type virus (Clem and Miller, 1993).
3. The defects in hemocoelic infectivity of mutants lacking *p35* can be completely suppressed by the Cp-*iap* gene (Clem *et al.*, 1994). The ability of this unrelated antiapoptotic gene to rescue infectivity *in vivo* is probably the best evidence to date that apoptosis can be an important defense against baculovirus infection.

Despite the fact that the mutation of *p35* greatly reduced infectivity in *S. frugiperda,* the time required to kill those larvae that did become infected was no different from that of viruses containing *p35* (Clem and Miller, 1993; Clem *et al.*, 1994). This result suggests that apoptosis may be important in preventing the establishment of infection, since once an infection is established, the time required for the infection to kill the larvae is normal. The infection itself is not normal, however, since the production of occluded virus is still dramatically affected by the lack of *p35.* There is also a lack of liquefaction observed in both *S. frugiperda* and *T. ni* larvae infected with *p35* mutants. Interestingly, Cp-*iap* is not able to rescue larval liquefaction, even though it does rescue infectivity, suggesting possible tissue-specific activity of Cp-IAP (Clem *et al.*, 1994).

2. Influence on Host Range

Most baculoviruses have relatively narrow host ranges, despite their ability to enter a variety of nonpermissive cells. The molecular basis of baculovirus host range is complex, with each virus–host pair probably involving a different combination of factors (see Chapter 9, this volume). Our work has shown that apoptosis can play a role in determining baculovirus host range. Given the severe crippling of infectivity and occluded virus production in *p35* mutant-infected *S. frugiperda* larvae, there can be little doubt that this insect would not be a natural host for AcMNPV if not for the presence of the *p35* gene in the viral genome.

The presence of *p35* and *p94* in AcMNPV and BmNPV, but not OpMNPV, indicates that these genes must have been acquired or lost sometime after the divergence of OpMNPV and the AcMNPV–BmNPV common ancestor. In this regard, it is interesting that *p94* is the site of integration of a host retrotransposon (Miller and Miller, 1982). In addition to their adjacent location on the genome and absence in OpMNPV, there is also functional evidence suggesting that these genes coevolved (Clem *et al.*, 1994). As discussed above, *p35* mutants are about 25-fold less infectious than wild-type virus in *S. frugiperda* larvae by the *per os* route of infection. Mutation of *p94* has no apparent effect on AcMNPV infectivity and replication either *in vivo* or *in vitro.* However, when *p35* and *p94* are both mutated, the resulting virus has normal *per os* infectivity, even though its infectivity by hemocoelic injection is the same as a *p35* mutant (1000-fold reduced). The simplest interpretation of these results is that *p94* expression induces apoptosis in tissues that are important for *per os* infectivity, so that P35 is not required to block apoptosis if *p94* is not expressed. Although the function of P94 is not known and P94 is not required for infection in *S. frugiperda* or *T. ni,* these results suggest it may be required for replication in other host species. If P94 expression also triggers apoptosis, then *p35* would be necessary to block the suicide response and allow viral replication to proceed. Thus, there appears to be selective pressure to acquire both of these genes together. This scenario could potentially explain the observation that the BmNPV *p35* gene is apparently not required for infectivity in *B. mori,* since most of the BmNPV *p94* gene has been deleted (Kamita *et al.,* 1993).

Additional insights into the influence of apoptosis on host range are provided by studies describing AcMNPV infection of the *S. littoralis* and *C. fumiferana* cell lines SL2 and CF-203 (Chejanovsky and Gershburg, 1995; Palli *et al.,* 1996). These species are nonpermissive hosts for AcMNPV; the virus does not kill larvae of these species and does not replicate efficiently in cell lines derived from them. On the other hand, these species are both permissive hosts for other baculoviruses (SlNPV and CfNPV, respectively). The reason why these species are nonpermissive hosts for AcMNPV may be due to apoptosis, which is triggered by infection of SL2 and CF-203 cells with AcMNPV but not with SlNPV or CfNPV (Chejanovsky and Gershburg, 1995; Palli *et al.,* 1996). The consequences of apoptosis in these cells appears to be similar to that of *p35* mutant infection of SF-21 cells, including a large decrease in the yield of budded virus and a lack of occluded virus production. Even though AcMNPV triggers apoptosis in these cells, the expression of the *p35* gene is insufficient to block the apoptotic response. Given the ability of P35 to block apoptosis in phylogenetically distant eukaryotes, there must be a hindrance to the production or accumulation of P35 protein in these cell lines. For example, in CF-203 cells it was shown that the appearance of *p35* transcripts is delayed (Palli *et al.,* 1996). These systems thus provide examples of natural situations (involving the wild-type virus instead of virus mutants) where apoptosis may be an important determinant of baculovirus host range.

V. CONCLUSIONS

Apoptosis is a fundamentally important process, both in the normal functioning of multicellular organisms and in the interplay between viruses and their hosts. As a normal consequence of infection, at least some (and presumably many) baculoviruses induce apoptosis. In most cases, the resulting premature death of the host cell is detrimental to viral replication, since viral gene expression, DNA replication, and production of progeny virus are all severely affected. The ability of individual cells to respond to baculovirus infection by committing suicide appears to inhibit establishment of infection at the organismal level, and there is good evidence that it also affects the range of insects that an individual baculovirus can productively infect. In order to replicate and disseminate, the baculoviruses have had to evolve mechanisms to block apoptosis, and they have acquired genes that allow them to accomplish this task. One such gene is the ICE family protease inhibitor *p35,* which is the most ubiquitously acting antiapoptotic gene known. Studies utilizing *p35* have provided some of the strongest evidence to date that ICE proteases are highly conserved components of the death pathway and that they lie at or very near the culmination point of these pathways. Since P35 is a novel type of protease inhibitor, information gleaned from further study promises to reveal new insights into the catalytic mechanisms of this important family of proteases. The other known antiapoptotic baculovirus genes are the *iap* genes, which have cellular counterparts that are also involved in the regulation of apoptosis and whose functions are just beginning to be studied. Although there are hints that the IAPs may act at a point more upstream in the death pathway than P35, their mode of action is still unknown. A detailed understanding of the function of the human *iap* genes may lead to advancements in the treatment of human diseases such as spinal muscular atrophy. Thus, the study of baculovirus-induced apoptosis has already contributed greatly to our understanding of the molecular mechanisms of cell death.

In addition, the study of baculovirus-induced apoptosis has also emerged as an important factor in understanding baculovirus replication, virulence, and host range. The future holds great promise for research on baculoviruses and their ability to manipulate the apoptotic pathways of their insect hosts.

REFERENCES

Ayres, M. D., Howard, S. C., Kuzio, Lopez-Ferber, J. M., and Possee, R. D., 1994, The complete DNA sequence of *Autographa californica* nuclear polyhedrosis virus, *Virology* **202:**586–605.

Barlow, P. N., Luisi, B., Milner, A., Elliott, M., and Everett, R., 1994, Structure of the C_3HC_4 domain by ^{1}H-nuclear magnetic resonance spectroscopy, *J. Molec. Biol.* **2337:**201–211.

Beidler, D. R., Tewari, M., Friesen, P. D., Poirier, G., and Dixit, V. M., 1995, The baculovirus p35 protein inhibits fas- and tumor necrosis factor-induced apoptosis, *J. Biol. Chem.* **270:**16256–16258.

Berg, J. M., and Shi, Y., 1996, The galvanization of biology: A growing appreciation for the roles of zinc, *Science* **271:**1081–1085.

Bertin, J., Mendrysa, S. M., LaCount, D. J., Gaur, S., Krebs, J. F., Armstrong, R. C., Tomaselli, K. J., and Friesen, P. D., 1996, Apoptotic suppression by baculovirus P35 involves cleavage by and inhibition of a virus-induced CED-3/ICE-like protease, *J. Virol.* **70:**6251–6259.

Birnbaum, M. J., Clem, R. J., and Miller, L. K., 1994, An apoptosis-inhibiting gene from a nuclear polyhedrosis virus encoding a peptide with cys/his sequence motifs, *J. Virol.* **68:**2521–2528.

Borden, K. L. B., Boddy, M. N., Lally, J., O'Reilly, N. J., Martin, S., Howe, K., Solomon, E., and Freemont, P. S., 1995, The solution structure of the RING finger domain from the acute promyelocytic leukaemia proto-oncoprotein PML, *EMBO J.* **14:**1532–1541.

Brooks, M. A., Ali, A. N., Turner, P. C., and Moyer, R. W., 1995, A rabbitpox virus serpin gene controls host range by inhibiting apoptosis in restrictive cells, *J. Virol.* **69:**7688–7698.

Bump, N. J., Hackett, M., Hugunin, M., Seshagiri, S., Brady, K., Chen, P., Ferenz, C., Franklin, S., Ghayur, T., Li, P., Licari, P., Mankovich, J., Shi, L., Greenberg, A. H., Miller, L. K., and Wong, W. W., 1995, Inhibition of ICE family proteases by baculovirus antiapoptotic protein p35, *Science* **269:**1885–1888.

Cartier, J. L., Hershberger, P. A., and Friesen, P. D., 1994, Suppression of apoptosis in insect cells stably transfected with baculovirus *p35:* Dominant interference by N-terminal sequences p35$^{1\text{-}76}$, *J. Virol.* **68:**7728–7737.

Chacon, M. R., Almazan, F., Nogal, M. L., Vinuela, E., and Rodriguez, J. F., 1995, The African swine fever virus IAP homolog is a late structural polypeptide, *Virology* **214:**670–674.

Chejanovsky, N., and Gershburg, E., 1995, The wild-type *Autographa californica* nuclear polyhedrosis virus induces apoptosis of *Spodoptera littoralis* cells, *Virology* **209:**519–525.

Cheng, E. H.-Y., Nicholas, J., Bellows, D. S., Hayward, G. S., Guo, H.-G., Reitz, M. S., and Hardwick, J. M., 1997, A Bcl-2 homolog encoded by Kaposi sarcoma-associated virus, human herpesvirus 8, inhibits apoptosis but does not heterodimerize with Bax or Bak. *Proc. Natl. Acad. Sci. USA* **94:**690–694.

Cheng, G., Cleary, A. M., Ye, Z.-s., Hong, D. I., Lederman, S., and Baltimore, D., 1995, Involvement of CRAF1, a relative of TRAF, in CD40 signalling, *Science* **267:**1494–1498.

Chiou, S.-K., Tseng, C.-C., Rao, L., and White, E., 1994, Functional complementation of the adenovirus E1b 19-kilodalton protein with Bcl-2 in the inhibition of apoptosis in infected cells, *J. Virol.* **68:**6553–6566.

Clem, R. J., and Miller, L. K., 1993, Apoptosis reduces both the *in vitro* replication and the *in vivo* infectivity of a baculovirus, *J. Virol.* **67:**3730–3738.

Clem, R. J., and Miller, L. K., 1994a, Control of programmed cell death by the baculovirus genes *p35* and *iap, Mol. Cell. Biol.* **14:**5212–5222.

Clem, R. J., and Miller, L. K., 1994b, Induction and inhibition of apoptosis by insect viruses, in: *Apoptosis II: The Molecular Basis of Cell Death* (F. O. Cope and L. D. Tomei, eds.), pp. 89–110, Cold Spring Harbor Laboratory Press, Plainview, NY.

Clem, R. J., Fechheimer, M., and Miller, L. K., 1991, Prevention of apoptosis by a baculovirus gene during infection of insect cells, *Science* **254:**1388–1390.

Clem, R. J., Robson, M., and Miller, L. K., 1994, Influence of infection route on the infectivity of baculovirus mutants lacking the apoptosis-inhibiting gene *p35* and the adjacent gene *p94, J. Virol.* **68:**6759–6762.

Clem, R. J., Duckett, C. S., Hudson, P., He, W. W., Thompson, C. B., and Hardwick, J. M., manuscript in preparation.

Clem, R. J., Hardwick, J. M., and Miller, L. K., 1996b, Anti-apoptotic genes of baculoviruses, *Cell Death Diff.* **3:**9–16.

Clouston, W. M., and Kerr, J. F. R., 1985, Apoptosis, lymphcytotoxicity and the containment of viral infections, *Med. Hypoth.* **18:**399–404.

Crook, N. E., Clem, R. J., and Miller, L. K., 1993, An apoptosis-inhibiting baculovirus gene with a zinc finger-like motif, *J. Virol.* **67:**2168–2174.

Debbas, M., and White, E., 1993, Wild-type p53 mediates apoptosis by E1A, which is inhibited by E1B, *Genes Dev.* **7:**546–554.

Dickson, J. A., and Friesen, P. D., 1991, Identification of upstream promoter elements mediating early transcription from the 35,000-molecular-weight protein gene of *Autographa californica* nuclear polyhedrosis virus, *J. Virol.* **65:**4006–4016.

Duckett, C. S., Nava, V. E., Gedrich, R. W., Clem, R. J., Van Dongen, J. L., Gilfillan, M. C., Shiels, H., Hardwick, J. M., and Thompson, C. B., 1996, A conserved family of cellular genes related to the baculovirus *iap* gene and encoding apoptosis inhibitors, *EMBO J.* **15:**2685–2694.

Duvall, E., and Wyllie, A. H., 1986, Death and the cell, *Immunol. Today* **7:**115–119.

Fadok, V. A., Voelker, D. R., Campbell, P. A., Cohen, J. J., Bratton, D. L., and Henson, P. M., 1992, Exposure of phosphatidylserine on the surface of apoptotic lymphocytes triggers specific recognition and removal by macrophages, *J. Immunol.* **148:**2207–2216.

Farrow, S. N., and Brown, R., 1996, New members of the Bcl-2 family and their protein partners, *Curr. Opin. Gen. Dev.* **6:**45–49.

Fraser, A., and Evan, G., 1996, A license to kill, *Cell* **85:**781–784.

Freemont, P. S., Hanson, I. M., and Trowsdale, J., 1991, A novel cysteine-rich sequence motif, *Cell* **64:**483–484.

Friesen, P. D., and Miller, L. K., 1987, Divergent transcription of early 35 and 94-kilodalton protein genes encoded by the *Hind*-III K genome fragment of the baculovirus *Autographa californica* nuclear polyhedrosis virus, *J. Virol.* **61**:2264–2272.

Gombart, A. F., Blissard, G. W., and Rohrmann, G. F., 1989, Characterization of the genetic organization of the *Hind*III-M region of the multicapsid nuclear polyhedrosis virus of *Orgyia pseudotsugata* reveals major differences among baculoviruses, *J. Gen. Virol.* **70**:1815–1828.

Gong, M., and Guarino, L. A., 1994, Expression of the *39K* promoter of *Autographa californica* nuclear polyhedrosis virus is increased by the apoptotic suppressor P35, *Virology* **204**:38–44.

Grether, M. E., Abrams, J. M., Agapite, J., White, K., and Steller, H., 1995, The *head involution defective* gene of *Drosophila melanogaster* functions in programmed cell death, *Genes Dev.* **9**:1694–1708.

Hay, B. A., Wolff, T., and Rubin, G. M., 1994, Expression of baculovirus P35 prevents cell death in *Drosophila, Development* **120**:2121–2129.

Hay, B. A., Wassarman, D. A., and Rubin, G. M., 1995, *Drosophila* homologs of baculovirus inhibitor of apoptosis proteins function to block cell death, *Cell* **83**:1253–1262.

Henderson, S., Huen, D., Rowe, M., Dawson, C., Johnson, G., and Rickinson, A., 1993, Epstein–Barr virus-coded BHRF1 protein, a viral homologue of Bcl-2, protects human B cells from programmed cell death, *Proc. Natl. Acad. Sci. USA* **90**:8479–8483.

Hershberger, P. A., Dickson, J. A., and Friesen, P. D., 1992, Site-specific mutagenesis of the 35-kilodalton protein gene encoded by *Autographa californica* nuclear polyhedrosis virus: Cell line-specific effects on virus replication, *J. Virol.* **66**:5525–5533.

Hershberger, P. A., LaCount, D. J., and Friesen, P. D., 1994, The apoptotic suppressor P35 is required early during baculovirus replication and is targeted to the cytosol of infected cells, *J. Virol.* **68**:3467–3477.

Hsu, H., Shu, H.-B., Pan, M.-G., and Goeddel, D. V., 1996, TRADD–TRAF2 and TRADD–FADD interactions define two distinct TNF receptor 1 signal transduction pathways, *Cell* **84**:299–308.

Hu, M. H., O'Rourke, K., Boguski, M. S., and Dixit, V. M., 1994, A novel RING finger protein interacts with the cytoplasmic domain of CD40, *J. Biol. Chem.* **269**:30069–30072.

Ink, B. S., Gilbert, C. S., and Evan, G. I., 1995, Delay of vaccinia virus-induced apoptosis in nonpermissive Chinese hamster ovary cells by the cowpox virus *CHOhr* and adenovirus *E1B 19K* genes, *J. Virol.* **69**:661–668.

Kamita, S. G., Majima, K., and Maeda, S., 1993, Identification and characterization of the *p35* gene of *Bombyx mori* nuclear polyhedrosis virus that prevents virus-induced apoptosis, *J. Virol.* **67**:455–463.

Kerr, J. F. R., Wyllie, A. H., and Currie, A. R., 1972, Apoptosis: A basic biological phenomenon with wide-ranging implications in tissue kinetics, *Br. J. Cancer* **26**:239–257.

Koval, T. M., 1996, Moths: Myths and mysteries of stress resistance, *BioEssays* **18**:149–156.

Laurent-Crawford, A. G., Krust, B., Riviere, Y., Muller, S., Kieny, M. P., Dauget, C., and Hovanessian, A. G., 1993, Membrane expression of HIV envelope glycoproteins triggers apoptosis in CD4 cells, *AIDS Res. Hum. Retroviruses* **9**:761–773.

Lee, J.-C., Chen, H.-H., Wei, H.-L., and Chao, Y.-C., 1993, Superinfection-induced apoptosis and its correlation with the reduction of viral progeny in cells persistently infected with Hz-1 baculovirus, *J. Virol.* **67**:6989–6994.

Li, C. J., Friedman, D. J., Wang, C., Metelev, V., and Pardee, A. B., 1995, Induction of apoptosis in uninfected lymphocytes by HIV-1 tat protein, *Science* **268**:429–431.

Liston, P., Roy, N., Tamai, K., Lefebvre, C., Baird, S., Cherton-Horvat, G., Farahani, R., McLean, M., Ikeda, J.-E., MacKenzie, A., and Korneluk, R. G., 1996, Suppression of apoptosis in mammalian cells by NAIP and a related family of IAP genes, *Nature* **379**:349–353.

Lockshin, R. A., 1985, Programmed cell death, in: *Comprehensive Insect Physiology, Biochemistry and Pharmacology* (G. A. Kerkut and L. I. Gilbert, eds.) Vol. 2, pp. 301–317, Pergamon Press, Elmsford, NY.

Lu, A., and Miller, L. K., 1995, The roles of eighteen baculovirus late expression factor genes in transcription and DNA replication, *J. Virol.* **69:**975–982.

Martin, S. J., and Green, D. R., 1995, Protease activation during apoptosis: Death by a thousand cuts? *Cell* **82:**349–352.

Martinou, I., Fernandez, P.-A., Missotten, M., White, E., Allet, B., Sadoul, R., and Martinou, J.-C., 1995, Viral proteins E1B19K and p35 protect sympathetic neurons from cell death induced by NGF deprivation, *J. Cell Biol.* **128:**201–208.

Martz, E., and Howell, D. M., 1989, CTL: Virus control cells first and cytolytic cells second? *Immunol. Today* **10:**79–86.

Miller, D. W., and Miller, L. K., 1982, A virus mutant with an insertion of a copia-like transposable element, *Nature* **299:**562–564.

Mosialos, G., Birkenbach, M., Yalamanchili, R., VanArsdale, T., Ware, C., and Kieff, E., 1995, The Epstein–Barr virus transforming protein LMP1 engages signalling proteins for the tumor necrosis factor receptor family, *Cell* **80:**389–399.

Nava, V. E., Cheng, E. H.-Y., Veliuona, M., Zou, S., Clem, R. J., Mayer, M., and Hardwick, J. M., 1996a, Herpesvirus saimiri encodes a functional homolog of the human *bcl-2* oncogene, *J. Virol.* **71:**4118–4122.

Nava, V. E., Clem, R. J., and Hardwick, J. M., 1996b, Manuscript in preparation.

Neilan, J. G., Lu, Z., Afonso, C. L., Kutish, G. F., Sussman, M. D., and Rock, D. L., 1993, An African swine fever virus gene with similarity to the proto-oncogene bcl-2 and the Epstein–Barr virus gene BHRF1, *J. Virol.* **67:**4391–4394.

Noteborn, M. H. M., Todd, D., Verschueren, C. A. J., De Gauw, H. W. F. M., Curran, W. L., Veldkamp, S., Douglas, A. J., McNulty, M. S., Van der Eb, A. J., and Koch, G., 1994, A single chicken anemia virus protein induces apoptosis, *J. Virol.* **68:**346–351.

Ooi, B. G., and Miller, L. K., 1988, Regulation of host RNA levels during baculovirus infection, *Virology* **166:**515–523.

Palli, S. R., Caputo, G. F., Sohi, S. S., Brownwright, A. J., Ladd, T. R., Cook, B. J., Primavera, M., Arif, B. M., and Retnakaran, A., 1996, CfMNPV blocks AcMNPV-induced apoptosis in a continuous midgut cell line, *Virology* **222:**201–213.

Prikhod'ko, E. A., and Miller, L. K., 1996, Induction of apoptosis by baculovirus transactivator IE1, *J. Virol.*, **70:**7116–7124.

Quan, L. T., Caputo, A., Bleackley, R. C., Pickup, D. J., and Salvesen, G. S., 1995, Granzyme B is inhibited by the cowpox serpin cytokine response modifier A, *J. Biol. Chem.* **270:**10377–10379.

Rabizadeh, S., LaCount, D. J., Friesen, P. D., and Bredesen, D. E., 1993, Expression of the baculovirus *p35* gene inhibits mammalian neural cell death, *J. Neurochem.* **61:**2318–2321.

Raff, M. C., Barres, B. A., Burne, J. F., Coles, H. S., Ishizaki, Y., and Jacobson, M. D., 1993, Programmed cell death and the control of cell survival: Lessons from the nervous system, *Science* **262:**695–700.

Rao, L., Debbas, M., Sabbatini, P., Hockenbery, D., Korsmeyer, S., and White, E., 1992, The adenovirus E1A proteins induce apoptosis, which is inhibited by the E1B 19-kDa and Bcl-2 proteins, *Proc. Natl. Acad. Sci. USA* **89:**7742–7746.

Ray, C. A., Black, R. A., Kronheim, S. R., Greenstreet, T. A., Sleath, P. R., Salvesen, G. S., and Pickup, D. J., 1992, Viral inhibition of inflammation: Cowpox virus encodes an inhibitor of the interleukin-1beta converting enzyme, *Cell* **69:**597–604.

Razvi, E. S., and Welsh, R. M., 1995, Apoptosis in viral infections, *Adv. Virus Res.* **45:**1–60.

Ribeiro, B., Hutchinson, K., and Miller, L. K., 1994, A mutant baculovirus with a temperature sensitive IE-1 transregulatory protein, *J. Virol.* **68:**1075–1084.

Rothe, M., Wong, S. C., Henzel, W. J., and Goeddel, D. V., 1994, A novel family of putative signal transducers associated with the cytoplasmic domain of the 75 kDa tumor necrosis factor receptor, *Cell* **78:**681–692.

Rothe, M., Pan, M.-G., Henzel, W. J., Ayres, T. M., and Goeddel, D. V., 1995a, The TNFR2–TRAF signalling complex contains two novel proteins related to baculoviral inhibitor of apoptosis proteins, *Cell* **83:**1243–1252.

Rothe, M., Sarma, V., Dixit, V. M., and Goeddel, D. V., 1995b, TRAF2-mediated activation of NF-kB by TNF receptor 2 and CD40, *Science* **269:**1424–1427.

Roy, N., Mahadevan, M. S., McLean, M., Shutler, G., Yaraghi, Z., Farahani, R., Baird, S., Besner-Johnston, A., Lefebvre, C., Kang, X., Salih, M., Aubry, A., Tamai, K., Guan, X., Ioannou, P., Crawford, T. O., de Jong, P. J., Surh, L., Ikeda, J. E., R. Korneluk, G., and MacKenzie, A., 1995, The gene for neuronal apoptosis inhibitory protein is partially deleted in individuals with spinal muscular atrophy, *Cell* **80:**167–178.

Shen, Y., and Shenk, T. E., 1995, Viruses and apoptosis, *Curr. Biol.* **5:**105–111.

Shi, L., Kam, C. M., Powers, J. C., Aebersold, R., and Greenberg, A. H., 1992, Purification of three cytotoxic lymphocyte granule serine proteases that induce apoptosis through distinct substrate and target cell interactions, *J. Exp. Med.* **176:**1521–1529.

Song, H. Y., Dunbar, J. D., Zhang, Y. X., Guo, D., and Donner, D. B., 1995, Identification of a protein with homology to hsp90 that binds the type 1 tumor necrosis factor receptor, *J. Biol. Chem.* **270:**3574–3581.

Steller, H., 1995, Mechanisms and genes of cellular suicide, *Science* **267:**1445–1449.

Steller, H., and Grether, M. E., 1994, Programmed cell death in *Drosophila, Neuron* **13:**1269–1274.

Suarez, P., Diaz-Guerra, M., Prieto, C., Esteban, M., Castro, J. M., Nieto, A., and Ortin, J., 1996, Open reading frame 5 of porcine reproductive and respiratory syndrome virus as a cause of virus-induced apoptosis, *J. Virol.* **70:**2876–2882.

Sugimoto, A., Friesen, P. D., and Rothman, J. H., 1994, Baculovirus *p35* prevents developmentally programmed cell death and rescues a *ced-9* mutant in the nematode *Caenorhabditis elegans, EMBO J.* **13:**2023–2028.

Takahashi, A., Alnemri, E. S. Y., Lazebnik, A., Fernandes-Alnemri, T., Litwack, G., Moir, R. D., Goldman, R. D., Poirier, G. G., Kaufmann, S. H., and Earnshaw, W. C., 1996, Cleavage of lamin A by Mch2alpha but not CPP32: Multiple interleukin 1beta-converting enzyme-related proteases with distinct substrate recognition properties are active in apoptosis, *Proc. Natl. Acad. Sci. USA* **93:**8395–8400.

Tartaglia, L. A., Ayres, T. M., Wong, G. H. W., and Goeddel, D. V., 1993, A novel domain within the 55 kd TNF receptor signals cell death, *Cell* **74:**845–853.

Tewari, M., and Dixit, V. M., 1996, Recent advances in tumor necrosis factor and CD40 signaling, *Curr. Opin. Gen. Dev.* **6:**39–44.

Thompson, C. B., 1995, Apoptosis in the pathogenesis and treatment of disease, *Science* **267:**1456–1462.

Todd, J., Passarelli, A. L., and Miller, L. K., 1995, Eighteen baculovirus genes, including *lef-11, p35,* 39K and *p47,* support late gene expression, *J. Virol.* **69:**968–974.

Todd, J. W., Passarelli, A. L., Lu, A., and L. K. Miller, 1996, Factors regulating baculovirus late and very late gene expression in transient-expression assays, *J. Virol.* **70:**2307–2317.

Tollefson, A. E., Scaria, A., Hermiston, T. W., Ryerse, J. S., Wold, L. J., and Wold, W. S. M., 1996, The adenovirus death protein (E3-11.6K) is required at very late stages of infection for efficient cell lysis and release of adenovirus from infected cells, *J. Virol.* **70:**2296–2306.

Truman, J. W., and Schwartz, L. M., 1984, Steroid regulation of neuronal death in the moth nervous system, *J. Neurosci.* **4:**274–280.

Tyler, K. L., Squier, M. K. T., Rodgers, S. E., Schneider, B. E., Oberhaus, S. M., Grdina, T. A., Cohen, J. J., and Dermody, T. S., 1995, Differences in the capacity of reovirus strains to induce apoptosis are determined by the viral attachment protein signal, *J. Virol.* **69:**6972–6979.

Ubol, S., Tucker, P. C., Griffin, D. E., and Hardwick, J. M., 1994, Neurovirulent strains of *Alphavirus* induce apoptosis in *bcl-2*-expressing cells: Role of a single amino acid change in the E2 glycoprotein, *Proc. Natl. Acad. Sci. USA* **91:**5202–5206.

Ucker, D. S., 1991, Death by suicide: One way to go in mammalian cellular development? *New Biol.* **3:**103–109.

Uren, A. G., Pakusch, M., Hawkins, C. J., Puls, K. L., and Vaux, D. L., 1996, Cloning and expression of apoptosis inhibitory protein homologs that function to inhibit apoptosis and/or bind tumor necrosis factor receptor-associated factors, *Proc. Natl. Acad. Sci. USA* **93:**4974–4978.

van Oers, M. M., Flipsen, J. T. M., Reusken, C. B. E. M., Sliwinsky, E. L., Goldbach, R. W., and Vlak, J. M., 1993, Functional domains of the p10 protein of *Autographa californica* nuclear polyhedrosis virus, *J. Gen. Virol.* **74:**563–574.

Vaux, D. L., and Strasser, A., 1996, The molecular biology of apoptosis, *Proc. Natl. Acad. Sci. USA* **93:**2239–2244.

Walker, N. I., Harmon, B. V., Gobe, G. C., and Kerr, J. F. R., 1988, Patterns of cell death, *Methods Achiev. Exp. Pathol.* **13:**18–54.

White, K., Grether, M. E., Abrams, J. M., Young, L., Farrell, K, and Steller, H., 1994, Genetic control of programmed cell death in *Drosophila, Science* **264:**677–683.

Williams, G. T., and Smith, C. A., 1993, Molecular regulation of apoptosis: Genetic controls on cell death, *Cell* **74:**777–779.

Williams, G. V., Rohel, D. Z., Kuzio, J., and Faulkner, P., 1989, A cytopathological investigation of *Autographa californica* nuclear polyhedrosis virus p10 gene function using insertion–deletion mutants, *J. Gen. Virol.* **70:**187–202.

Wold, W. S. M., and Gooding, L. R., 1991, Region E3 of adenovirus: A cassette of genes involved in host immunosurveillance and virus-cell interactions, *Virology* **184:**1–8.

Wyllie, A. H., 1980, Glucocorticoid-induced thymocyte apoptosis is associated with endogenous endonuclease activation, *Nature* **284:**555–556.

Wyllie, A. H., Kerr, J. F. R., and Currie, A. R., 1980, Cell death: The significance of apoptosis, *Int. Rev. Cytol.* **68:**251–306.

Xue, D., and Horvitz, H. R., 1995, Inhibition of the *Caenorhabditis elegans* cell-death protease CED-3 by a CED-3 cleavage site in baculovirus p35 protein, *Nature* **377:**248–251.

Zhu, H., Shen, Y., and Shenk, T., 1995, Human cytomegalovirus IE1 and IE2 proteins block apoptosis, *J. Virol.* **69:**7960–7970.

CHAPTER 11

Auxiliary Genes of Baculoviruses

DAVID R. O'REILLY

I. INTRODUCTION

A consideration of the genes encoded by viruses suggests that they can be divided into two broad groups based on functions. First, all viruses carry genes essential for their own propagation. In addition to genes encoding viral structural proteins, these also include genes whose function is to ensure the expression of viral genes, the replication of its genome, the assembly and release of progeny virus particles, and the effective transmission of the virus to a new host. Viruses with small genomes achieve many of these functions by subverting the machinery of the host cell. Many small viruses only carry these essential genes.

The evolution of larger genome size enables a virus to carry a wider range of genes. Many of these are still essential genes, often serving to reduce the dependency of the virus on its host cell. Thus, many larger viruses encode some or all of their own transcription machinery, genome replication machinery, and so forth. However, in addition to these extra essential genes, larger genome size enables the virus to carry genes of a second class, designated here as auxiliary genes. Such genes are not essential for viral replication, but nonetheless provide it with some selective advantage. Baculovirus genomes are among the largest of viral genomes, and thus can be expected to encode a large variety of auxiliary genes. It is the purpose of this chapter to review the functions of some of them.

The number of baculovirus auxiliary genes already identified precludes the discussion of all of them in a single chapter. In any case, for some of these genes, there is little or no information on possible function, and thus

---

DAVID R. O'REILLY • Department of Biology, Imperial College of Science, Technology, and Medicine, London SW7 2BB, England.

The Baculoviruses, edited by Lois K. Miller. Plenum Press, New York, 1997.

little to discuss. Therefore the scope of the chapter has been limited to genes known to be nonessential for which there are some clues to possible function and that are not discussed extensively elsewhere in this volume. The one exception to these criteria is the inclusion of the protein kinase 1 (*pk1*) gene, which may be essential. As of yet, it is not known whether this gene is nonessential. However, as the protein kinase 2 (*pk2*) and protein tyrosine phosphatase (*ptp*) genes are discussed, it was included for the sake of completeness. In addition to these three genes, the proliferating cell nuclear antigen (*pcna*), ubiquitin (*ubi*), superoxide dismutase (*sod*), p10 (*p10*), conotoxinlike peptide (*ctl*), cathepsin (*cath*), chitinase (*chiA*), and ecdysteroid UDP-glucosyltransferase (*egt*) genes will be reviewed.

The variety of potential functions of these genes makes it difficult to organize the discussion in a biological meaningful way. However, examination of their likely functions suggests they can be divided broadly into two categories. Proteins like the protein kinases and protein tyrosine phosphatase are likely to effect viral replication within an infected cell. In contrast, chitinase and cathepsin, which function to release progeny occluded virus (OV) from insect cadavers, influence viral propagation at the organismal level. Similarly, EGT functions by manipulating host physiology, again acting at the organismal level. The chapter therefore has been subdivided into two parts: genes likely to facilitate viral replication at the cellular level and those likely to function outside infected cells, i.e., at the organismal level. Note that for some genes addressed, however, function is not well enough understood to permit classification with any confidence. Thus, superoxide dismutase is included in the first section because it lacks a signal sequence and superoxide dismutases are generally cell associated. However, as discussed later, it might function in the extracellular environment after cell lysis. Conversely, the conotoxinlike peptide is included in the second category because it is secreted. This does not preclude the possibility that it exerts significant effects within infected cells.

Aside from this broad subdivision, the only other groupings made are of genes likely to function in related pathways. Thus, the protein kinases and phosphatase are grouped together. Similarly, chitinase and cathepsin are discussed in the same section.

II. GENES LIKELY TO FUNCTION AT THE CELLULAR LEVEL

A. Proliferating Cell Nuclear Antigen

1. Identification and Expression of PCNA

Initial studies of the *pcna* gene region in *Autographa californica* nuclear polyhedrosis virus (AcMNPV) were prompted by the observation that a single prominent transcript of 1.7 kilobases (kb) derived from the EcoRI T fragment of the viral genome was present early but not late in infection (Mainprize *et al.*, 1986). Crawford and Miller (1988) undertook to identify the gene(s) giving rise to this transcript and cloned and sequenced the EcoRI T fragment. Their data revealed the presence of three open reading frames (ORFs), designated *etl*, *etm*, and *ets* (for EcoRI T large, medium, and small). Subsequent database homology searches revealed that the protein encoded by *etl* displays considerable similarity to eukaryotic proliferating cell nuclear antigen (PCNA) (O'Reilly *et al.*, 1989). For example, it displays 42% amino acid identity and 63% similarity to mammalian PCNAs. Thus, *etl* has been renamed *pcna*. Currently, there is no evidence concerning the function(s) of *etm* and *ets*. The *pcna* gene is transcribed as a 1.7-kb transcript early in infection, presumably corresponding to the early transcript identified in previous studies (Crawford and Miller, 1988). It appears to be translated as a 28-kDa protein, detectable from 4 to 12 hr post infection (p.i.).

2. Functions of Cellular PCNA

PCNA plays a central role in eukaryotic DNA metabolism, being involved in DNA replication, repair, and possibly control of the cell cycle. It is a processivity factor for DNA polymerases δ and, under certain conditions, ϵ (Bambara and Jessee, 1991), acting as a sliding clamp that tethers the polymerase to the DNA. It also functions in nucleotide excision repair of DNA. Fractionation of cell extracts capable of carrying out nucleotide excision repair *in vitro* has shown that PCNA is required for the repair synthesis phase of the reaction (Nichols and Sancar, 1992; Shivji *et al.*, 1992). This was initially thought to be due to its activity as a polymerase accessory protein. However, more recent evidence indicates that the DNA replication and repair functions of PCNA are distinct. The protein Cip1, which is induced by DNA damage, inhibits the DNA replication function of PCNA but does not affect its repair activity (Shivji *et al.*, 1994). This presumably allows repair of damaged DNA while blocking replication until repair is complete. It is not known how this differential effect is achieved. PCNA may also be involved in the regulation of the cell cycle, since it is found in a complex with cyclin D and various cyclin-dependent kinases (Xiong *et al.*, 1992). PCNA is essential for viability in yeast (Waseem *et al.*, 1992) and for SV40 DNA replication *in vitro* (Prelich *et al.*, 1987).

3. Possible Functions of Viral PCNA

Based on the known functions of cellular PCNA, the most obvious hypothesis for the function of Ac-PCNA is that it acts as an accessory factor for the viral DNA polymerase. The phenotype of vETLβgal, a viral mutant lacking *pcna*, provides some support for this hypothesis. The most notable feature of cells infected by vETLβgal is that viral late gene expression is markedly delayed (Crawford and Miller, 1988). Such an effect could be due to defective DNA replication, since late gene expression is dependent on DNA replication. Examination of DNA replication in vETLβgal-infected cells revealed that it was delayed to a small extent relative to wild-type

AcMNPV-infected cells (O'Reilly *et al.*, 1989). However, several observations argue against a critical role for *pcna* in AcMNPV DNA replication. First, it is not essential, which would be surprising for a protein playing a central role in viral DNA replication. *Orgyia pseudotsugata* NPV (OpMNPV) PCNA is only 30% identical to Ac-PCNA (Ahrens *et al.*, 1997), suggesting a lack of strong functional constraints on the sequence. Indeed, *Bombyx mori* NPV (BmNPV) lacks a *pcna* gene entirely (S. Maeda, unpublished data). Second, there is some evidence that disruption of *Ac-pcna* may also affect early gene expression (Crawford and Miller, 1988). This would be difficult to reconcile with an effect primarily at the level of DNA replication. Third, purified AcMNPV DNA polymerase is reported to be highly efficient even in the absence of PCNA (L. A. Guarino, unpublished data). Finally, *pcna* is not one of the viral genes known to be involved in origin-dependent plasmid replication in infected insect cells (Kool *et al.*, 1994; reviewed in Chapter 7, this volume). Furthermore, deliberate cotransfection of an excess of a *pcna*-expressing plasmid had no effect on replication in this system.

What are other possible functions of Ac-PCNA? Given the marked delay in late gene expression in vETLβgal-infected cells, one possibility is that it plays some role in viral transcription rather than DNA replication. A precedent for such a role can be found in bacteriophage T4. The T4 gene 45 product functions as an accessory protein for the viral DNA polymerase acting as a sliding clamp, analogous to PCNA. This protein is also involved in late gene transcription, and Herendeen *et al.* (1989) have proposed that the DNA replication apparatus acts as a "sliding enhancer" of late gene expression. Arguing against a role for Ac-PCNA in late transcription is the fact that *pcna* has not been identified as one of the viral genes required for late gene expression in an *in vivo* transient expression system (Lu and Miller, 1995).

Another possibility is that Ac-PCNA plays a role in viral DNA repair. As discussed above, there is evidence that the DNA replication and repair functions of PCNA are separable (Shivji *et al.*, 1994). This suggests the novel possibility that the AcMNPV encodes PCNA to enhance the repair of its DNA. Such a mechanism could clearly be advantageous to the virus, given that it is often exposed to UV irradiation from sunlight during horizontal transmission from one insect to another. Furthermore, a defect in the repair of viral DNA could account for the observed delay in viral gene expression, since it is known that many DNA lesions block transcription (e.g., Sauerbier and Hercules, 1993, and references therein).

It is well known that host cell division is arrested following viral infection (Brown and Faulkner, 1975). As noted above, there is some evidence that PCNA is involved in the regulation of the cell cycle in mammalian cells, raising the possibility that Ac-PCNA plays some role in host cell arrest following infection. To investigate this, we have compared cell multiplication after infection with vETLβgal or with a control virus. Cell division was arrested in both cases (D. R. O'Reilly, unpublished data), indicating that Ac-PCNA does not play a role in the inhibition of host cell division.

Finally, it is still a formal possibility that Ac-PCNA does play an important role in viral DNA replication. The experiments outlined above were all done in a single cell line (SF21 cells) and were done in actively growing cells. It is conceivable that host PCNA can substitute for Ac-PCNA under these conditions, but that Ac-PCNA is required for replication in quiescent cells or in cells from a different tissue and/or host species.

Considerable work is still required to unravel the role of the *pcna* gene in AcMNPV replication. It is particularly difficult to understand how *pcna* can apparently play an important role in late gene expression in infected cells, and yet is not found to be important for viral transcription in transient expression assays. In this regard, it is important to bear in mind the caveat that the data from vETLβgal-infected cells may be misleading. In this mutant, the *pcna* gene is disrupted by the insertion of the *lacZ* gene in-frame at a site approximately two thirds through the gene so that a fusion protein including the N-terminal 166 amino acids of PCNA is produced (Crawford and Miller, 1988). It is conceivable that this fusion protein can carry out some PCNA functions, or alternatively can produce unexpected phenotypes by acting in a dominant negative fashion.

B. Protein Kinase 1 and 2

1. Importance of Phosphorylation in Baculovirus Infections

Phosphorylation of proteins is a common mechanism for the regulation of protein function at the posttranslational level. A central feature of this control mechanism is that, due to the interplay between protein kinases and protein phosphatases, it is readily reversible. Phosphorylation thus allows for a protein to be rapidly switched between different activity states. This form of control is widely used in biological systems, ranging from signal transduction, transcription, translation, and cell cycle progression to ion transport and other metabolic processes. There is little doubt that it also plays an important role in baculovirus infections. Many baculovirus proteins are known to be phosphorylated, including IE1, the core protein p6.9, 39K, gp64, vp39, ORF1629, the calyx protein, and possibly p10 (Choi and Guarino, 1995; Vialard and Richardson, 1993; Cheley *et al.*, 1992; Guarino *et al.*, 1992; Whitt and Manning, 1988; Kelly and Lescott, 1984; Maruniak and Summers, 1981). In some cases, there is evidence that phosphorylation is of functional importance. Choi and Guarino (1995) have reported that dephosphorylation of IE1 results in a loss of its DNA-binding activity. Phosphorylation of the *Plodia interpunctella* granulosis virus (PiGV) core protein may be required for release of viral DNA from the nucleocapsid at the start of the infection process (Funk and Consigli, 1993). Dephosphorylation of the protein would then allow packaging of progeny DNA molecules into virions at the end of the cycle. Cheley *et al.* (1992) have shown that phosphorylation of a serine at the C-terminus of p10 causes it to bind microtubules.

It has been known for some time that protein kinase activity is association with AcMNPV budded virus (BV) and OV (Miller *et al.*, 1983). Similarly, purified PiGV nucleocapsids have an associated kinase activity (Wilson and Consigli, 1985). AcMNPV is now known to encode at least three proteins that may participate in the control of phosphorylation: protein kinase 1 and 2 and protein tyrosine phosphatase.

2. Identification and Activity of Protein Kinase 1

Sequence analysis of the EcoRI I fragment of AcMNPV revealed an ORF predicted to encode a protein kinase of 272 amino acids (Ayres *et al.*, 1994; Reilly and Guarino, 1994; Possee *et al.*, 1991). Bischoff and Slavicek (1994) reported the cloning of a similar gene from the *Lymantria dispar* MNPV (LdMNPV). Both genes showed significant similarity to the catalytic domains of serine–threonine protein kinases. Such enzymes are known to have 11 catalytic domain consensus motifs (Hanks *et al.*, 1988). All these motifs are conserved in *Ac-* and *Ld-pk1* genes, suggesting they encode active enzymes. This was confirmed by assay of proteins produced by *in vitro* translation of these genes (Bischoff and Slavicek, 1994; Reilly and Guarino, 1994). The viral kinases are significantly smaller than related cellular serine–threonine kinases because they lack a large N-terminal regulatory domain. There may be some differences in the temporal control of *pk1* expression between the two viruses. Bischoff and Slavicek (1994) reported that *Ld-pk1* is expressed throughout infection, initially from an early promoter and subsequently from a late promoter. In contrast, Reilly and Guarino (1994) characterize *Ac-pk1* as a late gene only. The significance of this difference is not clear.

In addition to AcMNPV and LdMNPV, BmNPV, OpMNPV, and *Helicoverpa zea* (HzNPV) encode *pk1* genes, suggesting that *pk1* serves a functionally important role (Ahrens *et al.*, 1997; Maeda, 1996; Tribe *et al.*, 1994). Recent data from Weaver and colleagues have begun to shed some light on what this role might be. They report the characterization of several AcMNPV mutants with point mutations in *pk1* that are temperature sensitive for very late gene expression (Fan *et al.*, 1996). One of these mutants also has a temperature-sensitive defect in late gene expression and plaque production. Thus, *pk1* is likely to be essential. The effect on very late gene expression is manifested by a marked reduction in the steady-state levels of very late gene mRNA. However, the initiation of very late gene transcription is not reduced to the same extent, leading Fan *et al.* (1996) to suggest that the effect of *Ac-pk1* disruption is mediated partially at a posttranscriptional level. Note that *Ac-pk1* was not among the genes identified by Miller and colleagues in a transient expression assay for late and very late expression factors (Lu and Miller, 1995). The apparent discrepancy between these two sets of data remains to be resolved.

How protein kinase 1 affects late and/or very late gene expression is not yet understood. The identity of target protein(s), either viral or cellular, that it phosphorylates is unclear. However, Broussard *et al.* (1996) have presented evidence suggesting that a portion of AcMNPV 39K phosphorylation is mediated by a viral kinase. This is intriguing since the 39K protein has been identified as a late gene expression factor (Lu and Miller, 1995; see Chapter 8, this volume). Another clue to its mode of action has been provided by Weaver and colleagues, who have found that AcMNPV protein kinase 1 interacts with a viral protein they designate PKIP (protein kinase interacting protein) (X. Fan and R. F. Weaver, unpublished data). PKIP apparently stimulates the activity of protein kinase 1 *in vitro.* The *pkip* gene appears to be essential.

3. Identification and Characterization of Protein Kinase 2

The *Ac-pk2* gene was first identified during sequencing of the *lef7* gene region (Morris *et al.,* 1994). It is transcribed primarily as an early gene and is translated to yield a 25-kDa protein (Li and Miller, 1995a). Unlike *pk1, pk2* is missing the first five catalytic domain consensus motifs, strongly suggesting that it does not encode a functional enzyme. However, it displays significant similarity to eIF2α kinases. These enzymes inhibit protein synthesis in eukaryotes via the phosphorylation of the translation factor eIF2α (eukaryotic initiation factor 2α). These observations suggested the possibility that, although truncated, protein kinase 2 might be involved in the regulation of translation in infected cells (Morris *et al.,* 1994). Yeast eIF2α kinase mutants can block the normal down-regulation of translation due to eIF2α kinase. To test whether protein kinase 2 is implicated in the regulation of translation in infected cells, Li and Miller (1995a) generated an AcMNPV mutant lacking a functional *pk2* gene. This mutant displayed no noticeable phenotypic alterations compared to wild-type virus. Host cell and viral protein synthesis and phosphorylation were unchanged. Similarly, the infectivity and virulence of the mutant virus in *Spodoptera frugiperda* larvae were identical to wild-type AcMNPV.

Further evidence that *pk2* does not play a central role in baculovirus replication is the observation that it is not universally conserved among different baculoviruses. Whereas BmMNPV has a *pk2* gene (Maeda, 1996), it is lacking from the OpMNPV genome (Ahrens *et al.,* 1997). Elucidation of the role of this gene awaits further study.

C. Protein Tyrosine Phosphatase

1. Identification and Activity of PTP

The protein tyrosine phosphatases (PTPs) are an extremely diverse group of proteins that can be divided into two subfamilies—the receptor and nonreceptor PTPs—depending on whether they span the membrane. PTPs are identified by possession of the sequence (I/V)HCXAGXXR(S/T)G, known as the HC motif. The cysteine residue in this motif is essential for activity (Charbonneau and Tonks, 1992). Recently, it has become apparent that some nonreceptor PTPs have the ability to dephosphorylate both phosphotyrosine and phosphoserine/phosphothreonine residues (Keyse, 1995). The first such dual-specificity phosphatase identified was the vaccinia VH1 protein (Guan *et al.*, 1991), but the family is growing rapidly. Currently, there are thought to be at least three subfamilies of dual-specificity phosphatases: one comprising VH1, its poxvirus homologues, and mammalian MAP kinase phosphatases; a second including two putative PTPs, the products of the *cm01g6* gene of *Caenorhabditis elegans,* and the *CDC14* gene of *Saccharomyces cerevisiae,* respectively; and the third represented by the cell cycle inducer cdc25 and its homologues (Sheng and Charbonneau, 1993).

AcMNPV *orf1* encodes a protein of 168 amino acids containing the HC motif (Kim and Weaver, 1993; Possee *et al.*, 1991). Based on the lack of any apparent transmembrane motif, this protein appears to be a member of the nonreceptor PTP subfamily. Biochemical studies have shown that the encoded protein is in fact a dual-specificity phosphatase (Hakes *et al.*, 1993; Kim and Weaver, 1993; Sheng and Charbonneau, 1993).

Ac-ptp is more similar to *cm01g6* and *CDC14* than it is to *vh1,* suggesting that it belongs to the second subfamily of dual-specificity phosphatases discussed above (Sheng and Charbonneau, 1993). We recently identified a *ptp* gene in *Mamestra brassicae* NPV (MbMNPV) that bears little resemblance to *Ac-ptp* but is quite closely related to *vh1* (C. Phanis and D. R. O'Reilly, unpublished data). Thus, *Ac-ptp* and *Mb-ptp* appears to belong to different subfamilies of dual-specificity phosphatases, suggesting these viruses acquired them independently. Sequence analysis of OpMNPV has revealed the presence of two *ptp* homologues, one (*ptp-1*) resembling *Ac-ptp* and the other (*ptp-2*) being more similar to *vh1* (Ahrens *et al.*, 1997). It may be that MbMNPV encodes a second PTP that is the true homologue of Ac-PTP.

2. Expression and Inactivation of *Ac-ptp*

Kim and Weaver (1993) reported that *Ac-ptp* is transcribed as a very late gene. However, immunoblot analysis has shown that the protein can be detected in both the cytoplasm and nucleus of infected cells from about 12 hr p.i., an expression pattern more reminiscent of a late gene (Li and Miller, 1995b). PTP shows some association with fibrillar structures in infected cells. It can also be detected associated with both BV and OV, suggesting it is a component of the viral capsid (see also Chapter 2, this volume). Thus, it is also likely to be present at low levels early in infection. Comparison of phosphoprotein profiles in cells infected by wild-type virus or by a mutant lacking a functional *ptp* suggests that a 35-kDa phosphoprotein present at 24 hr p.i. may be dephosphorylated by Ac-PTP (Li and Miller, 1995c). Broussard *et al.* (1996) suggest that the 39K protein can be partially dephosphorylated by a viral phosphatase. It is unclear whether this corresponds to the 35-kDa protein identified by Li and Miller (1995c). Ac-PTP probably only recognizes a small number of specific substrates, since there was no evidence of global dephosphorylation in wild-type compared to mutant-infected cells (Li and Miller, 1995c).

The phenotype of vPTPdel, the AcMNPV mutant disrupted in the *ptp* gene, does not give many clues to gene function. In infected *S. frugiperda* larvae, this virus was indistinguishable from wild type in terms of lethal dose, survival time, and yield of progeny OV (Li and Miller, 1995c). Similarly, the mutant was identical to wild type during infection of *Trichoplusia ni* cells in culture. However, a partial phenotype was observed during infection of SF-21 cells in culture. There was a greater variance in the numbers of OV particles produced per infected cell than observed with wild-type virus, with some cells producing many OV particles and others producing few or none. The authors speculated that this heterogeneity might suggest that PTP functions in a cell-cycle-dependent manner. Production of BV was also reduced in vPTPdel infection of SF-21 cells.

3. Possible Functions of Baculovirus PTPs

It is instructive to consider the poxvirus PTPs when speculating on the function of the baculovirus enzymes, since they have a number of features in common. Baculovirus and poxvirus PTPs are all considerably smaller than cellular PTPs (Liu *et al.*, 1995; Mossman *et al.*, 1995). In the case of vaccinia VH1, it has been shown that the protein is expressed late in infection and, like Ac-PTP, is incorporated into the viral capsid (Liu *et al.*, 1995). The enzymes from both viruses are dual-specificity phosphatases. Liu *et al.* (1995) have shown that VH1 is required, either directly or indirectly, for efficient early gene expression. The vaccinia 11-kDa histonelike DNA binding protein was found to be a substrate of VH1, suggesting that hyperphosphorylation of this protein in the absence of VH1 is responsible for the early transcription defect (Liu *et al.*, 1995). It is tempting to speculate that PTPs play a similar role in baculovirus infections. However, Li and Miller (1995c) have reported that the kinetics of protein synthesis appear normal until approximately 24 hr p.i. in vPTPdel-infected cells, arguing against a role in early gene expression. Furthermore, unlike Ac-PTP, the poxvirus PTPs are essential for virus viability (Liu *et al.*, 1995; Mossman *et al.*, 1995). Nonetheless, it would be premature to conclude that baculovirus and poxvirus PTPs have unrelated roles in the viral life cycle. It could be, for example, that PTP contributes to dephosphorylation of the core protein prior to encapsidation of viral DNA (discussed above). Such a function might account for the defect in BV and OB production observed in vPTPdel-infected SF-21 cells.

Alternatively, baculovirus PTP function may be similar to that of some of the other PTPs. The MAP kinase phosphatases are involved in regulating signal transmission by dephosphorylating members of the mitogen-activated protein kinase (MAP kinase) family, whereas cdc25 removes inhibitory phosphates from cyclin-dependent kinases that drive the cell cycle. Thus, directly or indirectly, many PTPs are implicated in the control of cell division, and it is conceivable that baculovirus PTPs may have some cell-cycle-related role. In this context, it is interesting that Li and Miller (1995c) speculated that Ac-PTP functions in a cell-cycle-dependent manner.

D. Ubiquitin

1. Function of Cellular Ubiquitin

Ubiquitin is a 76 amino acid protein that is abundant in all eukaryotes. It is one of the most highly conserved proteins known, with animal and yeast ubiquitin differing by only three amino acid residues. Its principal function in the cell is to signal the degradation of proteins by the 26S proteosome (Hochstrasser, 1996). Degradation is signaled by the covalent linkage of a chain of ubiquitin molecules to the target protein. Initially, a ubiquitin monomer is attached via an isopeptide linkage between the C-terminal glycine of ubiquitin and a lysine in the target protein. The ubiquitin chain is then extended by links between the C-terminal glycine of one ubiquitin monomer and Lys_{48} of the next. Multiubiquitinated proteins are targeted to the proteosome, where the protein is degraded and the ubiquitin released.

In addition to its major role in protein degradation, ubiquitin may participate in other cellular functions (reviewed in Hochstrasser, 1996). In particular, it is required for reinternalization of the yeast cell surface receptor Ste2, and is likely to serve as a signal for receptor endocytosis in general. In addition, ubiquitination is in some way required for the activation of a kinase that phosphorylates the transcription factor inhibitor IκB.

2. Identification and Expression of Baculovirus Ubiquitin

The AcMNPV ubiquitin gene (*ubi*) is located immediately downstream of the *39k* gene (Guarino, 1990). AcMNPV ubiquitin is only 76% identical to animal ubiquitin, making it the most divergent ubiquitin known. The OpMNPV and *S. exigua* MNPV *ubi* genes have also been sequenced. The proteins they encode are 79 and 84% identical to animal ubiquitin, and 84 and 75% identical to the AcMNPV protein, respectively (van Strien *et al.*, 1996; Russell and Rohrmann, 1993). To determine whether this divergence was specific to baculovirus ubiquitins, the ubiquitin gene of *S. frugiperda* was cloned, sequenced, and found to be identical to animal ubiquitin, suggesting that baculovirus ubiquitins have diverged markedly since their acquisition by the virus (Guarino, 1990).

In eukaryotic cells, ubiquitin is generally encoded either as an in-frame fusion with another protein, often a ribosomal protein, or as a polyubiquitin gene, encoding from 5 to 18 ubiquitin monomers. In either case, C-terminal cleavage is necessary to release the ubiquitin monomer. The *Ac-ubi* gene encodes a 77 amino acid protein, i.e., with one additional amino acid after the usual C-terminal glycine residue, whereas *Se-ubi* comprises 80 amino acids, with three residues after the C-terminal glycine. In contrast, *Op-ubi* has a C-terminal extension of 17 amino acids. Presumably, in all cases the protein is processed to yield a 76 amino acid mature protein.

Ac-ubi is abundantly expressed late in infection (Guarino, 1990), suggesting that it might be a structural protein. This has been confirmed, with several forms of ubiquitin identified in BV particles (Guarino *et al.*, 1995). A number of capsid proteins seem to be conjugated with ubiquitin. In addition, substantial amounts of free ubiquitin monomers and dimers are present in the virus particle. Unconjugated ubiquitin is almost as abundant as the major envelope protein GP64 and accounts for approximately 2% of the weight of the virion. Both host- and virus-encoded ubiquitin are found in the virion. Some of this material is located between the nucleocapsid and the envelope, whereas the remainder is associated with the inner face of the envelope. The envelope-associated fraction is anchored by a unique phospholipid anchor, comprising a diacylphosphoglyceride joined via an ester linkage to a serine or threonine residue in the protein (Guarino *et al.*, 1995). Both host and viral ubiquitin proteins are modified in this way.

3. Functional Characterization of Baculovirus Ubiquitin

The extreme conservation of eukaryotic ubiquitins suggests that few amino acid changes can be tolerated without significant loss of function. The fact that baculovirus ubiquitins are not constrained in the same way is strong evidence that their function(s) differ significantly from cellular ubiquitin. A recent study of AcMNPV ubiquitin confirms that this is the case. In an *in vitro* system, the viral protein supports general protein degradation at only 40% the rate of eukaryotic ubiquitin (Haas *et al.*, 1996). This is attributed to a defect at the stage of conjugation of ubiquitin to the target proteins. Furthermore, multiubiquitination of certain target proteins was severely inhibited. The authors speculate that viral ubiquitin could act to block the degradation of selected proteins during viral infection.

The initial demonstration that *ubi* is nonessential for viral replication was made by Fraser *et al.* (1995), who mapped a transposon insertion event to within *Ac-ubi.* Recently, Reilly and Guarino (1996) have also described the generation of an AcMNPV mutant with a frame-shift mutation in *ubi.* The phenotype of this mutant indicates that the protein plays a role in the formation of progeny BV. The total yield of progeny virus particles was five- to tenfold lower from cells infected with Vubi-FS, the virus carrying the frame-shift mutation, compared to cells infected with wild-type virus. Surprisingly, this deficiency in cell culture was not manifested in any difference in virus infectivity or virulence during infection of insect larvae. However, these tests are difficult to evaluate since they were carried out in a different species to the cell culture studies.

A role in the assembly or budding of virions from infected cells might be consistent with the location of ubiquitin to the inner face of the cytoplasmic membrane and to the viral capsid. However, progeny virus budded from Vubi-FS-infected cells still contains host ubiquitin, so that it is difficult to identify a specific role for viral ubiquitin. As noted above, Haas *et al.* (1996) speculate that viral ubiquitin might function to block the degradation of selected proteins. This would first require the conjugation of viral ubiquitin to such proteins. The identity of any target protein specifically conjugated by viral ubiquitin is not yet known. No differences were observed between the ubiquitination profiles of cells infected with Vubi-FS or wild-type virus (Reilly and Guarino, 1996). Some evidence appears to indicate that there may be significant differences in ubiquitination between OpMNPV- and AcMNPV-infected cells. In OpMNPV-infected cells, the major ubiquitinated protein was a virus-induced 73-kDa protein. There was little or no free ubiquitin in evidence (Russell and Rohrmann, 1993). In contrast, a huge range of proteins are ubiquitinated in AcMNPV-infected cells (Reilly and Guarino, 1996). A virus-induced protein of about 70 kDa is observed, but is present in relatively small amounts. Approximately half of the total ubiquitin in these cells is present in an unconjugated form. Neither experiment distinguished between host and viral ubiquitin. It remains to be seen whether these represent bona fide differences between the two viruses or simply reflect procedural differences in the two studies.

Although baculoviruses are the only viruses known to encode their own ubiquitin, the protein appears to play a significant role in many viral systems. Some cytopathogenic variants of bovine viral diarrhea virus have incorporated a host ubiquitin gene into their genome (Meyers *et al.*, 1991). The *Chlorella* virus PBCV-1 contains a gene with similarity to ubiquitin carboxy-terminal hydrolase (Li *et al.*, 1995), whereas African swine fever virus encodes a ubiquitin-conjugating enzyme (Dixon *et al.*, 1994). Ubiquitin is also a common component of virus particles. African swine fever virus contains a number of ubiquitin-conjugated proteins and may contain a modified form of unconjugated ubiquitin like that found in AcMNPV (Hingamp *et al.*, 1995). Ubiquitin is also conjugated to the coat proteins of several plant viruses (Dunigan *et al.*, 1988), and avian leukosis virus includes free ubiquitin (Putterman *et al.*, 1980). Unfortunately, notwithstanding this accumulating evidence that ubiquitin is important to a variety of viruses, we have as yet no clear understanding of its precise function in any of these cases.

E. p10

1. Structure of p10 Proteins

The p10 protein is perhaps the most enigmatic of the proteins discussed in this chapter, despite the fact that it was identified in the early 1980s and has been studied intensively since. It initially attracted attention because of its high expression levels late in the infection process (Adang and Miller, 1982; Smith *et al.*, 1982), and the AcMNPV *p10* gene was cloned and sequenced soon thereafter (Kuzio *et al.*, 1984). Transcription studies confirmed that *p10* is expressed as a very late gene, with kinetics similar to but not identical to polyhedrin (Roelvink *et al.*, 1992; Rankin *et al.*, 1986). The *p10* promoter appears to be of similar strength to the polyhedrin promoter and has found considerable use in baculovirus expression vectors (see Chapter 13, this volume).

A feature of baculovirus p10 proteins is their relative lack of sequence similarity. Thus, OpMNPV and SeMNPV p10s are only 41 and 26% identical at the amino acid sequence level to AcMNPV p10, in contrast to polyhedrins, which display 90 and 84% identity, respectively (Zuidema *et al.*, 1993; Leisy *et al.*, 1986). This lack of sequence conservation has hampered the identification of p10 homologues in more distantly related viruses. No direct p10 homologue has been identified in any GV yet (but see Section II.E.2).

Despite the lack of primary amino acid sequence similarity, p10 proteins seem to share a number of structural features (van Oers and Vlak, 1996; Wilson *et al.*, 1995; van Oers *et al.*, 1993; Zuidema *et al.*, 1993). The N-terminal half of the protein includes a heptad repeat of hydrophobic amino acids (with hydrophobic residues at the first and fourth positions of the repeat) that is characteristic of sequences predicted to form an α-helical coiled–coil structure. This is followed by a hydrophilic region, rich in proline residues, whereas the extreme C-terminus is rich in basic amino acids. The p10 proteins tend to lack cysteine, tryptophan, and histidine residues.

There is some evidence that p10 may represent a minor component of OV (Quant-Russell *et al.*, 1987). However, it is not clear that this is a functional association. It may simply reflect small amounts of protein trapped during the occlusion process (see Chapter 2, this volume). Immunoelectron microscopy has shown that p10 is associated with the extensive fibrillar structures observed in the nucleus and cytoplasm of NPV-infected cells (Russell *et al.*, 1991; van der Wilk *et al.*, 1987). Characterization of mutants in which the *p10* gene has been disrupted has confirmed that p10 is essential for the formation of the fibrillar structures, and it is likely to be the major constituent of these structures (Williams *et al.*, 1989; Vlak *et al.*, 1988; Gonnet and Devauchelle, 1987). Two domains within the p10 protein appear to be required for fibrillar body formation: the N-terminal coiled–coil domain and the basic C-terminal tail. van Oers *et al.* (1993) found that the coiled–coil domain is required for aggregation of p10 proteins (presumably via the coiled–coil motif) but that the C-terminal basic amino acids are additionally required for assembly into fibrillar bodies. Cheley *et al.* (1992) have reported that overexpression of the catalytic subunit of a cAMP-dependent protein kinase by AcMNPV results in the phosphorylation of p10 at a C-terminal serine residue. This phosphorylated p10 functioned as a microtubule-binding protein, suggesting that microtubule association might be involved in the formation of fibrillar bodies. However, there is little evidence that p10 is phosphorylated in this manner during normal infections. Furthermore, AcMNPV p10 mutants in which either or both of the two C-terminal serine residues were mutated to alanine have been generated (van Oers and Vlak, 1996). Fibrillar body formation occurred normally in cells infected by any of these mutants, apparently excluding a function for phosphorylation of the p10 C-terminus in the formation of these bodies.

2. Possible Functions of p10

Although a large number of *p10* mutant viruses have now been generated, its role during viral infection is still unclear. Disruption of *p10* has little effect on viral virulence or infectivity (Vlak *et al.*, 1988). While it is well established that p10 is involved in the formation of fibrillar bodies in infected cells, there is little understanding of the significance of these bodies to the infection process. Currently, data from the characterization of *p10* mutant viruses suggest two possible functions for p10. First, there is some evidence that p10–fibrillar bodies are involved in occlusion body morphogenesis, specifically in the formation of the calyx around the occlusion bodies. Similarly, a role in the formation of electron-dense spacers, which may be calyx precursors, is proposed. However, the evidence for a role in calyx formation is quite ambiguous, with some *p10* mutants apparently possessing electron-dense spacers and/or occlusion bodies with a partial or complete calyx. These issues are discussed in more detail in Chapter 2 (this volume). Second, p10 appears to be involved in the rupture of the nucleus late in the infection process. Williams *et al.* (1989) reported that *p10* mutants failed to liberate polyhedra late in infection, suggesting a role for p10 in cell lysis. van Oers *et al.* (1993) demonstrated that there was no difference in the release of cytoplasmic contents from wild-type or *p10*-mutant-infected cells and proposed that p10 acts specifically in the disruption of the nuclear membrane. This function of p10 maps to the central, proline-rich region of the protein and is not dependent on the ability to form fibrillar bodies. It is assumed that nuclear lysis contributes to the release of OV from infected insects, and hence their effective dissemination in the environment, but this has not been directly demonstrated.

To date, p10 proteins have only been identified in NPVs. However, an entomopoxvirus protein with some similarity to p10 was described recently (Alaoui-Ismaili and Richardson, 1996). This protein, designated the filament-associated late protein of entomopoxviruses (FALPE) does not display significant amino acid sequence identity with p10 and is substantially larger. However, it possesses an N-terminal region that is predicted to form an amphipathic helix, a central region that is extremely proline rich, and a basic C-terminal tail. In addition to these structural similarities, there are also functional similarities. This protein is associated with an extensive fibrillar network in infected cells. These filaments are closely associated with viral occlusion bodies, reminiscent of the suggested role of p10 in baculovirus occlusion body morphogenesis. Recently, we have identified an ORF in CpGV that may also encode a protein of this type (W. Kang *et al.*, unpublished data). CpGV ORF17R appears to encode a late protein that has a large domain predicted to adopt an α-helical coiled–coil structure. A proline-rich region is also present, although this is N-terminal to the coiled–coil domain in this case. Perhaps we are seeing the emergence of a diverse family of filament-forming proteins that are only loosely related at the sequence level.

F. Superoxide Dismutase

1. Function of Cellular Superoxide Dismutases

Superoxide dismutases (SODs) catalyze the dismutation of the superoxide radical O_2- into H_2O_2 and O_2 (reviewed in Pardini, 1995). The O_2^- anion, generated as a by-product of oxygen utilization of aerobic organisms, is directly or indirectly responsible for most of the cellular toxicity due to oxygen. It can act both as an oxidant or reductant itself. In addition, it is readily converted into more toxic metabolites, such as the strong oxidants $HO_2\cdot$ and $HO\cdot$. The superoxide radical and its metabolites cause damage to most cellular constituents, including nucleic acids, carbohydrates, proteins, and lipid membranes. SODs represent part of the enzymatic defense against oxygen toxicity and are present in almost all aerotolerant organisms. Two unrelated families of SODs are known. The manganese-containing SODs (MnSODs) and iron-containing SODs (FeSODs) belong to one family and are found in prokaryotes and in the mitochondria of eukaryotes. The second family comprises the copper- and zinc-containing SODs (Cu,ZnSODs), found in the cytoplasm and chloroplasts of eukaryotes.

2. Identification, Activity, and Possible Functions of Baculovirus SODs

AcMNPV encodes a Cu,ZnSOD homologue that is expressed as a low-abundance protein late in infection (Tomalski *et al.*, 1991). The predicted protein is up to 54% identical with eukaryotic Cu,ZnSODs. Cu,ZnSOD homologues have also been identified in the genomes of other baculoviruses, including BmNPV and OpMNPV (G. F. Rohrmann, unpublished data; Maeda, 1996), as well as other large DNA viruses including vaccinia (Johnson *et al.*, 1993) and *Chlorella* PBCV-1 virus (Lu *et al.*, 1996).

The function of the baculovirus *sod* gene (or any viral *sod*) is unknown. *Ac-sod* has been deleted with no apparent effect on viral growth either in cell culture or in insects (Tomalski *et al.*, 1991). Similarly, the viral SOD did not appear to facilitate viral growth when elevated levels of superoxide anions were induced experimentally. Treatment of cell cultures with paraquat, an inducer of superoxide radical production, caused a similar reduction in the replication of wild-type and *sod*-mutant AcMNPV. It has not been formally proven that *Ac-sod* encodes a functional enzyme. SOD assays of infected cells or insects revealed high levels of endogenous SOD but no virus-specific activity. The amino acid sequence of AcMNPV SOD retains all residues known to be essential for metal coordination, and it seems most likely that the high endogenous activity simply swamped virus specific activity. Nonetheless, some doubt must remain whether *Ac-sod* encodes a functional dismutase.

Given that high levels of host SOD appear to be present during baculovirus infection, under what conditions might a viral-encoded enzyme be advantageous? Host SOD activity is likely to be retained almost exclusively within cells, so perhaps the viral SOD serves to protect the virus in extracellular environments. The production of a burst of superoxide radicals is a common mechanism whereby phagocytic cells attack engulfed material. Perhaps SOD functions to protect the virus against the host immune response. The available evidence suggests that this is not the case. First, as noted above, deletion of *sod* has no effect on the *in vivo* infectivity or virulence of AcMNPV. This is true whether the insects are infected by feeding or by direct injection of BV into the hemocoel (D. R. O'Reilly, unpublished data). Second, SOD is not a component of AcMNPV BV particles (Tomalski *et al.*, 1991).

Alternatively, SOD might serve to protect OV in the environment from superoxide radicals generated by exposure to sunlight. Ignoffo and Garcia (1994) have demonstrated that superoxide radicals contribute to sunlight-induced damage of OV in the environment and showed that addition of exogenous SOD afforded some protection. Tomalski *et al.* (1991) investigated whether SOD was associated with OV, but this study was inconclusive, due to the presence of many proteins of the approximate size of SOD in OV preparations. Thus, it remains a possibility that a function of baculovirus SOD is to protect OV from sunlight-induced superoxide radicals in the environment.

III. GENES LIKELY TO FUNCTION AT THE ORGANISMAL LEVEL

A. Conotoxinlike Peptide

1. Identification and Expression of Conotoxinlike Peptide

The AcMNPV conotoxinlike peptide (CTL) is a small, cysteine-rich peptide identified based on its sequence similarity to the ω-conotoxins, a family of calcium channel antagonists found in predatory marine snails (Eldridge *et al.*, 1992a). The *ctl* gene comprises 53 codons, with the conotoxin homology restricted to the C-terminal 30 amino acids. The N-terminal

```
GVIA            -CKSPGSSCSPTSYNCCRS-CNPYTKRCY
                 |   |  |      ||   | |    |
Ac-CTL          ACAETGAVCVHNDE-CCSGACSPIFNYCLPQ
                 |   |  |   || || | |      |
Aptotoxin III   -CNSKGTPCTNADE-CCGGKCAYNVWNCIGGGCSKTCGY
```

FIGURE 1. Alignment of the peptide sequences of Ac-CTL (after the putative signal peptide), ω-conotoxin GVIA (Olivera *et al.*, 1984) and aptotoxin III (Skinner *et al.*, 1992). Identical residues are indicated by vertical lines, and hyphens denote gaps introduced to achieve an optimum alignment.

part of the protein is quite hydrophobic, and Eldridge *et al.* (1992a) speculate that it is a signal sequence for secretion. In support of this, they report that CTL can be detected in the extracellular fluid of virus-infected cells. The peptide migrates with an apparent mass of 12 kDa on sodium dodecyl sulfate–polyacrylamide gel electrophoresis (SDS-PAGE) gels, although its predicted size is only 3 kDa. Derivatization of the thiol groups reduces the apparent mass to 5 kDa, indicating that the high cysteine content causes the anomalous migration on SDS-PAGE gels. The *ctl* gene is transcribed late in infection as the first *orf* in a bicistronic mRNA. CTL appears to be produced in very small quantities, and it was not possible to prepare enough sample to carry out N-terminal sequence analysis. Thus, there is no direct evidence that the putative signal sequence is cleaved from the peptide.

The ω-conotoxins are a family of 25- to 30-residue peptides that share a highly conserved patterns of cysteine and glycine residues: C-X3-G-X2-C-X6-CC-X2-3-C-X4-6-C (X = any amino acid) (Olivera *et al.*, 1991). AcMNPV CTL displays even greater similarity to a family of spider toxins known as aptotoxins (D. R. O'Reilly, unpublished data). A multiple alignment of CTL (without the predicted signal peptide) with aptotoxin III (Skinner *et al.*, 1992) and ω-conotoxin GVIA (Olivera *et al.*, 1984) is presented in Fig. 1. BmNPV does not encode any CTL homologue (Maeda, 1996). In contrast, OpMNPV encodes two CTL homologues (Ahrens *et al.*, 1997).

2. Possible Function of CTL

The biochemical activity of CTL is purely speculative at present. The sequence similarity with the ω-conotoxins suggests that CTL might participate in the regulation of calcium levels during infection, but there is little direct evidence of this. As shown in Fig. 1, the homology between CTL and the ω-conotoxins is essentially limited to the conserved cysteine and glycine spacing (Eldridge *et al.*, 1992a). However, this is also true of the ω-conotoxins themselves, which display extreme sequence hypervariability between the conserved cysteine and glycine residues, even among toxins known to interact with the same calcium channel subtype (Olivera *et al.*, 1991). The aptotoxins are insecticidal peptides, causing rapid paralysis and death within 24 hr (Skinner *et al.*, 1992). Presumably they also act as calcium channel antagonists. It seems likely therefore that CTL is a calcium channel antagonist, although this has not been demonstrated directly.

The biological function of CTL remains unknown at present. The most obvious hypothesis, since both the conotoxins and aptotoxins are paralytic toxins, is that CTL might be responsible for inducing some form of paralysis in infected insects, e.g., the lethargic state often observed late in the infection process. Alternatively, one might expect other effects that could be attributed to a disruption of calcium transport across membranes. Since it is not clear what proportion of CTL is secreted, such effects might be mediated intracellularly and/or extracellularly. A recombinant AcMNPV lacking a functional *ctl* has been generated and its biological properties examined thoroughly (Eldridge *et al.*, 1992a). No defects were observed in viral growth in cell culture or infectivity and virulence in insect larvae. In addition, infected insects were examined for a wide variety of possible behavioral alterations including decreased lethargy, increased general mobility, modified climbing, phototaxis, or aggression. No differences between wild-type AcMNPV and the *ctl*-minus virus were observed in any of the assays. Thus, the function of this fascinating small gene remains obscure.

B. Cathepsin and Chitinase

1. Identification and Biochemical Activity of Cathepsin and Chitinase

The viral chitinase (*chiA*) and cathepsin (*cath*) genes are considered together in this section because available evidence indicates they act together. In addition, they are located adjacent to each other in many baculovirus genomes. They are generally located in a head-to-head arrangement and are, at least in the case of AcMNPV, expressed late in infection (Hawtin *et al.*, 1995; Slack *et al.*, 1995). They now have been identified in several NPVs and also in CpGV (W. Kang *et al.*, unpublished data), suggesting they are widely distributed in baculoviruses. The evolutionary origin of these genes is discussed in Chapter 5 (this volume). Although both genes were first identified based on sequence homologies, biochemical studies have confirmed that the encoded enzymes possess the predicted activities. The *cath* gene was first identified based on its homology to the cysteine proteases papain and the cathepsins (Hill *et al.*, 1995; Rawlings *et al.*, 1992). AcMNPV cathepsin is insensitive to serine, aspartate, or metalloprotease inhibitors, but is strongly inhibited by a variety of cysteine protease inhibitors, confirming that it is a cysteine protease (Brömme and Okamoto, 1995; Slack *et al.*, 1995; Ohkawa *et al.*, 1994). Based on sequence similarities, Ohkawa *et al.* (1994) and Slack *et al.* (1995) suggest that the viral cathepsins most closely resemble cathepsins H and/or L. However, detailed characterization of the substrate specificity of AcMNPV cathepsin has shown that it strongly favors a dibasic amino acid cleavage site, similar to cathepsin B (Brömme and Okamoto, 1995). This probably explains the earlier failure of Rawlings *et al.* (1992) to detect any virus-specific cysteine protease activity, since the target cleavage site they used (Phe-Arg) is not cleaved efficiently by the viral cathepsin (Brömme and Okamoto, 1995). The enzyme has a strong pH optimum at pH 5.0–5.5 (Brömme and Okamoto, 1995).

AcMNPV chitinase possesses both exo- and endochitinase activity against a variety of substrates (Hawtin *et al.*, 1995). This contrasts with bacterial, plant, and fungal chitinases, where multiple chitinases with different, narrow specificities tend to be produced. Low levels of chitinase activity are secreted from uninfected SF cells (Hawtin *et al.*, 1995). The viral chitinase is readily distinguished from this host enzyme based on its relative insensitivity to allosamidin, a chitinase inhibitor (R. E. Hawtin *et al.*, unpublished data). It is active over a very broad pH range, from at least pH 3 to 10.

2. Structure and Subcellular Localization of Cathepsin and Chitinase

Immunoblotting studies of AcMNPV cathepsin reveal three bands of 27 kDa, 32 kDa, and 35 kDa (Slack *et al.*, 1995). Based on tunicamycin studies and comparison with other cysteine proteases, Slack *et al.* (1995) suggest that the enzyme is first produced as a preproprotein from which an N-terminal signal sequence is cleaved to give the 32-kDa proprotein. This is glycosylated to give the 35-kDa form, which is then cleaved to yield the 27-kDa mature glycosylated enzyme. They predict that cleavage of the proenzyme takes place at Pro_{113} (Slack *et al.*, 1995), but N-terminal sequence analysis of the mature enzyme reveals that it is cleaved at Asn_{97} (Brömme and Okamoto, 1995). A single site for N-linked glycosylation is present in the mature enzyme at Asn_{158}. AcMNPV cathepsin does not appear to be secreted from infected cells. Instead, subcellular fractionation experiments suggest it is found in a particulate fraction of cell lysates that includes nuclei, mitochondria, and lysosomes (Slack *et al.*, 1995). Since the cathepsins are normally lysosomal enzymes, it seems plausible that the viral cathepsin is also localized to the lysosomes. However, further studies are required to confirm this.

The structure and subcellular localization of the viral chitinase are less clear. Immunoblotting experiments have shown that AcMNPV chitinase is expressed as a protein of approximately 58 kDa (Hawtin *et al.*, 1995). The nucleotide sequence predicts a protein of 60.9 kDa, suggesting that the protein is not processed to a great extent. Although the predicted amino acid sequence appears to include a signal sequence, more than 90% of enzyme activity is retained with the infected cell. Immunofluorescence studies show that AcMNPV chitinase is confined to the cytoplasm of infected cells, apparently aggregating in foci (Hawtin *et al.*, 1995). Whether these foci represent accumulation in some subcellular organelle is not known.

3. Function of Cathepsin and Chitinase

A variety of studies have now examined the effects of deleting either or both the *cath* and *chiA* genes from AcMNPV and BmNPV, and a relatively coherent picture of the function of the encoded enzymes has emerged. Initial studies deleting either gene individually resulted in very similar phenotypic effects, the most notable being that insects infected by such viruses failed to liquefy and retained their cuticular integrity after death (R. E. Hawtin *et al.*, unpublished data; Slack *et al.*, 1995; Ohkawa *et al.*, 1994). Insect cuticle is composed mainly of chitin fibers embedded in a protein matrix and its degradation requires the synergistic action of both proteinases and chitinases (Samuels and Paterson, 1995). Coinfection of *cath*-minus and *chiA*-minus mutants of AcMNPV results in normal host cuticular breakdown and tissue liquefaction, confirming that both enzymes are required for this process in infected larvae (R. E. Hawtin *et al.*, unpublished data). Thus, it appears that baculoviruses encode these enzymes to facilitate host cuticle breakdown after death, causing release of the progeny OV into the environment. This presumably results in a significant advantage to the virus in terms of more efficient dissemination of progeny OV, and hence more efficient horizontal spread. In addition to cuticular degradation, there is evidence that the viral cathepsin also participates in the degradation of internal tissues of the insect, which would also facilitate liquefaction and the release of progeny OV (Ohkawa *et al.*, 1994).

Other functions for cathepsin and chitinase have also been considered. The fact that AcMNPV chitinase is active at high pH and that the peritropic membrane lining the host midgut is composed partially of chitin suggests the possibility that chitinase associated with OV could facilitate penetration of the virus through the peritropic membrane. However, although chitinase is detected in association with OV, there is no difference in the infectivity of wild-type and *chiA*-minus virus, showing that viral-encoded chitinase does not play a major role during the initial stages of infection (R. E. Hawtin *et al.*, unpublished data). Lanier *et al.* (1996) have reported that AcMNPV cathepsin is associated with BV. They also report that the enzyme can cleave actin *in vitro*, and they speculate that it might be responsible for actin filament reorganization early in infection. However, infection with a deletion mutant lacking *cath* did not support this possibility. Currently, therefore, the available evidence only indicates that cathepsin and chitinase act together to facilitate the degradation of the infected host and release of progeny OV at the end of the infection process.

C. Ecdysteroid UDP–Glucosyltransferase

1. The *egt* Gene, Protein, and Sequence Homologies

The ecdysteroid UDP–glucosyltransferase (*egt*) gene was first identified in AcMNPV during a study of regions of the genome known to be nonessential for viral replication (O'Reilly and Miller, 1989). Studies by Kumar and Miller (1987) had concluded that a region within the PstI G fragment of the AcMNPV genome is preferentially lost upon serial passage in cell culture.

Characterization of this region revealed that two ORFs were disrupted by these deletions: the *egt* and *da26* genes (O'Reilly and Miller, 1990; O'Reilly *et al.*, 1990). The function of *da26* is not yet known (O'Reilly *et al.*, 1990). The *egt* gene encodes a protein of 506 amino acids and is expressed early in infection as two 5′ coterminal transcripts of 1.8 and 3 kb, initiating 43 nucleotides upstream of the translational start site (O'Reilly and Miller, 1990). The EGT protein is secreted from infected cells, and N-terminal sequence analysis of EGT has shown that a signal sequence of 18 amino acids is cleaved off during export (O'Reilly *et al.*, 1992). The mature protein has an apparent mass of 60 kDa on SDS-PAGE gels (O'Reilly and Miller, 1990). We have recently found that EGT is N-glycosylated, with approximately 9 kDa of sugar residues on each protein molecule (O. P. Evans and D. R. O'Reilly, unpublished data). There are seven potential N-linked glycosylation sites in the AcMNPV EGT amino acid sequence. Removal of the sugar residues by glycosidase treatment has shown that N-linked glycosylation is not essential for enzyme activity (O.P. Evans and D. R. O'Reilly, unpublished data).

egt homologues have now been identified in a large number of baculovirus genomes, including representatives of both baculovirus subgroups, and we suspect it represents an ancestral gene that was present in the most recent common ancestor of baculoviruses (Clarke *et al.*, 1996). It seems likely that most or all baculoviruses will be found to encode an *egt* gene. The homologies between baculovirus EGTs have been discussed in detail in Clarke *et al.* (1996) and O'Reilly (1995). In all cases, they display high similarity throughout the lengths of their sequence, with the exception of short sequences at the N-terminus and in the center of the protein (corresponding to AcMNPV EGT amino acids 1–18 and 270–290). The nonconserved region in the center of the protein is discussed later. The N-terminal region corresponds to the signal sequence of AcMNPV. Although different viral EGTs do not share high levels of sequence identity in this region, they all contain a high proportion of hydrophobic residues, suggestive of a signal sequence. In addition, the homology between the viral EGTs becomes high precisely at the position corresponding to the first amino acid of mature AcMNPV EGT. AcMNPV EGT is known to lack a signal at its C-terminus for retention of the protein in the endoplasmic reticulum (O'Reilly and Miller, 1989). Related mammalian proteins are found in the lumen of the endoplasmic reticulum and possess a short region at their extreme C-terminus that is rich in lysine residues and is thought to act as a stop–transfer sequence. This motif appears to be absent from all the baculovirus EGTs (O'Reilly, 1995). Thus, it seems likely that all baculovirus EGTs are secreted.

EGT belongs to a large family of UDP–glycosyltransferases including enzymes from plants, bacteria, and nematodes as well as from mammals (O'Reilly, 1995). These all share the ability to catalyze the transfer of sugar from a UDP–sugar to an aglycone substrate. Many of these sequences are quite divergent, but nonetheless seven amino acids are absolutely conserved among all of them (O'Reilly, 1995). All the conserved residues are found in the C-terminal half of these proteins (corresponding to AcMNPV residues Gly_{298}, Trp_{349}, Gln_{352}, Gly_{369}, Glu_{375}, Pro_{382}, and Pro_{387}). In addition, a serine residue at the N-terminus of the protein (corresponding to AcMNPV Ser_{31}) is present in many but not all of these enzymes. The absolute conservation of the seven residues in the C-terminal half of the protein suggests they are essential for enzyme activity. There is some evidence for this. Replacement of Gly_{309} of human bilirubin UDP–glucuronosyltransferase (corresponding to AcMNPV Gly_{298}) with glutamic acid abolishes enzyme activity (Erps *et al.*, 1994). Mackenzie (1990) has proposed that the C-terminal half of the mammalian enzymes is involved in UDP–sugar binding, whereas the N-terminal part of the protein determines aglycone binding. It seems likely that the conserved residues in the C-terminal half of the protein are critical for UDP–sugar binding. The nonconserved region in the center of the viral EGTs may be a linker between an N-terminal ecdysteroid-binding domain and the C-terminal UDP–sugar binding domain.

2. EGT Enzyme Activity

The enzyme activity of EGT was first elucidated because of the protein's homology to several mammalian UDP–glucuronosyltransferases (O'Reilly and Miller, 1989). These enzymes catalyze the conjugation of a variety of exogenous and endogenous lipophilic substrates with glucuronic acid (Burchell and Coughtrie, 1989). This led to the speculation that EGT catalyzed a similar conjugation reaction. Assay of various substrates led to the initial discovery that EGT catalyzed the conjugation of ecdysteroids with UDP–glucose (O'Reilly and Miller, 1989). The ecdysone–glucoside formed by the AcMNPV EGT-catalyzed conjugation of ecdysone and UDP–glucose has been purified and its structure determined (O'Reilly *et al.*, 1991). The glucose is linked to the hydroxyl at position C-22 of the ecdysone molecule, yielding the conjugate ecdysone 22-O-β-D-glucopyranoside. In addition to ecdysone, 20-hydroxyecdysone, 26-hydroxyecdysone, and makisterone A can be conjugated with glucose by AcMNPV EGT (O'Reilly *et al.*, 1991). In contrast, any ecdysteroid lacking a hydroxyl group at C-22 is not conjugated. These data strongly suggest that the presence of a hydroxyl moiety at position C-22 of the ecdysteroid is a primary requirement for EGT-catalyzed conjugation. The conjugation of 20-hydroxyecdysone is of note, as this is believed to be the most active form of the hormone *in vivo*. Similar ecdysteroid substrate specificity studies have been carried out for MbMNPV EGT (Clarke *et al.*, 1996) that show that it closely resembles AcMNPV EGT. Such information is not available for other baculovirus EGTs, but it seems likely they are also similar.

A wide range of nonecdysteroids have also been tested as potential substrates for AcMNPV EGT (O'Reilly and Miller, 1989). These include cholesterol (the precursor of ecdysteroids in insects), selected mammalian steroid hormones, and a variety of other typical substrates of the mammalian UDP–glucuronosyltransferases. No conjugation to any of these substrates was observed, supporting the idea that EGT is specific for ecdysteroids.

The specificity of baculovirus EGTs for the UDP–sugar used as donor has also been investigated. In addition to UDP–glucose, UDP–galactose was found to function as a sugar donor for AcMNPV EGT (O'Reilly *et al.*, 1992). However, no other UDP–sugars were utilized. LdMNPV and MbMNPV EGTs can also use both UDP–glucose and UDP–galactose (Clarke *et al.*, 1996; Kelly *et al.*, 1995).

3. EGT Conjugation of Ecdysteroids *in Vivo*

Although the *in vitro* enzyme assays described above indicated that EGT can use either UDP–glucose or UDP–galactose as a sugar donor, there may be some specificity of UDP–sugar utilization *in vivo*. Characterization of the conjugation reaction in AcMNPV-infected *S. frugiperda* hemolymph demonstrated that only ecdysteroid–galactosides are formed (O'Reilly *et al.*, 1992). In contrast, the ecdysteroid conjugates formed in the hemolymph of *L. dispar* infected with LdMNPV are exclusively glucosides (Kelly *et al.*, 1995). Whether these observations are of biological significance remains to be determined. The relative affinity of either enzyme for each UDP–sugar has not been measured, and it is not clear whether these differences reflect a differential affinity for the UDP–sugars on the part of the enzyme or simply reflect which UDP–sugar is present in the hemolymph of the insect host. Furthermore, it is not clear if there are any functional differences between ecdysteroid glucosides and galactosides. Ecdysteroid galactosides are not known from any other insect system.

There seems to be some variation in the effects of baculoviral EGTs on host ecdysteroid levels following infection. Early experiments indicated that ecdysteroid levels never rise above basal levels in final instar *S. frugiperda* larvae infected with AcMNPV (O'Reilly *et al.*, 1992), suggesting that conjugated ecdysteroids are cleared very rapidly from the hemolymph of the infected insect. More recently, however, similar experiments detected high levels of ecdysteroids in the infected insects. These ecdysteroids were all conjugated (O'Reilly *et al.*, 1995). Thus, the rate at which the conjugated ecdysteroids are cleared from the hemolymph may be variable. The factor(s) underlying this variability in the AcMNPV–*S. frugiperda* system are unknown. In *L. dispar* the developmental stage of the insect at the time of infection may be important. Park *et al.* (1993) reported that ecdysteroids are undetectable in the hemolymph of final instar *L. dispar* infected by LdMNPV. They also found that ecdysteroid levels are higher in LdMNPV-infected fourth instar *L. dispar* larvae than in uninfected larvae of the same stage. Again, these ecdysteroids are all conjugated (Park *et al.*, 1993).

4. EGT Effects on Host Development and Survival

There is now substantial evidence that sugar conjugation of ecdysteroids by EGT causes suppression of host molting. This was first demonstrated by generating a mutant AcMNPV lacking a functional *egt* (O'Reilly and Miller, 1989). Comparison of the development of penultimate or final instar *S. frugiperda* larvae infected with wild-type AcMNPV or with the *egt* mutant confirmed that *egt* expression by wild-type AcMNPV arrested the development of the infected insect (O'Reilly and Miller, 1989). Similar results have been obtained in various other systems, including AcMNPV-infected *T. ni* (Eldridge *et al.*, 1992b) and *Heliothis virescens* (D. R. O'Reilly and T. J. Kelly, unpublished data) and LdMNPV-infected *L. dispar* (Burand and Park, 1992). Molting is blocked irrespective of whether the conjugated ecdysteroids are cleared from the hemolymph or whether the ecdysteroids are conjugated with glucose or galactose (O'Reilly *et al.*, 1995; Park *et al.*, 1993).

In the experiments discussed above, the insects were infected with high doses of virus. In recent studies, we have found that the proportion of insects whose development is arrested is directly dependent on the virus dose received (D. R. O'Reilly and T. J. Kelly, unpublished data). Furthermore, the proportion of insects arrested is inversely dependent on the time of infection within the instar, i.e., the later in the instar an insect is infected, the less likely it is to be developmentally arrested. These data show that, not unexpectedly, EGT has to accumulate to some threshold level to arrest development of the infected insect. The lower the amount of the inoculum or the later in the instar it is received, the less likely it is that EGT levels will reach this threshold before the insect is committed to ecdysis.

Analysis of the development of insects infected by *egt*-minus AcMNPV showed that, in effect, the function of *egt* is to lengthen the time after infection the insect feeds. Uninfected insects normally cease feeding during a larval–larval molt and prior to pupation. Whereas insects infected by a virus lacking *egt* also experience this feeding cessation (because their development is not disrupted), those infected by wild-type AcMNPV (encoding an active *egt* gene) do not arrest feeding, and therefore feed for a longer period of time than they otherwise would. Insects infected by *egt*-minus AcMNPV also succumb to the viral infection sooner than those infected by wild-type virus. For both *S. frugiperda* and *T. ni*, the 50% survival time (ST_{50}) for insects infected by *egt*-minus AcMNPV is 20–30% shorter than for insects infected with wild-type virus (Eldridge *et al.*, 1992b; O'Reilly and Miller, 1991). In contrast, the dose–mortality response does not seem to be significantly altered. The mechanistic basis for the accelerated mortality is not clear. Visual observation indicates that many *egt*-minus-infected insects die during or soon after a molt, suggesting that stresses associated with molting may not be well supported by an infected insect. Alternatively, data from Flipsen *et al.* (1995) suggest that precocious degeneration of the Malpighian tubules may accelerate mortality. In *S. exigua* larvae infected by AcMNPV, the Malpighian tubules degenerate more rapidly after infection with an *egt*-minus virus than with wild-type virus. The basis for this effect is not known (but see Section III.C.6).

5. Significance of EGT for Development of Baculovirus Pesticides

Much of the interest in baculoviruses has been due to their potential as biological insect control agents. A major goal of recent research has been to genetically engineer baculoviruses that can cause infected insects to cease feeding more rapidly after infection (see Chapter 13, this volume). An implication of the data discussed above is that deletion of *egt* actually improves the pesticidal properties of the virus. Insects infected by an *egt*-minus virus display reduced feeding and earlier death than those infected by wild-type virus. Final instar *S. frugiperda* infected with *egt*-minus AcMNPV consume approximately 30–40% less diet than those infected with the wild-type parent virus (O'Reilly and Miller, 1991). Thus, deletion of the *egt* gene is likely to be a common first step in the generation of novel recombinant baculovirus pesticides (see Chapter 13, this volume). The prospect of generating improved baculovirus pesticides accounts for the interest that has been shown in identifying the *egt* genes of other baculoviruses (see Section III.C.1).

6. Biological Role of EGT

The critical question to address concerning the role of *egt* in the baculovirus life cycle is what benefit the virus acquires by expressing this gene. Is it advantageous to arrest development of the host insect? Does *egt* enhance viral fitness in any other way? We have shown that one benefit of *egt* expression is that the yield of progeny OV is increased. Final instar *S. frugiperda* infected by *egt*-minus AcMNPV yielded approximately 23% fewer OV than insects infected by wild-type virus (O'Reilly and Miller, 1991). This increased yield from wild-type-infected insects was correlated with their increased size and feeding due to the inhibition of their development to a pharate pupal stage. These initial data were derived from precisely staged insects infected with large doses of virus. More recently, we have found that an *egt*-mediated increase in progeny OV yield is also observed in insects infected in earlier instars (E. E. Clarke *et al.*, unpublished data). These data indicate that a function of *egt* is to increase the yield of progeny virus per insect by arresting host development, and thereby enhance the spread of the virus through host insect populations. As noted above, whether developmental arrest is observed following infection depends both on virus dose and on the precise time of infection within the instar. We therefore expect that the size of the yield increase will be highly variable from insect to insect, depending on the dose received and the developmental stage at infection. On the other hand, *egt* expression delays insect death, and therefore delays the release of progeny virus. This is likely to be disadvantageous to the virus in terms of spread through a host population. Thus, there will be some trade-off between the increased yield and the delayed release of progeny virus due to *egt*. We are currently investigating the parameters governing this trade-off to obtain an estimate of the actual increase in fitness gained by the virus through expression of *egt*.

Currently, we cannot exclude the possibility that *egt* expression serves a function(s) other than directly increasing the yield of progeny virus. One hypothesis, which remains untested, is that a function of *egt* is to change the behavior of the infected insect. Many lepidopterans leave the host plant and burrow into the soil prior to pupation. Such behavior would likely be detrimental to the spread of the virus through an insect population. Thus, an additional function of *egt* may be to keep final instar insects (which yield the greatest number of progeny OV) on the food plant.

Both of the above possibilities depend on the ability of *egt* to block the development of the infected insect. However, accumulating evidence suggests this need not necessarily be the case. In the experiments of Flipsen *et al.* (1995) examining the fate of the Malpighian tubules, essentially all the infected larvae molted into the next instar, irrespective of whether they were infected by wild-type or *egt*-minus AcMNPV. Nonetheless, precocious degeneration of the Malpighian tubules were only observed in insects infected by the *egt*-minus virus. These insects also succumbed to the infection approximately 30 hr earlier than those infected by wild-type virus. Thus, *egt* seems to impact on the course of the infection even if molting is not prevented. The fact that *egt*-specific effects are observed even when molting is not blocked (and therefore significant levels of ecdysteroids are present in the infected insect) would appear to suggest that the role of this enzyme in the viral life cycle does not depend on the complete inactivation of the insects ecdysteroids. Nevertheless, there is reason to believe that the virus-induced degeneration of *S. exigua* Malpighian tubules is ecdysteroid mediated. In particular, ecdysteroids have been shown to induce Malpighian tubule remodeling during pupation (but not during other molts) in certain species (Ryerse, 1980). The fact that EGT can block virus-induced degeneration of this tissue even though it has not accumulated to levels sufficient to block molting may indicate that rather high levels of ecdysteroids are required to induce tubule degeneration. Alternatively, it may be that ecdysteroid glycosides have some direct effect on Malpighian tubules. Clearly, further research is needed to elucidate the relationship between EGT and Malpighian tubule survival.

There is other evidence that suggests that EGT-mediated reduction of active ecdysteroid titers is of direct importance to viral replication in the insect. We have constructed a series of viruses that overproduce insect prothoracicotropoic hormone (PTTH) (O'Reilly *et al.*, 1995). PTTH stimulates the prothoracic glands to produce ecdysteroids, and insects infected by these viruses displayed elevated ecdysteroid titers at inappropriate developmental times. The infectivity of the *egt*-minus virus overexpressing PTTH was approximately 100-fold reduced compared to either *egt*-minus AcMNPV (not expressing PTTH) or *egt*-plus AcMNPV overexpressing PTTH. This was not related to whether molting was arrested, since neither *egt*-minus virus blocked host molting. Although elevated levels of ecdysteroids were observed in insects infected by both viruses overexpressing PTTH (*egt*-plus and -minus), they were all conjugated in insects infected by the *egt*-plus virus (O'Reilly *et al.*, 1995). The most likely hypothesis to explain these observations is that the elevated ecdysteroid titers are somehow detrimental to virus replication. This effect would not be seen in insects infected with the *egt*-plus virus overexpressing PTTH because the excess ecdysteroids are all inactivated.

Other reports in the literature support the idea that ecdysteroids have an adverse effect on baculovirus replication *in vivo.* Keeley and Vinson (1975) reported that injection of 20-hydroxyecdysone into baculovirus-infected *H. virescens* resulted in a delay in the onset of virus-induced pathology as well as a decrease in the mortality of the infected insects. More indirectly, several workers have suggested that baculovirus replication is disrupted during metamorphosis (Murray *et al.*, 1991, and references therein). Ecdysteroids are known to induce a broad spectrum of responses in different tissues (Riddiford, 1985), including the effects on Malpighian tubules mentioned above. These effects are mediated via the control of transcription, and the expression of numerous host genes can be regulated by ecdysteroids. In addition, depending on the tissue and developmental stage, exposure to ecdysteroids may result in either an induction or suppression of cell division. At certain developmental stages, an increase followed by a decrease in ecdysteroid titers is the signal for the programmed death of selected cells (Robinow *et al.*, 1993; Schwartz and Truman, 1983). Many of these responses could be detrimental to viral replication in the insect. It seems possible that a significant function of *egt* expression is to counteract adverse effects of ecdysteroids on viral replication at the cellular level.

While there are some unanswered questions concerning the precise functions of *egt*, it is nonetheless possible to provide a reasonably complete and coherent account of the role of this gene in the baculovirus life cycle. Following infection of an insect larva, the gene is expressed rapidly in each infected cell. The protein is secreted into the insect's hemocoel. As infection spreads through the insect and more cells become infected, EGT continues to accumulate in the hemolymph. Here it catalyzes the conjugation of circulating ecdysteroids with glucose or galactose, thereby inactivating them. If the insect received a sufficient dose of virus early enough in the instar, EGT activity in the hemolymph will rise to the point whereby host ecdysteroids are largely inactivated and active ecdysteroid titers are prevented from reaching levels sufficient to commit the insect to ecdysis. The host becomes developmentally arrested at this point but continues to feed, leading to an increase in the production of progeny virus.

IV. CONCLUDING REMARKS

A sobering feature of this discussion of baculovirus auxiliary genes has been our lack of understanding of the functions of most of them. A clear picture has only emerged for three of these genes, *chiA*, *cath*, and *egt*. In many cases, inactivation of the gene results in no detectable phenotypic alteration, and the only clue to function derives from sequence homology to other proteins in the database. It is worth recalling that there are other auxiliary genes, not discussed in this chapter, for which we do not have even this clue to function. It is reasonable to assume that all these genes have some significant role in the viral life cycle. Experience with other viruses has shown that they will not expend the cost of replicating and expressing a gene unless there is some compensating benefit to the virus in terms of increased fitness. The reason why functions for these genes have not been identified yet most likely reflects the difficulty in devising assays that adequately test their possible roles in the virus life cycle. The assays used to date are likely to be inadequate in a number of ways. Most obviously, they generally only test for effects in one or a small number of host species. There is considerable evidence of particular genes functioning in a host-specific manner. The *ptp* gene addressed above, in which a phenotype was only observed in *S. frugiperda* (SF) cells, is one example. More dramatic examples addressed elsewhere in this volume include the *p35* gene, which is not required for replication in *T. ni* cells, and *hcf-1*, which conversely is not required in SF cells (see Chapter 9, this volume). In this regard, it is perhaps unfortunate that much of the basic baculovirus research carried out to date has been on AcMNPV, one of the broadest host range baculoviruses. Perhaps studies of narrower host range viruses will prove to be more informative.

A second possible difficulty in establishing the biological function of a particular gene is the sensitivity of the assay. A newly acquired gene will be fixed in the viral population if the increase in fitness it confers outweighs the costs of its replication and expression. It is possible that an evolutionarily significant increase in fitness might nonetheless be too small to be detected reliably with current laboratory assays.

Finally, one must also consider precisely what parameters are generally assayed. What contributes to viral fitness? In addition to efficient production of progeny virus in the cell and in the insect, the virus must be able to disseminate effectively through host populations and survive, perhaps for long periods, in the absence of high densities of its host. Parameters that might be important in this regard include host behavior, OV stability, ability of the virus to survive as a latent infection, and other factors that might affect vertical transmission. Such aspects of viral fitness are rarely assayed.

Thus, gaining a complete understanding of how many of these auxiliary genes contribute to viral fitness may require a substantial amount of work. However, the functions already elucidated for *chiA*, *cath*, and *egt* suggest that it will be worth the effort. One cannot but be impressed at the sophistication with which the virus manipulates and controls its environment. We can be assured of much fascinating biology yet to come as the functions of other auxiliary genes are finally unraveled.

ACKNOWLEDGMENTS. The author would like to thank S. Maeda, R. Possee, G. Rohrmann, J. Vlak, and R. Weaver for kindly providing data before publication, and L. Guarino, A. Lu, L. Miller, R. Possee, G. Rohrmann, and R. Weaver for comments on the manuscript.

REFERENCES

Adang, M. J., and Miller, L. K., 1982, Molecular cloning of DNA complementary to messenger RNA of the baculovirus *Autographa californica* nuclear polyhedrosis virus location and gene products of RNA transcripts found late in infection, *J. Virol.* **44:**782.

Ahrens, C. H., Russell, R. L. Q., Funk, C. J., Evans, J. T., Harwood, S. H., and Rohrmann, G. F., 1997, The sequence of the *Orgyia pseudotsugata* multicapsid nuclear polyhedrosis virus genome, *Virology* **229:**381.

Alaoui-Ismaili, M. H., and Richardson, C. D., 1996, Identification and characterization of a filament-associated protein encoded by *Amsacta moorei* entomopoxvirus, *J. Virol.* **70:**2697.

Ayres, M. D., Howard, S. C., Kuzio, J., Lopez-Ferber, M., and Possee, R. D., 1994, The complete DNA sequence of *Autographa californica* nuclear polyhedrosis virus, *Virology* **202:**586.

Bambara, R. A., and Jessee, C. B., 1991, Properties of DNA polymerases δ and ϵ, and their roles in eukaryotic DNA replication, *Biochim. Biophys. Acta* **1088:**11.

Bischoff, D. S., and Slavicek, J. M., 1994, Identification and characterization of a protein kinase gene in the *Lymantria dispar* multinucleocapsid nuclear polyhedrosis virus, *J. Virol.* **68:**1728.

Brömme, D., and Okamoto, K., 1995, The baculovirus cysteine-protease has a cathepsin B-like S2-subsite specifity, *Biol. Chem. Hoppe-Seyler* **376:**611.

Broussard, D. R., Guarino, L. A., and Jarvis, D. L., 1996, Dynamic phosphorylation of *Autographa californica* nuclear polyhedrosis virus pp31, *J. Virol.* **70:**6767.

Brown, M., and Faulkner, P., 1975, Factors affecting the yield of virus in a cloned cell line of *Trichoplusia ni* infected with a nuclear polyhedrosis virus, *J. Invertebr. Pathol.* **26:**251.

Burand, J., and Park, E., 1992, Effect of nuclear polyhedrosis virus infection on the development and pupation of gypsy moth larvae, *J. Invertebr. Pathol.* **60:**171.

Burchell, B., and Coughtrie, M., 1989, UDP-glucuronosyltransferases, *Pharmacol. Ther.* **43:**261.

Charbonneau, H., and Tonks, N. K., 1992, 1002 protein phosphatases? *Annu. Rev. Cell Biol.* **8:**463.

Cheley, S., Kosik, K. S., Paskevich, P., Bakalis, S., and Bayley, H., 1992, Phosphorylated baculovirus p10 is a heat-stable microtubule-associated protein associated with process formation in Sf9 cells, *J. Cell Sci.* **102:**739.

Choi, J., and Guarino, L. A., 1995, Expression of the IE1 transactivator of *Autographa californica* nuclear polyhedrosis virus during viral infection, *Virology* **209:**99.

Clarke, E. E., Tristem, M., Cory, J. S., and O'Reilly, D. R., 1996, Characterization of the ecdysteroid UDP-glucosyltransferase gene from *Mamestra brassicae* nucleopolyhedrosis virus, *J. Gen. Virol.* **77:**2865.

Crawford, A. M., and Miller, L. K., 1988, Characterization of an early gene accelerating expression of late genes of the baculovirus *Autographa californica* nuclear polyhedrosis virus, *J. Virol.* **62:**2773.

Dixon, L. K., Twigg, S. R. F., Baylis, S. A., Vydelingum, S., Bristow, C., Hammond, J. M., and Smith, G. L., 1994, Nucleotide sequence of a 55 kbp region from the right end of the genome of a pathogenic African swine fever virus isolate (Malawi LIL20/1), *J. Gen. Virol.* **75:**1655.

Dunigan, D. D., Dietzgen, R. G., Schoelz, J. E., and Zaitlin, M., 1988, Tobacco mosaic virus particles contain ubiquitinated coat protein subunits, *Virology* **165:**310.

Eldridge, R., Li, Y., and Miller, L. K., 1992a, Characterization of a baculovirus gene encoding a small conotoxin-like polypeptide, *J. Virol.* **66:**6563.

Eldridge, R., O'Reilly, D. R., Hammock, B. D., and Miller, L. K., 1992b, Insecticidal properties of genetically engineered baculoviruses expressing an insect juvenile hormone esterase gene, *Appl. Environ. Microbiol.* **58:**1583.

Erps, L. T., Ritter, J. K., Hersh, J. H., Blossom, D., Martin, N. C., and Owens, I. S., 1994, Identification of two single base substitutions in the UGT1 gene locus which abolish bilirubin uridine diphosphate glucuronosyltransferase activity *in vitro, J. Clin. Invest.* **93:**564.

Fan, X., Thirunavukkarasu, K., and Weaver, R. F., 1996, Temperature-sensitive mutations in the protein kinase-1 (*pk-1*) gene of the *Autographa californica* nuclear polyhedrosis virus that block very late gene expression, *Virology* **224:**1.

Flipsen, J. T. M., Mans, R. M. W., Kleefsman, A. W. F., Knebel-Mörsdorf, D., and Vlak, J. M., 1995, Deletion of the baculovirus ecdysteroid UDP–glucosyltransferase gene induces early degeneration of Malpighian tubules in infected insects, *J. Virol.* **69:**4529.

Fraser, M. J., Cary, L., Boonvisudhi, K., and Wang, H.-G. H., 1995, Assay for movement of lepidopteran transposon IFP2 in insect cells using a baculovirus genome as a target DNA, *Virology* **211:**397.

Funk, C. J., and Consigli, R. A., 1993, Phosphate cycling on the basic protein of *Plodia interpunctella* granulosis virus, *Virology* **193:**396.

Gonnet, P., and Devauchelle, G., 1987, Obtention par recombinaison dans le gène du polypeptide p10 d'un baculovirus exprimant le gène de résistance à la néomycine dans les cellules d'insecte, *C. R. Acad. Sci. Paris* **305** (Serie III):111.

Guan, K., Broyles, S. S., and Dixon, J. E., 1991, A Tyr/Ser protein phosphatase encoded by vaccinia virus, *Nature* **350:**359.

Guarino, L. A., 1990, Identification of a viral gene encoding a ubiquitin-like protein, *Proc. Natl. Acad. Sci. USA* **87:**409.

Guarino, L. A., Dong, W., Xu, B., Broussard, D., Davis, R. W., and Jarvis, D. L., 1992, Baculovirus phosphoprotein pp31 is associated with virogenic stroma, *J. Virol.* **66:**7113.

Guarino, L. A., Smith, G., and Dong, W., 1995, Ubiquitin is attached to membranes of baculovirus particles by a novel type of phospholipid anchor, *Cell* **80:**301.

Haas, A. L., Katzung, D. J., Reback, P. M., and Guarino, L. M., 1996, Functional characterization of the ubiquitin variant encoded by the baculovirus *Autographa californica, Biochemistry* **35:**5385.

Hakes, D. J., Martell, K. J., Zhao, W.-G., Massung, R. F., Esposito, J. J., and Dixon, J. E., 1993, A protein phosphatase related to the vaccinia virus VH1 is encoded in the genomes of several orthopoxviruses and a baculovirus, *Proc. Natl. Acad. Sci. USA* **90:**4017.

Hanks, S. K., Quinn, A. M., and Hunter, T., 1988, The protein kinase family: Conserved features and deduced phylogeny of the catalytic domains, *Science* **241:**42.

Hawtin, R. E., Arnold, K., Ayres, M. D., Zanotto, P. M. D. A., Howard, S. C., Gooday, G. W., Chappell, L. H., Kitts, P. A., King, L. A., and Possee, R. D., 1995, Identification and preliminary characterization of a chitinase gene in the *Autographa californica* nuclear polyhedrosis virus genome, *Virology* **212:**673.

Herendeen, D., Kassavetis, G., Barry, J., Alberts, B., and Geiduschek, E., 1989, Enhancement of bacteriophage T4 late transcription by components of the T4 DNA replication apparatus, *Science* **245:**952.

Hill, J. E., Kuzio, J., and Faulkner, P., 1995, Identification and characterization of the v-cath gene of the baculovirus CfMNPV, *Biochim. Biophys. Acta* **1264:**275.

Hingamp, P. M., Leyland, M. L., Webb, J., Twigger, S., Mayer, R. J., and Dixon, L. K., 1995, Characterization of a ubiquitinated protein which is externally located in African swine fever virions, *J. Virol.* **69:**1785.

Hochstrasser, M., 1996, Protein degradation or regulation: Ub the judge, *Cell* **84:**813.

Ignoffo, C. M., and Garcia, C., 1994, Antioxidant and oxidant enzyme effects of the inactivation of inclusion bodies of the *Heliothis* baculovirus by simulated sunlight UV, *Environ. Entomol.* **23:**1025.

Johnson, G. P., Goebel, S. J., and Paoletti, E., 1993, An update on the vaccinia virus genome, *Virology* **196:**381.

Keeley, L. L., and Vinson, S. B., 1975, Beta-ecdysone effects on the development of nucleopolyhedrosis in *Heliothis* spp, *J. Invertebr. Pathol.* **26:**121.

Kelly, D. C., and Lescott, T., 1984, Baculovirus replication: Phosphorylation of polypeptides synthesised in *Trichoplusia ni* nuclear polyhedrosis virus-infected cells, *J. Gen. Virol.* **65:**1183.

Kelly, T. J., Park, E. J., Masler, C. A., and Burand, J. P., 1995, Characterization of the glycosylated ecdysteroids in the hemolymph of baculovirus-infected gypsy moth larvae and cells in culture, *Eur. J. Entomol.* **92:**51.

Keyse, S. M., 1995, An emerging family of dual specificity MAP kinase phosphatases, *Biochim. Biophys. Acta* **1265:**152.

Kim, D., and Weaver, R. F., 1993, Transcription mapping and functional analysis of the protein tyrosine/serine phosphatase (PTPase) gene of the *Autographa californica* nuclear polyhedrosis virus, *Virology* **195:**587.

Kool, M., Ahrens, C. H., Goldbach, R. W., Rohrmann, G. F., and Vlak, J. M., 1994, Identification of genes involves in DNA replication of *Autographa californica* baculovirus, *Proc. Natl. Acad. Sci. USA* **91:**11212.

Kumar, S., and Miller, L. K., 1987, Effects of serial passage of *Autographa californica* nuclear polyhedrosis virus in cell culture, *Virus Res.* **7:**335.

Kuzio, J., Rohel, D. Z., Curry, C. J., Krebs, A., Carstens, E. G., and Faulkner, P., 1984, Nucleotide sequence of the p10 polypeptide gene of *Autographa californica* nuclear polyhedrosis virus, *Virology* **139:**414.

Lanier, L. M., Slack, J. M., and Volkman, L. E., 1996, Actin binding and proteolysis by the baculovirus AcMNPV: The role of virion-associated V-CATH, *Virology* **216:**380.

Leisy, D. J., Rohrmann, G. F., Nesson, M., and Beaudreau, G. S., 1986, Nucleotide sequencing and transcriptional mapping of the *Orgyia pseudotsugata* multicapsid nuclear polyhedrosis virus p10 gene, *Virology* **153:**157.

Li, Y., and Miller, L. K., 1995a, Expression and functional analysis of a baculovirus gene encoding a truncated protein kinase homolog, *Virology* **206:**314.

Li, Y., and Miller, L. K., 1995b, Expression and localization of a baculovirus protein phosphatase, *J. Gen. Virol.* **76:**2941.

Li, Y., and Miller, L. K., 1995c, Properties of a baculovirus mutant defective in the protein phosphatase gene, *J. Virol.* **69:**4533.

Li, Y., Lu, Z., Burbank, D. E., Kutish, G. F., Rock, D. L., and Van Etten, J. L., 1995, Analysis of 43 kb of the *Chlorella* virus PBCV-1 330-kb genome: Map positions 45 to 88, *Virology* **212:**134.

Liu, K., Lemon, B., and Traktman, P., 1995, The dual-specificity phosphatase encoded by vaccinia virus, VH1, is essential for viral transcription *in vivo* and *in vitro*, *J. Virol.* **69:**7823.

Lu, A., and Miller, L. K., 1995, The roles of eighteen baculovirus late expression factor genes in transcription and DNA replication, *J. Virol.* **69:**975.

Lu, Z., Li, Y., Que, Q., Kutish, G., Rock, D. L., and Van Etten, J. L., 1996, Analysis of 94kb of the *Chlorella* virus PBCV-1 330-kb genome: Map positions 88 to 182, *Virology* **216:**102.

Mackenzie, P., 1990, Expression of chimeric cDNAs in cell culture defines a region of UDP glucuronosyltransferase involved in substrate selection, *J. Biol. Chem.* **265:**3432.

Maeda, S., 1996, Genbank entry, Accession number L33180.

Mainprize, T. H., Lee, K., and Miller, L. K., 1986, Variation in the temporal expression of overlapping baculovirus transcripts, *Virus Res.* **6:**85.

Maruniak, J. E., and Summers, M. D., 1981, *Autographa californica* nuclear polyhedrosis virus phosphoproteins and synthesis of intra cellular proteins after virus infection, *Virology* **109:**25.

Meyers, G., Tautz, N., Dubovi, E. J., and Thiel, H. J., 1991, Viral cytopathogenicity correlated with integration of ubiquitin-coding sequences, *Virology* **180:**602.

Miller, L. K., Adang, M. J., and Browne, D., 1983, Protein kinase activity associated with the extracellular and occluded forms of the baculovirus *Autographa californica* nuclear polyhedrosis virus, *J. Virol.* **46:**275.

Morris, T. D., Todd, J. W., Fisher, B., and Miller, L. K., 1994, Identification of *lef-7:* A baculovirus gene affecting late gene expression, *Virology* **200:**360.

Mossman, K., Ostergaard, H., Upton, C., and McFadden, G., 1995, Myxoma virus and Shope fibroma virus encode dual-specificity tyrosine/serine phosphatases which are essential for virus viability, *Virology* **206:**572.

Murray, K. D., Shields, K. S., Burand, J. P., and Elkinton, J. S., 1991, The effect of gypsy moth metamorphosis on the development of nuclear polyhedrosis virus infection, *J. Invertbr. Pathol.* **57:**352.

Nichols, A. F., and Sancar, A., 1992, Purification of PCNA as a nucleotide excision repair protein, *Nucleic Acids Res.* **20:**2441.

Ohkawa, T., Majima, K., and Maeda, S., 1994, A cysteine protease encoded by the baculovirus *Bombyx mori* nuclear polyhedrosis virus, *J. Virol.* **68:**6619.

Olivera, B. M., McIntosh, J. M., Cruz, L. J., Luque, F. A., and Gray, W. R., 1984, Purification and sequence of a presynaptic peptide toxin from *Conus geographus* venom, *Biochemistry* **23:**5087.

Olivera, B. M., Rivier, J., Scott, J. K., Hillyard, D. R., and Cruz, L. J., 1991, Conotoxins, *J. Biol. Chem.* **266:**22067.

O'Reilly, D. R., 1995, Baculovirus-encoded ecdysteroid UDP–glucosyltransferases, *Insect Biochem. Mol. Biol.* **25:**541.

O'Reilly, D. R., and Miller, L. K., 1989, A baculovirus blocks insect molting by producing ecdysteroid UDP–glucosyltransferase, *Science* **245:**1110.

O'Reilly, D. R., and Miller, L. K., 1990, Regulation of expression of a baculovirus ecdysteroid UDP–glucosyltransferase gene, *J. Virol.* **64:**1321.

O'Reilly, D. R., and Miller, L. K., 1991, Improvement of a baculovirus pesticide by deletion of the *egt* gene, *Bio/Technology* **9:**1086.

O'Reilly, D. R., Crawford, A. M., and Miller, L. K., 1989, Viral proliferating cell nuclear antigen, *Nature* **337:**606.

O'Reilly, D. R., Passarelli, A. L., Goldman, I. F., and Miller, L. K., 1990, Characterization of the DA26 gene in a hypervariable region of the *Autographa californica* nuclear polyhedrosis virus genome, *J. Gen. Virol.* **7:**1029.

O'Reilly, D. R., Howarth, O. W., Rees, H. H., and Miller, L. K., 1991, Structure of the ecdysone glucoside formed by a baculovirus ecdysteroid UDP–glucosyltransferase, *Insect Biochem.* **21:**795.

O'Reilly, D. R., Brown, M. R., and Miller, L. K., 1992, Alteration of ecdysteroid metabolism due to baculovirus infection of the fall armyworm *Spodoptera frugiperda:* Host ecdysteroids are conjugated with galactose, *Insect Biochem. Mol. Biol.* **22:**313.

O'Reilly, D. R., Kelly, T. J., Masler, E. P., Thyagaraja, B. S., Robson, R. M., Shaw, T. C., and Miller, L. K., 1995, Overexpression of *Bombyx mori* prothoracicotropic hormone using baculovirus vectors, *Insect Biochem. Mol. Biol.* **25:**475.

Pardini, R. S., 1995, Toxicity of oxygen from naturally occurring redox-active pro-oxidants, *Arch. Insect Biochem. Physiol.* **29:**101.

Park, E. J., Burand, J. P., and Yin, C.-M., 1993, The effect of baculovirus infection on ecdysteroid titer in gypsy moth larvae, *J. Insect Physiol.* **39:**791.

Possee, R., Sun. T.-P., Howard, S., Ayres, M., Hill-Perkins, M., and Gearing, K., 1991, Nucleotide sequence of the *Autographa californica* nuclear polyhedrosis 9.4 kbp EcoRI-I and -R (polyhedrin gene) region, *Virology* **185:**229.

Prelich, G., Tan, C. K., Kostura, M., Mathews, M. B., So, A. G., Downey, K. M., and Stillman, B., 1987, Functional identity of proliferating cell nuclear antigen and a DNA polymerase-δ auxiliary protein, *Nature* **326:**517.

Putterman, D., Pepinsky, R. B., and Vogt, V. M., 1990, Ubiquitin in avian leukosis virus particles, *Virology* **176:**633.

Quant-Russell, R. L., Pearson, M. N., Rohrmann, G. F., and Beaudreau, G. S., 1987, Characterization of baculovirus p10 synthesis using monoclonal antibodies, *Virology* **160:**9.

Rankin, C., Ladin, B. F., and Weaver, R. F., 1986, Physical mapping of temporally regulated, overlapping transcripts in the region of the 10K protein gene in *Autographa californica* nuclear polyhedrosis virus, *J. Virol.* **57:**18.

Rawlings, N. D., Pearl, L. H., and Buttle, D. J., 1992, The baculovirus *Autographa californica* nuclear polyhedrosis virus genome includes a papain-like sequence, *Biol. Chem. Hoppe-Seyler* **373:**1211.

Reilly, L. M., and Guarino, L. A., 1994, The pk-1 gene of *Autographa californica* nuclear polyhedrosis virus encodes a protein kinase, *J. Gen. Virol.* **75:**2999.

Reilly, L. M., and Guarino, L. A., 1996, The viral ubiquitin gene of *Autographa californica* nuclear polyhedrosis virus is not essential for viral replication, *Virology* **218:**243.

Riddiford, L. M., 1985, Hormone action at the cellular level, in: *Comprehensive Insect Physiology, Biochemistry and Pharmacology* (G. A. Kerkut and L. I. Gilbert, eds.), Vol. 8, pp. 37–48, Pergamon Press, Oxford, England.

Robinow, S., Talbot, W. S., Hogness, D. S., and Truman, J. W., 1993, Programmed cell death in the *Drosophila* CNS is ecdysone-regulated and coupled with a specific ecdysone receptor isoform, *Development* **119:**1251.

Roelvink, P., Van Meer, M., De Kort, C., Possee, R., Hammock, B., and Vlak, J., 1992, Dissimilar expression of *Autographa californica* multiple nucleocapsid nuclear polyhedrosis virus polyhedrin and p10 genes, *J. Gen Virol.* **73:**1481.

Russell, R. L. Q., and Rohrmann, G. F., 1993, Nucleotide sequence of the ubiquitin-39K gene region from the *Orgyia pseudotsugata* multinucleocapsid nuclear polyhedrosis virus genome, *J. Gen. Virol.* **74:**1191.

Russell, R. L. Q., Pearson, M. N., and Rohrmann, G. F., 1991, Immunoelectron microscopic examination of *Orgyia pseudotsugata* multicapsid nuclear polyhedrosis virus-infected *Lymantria dispar* cells: Time course and localization of major polyhedron-associated proteins, *J. Gen. Virol.* **72:**275.

Ryerse, J. .S., 1980, The control of Malpighian tubule developmental physiology by 20-hydroxyecdysone and juvenile hormone, *J. Insect Physiol.* **26:**449.

Samuels, R. I., and Paterson, I. C., 1995, Cuticle degrading proteases from insect moulting fluid and culture filtrates of entomopathogenic fungi, *Comp. Biochem. Physiol.* **110B:**661.

Sauerbier, W., and Hercules, W., 1993, Gene and transcription unit mapping by radiation effects, *Annu. Rev. Genet.* **12:**329.

Schwartz, L. M., and Truman, J. W., 1983, Hormonal control of rates of metamorphic development in the tobacco hornworm *Manduca sexta*, *Dev. Biol.* **99:**103.

Sheng, Z., and Charbonneau, H., 1993, The baculovirus *Autographa californica* encodes a protein tyrosine phosphatase, *J. Biol. Chem.* **268:**4728.

Shivji, M., Kenny, M., and Wood, R., 1992, Proliferating cell nuclear antigen is required for DNA excision repair, *Cell* **69:**367.

Shivji, M. K. K., Grey, S. J., Strausfeld, U. P., Wood, R. D., and Blow, J. J., 1994, Cip1 inhibits DNA replication but not PCNA-dependent nucleotide excision-repair, *Curr. Biol.* **4:**1062.

Skinner, W. S., Dennis, P. A., Li, J. P., and Quistad, G. B., 1992, Identification of insecticidal peptides from venom of the trap-door spider, *Aptostichus ichlingeri (Ctenizidae)*, *Toxicon* **30:**1043.

Slack, J. M., Kuzio, J., and Faulkner, P., 1995, Characterization of *v-cath*, a cathepsin L-like proteinase expressed by the baculovirus *Autographa californica* multiple nuclear polyhedrosis virus, *J. Gen. Virol.* **76:**1091.

Smith, G. E., Vlak, J. M., and Summers, M. D., 1982, *In vitro* translation of *Autographa californica* nuclear polyhedrosis virus early and late mRNAs, *J. Virol.* **44:**189.

Tomalski, M. D., Eldridge, R., and Miller, L. K., 1991, A baculovirus homolog of a Cu/Zn superoxide dismutase gene, *Virology* **184:**149.

Tribe, D., Bulach, D. M., Goodge, K., Robertson, A. P. S., Wu, T., Lee, H., McAdams, A., and Cowan, P. J., 1994, Genbank entry, Accession number L36721.

van der Wilk, F., van Lent, J. W. M., and Vlak, J. M., 1987, Immunogold detection of polyhedrin, p10 and virion antigens in *Autographa californica* nuclear polyhedrosis virus-infected *Spodoptera frugiperda* cells, *J. Gen. Virol.* **68:**2615.

van Oers, M. M., Flipsen, J. T. M., Reusken, C. B. E. M., Sliwinsky, E. L., Goldbach, R. W., and Vlak, J. M., 1993, Functional domains of the p10 protein of *Autographa californica* nuclear polyhedrosis virus, *J. Gen. Virol.* **74:**563.

van Oers, M. M., and Vlak, J. M., The baculovirus 10 kilodalton protein, *J. Invertebr. Pathol.*, in press.

van Strien, E. A., Jansen, B. J. H., Mans, R. M. W., Zuidema, D., and Vlak, J. M., 1996 Sequence and transcriptional analysis of the ubiquitin gene cluster in the genome of *Spodoptera exigua* nucleopolyhedrovirus, *J. Gen. Virol.* **77:**2311.

Vialard, J. E., and Richardson, C. D., 1993, The 1,629-nucleotide open reading frame located down stream of the *Autographa californica* nuclear polyhedrosis virus polyhedrin gene encodes a nucleocapsid-associated phosphoprotein, *J. Virol.* **67:**5859.

Vlak, J. M., Klinkenberg, F. A., Zaal, K. J. M., Usmany, M., Klinge-Roode, E. C., Geervliet, J. B. F., Roosien, J., and Van Lent, J. W. M., 1988, Functional studies on the p10 gene of *Autographa californica* nuclear polyhedrosis virus using a recombinant expressing a p10-β-galactosidase fusion gene, *J. Gen. Virol.* **69:**765.

Waseem, N. H., Labib, L., Nurse, P., and Lane, D. P., 1992, Isolation and analysis of the fission yeast gene encoding polymerase δ accessory protein PCNA, *EMBO J.* **11:**5111.

Whitt, M. A., and Manning, J. S., 1988, A phosphorylated 34-kDa protein and a subpopulation of polyhedrin are thiol linked to the carbohydrate layer surrounding a baculovirus occlusion body, *Virology* **163:**33.

Williams, G. V., Rohel, D. Z., Kuzio, J., and Faulkner, P., 1989, A cytopathological investigation of *Autographa californica* nuclear polyhedrosis virus p10 gene function using insertion-deletion mutants, *J. Gen. Virol.* **70:**187.

Wilson, J. A., Hill, J. E., Kuzio, J., and Faulkner, P., 1995, Characterization of the baculovirus *Choristoneura fumiferana* multicapsid nuclear polyhedrosis virus p10 gene indicates that the polypeptide contains a coiled–coil domain, *J. Gen. Virol.* **76:**2923.

Wilson, M. E., and Consigli, R. A., 1985, Functions of a protein kinase activity associated with purified capsids of the granulosis virus infecting *Plodia interpunctella*, *Virology* **143:**526.

Xiong, Y., Zhang, H., and Beach, D., 1992, D type cyclins associate with multiple protein kinases and the DNA replication and repair factor PCNA, *Cell* **71:**505.

Zuidema, D., van Oers, M. M., van Strien, E. A., Caballero, P. C., Klok, E.-J., Goldbach, R. W., and Vlak, J. M., 1993, Nucleotide sequence and transcriptional analysis of the p10 gene of *Spodoptera exigua* nuclear polyhedrosis virus, *J. Gen. Virol.* **74:**1017.

CHAPTER 12

Baculovirus Ecology

JENNY S. CORY, ROSEMARY S. HAILS, AND STEVEN M. SAIT

I. INTRODUCTION

While baculoviruses continue to be the most-studied group of insect viruses, we still know surprisingly little about their ecology. As research into their molecular characterization has continued apace, less interest, or perhaps fewer resources, have been devoted to the study of their behavior in nature, despite this being the key to our understanding of the effective use of baculoviruses as biological control agents and the assessment of any perceived risks attached to their release into the environment. Part of this problem may have originated from the difficulty in identifying these organisms, particularly since baculoviruses appear to be notoriously scarce in many natural populations. Since the last major reviews of this area (e.g., Evans, 1986; Entwistle and Evans, 1985), there have been significant advances in two key areas, which has meant that (baculo)virus ecology has moved into a more interesting and challenging phase. The first of these advances has been the expanding use of sophisticated molecular techniques. This has allowed the identification of individual isolates and the monitoring of the structure and changes in viral populations. In addition, DNA sequencing information on the baculovirus genome is beginning to highlight the variety of mechanisms employed by a baculovirus to manipulate the host for its own survival.

JENNY S. CORY, ROSEMARY S. HAILS, AND STEVEN M. SAIT • Ecology and Biocontrol Group, NERC Institute of Virology and Environmental Microbiology, Oxford OX1 3SR, England.

The Baculoviruses, edited by Lois K. Miller. Plenum Press, New York, 1997.

The second major advance has been in the development of mathematical models and the construction of a conceptual framework in which to place baculovirus ecology. While the theoretical basis for disease interaction in vertebrates has received considerable attention, the role of disease in invertebrate population dynamics has been less studied. Anderson and May (1981) published a seminal work on models relevant to invertebrate–microparasite interactions that has since formed a basis for many theoretical developments (see Briggs *et al.*, 1995, for a review). While there are some 'areas that have yet to benefit from a formal quantitative framework, the parameterization and empirical testing of existing models has lagged behind the theoretical developments in baculovirus ecology.

This chapter is not intended to be an exhaustive review. Instead, we have brought together the recent advances in the study of baculoviruses and synthesized an account of what we understand about their ecology, from the individual, to the population, and finally to the community level. We begin by outlining the factors that influence both the host and baculovirus within individuals, such as variation in viral genotype and host range. We move beyond the individual to explore the host–pathogen interaction at the population level and the dynamic patterns observed in natural systems. Finally, we take a broader ecosystem perspective and discuss multitrophic associations beyond the central host–pathogen interaction and assess the role of baculoviruses in species coexistence and community structure.

II. INDIVIDUALS

The relationship between a baculovirus and its host at the individual level is governed by many factors. It is widely recognized, for example, that environmental variables such as temperature affect the response of the insect to infection. Other factors, like host age and the concentration of the virus inoculum, are known to affect virus pathogenicity and the time it takes for hosts to die. These particular issues have a solid foundation of many years of empirical study and are firmly established in the baculovirus literature, and so will not be repeated here.

A. Virus Variation

Nuclear polyhedrosis viruses (NPVs) and granulosis viruses (GVs) have been isolated from hundreds of insect species, mainly Lepidoptera. Bioassays and techniques based on electron microscopy and serology indicated that baculoviruses were not identical. The advent and widespread use of DNA-based techniques, particularly restriction endonuclease profiles, allowed these differences to be defined on a molecular level. The suitability of naming a baculovirus after the species from which it was isolated is open to debate because of the observed patterns of variability. Comparison of NPVs and GVs from the same species in different locations has shown considerable variation in genetic structure (e.g., Shapiro *et al.*, 1991; Cherry and Summers, 1985; Crook *et al.*, 1985; Gettig and McCarthy, 1982) and biological properties (e.g., Hatfield and Entwistle, 1988; Allaway and Payne, 1983; Hughes *et al.*, 1983). Conversely, some studies have shown that baculoviruses collected from different species in the same area can be very similar (Goto *et al.*, 1992).

More interesting than the regional differences described above is the variation found within baculovirus isolates. (The term *isolate* tends to cover any baculovirus that is isolated from one insect or a group of insects from one species at one time and location.) Restriction endonuclease profiles of field-collected isolates frequently show submolar bands, indicating the presence of different genotypes. *In vitro* and, more importantly, *in vivo* cloning of these wild-type isolates have demonstrated the presence of a wide array of genotypically distinct strains (e.g., Maeda *et al.*, 1990; Smith and Crook, 1988; Knell and Summers, 1981; Lee and Miller, 1978), but the phenotypic consequences of this variation have received little attention. Plaque-purified clones of gypsy moth *Lymantria dispar* NPV have been shown to vary in infectivity and productivity in cell culture (Lynn *et al.*, 1993). Clones of the pine beauty moth *Panolis flammea* NPV, derived by *in vivo* techniques, also exhibit significant differences in LD_{50}, speed of kill, productivity, and host range (Paul, 1997). Differences in rates of melanization and lysis at death have also been recorded (Corsaro and Fraser, 1987; Hamm and Styer, 1985), which may implicate the recently discovered chitinase and cathepsin genes (see Section II.B).

The presence of this variation raises some interesting questions, particularly with regard to the maintenance of baculovirus diversity and its affect on baculovirus–host interactions. Recent work has indicated that closely related baculoviruses undergo high levels of recombination (Smith and Crook, 1993; Crozier and Ribiero, 1992), which may produce the large numbers of variants observed. Alternatively, genotypic variation may be generated by small mutations, sequence duplication, or acquisition of cellular–host DNA (Brown *et al.*, 1985). Most importantly, the phenotypic differences between these variants will alter baculovirus–host population dynamics. For example, where mixed wild-type isolates are collected from the field each year for reapplication and so are continually being repassaged, the selection of particular variants will be promoted. This may reduce the effectiveness of the baculovirus in the field, since characteristics that promote baculovirus survival, such as maximized yield, may act against the selection of features that enhance its abilities as a bioinsecticide, such as rapid speed of action. This suggests that the clonal diversity of isolates needs to be determined, and those with defined activity profiles selected for specific biocontrol programs. Moreover, multiple passaging of virus should be used with care until we understand more about what determines the selection of particular variants.

B. Genes to Ecology

The sequencing of the baculovirus genome has revealed that viruses possess genes that appear to manipulate their hosts in a number of subtle ways that may influence virus survival. If it is possible to attach phenotypic traits to specific genes, their selective advantage at the population level can be assessed. The identification, biochemical activity, and function of these genes, particularly the cathepsin and chitinase genes and the ecdysteroid UDP–glucosyltransferase (*egt*) gene, are covered elsewhere (Chapter 11, this volume), but it is worth discussing briefly their ecological implications. The primary function of both the chitinase and cathepsin genes appears to be the enhancement of the release of occlusion bodies from the insect cadaver (Hawtin *et al.*, 1995; Slack *et al.*, 1995; Ohkawa *et al.*, 1994). The subsequent liquefaction of the infected host may increase transmission through the rapid release of virus to healthy conspecifics and the wider contamination of the environment, although this has yet to be demonstrated in the field.

The expression of EGT appears to have a more complex effect on the insect larvae (O'Reilly and Miller, 1989), and we are still investigating how this gene affects viral fitness. There is substantial evidence that the insect lives longer, grows larger, and so produces more virus, although this will have the trade-off in delaying the release of progeny virus (E. E. Clarke *et al.*, unpublished data). While we do not fully understand the effects of EGT expression, it is evident that it alters several parameters that are likely to influence the dynamics of the pathogen and its host, and thus offers a fascinating opportunity to investigate the effects of this gene at a population level and the conditions under which it would invade and persist in baculovirus populations.

C. Host Variation and Disease Resistance

Variation in host response and the development of resistance to pathogens are key issues in the long-term use of baculoviruses for pest control, yet these topics have received surprisingly little attention (but see reviews by Fuxa, 1993, Briese, 1986). Populations of the same species collected from different regions often differ in their susceptibility. For example, Fuxa (1987) showed that *Spodoptera frugiperda* populations in North, Central, and South America differed significantly in their susceptibility to NPV, and Briese and Mende (1981) demonstrated that the LD_{50} of 16 populations of the potato tuber moth, *Phthorimaea operculella,* differed by a factor of almost 12. One explanation for this variation is that previous exposure to virus in different localities has selected for resistant populations. It may be possible to monitor such changes as a result of pest control operations, but few programs have been running for long enough to detect any changes in susceptibility. The exception is perhaps the velvet bean caterpillar, *Anticarsia gemmatalis,* control program in Brazil, which has been in progress for 10 years. A survey of populations from sites that have been treated regularly with NPV showed that the LD_{50} tended to increase with the numbers of years that the site had been treated (Abot *et al.*, 1995).

A number of studies have shown that it is possible to generate "resistance" to baculoviruses in different host strains. Briese and Mende (1983) obtained a 140-fold increase in resistance in a field population of *P. operculella* after six generations, while Fuxa *et al.* (1988) quadrupled the LD_{50} of a laboratory culture of *S. frugiperda* in seven generations. In contrast, Boots and Begon (1993) only measured a 1.96-fold difference in susceptibility in the stored product pest, *Plodia interpunctella,* after approximately 15 generations in which the host and virus populations interacted continuously (see Sait *et al.*, 1994b). However, it has not proved possible to generate resistance in all species. For example, Kaomini and Roush (1988) did not record any change in the mortality response of *Heliothis virescens* after 15 generations of exposure to the *Heliothis zea* SNPV.

It is commonly accepted that for polymorphism in disease susceptibility to be maintained in the host population, the development of resistance should have a cost in the absence of the pathogen (e.g., Bowers *et al.*, 1994). However, these costs have only been investigated in a few systems. Two studies on *P. interpunctella* and its GV showed that the most resistant populations have significantly longer development times (Boots and Begon, 1993; Vail and Tebbets, 1990). Boots and Begon (1993) also demonstrated a reduction in egg viability in more resistant populations. Fuxa and Richter (1989) showed that resistant *S. frugiperda* populations had reduced fecundity and egg viability, although these insects had a shorter life span. Resistance tends to be lost once the selection pressure is removed. For example, after one generation in the absence of the pathogen, the LD_{50} of the selected *S. frugiperda* population was no longer significantly different from the control colony (Fuxa and Richter, 1989). There is also evidence that resistance to *S. frugiperda* NPV provides some cross-resistance to other baculoviruses (Fuxa and Richter, 1990).

As discussed below, disease is thought to be a factor that causes the population cycles observed in many species of forest Lepidoptera. Martignoni and Schmid (1961) suggested that the susceptibility of Lepidoptera to baculoviruses may change at different stages of the population cycle. Myers (1988, 1990) put forward the "disease defense hypothesis," which suggests that if a reduction in fecundity was a trade-off for increased resistance to disease, populations exposed to virus epizootics at high densities would be expected to decline. When the selection pressure is removed, the susceptibility of the population should increase together with its fecundity, thereby creating cycles. One problem with the investigation of baculovirus resistance is that few, if any, studies have ruled out sublethal infection as a cause. As discussed in Section II.E, many of the consequences of sublethal infections, particularly altered development time and reduced fecundity, are the same as the proposed trade-offs for disease resistance. These two possibilities have been compared in field studies on the Western tent caterpillar, *Malacosoma californicum pluviale,* and its NPV. Several populations were surveyed over a number of years and their fecundity related to different phases of population fluctuations. The shifts in frequencies of large and small egg masses and mean fecundity changed rapidly between different sites (Myers and Kukan, 1995). Viral disease was also more common in small egg masses. When the susceptibility of caterpillars from small and large egg masses was compared, insects from the smaller masses were not found to be more resistant to disease, and, if anything, resistance tended to increase with size of egg mass (Rothman and Myers, 1996b). The authors concluded that the observed effects were more consistent with the effects of a sublethal infection rather than disease-induced resistance. Clearly, susceptibility to baculoviruses is a complex issue that requires more detailed study, particularly in field populations.

D. Host Range

While there is an acknowledged difficulty in deciding when different baculovirus isolates comprise separate "species," there is little doubt that baculoviruses from different hosts vary in the number of species they infect. Host range research has been fraught with problems, primarily due to the difficulties in assessing true cross-infection. This has been alleviated to some extent with the use of techniques, such as restriction endonuclease analysis of viral DNA, to confirm the identity of both inoculum and progeny virus. However, this does mean that some doubt must be cast over many host range studies in which progeny virus has not been checked. There is, in fact, very little detailed host range data for baculoviruses, with most studies involving only a handful of species. Varying bioassay techniques make data comparison difficult, and there is an urgent need for standardization in this area. The larger host range studies have been biased toward Lepidoptera, primarily noctuids, which are easier both to collect and rear. Probably the only overall trend that can be recognized is the narrow host range of Lymantriid NPVs, such as those from *L. dispar, Orgyia antiqua,* and *Euproctis chrysorrhoea* (Barber *et al.,* 1993; Richards, 1993; J. S. Cory *et al.,* unpublished data), most of which only appear to infect single species, compared with those isolated from the noctuids such as *Autographa californica, Anagrapha falcifera,* and *Mamestra brassicae* (Bishop *et al.,* 1995; Hostetter and Puttler, 1991; Doyle *et al.,* 1990; Payne, 1986), which infect dozens of insect species and from more than one lepidopteran family. This trend does not extend across the Noctuidae. Viruses from other species, such as those from various *Spodoptera* species, also appear to be very narrow in their host range (I. C. Hauxwell and J. S. Cory, unpublished data). No patterns can be discerned in terms of the phylogenetic relationships of the species that can be infected, with two host species within a single genus often having totally different responses. The host range of GVs and also the hymenopteran NPVs are reportedly narrower than the lepidopteran NPVs (e.g., Zethner and Øgaard, 1982; Cunningham and Entwistle, 1981; Ignoffo and Couch, 1981), although wider testing is needed.

Interest in baculovirus host range has developed primarily as a result of requirements for the registration of baculovirus biopesticides. This topic has expanded with the recent development of genetically modified baculoviruses and the need to assess their environmental impact. Perhaps the main issue this has highlighted is that not all species are equally susceptible to a particular virus, with a continuous gradient of susceptibility being exhibited. To distinguish between highly susceptible and less susceptible species, an arbitrary definition for "permissive," "semipermissive," and "nonpermissive" species was proposed by Bishop *et al.* (1995). Laboratory host range testing will give an index of which species could become infected under ideal conditions and is therefore a valuable tool for filtering out species that are unlikely to be infectable. However, so many other factors will influence (and probably reduce) the risk of infection in the field, such as the degree of niche overlap between permissive and semipermissive species, that more realistic studies need to be performed to determine the "ecological" host range of any virus.

E. Sublethal Effects

Interest in host–baculovirus infections has tended to focus on host mortality and its immediate effects on pest populations. However, there is considerable evidence that baculoviruses can also cause sublethal effects on their hosts that, if deleterious, could further enhance the baculovirus's impact on its host population (see review in Rothman and Myers, 1996a). Typical sublethal effects include alterations in development time (e.g., Goulson and Cory, 1995; Sait *et al.*, 1994a; Patil *et al.*, 1989), reductions in fecundity (e.g., Rothman and Myers, 1994; Sait *et al.*, 1994a; Patil *et al.*, 1989; Shapiro and Robertson, 1987; Young and Yearian, 1982), reduced egg viability (Sait *et al.*, 1994a; Young, 1990; Patil *et al.*, 1989; Santiago-Alvarez and Vargas-Osuna, 1988; Melamed-Madjar and Raccah, 1979), and changes in sex ratio (Santiago-Alvarez and Vargas-Osuna, 1986; Melamed-Madjar and Raccah, 1979).

The mechanisms by which sublethal effects might be exerted are poorly understood. It is not clear, for example, whether the virus is still within the host or, if it is no longer present, whether the effect observed is simply the result of the initial virus challenge. There is a need to determine the location of the virus, its possible activity, and under what circumstances it may be transmitted to other hosts. One potential mechanism for removing invading virus, the shedding of infected midgut cells (Engelhard and Volkman, 1995; Keddie *et al.*, 1989), may have a cost to the larva in terms of diversion of resources that might otherwise have been directed toward host growth and other functions. The passage of virus to progeny is the best evidence that the baculovirus remains in some species after virus challenge (e.g., Fuxa and Richter, 1993; Smits and Vlak, 1988a), and it is often assumed that the host constrains the infection to a low level. Hormonal and enzyme changes associated with virus infections (e.g., Burand and Park, 1992; O'Reilly and Miller, 1989; Subrahmanyam and Ramakrishnan, 1980, 1981) may affect host development or the production or development of viable eggs and sperm (Gelbic and Metwally, 1981; Riddiford and Williams, 1967). Sait *et al.* (1997) examined the effects of a sublethal GV challenge on sperm numbers and sizes in *P. interpunctella,* for which a reduced fertility effect has been recorded (Sait *et al.*, 1994a). While infected males tended to produce fewer sperm, the nonsignificant reduction could not explain the decline in fertility observed previously for this host–pathogen interaction. Clearly, measuring mortality alone is likely to underestimate the influence of baculoviruses and we need more information on the distribution and prevalence of sublethal effects before we can understand their role in host–virus population dynamics.

III. POPULATIONS

In order to appreciate fully the association between baculoviruses and their hosts and to exploit them effectively in biological control programs, we need to consider the interaction at the population level. Population dynamics of host–baculovirus interactions has, for the most part, been addressed theoretically, but only recently have model predictions been tested empirically. Two broad theoretical approaches have been adopted: relatively simple strategic and often analytical models and more complex tactical simulation models. The latter are typically designed to investigate one system in detail, often including a detailed description of the biotic and abiotic environment, while the former tackle more general questions. Reviews of system-specific tactical simulation models are provided by Brown (1987) and Onstad and Carruthers (1990).

A. A Framework for Insect–Baculovirus Interactions

Much of the theoretical work on the temporal patterns of insect pathogens has developed from the basic model by Anderson and May (1981) (Fig. 1). This model has three dynamic variables: the density of susceptible (S) and infected hosts (I), and the density of infectious particles in the environment (W). This model considers a one host–one pathogen interaction and

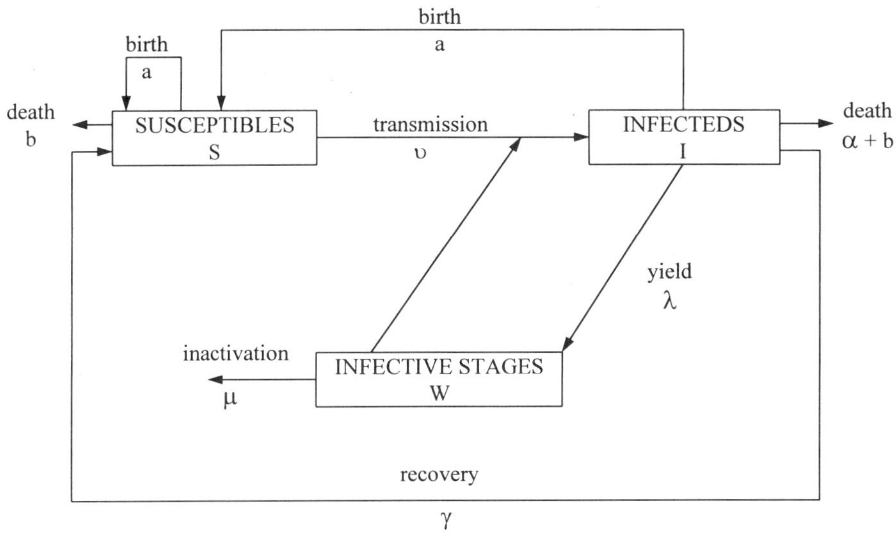

FIGURE 1. The basic model of Anderson and May (1981) has three dynamic variables: the density of susceptible hosts (S) and infected hosts (I), and the density of infectious particles in the environment (W). Other parameters are transmission coefficient (or rate at which infective stages successfully infect hosts) (ν), virulence (or disease induced mortality rate) (α), rate of production of infective stages, per host (λ), and mortality rate of the infectious particles in the environment (μ), host birth rate (a), natural mortality rate of hosts (b), and rate of host recovery from infection (γ).

assumes exponential host population growth in the absence of the pathogen. Although this model is clearly a simplification, it embodies the four basic processes of any insect host–baculovirus interaction at the population level (considered fully below), namely, transmission (ν), virulence (or speed of kill, α), yield (λ), and persistence of the infectious particles in the environment (μ), and it allows the definition of the ecological conditions under which the virus is likely to have the potential to regulate its host. Since these dynamic processes are fundamental to our understanding of the central host–baculovirus interaction, we will consider the ecological implications of these terms in detail.

1. Transmission

Transmission of a baculovirus between infected and susceptible hosts occurs via two routes. The most common route of infection occurs when a susceptible individual ingests sufficient occlusion bodies (OBs) from the environment, and is referred to as horizontal transmission. The release of virus in the form of OBs is a mechanism that promotes virus persistence in the environment outside the host. The alternative route of infection, termed vertical transmission, occurs across host generations (between parent and offspring), and may follow either contamination of the egg surface (transovum) or via virus contained within the egg (transovarial).

a. Horizontal Transmission

This process is central to the persistence of the virus in the host population and it can only be meaningfully quantified in the field where it is very sensitive to abiotic conditions. Models of disease transmission in insects have their origins in human epidemiology, in which the process is described as

$$\text{transmission} \propto SI$$

where S = density of susceptible individuals and I = density of infected individuals (e.g., Bailey, 1975). Quite simply, a new infection will occur each time an infected individual comes into contact with a susceptible individual. However, in insect–pathogen models, the main source of infection is the pool of OBs in the environment (represented as W), rather than infected larvae. Moreover, not every contact will lead to a successful infection, so a constant of proportionality or transmission parameter (ν) will reflect the ease with which the infection is acquired. Thus, the number of new infections per unit time may be expressed as νSW, where a *linear* relationship is assumed between the density of both susceptible hosts and infectious particles. In other words, the per capita rate of infection is constant at all densities of the host and increases linearly with pathogen density. The efficiency of transmission is given by ν, which embodies both the susceptibility of the host and the impact of the distribution of pathogen in the environment on the uptake of OBs by the host. The transmission of a baculovirus will only conform to this description if the assumption of homogenous mixing between host and pathogen is justified (the "mass action" assumption).

By making certain assumptions (which can be substantiated during the course of an experiment), and considering a short time period, the original Anderson and May (1981) model can be reduced and solved to provide an expression for ν:

$$\nu = \frac{-1}{tW_0} \ln \left[1 - \frac{I_t}{S_0} \right]$$

where W_0 = pathogen density at time 0; I_t = density of infected insects at time t; and S_0 = density of susceptible hosts at time 0. Therefore, by manipulating and monitoring host and pathogen density in the field, an estimate of ν can be obtained. Susceptibility to baculoviruses is known to be reduced dramatically in later instars, yet feeding rates increase. If the mass action assumption holds, it should be possible to predict relative rates of transmission in the field from dose response and feeding rate measurements in the laboratory. However, Goulson *et al.* (1995) found that estimates of ν in the field did not differ significantly between instars, despite the prediction from laboratory data that transmission rates should increase in older larvae.

A number of recent empirical studies have investigated the mass action assumption in the field for both baculoviruses and other pathogens (D'Amico *et al.*, 1996; Knell *et al.*, 1996; Thomas *et al.*, 1995; Dwyer, 1991; S. D. Vasconcelos *et al.*, unpublished), but in none of these cases was the assumption upheld. In two of these studies, insect cadavers were treated as infectious units (rather than OBs or equivalent). It was found that the infectivity of a cadaver changed with time as it broke up and released the pathogen, which subsequently decayed. For example, Thomas *et al.* (1995) found that infectivity was best described by a gamma function multiplied by a constant. In studies of baculoviruses, OBs are more frequently treated as the infectious units, but again the data do not conform to the mass action assumption. For example, D'Amico *et al.* (1996), studying *L. dispar* and its NPV, found that the transmission parameter declined as the density of susceptible hosts and infectious particles increased.

Empirical studies, therefore, suggest that the mass action assumption is not an adequate description of transmission dynamics in the field. Theoretical studies have investigated the consequences of contravening the mass action assumption. Hochberg (1989a) investigated a more general transmission term, incorporating density dependence, of the form

$$\upsilon S^{q+1} W^{p+1}$$

(see also Andreasen, 1989; Liu *et al.*, 1986). If $q < 0$, an increase in the density of OBs in the environment results in fewer "than expected" new infections in the host population (as observed by D'Amico *et al.*, 1996). This is likely if the OBs are distributed patchily in the environment, as occurs when they are liberated from cadavers. Infection is likely to be a function of the probability of encountering a cadaver rather than the number of OBs present in that corpse (provided that the number of cadavers has not increased), particularly if one encounter is sufficient to infect the host. The dynamic consequence of this nonlinearity is that the insect–pathogen interaction is more stable; it is as if at high pathogen densities, the host has a refuge from infection. A similar result is obtained if $p > 0$. An alternative approach was taken by Getz and Pickering (1983) who used a transmission term in which transmission was proportional to the fraction of susceptible hosts in the population (rather than the absolute number). In this case, the pathogen lost any regulatory power over the host population, though it continued to persist.

In summary, theory predicts that the absolute value of the transmission parameter will be important in determining host and pathogen abundance at any particular pathogen-regulated host equilibrium (Anderson and May, 1981), while the density-dependent form of the transmission term may influence the potential of the pathogen to regulate the host and the stability of the interaction (Hochberg, 1989a). Studies of the transmission process are a clear illustration of how models have aided the design of experiments, the results of which feed back into modifying the models.

b. Vertical Transmission

While horizontal transmission appears to be the most common route of virus transmission, there is also evidence that baculoviruses can be transmitted vertically. Data on this topic are highly variable, depending on host species, host age, and type of virus. For example, *S. frugiperda* dosed with NPV exhibited 14.3% mortality in their offspring, while, under the same conditions, *Anticarsia gemmatalis* NPV was not vertically transmitted in *A. gemmatalis* (Fuxa and Richter, 1993). Murray *et al.* (1991) found no evidence of vertical transmission following infection in fourth instar *L. dispar* larvae, but Shapiro and Robertson (1987) recorded up to 11.5% mortality in the progeny of second instar survivors of viral challenge. Evidence for vertical transmission has been found in other systems, including *Mythimna separata* NPV (Neelgund and Mathad, 1978), *Sesamia nonagroides* GV (Melamed-Madjar and Raccah, 1979), and *Spodoptera exigua* NPV (Smits and Vlak, 1988a). Fuxa and Richter (1991) were able to increase the rate of vertical transmission of *S. frugiperda* NPV by artificial selection, suggesting that the propensity to do so is determined genetically in that species. Vertical transmission enhances dispersal of the virus by enabling it to track its host during dispersal or migration, and may be important in a mobile pest species such as *S. frugiperda* where successive larval generations are unlikely to occupy the same site.

2. Virulence

Virulence is a term used in the literature to describe both the severity of the impact of a parasite on its host (often used interchangeably with pathogenicity) and the speed of that impact. Baculovirus infections are normally lethal, though some infections can be sublethal (see Section II.E) or possibly latent (Hughes *et al.*, 1993, 1997). (A latent infection is assumed to be one in which no symptoms are displayed, sublethal or otherwise; the host is not infectious, though it can be transmitted vertically within the host population.) It is also known that within one baculovirus"species" genotypically distinct clones can show considerable variability in speed of kill (see Section II.A). Baculoviruses therefore exhibit variability in both severity and speed of impact, and the selective forces that cause and maintain this variability are largely unknown.

Latent infections represent an extreme lack of virulence, albeit temporarily. Until recently, the evidence for latency has been largely circumstantial. For example, the appearance of infections in an apparently disease-free insect culture has frequently been ascribed to a latent virus having been triggered out of its nonvirulent state through stress (e.g., high densities, diet, rearing conditions) (Longworth and Cunningham, 1968), though the possibility of contamination can never be wholly excluded. Rather more intriguing is the observation that infection of an insect by one baculovirus in the laboratory can result in the death of that insect by a different baculovirus. This may be due to the first virus triggering the second out of its latent form, perhaps by providing vital functions that are missing in the latent virus (Hughes *et al.*, 1993; Longworth and Cunningham, 1968).

Polymerase chain reaction (PCR) amplification has identified sequences of *Mamestra brassicae* NPV (MbNPV) in a laboratory culture of *M. brassicae* (Hughes *et al.*, 1993). These sequences were found in all life stages of the insect, suggesting they were vertically transmitted from one generation to the next. However, it is not known whether these sequences were merely dysfunctional fragments of the viral genome, a low-level persistent infection, or a latent virus. Wild-type insects did die of an MbNPV-like infection after being fed fat body cells from the insects thought to carry the latent infection (Hughes *et al.*, 1997). Moreover, Hughes *et al.* (1997) found that the fat body cells contained mRNA specific for the polyhedrin gene and transcriptional factors capable of activating the baculovirus, providing indirect evidence that a latent MbNPV virus is maintained as a persistent infection in this culture.

Considering the speed with which the host is killed, there are obvious trade-offs between virulence and yield. If the baculovirus kills its host quickly, the number of replication cycles will be limited, with reduced yield as a consequence. This will happen if a baculovirus is engineered so that its speed of kill is increased artificially. For example, *Autographa californica* NPV (AcNPV) has been engineered to express an insect-selective scorpion toxin (AcNPV-ST3), causing a 25% reduction in time to death but a tenfold reduction in yield (Cory *et al.*, 1994; Stewart *et al.*, 1991). On the other hand, when observing time to death of insects given the same dose of wild-type AcNPV, those that take longer to die have greater yields (P. Hernández-Crespo *et al.*, unpublished data). This suggests that selection will act to reduce the speed of kill to maximize yield, and it may be why many baculoviruses have acquired the *egt* gene, which inhibits larval moulting (Section II.B). Given that these arguments suggest that baculoviruses might be expected to exploit their hosts fully, the existence of genotypically and phenotypically distinct isolates and the factors that cause and maintain this diversity remain important yet unanswered questions.

3. Yield

Baculovirus productivity in insects has been studied broadly with regard to optimization of *in vivo* production (Shapiro *et al.*, 1986), investigating the influence of virus dose, insect maintenance conditions, host species, age, instar, and sex (e.g., Vargas-Osuna *et al.*, 1995; Kelly and Entwistle, 1988; Smits and Vlak, 1988b). In natural populations, one of the most important influences on yield will be the host species and where it is found on the permissive–semipermissive continuum for that particular baculovirus. For example, the dynamics of AcNPV were compared in a semipermissive host, *M. brassicae,* and a permissive host *Trichoplusia ni.* The semipermissive host died more slowly after ingesting a given quantity of virus and produced more OBs on death as a result (P. Hernández-Crespo *et al.*, unpublished data). The relationship between virus yield, speed of kill, and virus dose was highly complex and differed between permissive and semipermissive hosts. Importantly, the difference in yield and speed of kill between AcNPV and AcNPV-ST3 that has been so clearly demonstrated in the permissive host (Cory *et al.*, 1994; Stewart *et al.*, 1991) is not apparant in the semipermissive host (P. Hernández-Crespo *et al.*, unpublished data). The dynamics of infection within the different classes of hosts are quite different; further studies should be conducted to ascertain whether a similar relationship is found in other permissive and semipermissive comparisons, and such data should be incorporated into a quantitative risk assessment.

4. Persistence

The ability to persist outside the host is one of the key features of NPVs and GVs. The rate at which baculovirus OBs become inactivated in the natural environment will undoubtedly influence the ability of the virus to persist though periods of low host density. When virus is delivered in the form of spray, persistence of the virus on foliage is vulnerable to UV irradiation (e.g., Jones *et al.*, 1993) and physical removal by wind and rain (Ignoffo and Hostetter, 1977). Plant architecture will also influence the rate at which degradation and removal occur (see Section IV.A), as well as various abiotic factors (see Evans, 1986). In contrast, virus in cadavers is afforded much better protection against degradation. For example, in a field study where NPV was released in the form of infected insects, there was no appreciable reduction in virus activity over a week (R. S. Hails *et al.*, unpublished data), compared to a 100-fold reduction with sprayed virus (P. Hernández-Crespo *et al.*, unpublished data). This contrasts with cadavers containing fungal pathogens, in which infectivity increases initially as the cadavers break up and release the pathogen, then decreases as the pathogen decays (Thomas *et al.*, 1995). Other environmental surfaces have been found to provide protection, such as the soil (McLeod *et al.*, 1982; Jaques, 1974a,b) and crevices in the bark of trees (Evans, 1986).

The presence of such reservoirs essentially provides the virus with a refuge (an alternative refuge is that provided by other hosts) (see Section IV.B). Hochberg (1989b) developed the basic model of Anderson and May (1981) and divided the population of OBs into two subpopulations by including a pathogen reservoir in which the pathogen is protected from degradation but is unavailable to infect the host. The rate of flow of the pathogen from protected to transmissible habitats is crucial in determining the type of dynamics displayed by the model. This heterogeneity in the pathogen population with the presence of a reservoir can give rise to more constant levels of host abundance (i.e., is a feature that stabilizes the host–pathogen interaction). Evidence that virus in refuges has any impact on the dynamics of the host is indirect. Fuxa and Geaghan (1983) found that the most significant factor affecting virus prevalence in *S. frugiperda* populations was overwintering virus in the soil. Additionally, soil application of baculoviruses can effectively control leaf-feeding pests, such as the cabbage looper (*T. ni*), if applied in high enough concentrations (Jaques 1970, 1974a,b). There is little doubt that abiotic factors move virus from cadavers on plants to the soil (e.g., D'Amico and Elkinton, 1995), but movement from the soil to the plant is far less studied. There is evidence that OBs may be transferred from soil to foliage by wind (Olofsson, 1988). However, in all these studies, there was no attempt to quantify the rate of translocation between protected and transmissible surfaces. A recent field experiment, using the cabbage moth *M. brassicae* and its NPV, found that persistence was moderate to low and rates of translocation back to leaf surfaces were sufficient to initiate epizootics and stabilize the interaction between host and baculovirus (Hails *et al.*, in review).

Other mechanisms may cause the reintroduction of the virus into the cycle of infection. The behavior of the insects themselves may bring them into contact with the soil, such as in the case of Lymantriid larvae, which often disperse from trees in their early instars (Dwyer and Elkinton, 1995; Richards, 1993). Larvae of *Wiseana* spp., a soil-dwelling pest of pastures that die from NPV infections, accumulate in the upper layers of soil, so forming a reservoir (Crawford and Kalmakoff, 1978). Subsequent grazing of the grasslands is thought to aid translocation of the NPV from the soil to the leaves, where susceptible caterpillars feed. When this routine is disturbed (e.g., when pastures are left ungrazed or ploughed over), there are frequent pest outbreaks (Crawford and Kalmakoff, 1977). This may be due to the disruption of the normal routes of translocation, which removes the stabilizing influence of the reservoir.

B. Temporal Patterns

Bringing together the four processes (transmission, virulence, yield, and persistence) in a theoretical framework (the basic model of Anderson and May, 1981), an insect–pathogen interaction will exhibit one of four types of dynamical behavior, which depends on the magnitude of the parameters describing these processes. If the replication rate of the virus is low, the virus will be unable to maintain itself in the host population. If virulence is low (i.e., the baculovirus takes a long time to kill its host), the virus is unlikely to regulate its host, though it may be able to persist in the host population. At intermediate levels of virulence and yield, the virus has the potential to regulate the host population. This may be either as a stable interaction, in which the host and virus are present at constant levels of abundance, or the system may display limit cycles. Stable limit cycles are most likely when considering viruses with fast speed of kill and persistent OBs.

This model was first tested against observational data from a number of insect–baculovirus associations in temperate forests that show long-term cycles in host population abundance. Estimates of the four parameters indicate that many of these associations fall into the area of parameter space of the model that would produce stable limit cycles. Since this model was first published, the debate has centered around whether cycles in pathogen abundance are caused by or cause the cycles in the host. Clearly, if an insect population cycles for reasons other than disease, so will the pathogen population, and observational data will not prove cause and effect. Modifications of the basic model, incorporating features that make it more biologically realistic, frequently result in the cycles becoming reduced in amplitude or disappearing altogether. For example, if the host population is regulated by density-dependent reproduction, cyclic behavior becomes less likely (Bowers *et al.*, 1993; Vezina and Peterman, 1985; but see Dwyer, 1994). However, other features are shown theoretically to make cyclic dynamics more likely, such as vertical transmission (Regnière, 1984) or a sublethal reduction in host production (Anderson, 1982).

Despite the theoretical demonstration of linked host-pathogen population cycles, the empirical evidence remains equivocal because of other biotic and abiotic factors that interact with the host (Berryman, 1966; Berryman *et al.*, 1990; Myers, 1988). In a laboratory study, where the confounding variables found in natural systems could be controlled or eliminated, Sait *et al.* (1994b) examined the long-term population dynamics of the Indian meal moth *P. interpunctella* and its GV. The host alone exhibited regular cycles, with a period of approximately one host generation. When interacting with the virus, however, the host still cycled but the period was increased by 3 or 4 days (Fig. 2). In addition, the regularity and integrity of these cycles were enhanced compared with the host alone, as a result of a reduction in variable larval developmental rates caused by greater GV-induced mortality in the early instars. Thus, in a population that already cycled, the pathogen strengthened the cycling behavior. The mechanism leading to the increase in cycle period in the host–pathogen populations was probably due to a complex interaction between sublethal or covert virus effects (Goulson and Hauxwell, 1995; Sait *et al.*, 1994b) and selection for resistant hosts (Begon and Boots, 1995; Boots and Begon, 1993; Vail and Tebbets, 1990), and served to illustrate that ecological and evolutionary processes will interact when populations are studied over intergenerational timescales.

Other empirical studies have attempted to fit theoretical predictions to experimental data. Dwyer and Elkinton (1993) used a reduced, within-

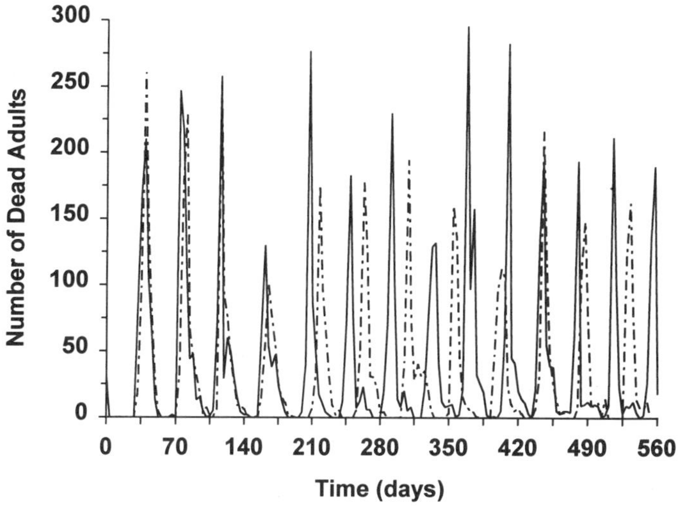

FIGURE 2. Typical trajectories of replicated *Plodia interpunctella*-alone populations (solid line) and *P. interpunctella* + GV populations (dot–dash line) illustrating the reduction in adult host abundance and the increase in the cycle period when the host was exposed continuously to the pathogen. Dead adult hosts were sampled twice weekly from the population cages. In the same time period adults in the host-alone population went through 13 generation cycles while those in the host-GV population went through 11 cycles (after Sait *et al.*, 1994a).

season model to describe the dynamics of the gypsy moth *L. dispar* and its NPV in field populations, estimating each of the parameters describing the four basic processes independently. The population trajectories were moderately accurate for five of the eight populations, predicting the general epizootic pattern. Inaccuracies in the remaining three populations were thought to be due to the transmission parameter not conforming to the mass action assumption (see Section III.A.2a). These results are encouraging, in that they suggest that this insect–virus system can be reduced to a small number of simple processes, yet still encapsulate the essence of the interaction.

These simple models may be adapted to include artificial manipulation of the pathogen population, and so investigate the potential success of a biological control program (Anderson, 1982). It is predicted that there is a threshold input of a biocontrol agent that is required to cause host extinction, but it is only close to that threshold that substantial reductions in host density are likely to be obtained. It should be noted that this particular model assumes homogeneous mixing of host and pathogen, with no spatial structure. In reality, the environment will be divided into patches (see Section III.D). Within one patch, such simple models may well be a reasonable approximation of the dynamics, thus allowing us to predict the thresholds required for *local* extinction of the host. This threshold is affected by biological features of the system, such as vertical transmission to the next generation. While theory and experiment have much to contribute to the design of biological control programs, it is rare that practitioners have the luxury of considering the ecological aspects. However, the integration of ecology and pest control will become more important, particularly when considering the use of genetically modified viruses.

The last few years has also seen the development of models of intermediate complexity, based on the original Anderson and May (1981) model but incorporating additional features of the insect–baculovirus interaction. An example of this is the development of a stage-structured model using time-delayed differential equations (Briggs and Godfray, 1995). This takes account of the fact that insects are not equally susceptible to baculoviruses at all stages of the life cycle; only the larval stages are susceptible, and early and late instars may differ in infectivity by orders of magnitude. This model, and variants of it (e.g., Briggs and Godfray, 1996), tends to be less stable than their simpler counterparts and, as yet, remain untested against field data.

C. Risk Assessment Issues

The assessment of the environmental impact of genetically modified organisms (GMOs) is currently an important issue, but without a more detailed understanding of baculovirus ecology, it will be difficult to assess the implications of widescale baculovirus release. Consideration should be given to both spatial and temporal processes; as such, all sections discussed in this chapter are relevant, such as the host range issue and the need to understand the dynamics of infections in permissive and semipermissive hosts, the prevalence of sublethal infections in natural host populations, the mechanisms by which viruses persist between epidemics, and the development of mathematical models that incorporate realistic features of baculovirus–host dynamics. Perhaps of most concern is the risk that GMOs and exotic natural baculoviruses may pose a threat to nontarget hosts (i.e., species not considered to be pests, but which may nevertheless become infected by the baculovirus). These nontarget hosts include lepidopteran species of conservation value.

It is not possible to quantify in absolute terms the probability that a novel virus will invade a host population. However, some progress can be made by considering the relative invasiveness of two viruses and the extent to which theory allows us to make some statements about relative rates of increase (particularly useful when considering an engineered virus and its wild-type counterpart). The Anderson and May (1981) model may be used to define the criteria required for a virus to invade a population of susceptible hosts. From this model (illustrated in Fig. 1), the change in the number of infected insects can be expressed as:

$$\frac{dI}{dt} = \upsilon SW - (\alpha + b + \gamma)I$$

where I = density of infected insects, S = density of susceptible insects, and all parameters are as explained for Fig. 1. The quantity $(\alpha + b + \gamma)$ represents the rate at which infected insects leave the infected class, and therefore the reciprocal of this represents the average length of time an insect is infected. Consider one infected insect. According to the model, throughout the time it is infected, it is releasing OBs at rate λ. However, with lepidopteran baculoviruses, all OBs are released at once upon death. If the total yield upon death is denoted Λ, this can be expressed as:

$$\Lambda = \frac{\lambda}{(\alpha + b + \gamma)}$$

Of the OBs that are released into the environment, some will be ingested by susceptible hosts, while others will become inactivated before they have had the chance to cause a new infection. The proportion that are successfully ingested before inactivation can be expressed as:

$$\frac{\upsilon H}{\mu + \upsilon H}$$

where H = density of insect host. Combining this information provides an expression for the basic reproductive rate of the pathogen (R_0), or the number of new infections expected from one primary infection. If one OB is ingested by a susceptible host, which then dies and yields Λ OBs (each of which have the above probability of being ingested again), then the expected number of new infections are:

$$\Lambda \frac{\upsilon H}{\mu + \upsilon H}$$

This quantity must be greater than 1, on average, for the virus to invade a population of susceptible hosts. Genetically modified baculoviruses (GMBVs) are engineered specifically to increase their speed of kill and this in turn reduces the yield of OBs from an infected insect (Λ). Additionally, in toxin-producing viruses, paralysis frequently occurs before death, resulting in the insect falling off the plant. This is likely to reduce the rate at which the virus is transmitted to other hosts (ν). It is evident from this simple model that such GMBVs will have lower R_0 than their wild-type counterparts. However, the shorter cycle of infection resulting from the increased speed of kill could mean that an engineered virus may go through more cycles within one season. Hence, its finite rate of increase on an annual basis may not be lower than the wild type. This is one area in which models could be developed and parameterized to answer these specific questions.

This same equation also illustrates another principle that has been referred to frequently throughout this chapter, namely the concept of a threshold host density (or critical community size) below which a virus is not able to persist in the population. Given that the above expression must be greater than 1 for virus persistence, this can be rearranged to give:

$$H_T > \frac{\mu}{(\Lambda - 1)\upsilon}$$

When the population falls below this threshold host density (H_T), horizontal transmission is insufficient for persistence of the virus. However, as discussed earlier, if a proportion of virus progeny are vertically transmitted, then that virus species can persist at lower host densities.

Clearly, GMBVs with lower yields and transmission rates will require higher critical community sizes to persist. The corollary to this is that rare species are less likely to sustain a virus population simply because they are rare. While this is a favorable conclusion for those wishing to release GMBVs (and those wishing to conserve endangered species), it should be remembered that the model from which this invasion criterion was derived is simple, deterministic, and assumes homogeneous mixing of host and pathogen. The three parameters involved (transmission, yield, and persistence) will vary considerably depending on the ecological and abiotic conditions, and therefore estimates of variability as well as expected values should be obtained. Future work would benefit from deriving and parameterizing similar invasion criteria from models that have incorporated important system-specific features that are likely to influence the dynamics. Incorporation of spatial heterogeneity into the theoretical framework frequently leads to unexpected results. It is this latter theme that is developed in the next section. We are therefore some way from being able to perform a rigorous quantitative risk assessment for GMBVs.

D. Spatial Processes

Studies of baculovirus epizootics have demonstrated clearly the capacity for baculoviruses to spread from the original focus of infection (Dwyer and Elkinton, 1995; Fuxa and Richter, 1994; Fuxa *et al.*, 1993; Dwyer, 1992; Entwistle *et al.*, 1983). The mechanisms by which viruses disperse are less clear. Parasites frequently affect the behavior of their hosts, which can have a significant impact on the fitness of both. However, quantitative studies on the behavioral effects of baculovirus infection and their ecological consequences have been lacking. Vasconcelos *et al.* (1996b) found that NPV infection increased the mobility of *M. brassicae* larvae in both laboratory and field experiments. Infected larvae also migrated to more exposed parts of the plant, in contrast to healthy larvae that were found under leaves. Predators and scavengers can act as passive dispersal agents, since baculoviruses tend to pass through their guts unharmed. On a larger scale, birds have been implicated as important agents of dispersal (e.g., Lautenschlager *et al.*, 1980; Entwistle *et al.*, 1977a,b). In forest sites that had been treated with NPV for control of the pine beauty moth *P. flammea*, up to 77% of birds captured were found to have fed on NPV-infected material during the larval period (Entwistle *et al.*, 1993). Arthropod predators may act as passive vectors on an intermediate scale, with their higher mobility compared to host larvae increasing the rate of inoculum dispersal (e.g., Vasconcelos *et al.*, 1996a; Fuxa and Richter, 1994; Fuxa *et al.*, 1993; Young and Yearian, 1987, 1992). However, while passive dispersal of baculoviruses is obviously feasible by innumerable routes, few studies have attempted to estimate quantitatively the contribution of different mechanisms by which the virus can spread in the environment.

It is only relatively recently that the importance of spatial processes in population models has been fully appreciated. There are a number of modeling frameworks commonly used to study spatial problems: reaction–diffusion models, cellular automata, and lattice models. Dwyer (1994) used a diffusion model to incorporate host movement in one dimension (both susceptible and infected hosts). While this is obviously a simplification, such models have been shown to be equivalent to their two-dimensional counterparts. At certain parameter values (specifically those that would cause limit cycles in the temporal dynamics), a traveling wave would form, consisting of a wave of susceptible hosts followed by a wave of infected hosts. Dwyer (1994) used this model to test the hypothesis advanced by Entwistle *et al.* (1983) that the disease spread in the European spruce sawfly *Gilpinia hercyniae* was so fast that it must require special mechanisms of transport (e.g., disperal by large predators). Using a conservative estimate of the diffusion coefficient (which encapsulates the speed of host movement), the model could easily explain the observed rate of disease spread. This suggested that simple diffusive host movement may be sufficient to explain data on the spread of epizootics; it is not necessary to invoke more complicated dispersal mechanisms. However, in another system, using *L. dispar*, Dwyer and Elkinton (1995) found that larval dispersal was not a sufficient explanation for the spatial spread of NPV, and they concluded that other mechanisms must be involved, suggesting that a parasitoid could be responsible for the enhanced dispersal.

Another group of models has been developed to study spatial problems that include the spatial element more explicitly. These comprise a grid of cells or patches in which individual populations are governed by a set of rules for one generation, at the end of which a fraction disperse to neighboring patches (Gilpin and Hanski, 1991). Such models have illustrated that temporal population models that are unstable and frequently go extinct locally can link up and persist over a larger spatial scale (e.g., Comins *et al.*, 1992). For example, models that assume homogenous mixing of host and pathogen predict that pathogens of high virulence will not persist. This is because they cause extinction of the host and effectively "burn themselves out." However, when these models are placed into a spatial framework (e.g., a lattice map), such pathogens can become endemic (Woods and Thomas, 1996). Because the healthy hosts are mobile, they can move from areas of high pathogen density and a proportion will escape infection. Thus, host extinction is avoided and the pathogen can persist. This clearly has implications for biological control and removes the historical distinction between "fast-acting biopesticides" and "slow-acting long-term biological control agents."

These models describing a number of small populations linked together spatially are referred to as *metapopulation models,* and the resulting dynamics produce a variety of spatial patterns. A similar framework has recently been applied to a host–pathogen system (White *et al.*, 1996) in which the environment is divided into a series of cells. At the end of each generation, a proportion of the host and pathogen populations disperse to neighboring patches. At certain parameter values a radial wave pattern would pulsate from the center, i.e., this is the two-dimensional equivalent to the reaction–diffusion model (though some details of the wave pattern differ). These two models produce a similar pattern due to the same underlying process. During an outbreak, pathogen levels become high and remain so for sometime afterward. The host remains at low densities until the pathogen has decayed sufficiently for the population to recover and create a focus from which it can spread out again. More interestingly, however, this model was capable of exhibiting two quasistable spatial patterns—spirals and radials—and the system would switch between the two with intermittent behavior in between. Thus, while the dynamics look consistent at one time scale, over a longer time scale the behavior is less predictable.

Incorporating spatial processes into models are providing new insights into host–pathogen dynamics, with implications for both biological control and risk assessment. There are many avenues yet to be developed, and this should prove to be a fruitful area over the next few years. One challenge will be to integrate spatial processes with multispecies models in an attempt to predict the impact of engineered viruses on ecological communities.

IV. MULTITROPHIC INTERACTIONS

For the most part, the study of baculovirus ecology has been concerned with understanding the central host–pathogen interaction alone. Clearly, this represents a simplification of the complex web of interactions that exists between species in natural populations. In considering multitrophic systems, we take an incremental step toward understanding the importance of baculoviruses in population and community ecology, and we begin to address some of the critical issues regarding the use of baculoviruses in biological control. Much of the exploration of host–pathogen interactions beyond two species has, to date, been theoretical. These models build on those developed at the population level and provide a framework in which to study and understand multispecies systems. They typically refer to generalized pathogens, but nonetheless possess elements that are applicable to baculoviruses, thus highlighting possible avenues for empirical studies. Broadly speaking, there are four main areas that have generated interest in multitrophic interactions: host–plant–pathogen, host–host–pathogen, host–pathogen–pathogen, and host–pathogen–predator (parasitoid) systems.

A. Host–Plant Interactions

Many biotic and abiotic factors influence the association between a host larva and its baculovirus in the field. One of these, the host food plant, can affect baculovirus ecology in two principle ways: through its structural architecture and its chemical constituents. Different foliar structures provide varying degrees of protection for the virus from degradation by UV irradiation and weather, thereby creating virus reservoirs on the plant and in the soil below. Plant structure may also affect the likelihood of contamination of the foliage with virus from the soil reservoir (see Section III.A.4). For example, although *Spodoptera frugiperda* was less susceptible on signalgrass in laboratory assays (Richter *et al.*, 1987), NPV prevalence in field populations was greater in signalgrass than other crops (Fuxa and Geaghan, 1983). Fuxa (1982) also found that NPV infection increased more slowly on corn and sorghum than in pastures even when the crop species were found in the same field. Host feeding site and mobility within a crop may affect the observed patterns of infection, but translocation of virus from a soil reservoir may also be greater in shorter, less dense crop types.

The effects of plant chemistry on the insect–baculovirus interaction is very complex, since the plant constituents also interact with each other (Duffey *et al.*, 1995). Laboratory studies have demonstrated that catechols (Felton *et al.*, 1987), additional casein or salts (Keating *et al.*, 1988a), and tannins (Young *et al.*, 1995; Keating *et al.*, 1988b) can reduce baculovirus infectivity. Similarly, exudates on the leaf surface of cotton, which are responsible for high pH and cation concentrations at the leaf surface, result in virus inactivation (Elleman and Entwistle, 1982, 1985; McLeod *et al.*, 1977). Studies in which virus infectivity has been compared on different types of foliage have shown repeatedly that LD_{50} varies considerably with food plant (Forschler *et al.*, 1992; Santiago-Alvarez and Ortiz-Garcia, 1992; Keating *et al.*, 1990; Keating and Yendol, 1987; Richter *et al.*, 1987). These laboratory studies give some indication of relative host susceptibilities on different crops, but, as shown above, they are not necessarily good indicators of the host–pathogen–plant interaction in the field where potential confounding factors can be found.

The reduced growth of an insect or the "stress" induced in the host by toxic plant chemicals may itself lead to greater susceptibility to disease rather than a direct effect of the plant constituents on infection. Several studies have attempted to tease apart these complexities by varying the dietary history of the insects and the substrate on which virus inoculum was given. Forschler *et al.* (1992) highlighted the importance of diet by comparing NPV-induced mortality of *H. zea* on a range of crops. Differences in mortality were found only when larvae were reared entirely on foliage. On the other hand, Schultz *et al.* (1992) found that the variation between mortality of gypsy moth *L. dispar* dosed on different tree foliage was maintained regardless of the larval diet before and after inoculation. Hydrolyzable tannins in leaves were shown to reduce the susceptibility of *L. dispar* to virus (Keating *et al.*, 1990; Schultz *et al.*, 1990). Furthermore, NPV infection in *L. dispar* larvae was negatively correlated with increasing levels of defoliation (Hunter and Schultz, 1993), suggesting that the elevated concentrations of tannis induced by herbivore damage can enhance the resistance of a pest insect in the presence of disease. These data illustrate that detailed laboratory and field investigations are necessary to understand more fully the intricacies of insect–plant–pathogen interactions.

B. Alternative Hosts

A wealth of potentially complex dynamic properties could arise out of host–host–pathogen interactions. To date, however, there are no experimental accounts of more than one host interacting with a common pathogen. These multispecies associations are directly relevant to our understanding of the potential use of natural and recombinant baculoviruses as biocontrol agents or biopesticides, where either alternative target or nontarget hosts will have a profound effect on the central target host–pathogen interaction. As we saw in Section II.D, several hosts may be classified as permissive or semipermissive according to some measure of pathogenicity. Semipermissive or nontarget hosts could function as baculovirus reservoirs when the primary host density is too low to support the pathogen so long as transmission occurs between them.

A number of theoretical approaches have considered the population dynamics, community structure, and biological control implications of host–host–pathogen systems. They possess features that are analogous to typical host–baculovirus interactions, and thus provide a framework into which the release of baculoviruses as bioinsecticides can be placed (see review in Begon and Bowers, 1995; Bowers and Begon, 1991; Holt and Pickering, 1985). Begon and Bowers (1994) developed a model, perhaps the most relevent to baculoviruses, incorporating infective stages external to the host (equivalent to horizontally transmitted baculovirus) and host self-regulation. The possible outcomes were varied and complex, but in relation to biological control, the interaction between what could be interpreted as permissive and semipermissive hosts showed that a nontarget host is unlikely to undermine pest control, through consumption and reduction of the virus source, or enhance pest control, though amplification of the virus beyond the level produced by target hosts. Conversely, they also showed that an alternative, nontarget host is at low risk of extinction, with the outcome being dependent primarily on its own interactions with other species.

These model predictions remain largely untested empirically, yet evidence suggests that the relationship between pathogenicity (or virulence), virus yield, and host species is complex at the individual level, with a number of implications for virus transmission and persistence. However, these data tell us very little about the potential three-species interaction. One possibility is apparent competition (Holt, 1977), in which one host species is reduced in abundance or excluded by another host species because of their shared pathogen. For example, the high yields produced by some semipermissive hosts may enhance the bioinsecticidal activity of the virus against the target species. The influence of pathogens on the coexistence of interacting hosts should be greater for closely related species because the overlap in susceptibility tends to increase as phylogenetic relatedness increases (Freeland, 1983).

C. Mixed Infections

A wealth of experimental studies have investigated mixed baculovirus infections in the same host (see review in Harper, 1986). Several possible outcomes of these host–pathogen–pathogen interactions have been documented, ranging from synergism (Arne and Nordin, 1995; Goto, 1990; Tanada and Hukuhara, 1971), to interference or antagonism (Ritter and Tanada, 1978). Another form of interaction has been termed *independent proliferation,* where there is neither a synergistic nor an interference effect on host mortality (Lowe and Paschke, 1968). The classification of these interactions is based typically on studies that provide overall mortality data, but this provides little information about the individual effect on each of the interacting pathogens. For example, independent proliferation must also constitute an interference interaction, because both pathogens are ultimately competing for the same limited resource—the host (Smith and Holt, 1996). An ecological notation, 0 0, 0 +, 0 −, and so on (where the symbols 0, +, and − represent mortality that is the same as, higher than, or lower than for each pathogen than if it were alone) may provide more information. Tanada and

Hukuhara (1971) report a synergistic effect of a mixed NPV-GV infection in *Pseudaletia unipuncta* based on an overall increase in host mortality, but that caused by GV is less than if it were acting alone, hence a (+/−) interaction. This kind of detailed information is needed to shed more light on mixed infections and their population dynamic consequences.

To date, there have been no experimental studies that have addressed the outcome of competition between viruses in mixed infections beyond the individual level, and once again the primary approach has been theoretical. A number of models have examined the evolutionary aspects of pathogen–pathogen interactions, such as the evolution of parasite virulence (Nowak and May, 1994; May and Anderson, 1983; Levin and Pimental, 1981). Hochberg and Holt (1990) explored the population dynamics of competing pathogens and showed that coexistence was determined by their relative transmission abilities and within-host interference (as with host–pathogen–parasitoid interactions described in Section IV.D). Andreasen and Pugliese (1995) showed that disease coexistence was possible with strong density-dependent regulation of the host, which also subsequently affected the relative transmission rates of the competing viruses.

The major drawback of these models is that they fail to include the possibility that hosts can be coinfected simultaneously. Since experimental evidence shows clearly that this is not the case, it would be a valuable exercise to incorporate this additional complexity and realism into the models. May and Nowak (1995) investigated, from an evolutionary viewpoint, the dynamics of different pathogen strains within the same host, analogous to coinfection of a host by two baculoviruses, and showed that virulence was the key parameter that determined whether different strains or species could coexist. Moreover, they showed that there was no limit to the number of pathogen variants that could evolve. As we saw in Section II.A, baculovirus isolates can consist of a wide range of genotypically distinct variants within the same host. Models dealing with coexistence of different pathogen strains could be applied directly to the phenotypic and population dynamic consequences of being a particular variant and to how baculovirus diversity is maintained.

D. Predators and Parasitoids

Understanding host–pathogen–predator interactions will have direct implications for the use of several natural enemies in integrated pest management strategies, as well as for concepts of species coexistence and community structure. The majority of interest has focused on the interaction between natural enemies at the individual level, but, common to most multitrophic systems, this issue has rarely been addressed at the population level.

1. Individual Level

Studies of baculovirus–predator interactions have shown that the predator is largely unaffected by the consumption of an infected prey (e.g., Vasconcelos *et al.*, 1996a; Young and Yearian, 1992). Anderson and May (1986) consider a theoretical interaction in which an established predator–prey system is invaded by a pathogen that only infects the prey. One outcome that is relevant to biocontrol suggests that if the prey is maintained below the threshold density necessary for pathogen persistence (see Section III.C), then quite clearly the pathogen cannot become established. On the other hand, the addition of the pathogen could depress the density of the prey population to levels that are unable to support the predator. These theoretical results have a parallel in many different fields where the superior competitor is the one that depresses the shared resource the most (e.g., plant competition theory (Tilman, 1990).

The interaction between baculoviruses and parasitoids, the most commonly investigated, is likely to be more intimate than simply one of predation, since the act of parasitism is also the means by which the parasitoid reproduces. Moreover, the baculovirus and developing parasitoid are in direct competition for a limited resource—the host—with such systems being analogous to the host–pathogen–pathogen interactions described above. The outcome of competition is determined largely by which parasite attacks first and the subsequent interval between coinvasion. When parasitism precedes pathogen infection, subsequent susceptibility to the disease may decrease (Murray *et al.*, 1995; Santiago-Alvarez and Caballero, 1990; Beegle and Oatman, 1974) or exhibit no change (Hochberg, 1991a; Santiago-Alvarez and Caballero, 1990; Eller *et al.*, 1988). Furthermore, successful parasitoid emergence from diseased hosts typically increases as the interval before virus infection increases (Murray *et al.*, 1995; Hochberg, 1991a; Caballero *et al.*, 1990; Cossentine and Lewis, 1986; Levin *et al.*, 1981; Beegle and Oatman, 1975; Irabagon and Brooks, 1974).

When pathogen infection precedes parasitism, the parasitoid generally fares badly, since a diseased host is typically physiologically unsuitable for parasitoid development, or the pathogen kills the host before the wasp is able to eclose (Hochberg, 1991a; Eller *et al.*, 1988; Cossentine and Lewis, 1986; Beegle and Oatman, 1975; Irabagon and Brooks, 1974). In the only studies to demonstrate a direct pathogenic effect of a baculovirus on an endoparasitoid larva, Kaya and Tanada (1972, 1973) found that a toxic factor was produced by certain strains of NPV- and GV-infecting *Pseudaletia unipuncta,* causing cessation of growth and eventual death in the braconid parasitoid *Apanteles militaris.*

Given the costs to the parasitoid of attacking infected prey, a number of species exhibit some degree of avoidance of infected hosts (Sait *et al.*, 1996; Caballero *et al.*, 1991; Versoi and Yendol, 1982), though in others there is no discrimination (Levin *et al.*, 1983; Beegle and Oatman, 1975). A parasitoid may become contaminated with and transmit the pathogen when it attacks infected hosts. Possible routes of transmission are by contamination of the host or its environment (e.g., Sait *et al.*, 1996; Hochberg, 1991b; Young and Yearian, 1990; Vail, 1981; Raimo *et al.*, 1977), or by direct injection of infectious virus particles through oviposition (e.g., Caballero *et al.*, 1990, 1991; Levin *et al.*, 1979, 1983; Beegle and Oatman, 1975; Irabagon and Brooks, 1974). Parasitoid-mediated pathogen transmission will enhance the dispersal of the pathogen in the environment in which the parasitoid forages for new (susceptible) prey. Such a process benefits the pathogen, since it will lead to a reduction in the host density threshold that is necessary for pathogen persistance (Onstad and Carruthers, 1990), but this may be offset by the increase in intensity of competition between the natural enemies.

The recent development of recombinant baculovirus pesticides requires concurrent assessment of possible adverse effects on nontarget parasitoids and predators. Moreover, natural enemies may play a crucial role in the dispersal of these viruses to nontarget hosts, perceived as a key issue in risk assessment. Not surprisingly, given the recent advent of recombinant viruses and the priority to understand the host–pathogen interaction first of all, there have been very few studies that investigate interactions with parasitoids and predators. McCutchen *et al.* (1996) recorded a reduction in the development time, and hence size, of *Microplitis croceipes* larvae developing in *Heliothis virescens* subsequently infected with toxin-expressing AcNPV. There was no effect on survival, but parasitoids were shown to transmit the recombinant virus mechanically. The population dynamic consequences of this phenomenon remain unexplored. Heinz *et al.* (1995) and McNitt *et al.* (1995) examined the effects of the same recombinant virus on a number of nontarget species and found no effect of inoculation or any oral toxicity. These laboratory studies do suggest that genetically modified baculoviruses may pose little threat to a diverse range of nontarget insects, but a great deal more investigation is undoubtedly required, particularly using controlled field studies.

2. Population Level

Evidence of the detrimental effects that pathogens can have on parasitoids in natural populations has come from investigations of epizootics in forest pests. Bird and Elgee (1957), for example, suggested that the disappearance of two introduced species of parasitoids that attacked the European spruce sawfly, *Gilpinia hercyniae,* was caused by a concurrent NPV epizootic. Further evidence has come following the use of baculoviruses as bioinsecticides, after which reduced rates of parasitism were recorded (Teakle *et al.*, 1985; Hamm and Hare, 1982; Vail *et al.*, 1972). Using a modeling approach, Berryman *et al.* (1990) examined the dynamic consequences of viral invasion of an established host–parasitoid interaction, based on data collected for the blackheaded budworm, *Acleris variana*. They showed that a virus epizootic destabilized the interaction and caused continuous outbreak cycles.

Also adopting a theoretical approach, Hochberg *et al.* (1990) explored a host–pathogen–parasitoid interaction at the population and community level and showed that a wealth of dynamic properties were exhibited by these complex interactions. Coexistence of natural enemies was enhanced if the superior ability of one to exploit prey (hosts) in the environment was balanced by the competitive superiority of the other within coinfected hosts. Experimental evidence suggests that parasitoids will be better at attacking and exploiting prey because of their active foraging strategies, while pathogens appear to hold the balance of power within hosts. In a laboratory study, Begon *et al.* (1996) investigated the long-term population dynamics of the Indian meal *P. interpunctella*, its GV, and the ichneumonid solitary endoparasitoid *Venturia canescens*. The component single- and two-species interactions fluctuated in cycles with a period of approximately one host generation, yet in the three-species system both the host and parasitoid exhibited multigeneration cycles, with periods of three to four host generations. As well as highlighting the complex link between natural enemy interactions and community structure, this study emphasized the vital role that baculoviruses can play as a dynamic force in population and community ecology, despite being notoriously inapparent in natural populations.

V. CONCLUSIONS

At an individual level we can formulate a reasonable picture of how baculoviruses interact with their hosts. However, investigation of several areas is still in its infancy, such as the mechanisms that underlie sublethal infections and the role of virus variation, which would benefit from a more systematic and rigorous approach as well as the application of molecular techniques. Much of what we understand about baculovirus ecology at the population level and in multitrophic interactions has been derived from the theoretical investigation of these complex systems and is far in advance of current empirical knowledge. However, the development of mathematical models has illustrated that they still have much to gain from the inclusion of more realism, such as host- and age-dependent pathogenicity and other transmission-related parameters of baculoviruses. Nonetheless, this review demonstrates that baculoviruses have the potential to exert profound and complex influences over species population dynamics, coexistence, and community structure. In order to test and support model predictions and to assess the true role of baculoviruses at all levels of the host–pathogen interaction, there is a clear need to develop experimental systems with an interdisciplinary approach, from the molecular level to the population level, in the field. This information is vital if we are to design more effective biological control programs based around the natural features of baculoviruses and also if we are to assess accurately any risks that might be attached to their use in a natural or genetically modified form.

ACKNOWLEDGMENTS. We should like to thank Judy Myers and Bob Possee for providing us with unpublished data, and Lois Miller, Greg Dwyer, and Mike Begon for helpful criticism of the chapter.

REFERENCES

Abot, A. R., Moscardi, F., Fuxa, J. R., Sosa-Gómez, D. R., and Richter, A. R., 1995, Susceptibility of populations of *Anticarsia gemmatalis* (Lepidoptera: Noctuidae) from Brazil and the United States to a nuclear polyhedrosis virus, *J. Entomol. Sci.* **30:**62.

Allaway, G. P., and Payne, C. C., 1983, A biochemical and biological comparison of three European isolates of nuclear polyhedrosis viruses from *Agrotis segetum, Arch. Virol.* **75:**43.

Anderson, R. M., 1982, Theoretical basis for the use of pathogens as biological control agents of pest species, *Parasitology* **84:**3.

Anderson, R. M., and May, R. M., 1981, The population dynamics of microparasites and their invertebrate hosts, *Philos. Trans. R. Soc. Lond. Biol. Sci.* **291:**451.

Anderson, R. M., and May, R. M., 1986, The invasion, persistence, and spread of infectious diseases within animal and plant communities, *Philos. Trans. R. Soc. Lond. Biol. Sci.* **314:**533.

Andreasen, V., 1989, Disease regulation of age-structured host populations, *Theor. Pop. Biol.* **36:**214.

Andreasen, V., and Pugliese, A., 1995, Pathogen coexistence induced by density-dependent host mortality, *J. Theor. Biol.* **177:**159.

Arne, C. N., and Nordin, G. L., 1995, Enhancement of indexes of viral infection by simultaneously administering *Helicoverpa zea* and *Autographa californica* nuclear polyhedrosis viruses to *Helicoverpa zea* (Boddie), *J. Invertebr. Pathol.* **66:**18.

Bailey, N. T. J., 1975, *The Mathematical Theory of Infectious Disease and Its Application,* 2nd ed., Griffin, London.

Barber, K. N., Kaupp, W. J., and Holmes, S. B., 1993, Specificity testing of the nuclear polyhedrosis virus of the gypsy moth, *Lymantria dispar* (L.) (Lepidoptera: Lymantriidae), *Can. Entomol.* **125:**1055.

Beegle, C. C., and Oatman, E. R., 1974, Differential susceptibility of parasitized and nonparasitized larvae of *Trichoplusia ni* to a nuclear polyhedrosis virus, *J. Invertebr. Pathol.* **24:**188.

Beegle, C. C., and Oatman, E. R., 1975, Effect of a nuclear polyhedrosis virus on the relationship between *Trichoplusia ni* (Lepidoptera: Noctuidae) and the parasite, *Hyposoter exiguae* (Hymenoptera: Ichneumonidae), *J. Invertebr. Pathol.* **25:**59.

Begon, M., and Boots, M., 1995, Covert infection or, simply, resistance? *Funct. Ecol.* **9:**549.

Begon, M., and Bowers, R. G., 1994, Host-host-pathogen models and microbial pest control: The effect of host self regulation, *J. Theor. Biol.* **169:**275.

Begon, M., and Bowers, R. G., 1995, Beyond host-pathogen dynamics, in: *Ecology of Infectious Diseases in Natural Populations* (B. T. Grenfell and A. P. Dobson, eds.), pp. 478–509, Cambridge University Press, Cambridge, England.

Begon, M., Sait, S. M., and Thompson, D. J., 1996, Predator–prey cycles with period shifts between two- and three-species systems, *Nature* **381:**311.

Berryman, A. A., 1996, What causes population cycles of forest Lepidoptera? *Trends Ecol. Evol.* **11:**28.

Berryman, A. A., Millstein, J. A., and Mason, R. R., 1990, Modelling Douglas-fir tussock moth population dynamics: The case for simple theoretical models, in: *Population Dynamics of Forest Insects* (A. D. Watt, S. R. Leather, M. D. Hunter, and N. A. C. Kidd, eds.), pp. 369–380, Intercept, Andover, Hampshire, England.

Bird, F. T., and Elgee, D. E., 1957, A virus and introduced parasites as factors controlling the European spruce sawfly, *Diprion hercyniae* (Htg), in central New Brunswick, *Can. Entomol.* **89**:371.

Bishop, D. H. L., Hirst, M. L., Possee, R. D., and Cory, J. S., 1995, Genetic engineering of microbes: Virus insecticides—A case study, in: *50 Years of Microbials* (G. K. Darby, P. A. Hunter, and A. D. Russell, eds.), pp. 249–277, Cambridge University Press, Cambridge, England.

Boots, M., and Begon, M., 1993, Trade-offs with resistance to a granulosis virus in the Indian meal moth, examined by a laboratory evolution experiment, *Funct. Ecol.* **7**:528.

Bowers, R. G., and Begon, M., 1991, A host-host-pathogen model with free living infective stages, applicable to microbial pest control, *J. Theor. Biol.* **148**:305.

Bowers, R. G., Begon, M., and Hodgkinson, D. E., 1993, Host–pathogen population cycles in forest insects? Lessons from simple models reconsidered, *Oikos* **67**:529.

Bowers, R. G., Boots, M., and Begon, M., 1994, Life-history trade-offs and the evolution of pathogen resistance: competition between host strains, *Proc. R. Soc. Lond.* **257**:247.

Briese, D. T., 1986, Host resistance to microbial control agents, in: *Biological Plant and Health Protection* (J. M. Franz, ed.), pp. 233–256, Fischer Verlag, Stuttgart.

Briese, D. T., and Mende, H. A., 1981, Differences in susceptibility to a granulosis virus between field populations of the potato moth, *Phthorimaea operculella* (Zeller) (Lepidoptera: Gelechiidae), *Bull. Entomol. Res.* **71**:11.

Briese, D. T., and Mende, H. A., 1983, Selection for increased resistance to a granulosis virus in the potato moth, *Phthorimaea operculella* (Zeller) (Lepidoptera: Gelechiidae), *Bull. Entomol. Res.* **73**:1.

Briggs, C. J., and Godfray, H. C. J., 1995, The dynamics of insect–pathogens in stage-structured populations, *Am. Nat.* **145**:855.

Briggs, C. J., and Godfray, H. C. J., 1996, The dynamics of insect–pathogen interactions in seasonal environments, *Theor. Pop. Biol.* **148**:807.

Briggs, C. J., Hails, R. S., Barlow, N. D., and Godfray, H. C. J., 1995. The dynamics of insect–pathogen interactions, in: *Ecology of Infectious Diseases in Natural Populations* (B. T. Grenfell and A. P. Dobson, eds.), pp. 478–509, Cambridge University Press, Cambridge, England.

Brown, G. C., 1987, Modeling, in: *Epizootiology of Insect Diseases* (J. R. Fuxa and Y. Tanada, eds.), pp. 43–68, John Wiley, New York.

Brown, S. E., Maruniak, J. E., and Knudson, D. L., 1985, Baculovirus (MNPV) genomic variants: Characterization of *Spodoptera exempta* MNPV DNAs and comparison with other *Autographa californica* MNPV DNAs, *J. Gen. Virol.* **66**:2431.

Burand, J. P., and Park, E. J., 1992, Effect of a nuclear polyhedrosis virus infection on the development and pupation of gypsy moth larvae, *J. Invertebr. Pathol.* **60**:171.

Caballero, P., Vargas-Osuna, E., and Santiago-Alvarez, C., 1990, Development of *Apanteles telengai* (Hym., Braconidae) and *Campoletis annulata* (Hym., Ichneumonidae) in granulosis virus (GV) infected *Agrotis segetum* (Lep., Noctuidae) larvae, *J. Appl. Entomol.* **10**:358.

Caballero, P., Vargas-Osuna, E., and Santiago-Alvarez, C., 1991, Parasitization of granulosis-virus infected and noninfected *Agrotis segetum* larvae and the virus transmission by three hymenopteran parasitoids, *Entomol. Exp. Appl.* **58**:55.

Cherry, C. L., and Summers, M. D., 1985, Genotypic variation among wild isolates of two nuclear polyhedrosis viruses isolated from *Spodoptera littoralis*, *J. Invertebr. Pathol.* **46**:289.

Comins, H. N., Hassell, M. P., and May, R. M., 1992, The spatial dynamics of a host–parasitoid system, *J. Anim. Ecol.* **61**:735.

Corsaro, B. G., and Fraser, M. J., 1987, Characterization of genotypic and phenotypic variation in plaque-purified strains of HzSNPV Elkar Isolate, *Intervirology* **28:**185.

Cory, J. S., Hirst, M. L., Williams, T., Hails, R. S., Goulson, D., Green, B. M., Carty, T. M., Possee, R. D., Cayley, P. J., and Bishop, D. H. L., 1994, Field trial of a genetically improved baculovirus insecticide, *Nature* **370:**138.

Cossentine, J. E., and Lewis, L. C., 1986, Potential of *Bonnetia comta* (Diptera: Tachinidae) in combination with a nuclear polyhedrosis virus to control *Agrotis ipsilon* (Lepidoptera: Noctuidae) larvae on corn, *Environ. Entomol.* **15:**360.

Crawford, A. M., and Kalmakoff, J., 1977, A host–virus interaction in a pasture habitat, *J. Invertebr. Pathol.* **29:**81.

Crawford, A. M., and Kalmakoff, J., 1978, Transmission of *Wiseana* spp. nuclear polyhedrosis virus in the pasture habitat, *N.Z. J. Agricult. Res.* **21:**521.

Crook, N. E., Spencer, R. A., Payne, C. C., and Leisy, D. J., 1985, Variation in *Cydia pomenella* granulosis virus isolates and physical maps of the DNA from three variants, *J. Gen. Virol.* **66:**2423.

Crozier, G., and Ribiero, H. C. T., 1992, Recombination as a possible major cause of genetic heterogeneity in *Anticarsia gemmatalis* nuclear polyhedrosis virus wild populations, *Virus Res.* **26:**183.

Cunningham, J. C., and Entwistle, P. F., 1981, Control of sawflies by baculoviruses, in: *Microbial Control of Pests and Plant Diseases 1970–1980* (H. D. Burges, ed.), pp. 379–408, Academic Press, London.

D'Amico, V., and Elkinton, J. S., 1995, Rainfall effects on the transmission of gypsy moth (Lepidoptera: Lymantriidae) nuclear polyhedrosis virus, *Environ. Entomol.* **24:**1144.

D'Amico, V., Elkinton, J. S., Dwyer, G., and Burand, J. P., 1996, Virus transmission in gypsy moths is not a simple mass action process, *Ecology* **77:**201.

Doyle, C. J., Hirst, M. L., Cory, J. S., and Entwistle, P. F., 1990, Risk assessment studies: Detailed host range testing of wild-type cabbage moth, *Mamestra brassicae* (Lepidoptera: Noctuidae), nuclear polyhedrosis virus, *Appl. Environ. Microbiol.* **56:**2704.

Duffey, S. S., Hoover, K., Bonning, B., and Hammock, B. D., 1995, The impact of host plant on the efficacy of baculoviruses, *Rev. Pest. Toxicol.* **3:**137.

Dwyer, G., 1991, The roles of density, stage, and patchiness in the transmission of an insect virus, *Ecology* **72:**559.

Dwyer, G., 1992, On the spatial spread of insect pathogens: theory and experiment, *Ecology* **73:**479.

Dwyer, G., 1994, Density-dependence and spatial structure in the dynamics of insect–pathogens, *Am. Nat.* **143:**533.

Dwyer, G., and Elkinton, J. S., 1993, Using simple models to predict virus epizootics in gypsy moth populations, *J. Anim. Ecol.* **62:**1.

Dwyer, G., and Elkinton, J. S., 1995, Host dispersal and the spatial spread of insect pathogens, *Ecology* **76:**1262.

Elleman, C. J., and Entwistle, P. F., 1982, A study of the glands on cotton responsible for the high pH and cation concentration of the leaf surface, *Ann. Appl. Biol.* **100:**553.

Elleman, C. J., and Entwistle, P. F., 1985, Inactivation of a nuclear polyhedrosis virus on cotton by the substances produced by the cotton leaf surface gland, *Ann. Appl. Biol.* **106:**83.

Eller, F. J., Boucias, D. G., and Tumlinson, J. H., 1988, Interactions between *Microplitis croceipes* (Hymenoptera: Braconidae) and a nuclear polyhedrosis virus of *Heliothis zea* (Lepidoptera: Noctuidae), *Environ. Entomol.* **17:**977.

Engelhard, E. K., and Volkman, L. E., 1995, Developmental resistance in fourth instar *Trichoplusia ni* orally inoculated with *Autographa californica* M nuclear polyhedrosis virus, *Virology* **209:**384.

Entwistle, P. F., and Evans, H. F., 1985, Viral control, in: *Comprehensive Insect Physiology, Biochemistry, and Pharmacology* (L. I. Gilbert and G. A. Kerkut, eds.), Vol. 12, pp. 347–412, Pergamon, Oxford, England.

Entwistle, P. F., Adams, P. H. W., and Evans, H. F., 1977a, Epizootiology of a nuclear polyhedrosis virus in European spruce sawfly (*Gilpinia hercyniae*): Birds as dispersal agents during winter, *J. Invertebr. Pathol.* **30:**15.

Entwistle, P. F., Adams, P. H. W., and Evans, H. F., 1977b, Epizootiology of a nuclear polyhedrosis virus in European spruce sawfly (*Gilpinia hercyniae*): The status of birds as dispersal agents of the virus during the larval season, *J. Invertebr. Pathol.* **29:**354.

Entwistle, P. F., Adams, P. H. W., Evans, H. F., and Rivers, C. F., 1983, Epizootiology of a nuclear polyhedrosis virus (Baculoviridae) in European spruce sawfly (*Gilpinia hercyniae*): Spread of disease from small epicentres in comparison with spread of baculovirus disease in other hosts, *J. Appl. Ecol.* **20:**473.

Entwistle, P. F., Forkner, A. C., Green, B. M., and Cory, J. S., 1993, Avian dispersal of nuclear polyhedrosis virus after induced epizootics in the pine beauty moth, *Panolis flammea* (Lepidoptera: Noctuidae), *Biol. Control* **3:**61.

Evans, H. F., 1986, Ecology and epizootiology of baculoviruses, in: *The Biology of Baculoviruses* (R. R. Granados and B. A. Federici, eds.), Vol. 2, pp. 89–132, CRC Press, Boca Raton, FL.

Felton, G. W., Duffey, S. S., Vail, P. V., Kaya, H. K., and Manning, J., 1987, Interaction of nuclear polyhedrosis virus with catechols: potential incompatibility for host–plant resistance against noctuid larvae, *J. Chem. Ecol.* **13:**945.

Forschler, B. T., Young, S. Y., and Felton, G. W., 1992, Diet and the susceptibility of *Helicoverpa zea* (Noctuidae: Lepidoptera) to a nuclear polyhedrosis virus, *Environ. Entomol.* **21:**1220.

Freeland, W. J., 1983, Parasites and the coexistence of animal host species, *Am. Nat.* **121:**223.

Fuxa, J. R., 1982, Prevalence of viral infections in populations of fall armyworm, *Spodoptera frugiperda,* in southeastern Louisiana, *Environ. Entomol.* **11:**239.

Fuxa, J. R., 1987, *Spodoptera frugiperda* susceptibility to nuclear polyhedrosis virus isolates with reference to insect migration, *Environ. Entomol.* **16:**218.

Fuxa, J. R., 1993, Insect resistance to viruses, in: *Parasites and Pathogens of Insects* (N. E. Beckage, S. N. Thompson, and B. A. Federici, eds.), vol. 2, *Pathogens,* pp. 197–209, Academic Press, San Diego, CA.

Fuxa, J. R., and Geaghan, J. P., 1983, Multiple regression analysis of factors affecting prevalence of nuclear polyhedrosis virus in *Spodoptera frugiperda* (Lepidoptera: Noctuidae) populations, *Environ. Entomol.* **12:**311.

Fuxa, J. R., and Richter, A. R., 1989, Reversion of resistance by *Spodoptera frugiperda* to nuclear polyhedrosis virus, *J. Invertebr. Pathol.* **53:**52.

Fuxa, J. R., and Richter, A. R., 1990, Response of nuclear polyhedrosis virus-resistant *Spodoptera frugiperda* larvae to other pathogens and to chemical insecticides, *J. Invertebr. Pathol.* **55:**272.

Fuxa, J. R., and Richter, A. R., 1991, Selection for an increased rate of vertical transmission of *Spodoptera frugiperda* (Lepidoptera: Noctuidae) nuclear polyhedrosis virus, *Environ. Entomol.* **20:**603.

Fuxa, J. R., and Richter, A. R., 1993, Lack of vertical transmission in *Anticarsia gemmatalis* (Lepidoptera: Noctuidae) nuclear polyhedrosis virus, a pathogen not indigenous to Louisiana, *Environ. Entomol.* **22:**425.

Fuxa, J. R., and Richter, A. R., 1994, Distance and rate of spread of *Anticarsia gemmatalis* (Lepidoptera: Noctuidae) nuclear polyhedrosis virus released into soybean, *Environ. Entomol.* **23:**1308.

Fuxa, J. R., Mitchell, F. L., and Richter, A. R., 1988, Resistance of *Spodoptera frugiperda* (Lep.: Noctuidae) to a nuclear polyhedrosis virus in the field and laboratory, *Entomophaga* **33:**55.

Fuxa, J. R., Richter, A. R., and Strother, M. S., 1993, Detection of *Anticarsia gemmatalis* nuclear polyhedrosis virus in predatory arthropods and parasitoids after viral release in Louisiana soybean, *J. Entomol. Sci.* **28:**51.

Gelbic, I., and Metwally, M. M., 1981, Changes in the development of male germinal cells in *Spodoptera littoralis* caused by effects of juvenoids (Lepidoptera, Noctuidae), *Acta Entomol. Bohemoslov.* **78:**10.

Gettig, R. R., and McCarthy, W. J., 1982, Genotypic variation among wild isolates of *Heliothis* spp nuclear polyhedrosis viruses from different regions, *Virology* **117**:245.

Getz, W. M., and Pickering, J., 1983, Epidemic models: thresholds and population regulation, *Am. Nat.* **121**:892.

Gilpin, M. E., and Hanski, I., 1991, *Metapopulation Dynamics: Empirical and Theoretical Investigations*, Academic Press, London.

Goto, C., 1990, Enhancement of a nuclear polyhedrosis virus (NPV) infection by a granulosis virus (GV) isolated from the spotted cutworm, *Xestia-c-nigrum* L. (Lepidoptera, Noctuidae), *Appl. Entomol. Zool.* **25**:135.

Goto, C., Minobe, Y., and Izuka, T., 1992, Restriction endonuclease analysis and mapping of the genomes of granulosis viruses isolated from *Xestia c-nigrum* and five other noctuid species, *J. Gen Virol.* **73**:1491.

Goulson, D., and Cory, J. S., 1995, Sublethal effects of baculovirus in the cabbage moth, *Mamestra brassicae, Biol. Control* **5**:361.

Goulson, D., and Hauxwell, C., 1995, Resistance or covert infection; baculovirus studies reexamined, *Funct. Ecol.* **9**:548.

Goulson, D., Hails, R. S., Williams, T., Hirst, M. L., Vasconcelos, S. D., Green, B. M., Carty, T. M., and Cory, J. S., 1995, Transmission dynamics of a virus in a stage-structured insect population, *Ecology* **76**:392.

Hails, R. S., Cory, J. S., Donnelly, C. A., Hirst, M. L., Speight, M. R., Goulson, D., and Williams, T., 1997, The role of environmental reservoirs in insect–pathogen interactions, *Proc. R. Soc. Lond. B*, in review.

Hamm, J. J., and Hare, W. W., 1982, Application of entomopathogens in irrigation water for control of the fall armyworms and corn earworms (Lepidoptera: Noctuidae) on corn, *J. Econ. Entomol.* **75**:1074.

Hamm, J. J., and Styer, E. L., 1985, Comparative pathology of isolates of *Spodoptera frugiperda* nuclear polyhedrosis virus in *S. frugiperda* and *S. exigua, J. Gen. Virol.* **66**:1249.

Harper, J. D., 1986, Interactions between baculoviruses and other entomopathogens, chemical pesticides, and parasitoids, in: *The Biology of Baculoviruses* (R. R. Granados and B. A. Federici, eds.), Vol. 1, pp. 133–135, CRC Press, Boca Raton, FL.

Hatfield, P. R., and Entwistle, P. F., 1988, Biological and biochemical comparison of nuclear polyhedrosis virus isolates pathogenic for the oriental armyworm, *Mythimna separata* (Lepidoptera: Noctuidae), *J. Invertebr. Pathol.* **52**:168.

Hawtin, R. E., Arnold, K., Ayres, M. D., Zanotto, P. M., Howard, S. C., Gooday, G. W., Chappell, L. H., Kitts, P. A., King, L. A., and Possee, R. D., 1995, Identification and preliminary characterization of a chitinase gene in the *Autographa californica* nuclear polyhedrosis virus, *Virology* **212**:673.

Heinz, K. M., McCutchen, B. F., Herrmann, R., Parrella, M. P., and Hammock, B. D., 1995, Direct effects of recombinant nuclear polyhedrosis viruses on selected nontarget organisms, *J. Econ. Entomol.* **88**:259.

Hochberg, M. E., 1989a, *The Population Dynamics of Host–Parasitoid–Pathogen Interactions*, PhD thesis, University of London.

Hochberg, M. E., 1989b, The potential role of pathogens in biological control, *Nature* **337**: 262.

Hochberg, M. E., 1991a, Intra-host interactions between a braconid endoparasitoid, *Apanteles glomeratus*, and a baculovirus for larvae of *Pieris brassicae, J. Anim. Ecol.* **60**:51.

Hochberg, M. E., 1991b, Extra-host interactions between a braconid endoparasitoid, *Apanteles glomeratus*, and a baculovirus for larvae of *Piers brassicae, J. Anim. Ecol.* **60**:65.

Hochberg, M. E., and Holt, R. D., 1990, The coexistence of competing parasites. I. The role of cross-species infection, *Am. Nat.* **136**:517.

Hochberg, M. E., Hassell, M. P., and May, R. M., 1990, The dynamics of host–parasitoid–pathogen interactions, *Am. Nat.* **135**:74.

Holt, R. D., 1977, Predation, apparent competition, and the structure of prey communities, *Theor. Pop. Biol.* **12**:197.

Holt, R. D., and Pickering, J., 1985, Infectious disease and species coexistence: A model in Lotka-Volterra form, *Am. Nat.* **126:**196.
Hostetter, D. L., and Puttler, B., 1991, A new broad host spectrum nuclear polyhedrosis virus isolated from a celery looper, *Anagrapha falcifera* (Kirby), (Lepidoptera: Noctuidae), *Environ. Entomol.* **20:**1480.
Hughes, D. S., Possee, R. D., and King, L. A., 1993, Activation and detection of a latent baculovirus resembling *Mamestra brassicae* nuclear polyhedrosis virus in *M. brassicae* insects, *Virology* **194:**608.
Hughes, D. S., Possee, R. D., and King, L. A., 1997, Evidence for the presence of a low-level, persistent baculovirus infection of *Mamestra brassicae* insects, *J. Gen. Virol.*, in press.
Hughes, P. R., Gettig, R. R., and McCarthy, W. J., 1983, Comparison of the time-mortality response of *Heliothis zea* to 14 isolates of *Heliothis* nuclear polyhedrosis virus, *J. Invertebr. Pathol.* **41:**256.
Hunter, M. D., and Schultz, J. C., 1993, Induced plant defences breached? Phytochemical induction protects an herbivore from disease, *Oecologia* **94:**195.
Ignoffo, C., and Couch, T., 1981, The nucleopolyhedrosis virus of *Heliothis* species as a microbial insecticide, in: *Microbial Control of Pests and Plant Diseases 1970–1980* (H. D. Burges, ed.), pp. 330–362, Academic Press, London.
Ignoffo, C. M., and Hostetter, D. L., 1977, Summary of the environmental stability of microbial insecticides, *Misc. Publ. Entomol. Soc. Am.* **10:**117.
Irabagon, T. A., and Brooks, W. M., 1974, Interaction of *Campoletis sonorensis* and a nuclear polyhedrosis virus in larvae of *Heliothis virescens, J. Econ. Entomol.* **67:**229.
Jaques, R. P., 1970, Application of viruses to soil and foliage for control of the cabbage looper and imported cabbageworm, *J. Invertebr. Pathol.* **15:**328.
Jaques, R. P., 1974a, Occurrence and accumulation of viruses of *Trichoplusia ni* in treated field plots, *J. Invertebr. Pathol.* **23:**140.
Jaques, R. P., 1974b, Occurrence and accumulation of the granulosis virus of *Pieris rapae* in treated field plots, *J. Invertebr. Pathol.* **23:**351.
Jones, K. A., Moawad, G., McKinley, D. J., and Grzywacz, D., 1993, The effect of natural sunlight on *Spodoptera littoralis* nuclear polyhedrosis virus, *Biocontrol Sci. Technol.* **3:**189.
Kaomini, M., and Roush, R. T., 1988, Absence of response to selection for resistance to nucleopolyhedrosis virus in *Heliothis virescens* (F.) (Lepidoptera: Noctuidae), *J. Entomol. Sci.* **23:**379.
Kaya, H. K., and Tanada, Y., 1972, Response of *Apanteles militaris* to a toxin produced in a granulosis-virus-infected host, *J. Invertebr. Pathol.* **19:**1.
Kaya, H. K., and Tanada, Y., 1973, Hemolymph factor in armyworm larvae infected with a nuclear polyhedrosis virus toxic to *Apanteles militaris, J. Invertebr. Pathol.* **2:**211.
Keating, S. T., and Yendol, W. G., 1987, Influence of selected host plants on gypsy moth (Lepidoptera: Lymantriidae) larval mortality caused by a baculovirus, *Environ. Entomol.* **16:**459.
Keating, S. T., McCarthy, W. J., and Yendol, W. G., 1988a, Gypsy moth (*Lymantria dispar*) larval susceptibility to a baculovirus affected by selected nutrients, hydrogen ions (pH), and plant allelochemicals in artificial diets, *J. Invertebr. Pathol.* **54:**165.
Keating, S. T., Yendol, W. G., and Schultz, J. C., 1988b, Relationship between susceptibility of gypsy moth larvae (Lepidoptera: Lymantriidae) to a baculovirus and host plant foliage constituents, *Environ. Entomol.* **17:**952.
Keating, S. T., Hunter, M. D., and Schultz, J. C., 1990, Leaf phenolic inhibition of gypsy moth nuclear polyhedrosis virus, *J. Chem. Ecol.* **16:**1445.
Keddie, B. A., Aponte, G. W., and Volkman, L. E., 1989, The pathway of infection of *Autographa californica* nuclear polyhedrosis virus in an insect host, *Science* **243:**1728.
Kelly, P. M., and Entwistle, P. F., 1988, *In vivo* mass production in the cabbage moth (*Mamestra brassicae*) of a heterologous (*Panolis*) and a homologous (*Mamestra*) nuclear polyhedrosis virus, *J. Virol. Methods* **19:**249.

Knell, J. D., and Summers, M. D., 1981, Investigation of genetic heterogeneity in wild isolates of *Spodoptera frugiperda* nuclear polyhedrosis virus by restriction endonuclease analysis of plaque-purified variants, *Virology* **112:**190.

Knell, R. J., Begon, M., and Thompson, D. J., 1996, Transmission dynamics of *Bacillus thuringiensis* infecting *Plodia interpunctella*: A test of the mass action assumption with an insect pathogen, *Proc. R. Soc. Lond. B* **263:**75.

Lautenschlager, R. A., Podgwaite, J. D., and Watson, D. E., 1980, Natural occurrence of the nucleopolyhedrosis virus of the gypsy moth, *Lymantria dispar* (Lep.: Lymantriidae) in wild birds and mammals, *Entomophaga* **25:**261.

Lee, H. H., and Miller, L. K., 1978, Isolation of genotypic variants of *Autographa californica* nuclear polyhedrosis virus, *J. Virol.* **27:**754.

Levin, D. B., Laing, J. E., and Jaques, R. P., 1979, Transmission of granulosis virus by *Apanteles glomeratus* to its host *Pieris rapae, J. Invertebr. Pathol.* **34:**317.

Levin, D. B., Laing, J. E., and Jaques, R. P., 1981, Interactions between *Apanteles glomeratus* (L.) (Hymenoptera: Braconidae) and granulosis virus in *Pieris rapae* (L.) (Lepidoptera: Pieridae), *Environ. Entomol.* **10:**65.

Levin, D. B., Laing, J. E., Jaques, R. P., and Corrigan, J. E., 1983, Transmission of the granulosis virus of *Pieris rapae* (Lepidoptera: Pieridae) by the parasitoid *Apanteles glomeratus* (Hymenoptera: Braconidae), *Environ. Entomol.* **12:**166.

Levin, S., and Pimental, D., 1981, Selection of intermediate rates of increase in parasite–host systems, *Am. Nat.* **117:**308.

Liu, W., Levin, S., and Iwasa, Y., 1986, Influence of nonlinear incidence rates upon the behavior of SIRS epidemiological models, *J. Math. Biol.* **23:**187.

Longworth, J. F., and Cunningham, J. C., 1968, The activation of occult nuclear-polyhedrosis viruses by foreign nuclear polyhedra, *J. Invertebr. Pathol.* **10:**361.

Lowe, R. E., and Paschke, J. D., 1968, Simultaneous infection with the nucleopolyhedrosis and granulosis viruses of *Trichoplusia ni, J. Invertebr. Pathol.* **12:**86.

Lynn, D. E., Shapiro, M., and Dougherty, E. M., 1993, Selection and screening of clonal isolates of the Abington strain of gypsy moth nuclear polyhedrosis virus, *J. Invertebr. Pathol.* **62:** 19.

Maeda, S., Mukohara, Y., and Kondo, A., 1990, Characteristically distinct isolates of the nuclear polyhedrosis virus from *Spodoptera litura, J. Gen. Virol.* **71:**2631.

Martignoni, M. E., and Schmid, P., 1961, Studies on the resistance to virus infections in natural populations of Lepidoptera, *J. Insect Pathol.* **3:**62.

May, R. M., and Anderson, R. M., 1983, Epidemiology and genetics in the coevolution of parasites and hosts, *Proc. R. Soc. Lond. B* **219:**281.

May, R. M., and Nowak, M. A., 1995, Coinfection and the evolution of parasite virulence, *Proc. R. Soc. Lond. B* **261:**209.

McCutchen, B. F., Herrmann, R., Heinz, K. M., Parrella, M. P., and Hammock, M. D., 1996, Effects of recombinant baculoviruses on a nontarget endoparasitoid of *Heliothis virescens, Biol. Control* **6:**45.

McLeod, P. J., Yearian, W. C., and Young, S. Y., 1977, Inactivation of *Baculovirus heliothis* by ultraviolet irradiation, dew and temperature, *J. Invertebr. Pathol.* **30:**237.

McLeod, P. J., Young, S. Y., and Yearian, W. C., 1982, Application of a baculovirus of *Pseudoplusia includens* to soybean: Efficacy and seasonal persistence, *Environ. Entomol.* **11:**412.

McNitt, L., Espelie, K. E., and Miller, L. K., 1995, Assessing the safety of toxin-producing baculovirus biopesticides to a nontarget predator, the social wasp *Polistes metricus* Say., *Biol. Control* **5:**267.

Melamed-Madjar, V., and Raccah, B., 1979, The transstadial and vertical transmission of a granulosis virus from the corn borer *Sesamia nonagroides, J. Invertebr. Pathol.* **33:**259.

Murray, D. A. H., Monsour, C. J., Teakle, R. E., Rynne, K. P., and Bean, J. A., 1995, Interactions between nuclear polyhedrosis virus and three larval parasitoids of *Helicoverpa armigera* (Hübner) (Lepidoptera: Noctuidae), *J. Aust. Entomol. Soc.* **34:**319.

Murray, K. D., Shields, K. S., Burand, J. P., and Elkinton, J. S., 1991, The effect of gypsy moth metamorphosis on the development of nuclear polyhedrosis infection, *J. Invertebr. Pathol.* **57:**352.

Myers, J. H., 1988, Can a general hypothesis explain population cycles of forest Lepidoptera? *Adv. Ecol. Res.* **18:**179.

Myers, J. H., 1990, Population cycles of western tent caterpillars: Experimental introductions and synchrony of fluctuations, *Ecology* **71:**986.

Myers, J. H., and Kukan, B., 1995, Changes in the fecundity of tent caterpillars: A correlated character of disease resistance or sublethal effects of disease? *Oecologia* **103:**475.

Neelgund, Y. F., and Mathad, S. B., 1978, Transmission of nuclear polyhedrosis virus in a laboratory population of the armyworm, *Mythimna (Pseudaletia) separata, J. Invertebr. Pathol.* **31:**143.

Nowak, M. A., and May, R. M., 1994, Superinfection and the evolution of parasite virulence, *Proc. R. Soc. Lond. B* **255:**81.

Ohkawa, T., Majima, K., and Maeda, S., 1994, A cysteine protease encoded by the baculovirus *Bombyx mori* nuclear polyhedrosis virus, *J. Virol.* **68:**6619.

Olofsson, E., 1988, Dispersal of the nuclear polyhedrosis virus of *Neodiprion sertifer* from soil to pine foliage with dust, *Entomol. Exp. Appl.* **46:**181.

Onstad, D. W., and Carruthers, R. I., 1990, Epizootiological models in insect diseases, *Ann. Rev. Entomol.* **35:**399.

O'Reilly, D. R., and Miller, L. K., 1989, A baculovirus blocks insect moulting by producing ecdysteroid UDP–glucosyltransferase, *Science* **245:**1110.

Patil, U. R., Savanurmath, C. J., Mathad, S. B., Aralaguppi, P. I., and Ingalhalli, S. S., 1989, Effects of nuclear polyhedrosis virus on the growth, development and reproduction in surviving generations of the armyworm *Mythimna (Pseudaletia) separata* (Walker), *J. Appl. Ent.* **108:**527.

Paul, R. D., 1997, *Evolution and Interaction of insect pathogens,* Ph.D. thesis, University of Reading, Reading, England.

Payne, C. C., 1986, Insect pathogenic viruses as pest control agents, in: *Biological and Plant Health Protection* (J. M. Franz, ed.) pp. 183–200, Fischer Verlag, Stuttgart.

Raimo, B., Reardon, R. C., and Podgwaite, J. D., 1977, Vectoring gypsy moth nuclear polyhedrosis virus by *Apanteles melanoscelus* (Hym.: Braconidae), *Entomophaga* **22:**207.

Regnière, J., 1984, Vertical transmission of diseases and population dynamics of insects with discrete populations: a model, *J. Theor. Biol.* **107:**287.

Richards, A. R., 1993, *Studies into the Ecology of a Lymantriid–Nuclear Polyhedrosis Virus Interaction,* D. Phil thesis, University of Oxford, Oxford, England.

Richter, A. R., Fuxa, J. R., and Abdel-Fattah, M., 1987, Effect of host plant on the susceptibility of *Spodoptera frugiperda* (Lepidoptera:Noctuidae) to a nuclear polyhedrosis virus, *Environ. Entomol.* **16:**1004.

Riddiford, L. M., and Williams, C. M., 1967, The effects of juvenile hormone analogues on the embryonic development of silkworms, *Proc. Natl. Acad. Sci. USA* **57:**595.

Ritter, K. S., and Tanada, Y., 1978, Interference between two nuclear polyhedrosis viruses of the armyworm, *Pseudaletia unipuncta* (Lep.:Noctuidae), *Entomophaga* **23:**349.

Rothman, L. D., and Myers, J. H., 1994, Nuclear polyhedrosis virus treatment effect on reproductive potential of western tent caterpillar (Lepidoptera: Lasiocampidae), *Environ. Entomol.* **23:**864.

Rothman, L. D., and Myers, J. H., 1996a, Debilitating effects of viral diseases on host Lepidoptera, *J. Invertebr. Pathol.* **67:**1.

Rothman, L. D., and Myers, J. H., 1996b, Is fecundity correlated with resistance to viral disease in the western tent caterpillar? *Ecol. Ent.* **21:**396.

Sait, S. M., Begon, M., and Thompson, D. J., 1994a, The effects of a sublethal infection in the Indian meal moth, *Plodia interpunctella, J. Anim. Ecol.* **63:**541.

Sait, S. M., Begon, M., and Thompson, D. J., 1994b, Long term population dynamics of the Indian meal moth *Plodia interpunctella* and its granulosis virus, *J. Anim. Ecol.* **63:**861.

Sait, S. M., Begon, M., Thompson, D. J., and Harvey, J. A., 1996, Parasitism of baculovirus-infected *Plodia interpunctella* by *Venturia canescens* and subsequent virus transmission, *Funct. Ecol.* **10:**586.

Sait, S. M., Gage, M. J. G., and Cook, P. A., 1997, Effects of a fertility-reducing baculovirus on sperm numbers and sizes in the Indian meal moth, *Plodia interpunctella, Funct. Ecol.,* in press.

Santiago-Alvarez, C., and Caballero, P., 1990, Susceptibility of parasitized *Agrotis segetum* larvae to a granulosis virus, *J. Invertebr. Pathol.* **56:**128.

Santiago-Alvarez, C., and Ortiz-Garcia, R., 1992, The influence of host plant on the susceptibility of *Spodoptera littoralis* (Boisd.) (Lep., Noctuidae) larvae to *Spodoptera littoralis* NPV (Baculoviridae, Baculovirus), *J. Appl. Entomol.* **114:**124.

Santiago-Alvarez, C., and Vargas-Osuna, E., 1986, Differential mortality between male and female *Spodoptera littoralis* larvae infected with a baculovirus, *J. Invertebr. Pathol.* **47:**374.

Santiago-Alvarez, C., and Vargas-Osuna, E., 1988, Reduction of reproductive capacity of *Spodoptera littoralis* males by a nuclear polyhedrosis virus (NPV), *J. Invertebr. Pathol.* **52:**142.

Schultz, J. C., Foster, M. A., and Montgomery, M. E., 1990, Host plant-mediated impacts of a baculovirus on gypsy moth populations, in: *Population Dynamics of Forest Insects* (A. D. Watt, S. R. Leather, M. D. Hunter, and N. A. C. Kidd, eds), pp. 303–313, Intercept, Andover, Hampshire, England.

Schultz, J. C. Hunter, M. D., and Appel, H. M., 1992, Antimicrobial activity of polyphenols mediates plant–herbivore interaction, in: *Plant Polyphenols, Synthesis Properties and Significance* (R. W. Hemingway and P. E. Laker, eds.), pp. 1–17, Plenum Press, New York.

Shapiro, M., and Robertson, J. L., 1987, Yield and activity of gypsy moth (Lepidoptera: Lymantriidae) nucleopolyhedrosis virus recovered from survivors of viral challenge, *J. Econ. Entomol.* **80:**901.

Shapiro, M., Robertson, J. L., and Bell, R. A., 1986, Quantitative and qualitative differences in gypsy moth (Lepidoptera: Lymantriidae) nucleopolyhedrosis virus produced in different-aged larvae, *J. Econ. Entomol.* **79:**1174.

Shapiro, M., Fuxa, J. R., Braymer, H. D., and Pashley, D. P., 1991, DNA restriction polymorphism in wild isolates of *Spodoptera frugiperda* nuclear polyhedrosis virus, *J. Invertebr. Pathol.* **58:**96.

Slack, J. M., Kuzio, J., and Faulkner, P., 1995, Characterization of *v-cath,* a cathepsin L-like proteinase expressed by the baculovirus *Autographa californica* multiple nuclear polyhedrosis virus, *J. Gen. Virol.* **76:**1091.

Smith, I. R. L., and Crook, N. E., 1988, *In vivo* isolation of baculovirus genotypes, *Virology* **166:**240.

Smith, I. R. L., and Crook, N. E., 1993, Characterization of new baculovirus genotypes arising from inoculation of *Pieris brassicae* with granulosis viruses, *J. Gen. Virol.* **74:**415.

Smith, V. H., and Holt, R. D., 1996, Resource competition and within-host disease dynamics, *Trends Ecol. Evol.* **11:**386.

Smits, P. H., and Vlak, J. M., 1988a, Biological activity of *Spodoptera exigua* nuclear polyhedrosis virus against *S. exigua* larvae, *J. Invertebr. Pathol.* **51:**107.

Smits, P. H., and Vlak, J. M., 1988b, Quantitative and qualitative aspects in the production of a nuclear polyhedrosis virus in *Spodoptera exigua* larvae, *Ann. Appl. Biol.* **112:**249.

Stewart, L. M. D., Hirst, M., Ferber, M. L., Merryweather, A. T., Cayley, P. J., and Possee, R. D., 1991, Construction of an improved baculovirus insecticide containing an insect-specific toxin gene, *Nature* **352:**85.

Subrahmanyam, B., and Ramakrishnan, N., 1980, The alteration of juvenile hormone titre in the haemolymph of *Spodoptera litura* (F.) due to baculovirus infection, *Experientia* **36:**471.

Subrahmanyam, B., and Ramakrishnan, N., 1981, Influence of a baculovirus infection on moulting and food consumption by *Spodoptera litura, J. Inverbebr. Pathol.* **38:**161.

Tanada, Y., and Hukuhara, T., 1971, Enhanced infectivity of a nuclear-polyhedrosis virus in larvae of the armyworm, *Pseudaletia unipuncta,* by a factor in the capsule of a granulosis virus, *J. Invertebr. Pathol.* **17:**116.

Teakle, R. E., Jensen, J. M., and Mulder, J. C., 1985, Susceptibility of *Heliothis armigera* (Lepidoptera: Noctuidae) on sorghum to nuclear polyhedrosis virus, *J. Econ. Entomol.* **78:**1373.

Thomas, M. B., Wood, S. N., and Lomer, C. J., 1995, Biological control of locusts and grasshoppers using a fungal pathogen: The importance of secondary cycling, *Proc. R. Soc. Lond. B* **259:**265.

Tilman, D., 1990, Mechanisms of plant competition for nutrients: the elements of a predictive theory of competition, in: *Perspectives on Plant Competition* (J. B. Grace and D. Tilman, eds.), pp. 117–142, Academic Press, San Diego.

Vail, P. V., 1981, Cabbage looper nuclear polyhedrosis virus–parasitoid interactions, *Environ. Entomol.* **10:**517.

Vail, P. V., and Tebbets, J. S., 1990, Comparative biology and susceptibility of *Plodia interpunctella* (Lepidoptera: Pyralidae) populations to a granulosis virus, *Environ. Entomol.* **19:**791.

Vail, P. V., Soo Hoo, C. F., Seay, R. S., Killinen, R. G., and Wolf, W. W., 1972, Microbial control of lepidopterous pests of fall lettuce in Arizona and effects of chemicals and microbial pesticides on parasitoids, *Environ. Entomol.* **1:**780.

Vargas-Osuna, E., Carranza, P., Aldebis, H. K., and Santiago-Alvarez, C., 1995, Production of *Agrotis segetum* granulosis virus in different larval instars of *A. segetum* (Lep., Noctuidae), *J. Appl. Entomol.* **119:**695.

Vasconcelos, S. D., Williams, T., Hails, R. S., and Cory, J. S., 1996a, Prey selection and baculovirus dissemination by carabid predators of Lepidoptera, *Ecol. Entomol.* **21:**98.

Vasconcelos, S. D., Cory, J. S., Wilson, K. R., Sait, S. M., and Hails, R. S., 1996b, Modified behavior in baculovirus-infected lepidopteran larvae and its impact on the spatial distribution of inoculum, *Biol. Control* **7:**299.

Versoi, P. L., and Yendol, W. G., 1982, Discrimination by the parasite, *Apanteles melanoscelus*, between healthy and virus-infected gypsy moth larvae, *Environ. Entomol.* **11:**42.

Vezina, A., and Peterman, R., 1985, Tests of the role of nuclear polyhedrosis virus in the population dynamics of its host, Douglas-fir tussock moth, *Orgyia pseudotsugata* (Lepidoptera: Lymantriidae), *Oecologia* **67:**260.

White, A., Begon, M., and Bowers, R. G., 1996, Host-pathogen systems in a spatially patchy environment, *Proc. R. Soc. Lond. B* **263:**325.

Woods, S. N., and Thomas, M. B., 1996, Space, time and persistence of virulent pathogens, *Proc. R. Soc. Lond. B* **263:**673.

Young, S. Y., 1990, Effect of a nuclear polyhedrosis virus infection in *Spodoptera ornithogalli* larvae on post larval stages and dissemination by adults, *J. Invertebr. Pathol.* **55:**69.

Young, S. Y., and Yearian, W. C., 1982, Nuclear polyhedrosis infection of *Pseudoplusia includens* (Lep.: Noctuidae) larvae: Effect on post larval stages and transmission, *Entomophaga* **27:**61.

Young, S. Y., and Yearian, W. C., 1987, *Nabis roseipennis* adults (Hemiptera: Nabidae) as disseminators of nuclear polyhedrosis virus to *Anticarsia gemmatalis* (Lepidoptera: Noctuidae) larvae, *Environ. Entomol.* **16:**1330.

Young, S. Y., and Yearian, W. C., 1990, Transmission of nuclear polyhedrosis virus by the parasitoid *Microplitis croceipes* (Hymenoptera: Braconidae) to *Heliothis virescens* (Lepidoptera: Noctuidae) on soybean, *Environ. Entomol.* **19:**251.

Young, S. Y., and Yearian, W. C., 1992, Movement of nuclear polyhedrosis virus into velvetbean caterpillar (Lepidoptera: Noctuidae) larval populations on soybean by *Nabis roseipennis* (Heteroptera: Nabidae) nymphs, *J. Entomol. Sci.* **27:**126.

Young, S. Y., Yang, J. G., and Felton, G. W., 1995, Inhibitory effects of dietary tannins on the infectivity of a nuclear polyhedrosis virus to *Helicoverpa zea* (Noctuidae: Lepidoptera), *Biol. Control* **5:**145.

Zethner, O., and Øgaard, L., 1982, Studies on the specificity of *Agrotis segetum* granulosis virus fed to larvae of 17 species of noctuids, *Tidsskr. Planteavl.* **86:**471.

CHAPTER 13

Commercialization of Baculoviral Insecticides

BRUCE C. BLACK, LYNN A. BRENNAN, PETER M. DIERKS, AND IVAN E. GARD

I. INTRODUCTION

The commercialization of baculoviruses has had a long and checkered history. Early insect virologists were quick to see the potential of these highly infectious and selective pathogens. This, in addition to the safety of these agents, proved irresistibly attractive for the development of these viruses as insecticides. During the 1970s, the first baculoviral product was introduced into the commercial arena, Elcar [*Helicoverpa zea* nuclear polyhedrosis virus (NPV)]. This product was accompanied by three noncommercial preparations produced by the US Forestry Service, namely Gypcheck (*Lymantria dispar* NPV), TM BioControl-1 (*Orgyia pseudotsuga* NPV), and Neocheck-S (*Neodiprion sertifer* NPV). During these early days, products suffered from a variety of problems, including variable potency, high production costs, poor formulation, and generally inferior field performance when compared to chemical alternatives. Eventually, production of these viral insecticides ceased with the exception of Gypcheck.

Since the 1970s, knowledge pertaining to the biology and application of these biopesticides has progressed steadily. By the 1990s, molecular biology and the development of baculovirus expression systems set the stage for a new era in the development of baculoviral insecticides. Most important

---

BRUCE C. BLACK, LYNN A. BRENNAN, PETER M. DIERKS, AND IVAN E. GARD • American Cyanamid Company, Agricultural Products Research Division, Insecticide Directed Basic Research, Princeton, New Jersey 08543.

The Baculoviruses, edited by Lois K. Miller. Plenum Press, New York, 1997.

have been the construction of fast-killing recombinant baculoviruses that express insect-specific toxins. For the first time, recombinant insect viruses have been constructed that have the potential to rival the efficacy of chemicals. In this review, we discuss many of the important issues that will determine the success of viral insecticides in the marketplace.

II. MARKET ISSUES AND DEVELOPMENT OF IMPROVED VIRUSES

The ability of foliar-applied baculoviruses to protect crops from insect damage is dependent on two fundamental issues: (1) effective dose acquisition, and (2) the speed of action of the acquired dose. Effective dose acquisition is critical to the success of any insecticide. For chemical insecticides that are contact poisons, it can be achieved when a target insect simply walks across a treated leaf surface. For a baculovirus, however, effective dose acquisition is achieved when the targeted insect(s) ingests a dose of virus that is sufficient to initiate a productive systemic infection. This is a relatively complex process that is affected by the rate of baculovirus application, the stability (residual activity) of the applied dose on the treated leaf surface over time, the feeding behavior of the pests that need to be controlled, and (most importantly) how well the host range of the baculovirus matches the composition of the pest complex at the time of application.

The second critical parameter is the amount of time required for the virus to curb the feeding activity of the infected pest insect. Historically, speed of action has been one of the most visible weaknesses of baculovirus-based insecticides. Most naturally occurring baculoviruses take from 4 to 7 days to kill their hosts and some, such as the gypsy moth virus (*Lymantria dispar* NPV), can take up to 18 to 21 days. During this time the infected insects continue to feed and may cause substantial crop damage. It is possible, therefore, to apply enough virus to control the entire pest population but still lose the crop to feeding damage.

Fortunately, there is reason to believe that most of the parameters affecting baculovirus performance can be optimized, and that once this is accomplished, baculovirus-based insecticides will become a significant and effective tool for insect pest management. Some of the needed technology, such as the ability to modify the "economic" host range of baculoviruses through genetic manipulation, is still in the earliest stages of development. Other components—including the development of much faster acting baculoviruses—are already a reality.

A. Improving Baculovirus Speed of Action: Gene Deletion

One strategy for improving the speed of action of baculovirus insecticides is to identify and remove from the viral genome any nonessential genes that might act to prolong the life of the host. Many viruses, particularly large DNA viruses, contain genes that alter the host organism's physiology for the purpose of maximizing the yield or transmission of viral progeny at the end of the infection cycle. These genes are generally not required for virus growth per se, but may play an important role in the evolutionary "fitness" of the virus in its natural habitat. One gene that appears to fall into this category is the viral ecdysteroid glucosyltransferase gene (*egt*), which was initially discovered in *Autographa californica* NPV (AcMNPV) (O'Reilly and Miller, 1989, 1990).

The *egt* gene encodes an enzyme that specifically inactivates ecdysteroid hormones by conjugating the C-22 hydroxyl with glucose or galactose (O'Reilly *et al.*, 1991, 1992). Hormonally controlled increases in ecdysteroid levels at specific times during development bring about a complex series of changes that result in molting and pupation. Infection of fall armyworm (*Spodoptera frugiperda*) larvae or cabbage looper (*Trichoplusia ni*) larvae with high levels of wild-type AcMNPV blocks the induction of ecdysteroid titers and causes a developmental arrest in which molting and pupation are suppressed (O'Reilly and Miller, 1989, 1990). This effect requires a functional *egt* gene.

An interesting observation made while studying the *egt* gene was that larvae infected with *egt*$^-$ mutants of AcMNPV die faster—up to 30% faster in some cases—than those infected with the wild-type virus (O'Reilly and Miller, 1991). The accelerated speed of action of the *egt*$^-$ viruses is coupled with a significant reduction in the amount of food consumed and a 20–30% reduction in the number of viral polyhedra produced at the time of death. While the role of the *egt* gene is clear (i.e., inactivation of host ecdysteroids), the mechanisms responsible for the accelerated speed of action of the *egt*$^-$ viruses are not so well defined (see O'Reilly, 1995). One possible explanation is that virus-infected larvae are unable to handle the physiological stress associated with molting. The weakness of this explanation, however, is that molting and molting behavior are not inhibited in all situations where mortality is accelerated by *egt*$^-$ viruses. For example, Vlak and co-workers have reported that although wild-type AcMNPV induces developmental arrest in fourth-instar beet armyworm (*Spodoptera exigua*) larvae, it has little impact on molting in early second-instar larvae (Flipsen *et al.*, 1995). As a result, the fraction of larvae molting from the second to the third instar in this experiment, and the timing of the molt, were the same for both the *egt*$^+$ and *egt*$^-$ viruses. Nevertheless, larvae infected with the *egt*$^-$ virus still died 30–35% faster than those infected with wild-type AcMNPV. These same workers also observed that the Malpighian tubules undergo early degeneration in larvae infected with *egt*$^-$ virus, and suggest that this may be responsible for the acceleration of mortality seen with the *egt*$^-$ mutants (Flipsen *et al.*, 1995). Additional work is required to see if this finding can be extended to other virus–insect combinations.

Although the impact of the *egt* deletion on the speed of action of

AcMNPV is relatively modest, the difference is significant enough to have a measurable effect on insect feeding damage in laboratory bioassays. For example, in an unpublished study conducted by American Cyanamid, cabbage leaf consumption by individual *Trichoplusia ni* larvae was reduced by 43% if the larvae were infected with an LC_{99} dose of wild-type AcMNPV *prior* to being placed on the cabbage leaves, and by 77% if they were infected by an *egt*$^-$ mutant of AcMNPV (designated vEGTDEL). This type of assay is useful as a laboratory benchmark. To understand how this virus performs in a more realistic setting, American Cyanamid conducted a series of greenhouse and small-scale field tests in 1993 and 1994 in which the effectiveness of vEGTDEL was compared with wild-type AcMNPV and two commercial *Bacillus thuringiensis* (Bt) bioinsecticides. These tests were conducted primarily in cotton and leafy vegetables and were directed at controlling cabbage loopers (*Trichoplusia ni*) and tobacco budworms (*Heliothis virescens*). In the greenhouse experiments, the hastened speed of action of vEGTDEL led to improved plant protection when compared to wild-type AcMNPV and in one instance resulted in a 60% reduction in damage to cotton flower buds (Treacy *et al.*, 1996). In the field trials, however, vEGTDEL tended to provide better control of cabbage loopers than wild-type AcMNPV, but the differences were less significant than those observed in the laboratory or the greenhouse (Treacy *et al.*, 1996). Similarly, there was a trend for vEGTDEL to be slightly less effective than the Bt standard in many of the tests, but the differences in the two treatments were frequently not significant.

B. Improving Baculovirus Speed of Action: Gene Insertion

While an *egt*$^-$ AcMNPV may be competitive with Bt insecticides in some situations, it is still too slow acting to provide the level of crop protection that growers have come to expect from conventional insecticides. Furthermore, the degree of improvement that can be achieved by gene deletion alone appears limited. For these reasons, researchers in academic, government, and commercial laboratories have been exploring the use of gene insertion technology as a means of achieving more substantial improvements in the performance of viral insecticides (Bonning and Hammock, 1992, 1996; Hammock *et al.*, 1993; Possee *et al.*, 1993a; Miller, 1995). In this approach the baculovirus is used primarily as a delivery system for a foreign gene whose expression in the targeted insect will interfere with some critical aspect of its physiology and result in reduced feeding or death. This endows the baculovirus with a second, faster mode of action, but does not otherwise interfere with its ability to replicate and cause disease.

A number of factors must be considered when deciding whether a gene is appropriate for enhancing the insecticidal properties of AcMNPV (Table I). At the top of the list is the requirement that the inserted gene have an insect-specific target or mode of action. Although the delivery system (the

TABLE I. Desirable Genetic Traits for Insecticidal Genes

Insect-specific target or mode of action
Active at very low levels
Fast acting
Lethal to host but not cytotoxic to cells in culture
Specialized processing (e.g., α-amidation) not required for activity

baculovirus) is itself insect specific, restricting the list of candidate genes to those that have an insect-specific target or mode of action helps alleviate concerns about the potential consequences of gene movement or inadvertent expression in a nontarget host. Another important requirement is that the gene must not target an aspect of intermediary metabolism or cell physiology that is essential for the short-term survival of the infected cells, since this would make it difficult or impossible to produce the virus product. Therefore, systems that operate at the organismal level make the most attractive targets. It is also likely that the best performance will be achieved with a gene product that is active at low levels, requires little or no specialized processing for activity, and has an immediate impact on the host.

1. Insertion of Insect Genes

One approach for implementing the gene insertion strategy is to use the baculovirus as a vehicle for promoting inappropriate expression of insect genes that normally regulate key aspects of insect physiology or development. Maeda (1989) initially tested this approach by inserting a synthetic gene for the diuretic hormone of *Manduca sexta* (tobacco hornworm) into *Bombyx mori* NPV (BmNPV). Diuretic hormone is a small (41 amino acids) peptide hormone that plays an important role in maintaining water balance in insects. To promote high-level expression of diuretic hormone, the synthetic gene was placed under the control of the very strong viral polyhedrin gene (*polh*) promoter and supplied with all the ancillary sequences needed for hormone secretion and α-amidation. Injection of a recombinant virus containing the synthetic diuretic hormone gene into *Bombyx mori* (silkworm) larvae resulted in the formation of biologically active diuretic hormone and caused a 30% reduction in the hemolymph volume compared to larvae infected with wild-type BmNPV. Unfortunately, the speed of action of the recombinant virus was only marginally faster than wild-type BmNPV.

Another potential target for disruption is the insect endocrine system. As mentioned earlier, molting is an essential part of insect development and is controlled by ecdysteroid hormones, such as 20-hydroxyecdysone. The synthesis of ecdysteroids during larval and pupal development occurs mainly in the prothoracic gland and is stimulated by the action of prothoracicotropic hormone (PTTH), a small neuropeptide. Ecdysteroid catabolism is controlled by a complex assortment of enzymes, some of which inactivate the hormones by glucosylation, similar to the action of the *egt* gene in AcMNPV. One might expect, therefore, that overexpression of PTTH from an *egt*$^-$ baculovirus could be used to artificially elevate ecdysteroid levels and induce precocious molting, which could in turn disrupt insect feeding behavior or survival. However, insertion of the *Bombyx mori* PTTH gene into AcMNPV (with or without *egt*) did not improve its insecticidal properties (O'Reilly *et al.*, 1995). In fact, the PTTH recombinant in the *egt*$^-$ background was approximately 100-fold less infectious than wild-type AcMNPV. This observation raises some interesting questions about the impact of ecdysteroids on baculovirus replication (see Chapter 11, this volume, and O'Reilly, 1995, for discussion), but does not solve the immediate problem of improving the crop protection properties of AcMNPV.

An alternative approach for disrupting the insect's endocrine system focuses on juvenile hormone (JH) metabolism. JH is a terpenoid hormone that controls the state of differentiation of insect tissues. It is synthesized by the corpora allata under the control of peptide hormones (allatostatin and allatatropin) and acts to keep tissues in a "juvenile" or larval state of differentiation. Production of JH during the early instars of lepidopteran development causes the insect to remain committed to larval development as it molts. During the last larval instar, JH largely disappears from the hemolymph and the larva begins the process of pupation and adult metamorphosis.

One of the enzymes responsible for the turnover of JH is juvenile hormone esterase (JHE), which cleaves the methyl ester of JH to yield the biologically inactive JH acid. It has been postulated that overexpression of JHE from a baculovirus vector should lead to a reduction in JH titer and precipitate premature pupation. However, this expectation has not been borne out by experimentation. A number of derivatives of AcMNPV have been constructed that contain the JHE gene of *Heliothis virescens* under the control of various late and very late baculovirus promoters (Hammock *et al.*, 1990; Bonning and Hammock, 1992; Eldridge *et al.*, 1992a). Although JHE titers were elevated by as much as 70-fold compared to mock infected insects, there was no consistent evidence that JHE expression had any significant impact on the time of death, feeding behavior, or developmental phenotype of the infected insects. By comparing the properties of *egt*$^+$ and *egt*$^-$ derivatives of AcMNPV containing the JHE gene, Eldridge *et al.* (1992a) effectively ruled out the possibility that *egt*-mediated inactivation of the ecdysteroid pool could have masked the effects of the JHE gene. Other attempts to account for the unexpectedly poor performance of the JHE constructs have focused on the *in vivo* stability and possible sequestration of the enzyme (Bonning and Hammock, 1992, 1996; Hammock *et al.*, 1993). Based on these considerations, the JHE gene was modified by site-directed mutagenesis and several derivatives were identified that improved the speed of action of AcMNPV by about 20–35% (Bonning and Hammock, 1992,

1996). At least one of these, however, is no longer able to cleave JH and appears to kill its host by a non-JH mechanism (Bonning *et al.*, 1995). This raises obvious questions about the target specificity of the native JHE enzyme and the mode of action of these JHE derivatives.

Attempts to use other developmental hormones, such as eclosion hormone, to improve the speed of action of AcMNPV also have proven unsuccessful (Eldridge *et al.*, 1992b). It appears, therefore, that while it may eventually be possible to use inappropriate expression of insect hormones as a strategy for improving the insecticidal properties of baculoviruses, our current understanding of insect development and the insect endocrine system is not sufficient to achieve a predictable outcome. On a more positive note, Gopalakrishnan *et al.* (1995) showed that *Spodoptera frugiperda* larvae injected with a nonoccluded derivative of AcMNPV expressing an insect chitinase died 20% faster than those injected with wild-type AcMNPV. Although the mode of action of this virus was not investigated, the results are consistent with the hypothesis that inappropriate expression of insect chitinase might prove insecticidal because of its impact on chitin metabolism in the gut or cuticle. Further research is needed to see if the effect can be reproduced with an orally administered occlusion-positive recombinant virus.

2. Insertion of Genes that Encode Insect-Specific Toxins

Of the strategies that have been explored to date, the insertion of insect-specific toxins, particularly insect-specific neurotoxins, holds the greatest promise for delivering a commercially viable baculovirus insecticide. Nature is replete with insect predators and insect parasites that use venoms to immobilize their prey. Although arthropod venoms contain a mixture of toxins and often have activity against organisms other than insects (including vertebrates), most contain one or more components that specifically target the insect nervous system. The first test of this strategy involved the use of an insect-specific toxin, insectotoxin-1, from the scorpion *Buthus eupeus* (Carbonell *et al.*, 1988). In these experiments, a synthetic coding region for the mature insectotoxin-1 peptide (36 amino acids) was placed under the control of the AcMNPV polyhedrin gene promoter in the presence and absence of a signal peptide. Although a peptide of the expected size was observed in virus-infected cells, no biologically active toxin was detected and the speed of action of the recombinant AcMNPV was not improved.

In contrast, significant improvements in virus efficacy have been obtained using other scorpion toxin genes. One gene that has attracted considerable interest is the AaIT* gene, which encodes the insect-specific venom component of the North African (Algerian) scorpion *Androctonus australis*

*The acronyms AaHIT and AaIT have been used extensively in the literature to refer to the same toxin.

(Hector) (Zlotkin *et al.*, 1971; Darbon *et al.*, 1982). AaIT is a small (70 amino acids) cysteine-rich peptide that causes rapid contractile paralysis in insects. It interacts specifically with voltage-dependent sodium channels in insect nervous tissue and causes sustained repetitive generation of action potentials (Walther *et al.*, 1976; Zlotkin *et al.*, 1985). AaIT has no activity on vertebrate nervous tissue (Teitelbaum *et al.*, 1979; Zlotkin, 1983) and is nontoxic to mice (Zlotkin, 1986). It also has no apparent effect on noninsect arthropods, such as crustaceans or scorpions (Zlotkin, 1986). In addition, unlike many vertebrate active scorpion toxins and depressant-type insect-selective scorpion toxins, the natural pathway for AaIT formation does not involve C-terminal processing of a propeptide (Bougis *et al.*, 1989; Zilberberg *et al.*, 1991). In summary, this gene satisfies all the criteria listed in Table I.

When susceptible larvae are infected with an AaIT-expressing recombinant baculovirus (either AcMNPV or BmNPV), they experience a progressive loss of neuromuscular control that culminates in complete or nearly complete contractile paralysis (Stewart *et al.*, 1991; Maeda *et al.*, 1991; McCutchen *et al.*, 1991). Death is not immediate and may be delayed after the onset of paralysis for as long as 36 hr; however, the time to death is inconsequential since the paralyzed larvae are unable to feed and tend to fall off the plants (Hoover *et al.*, 1995). For recombinant viruses in which the AaIT gene is under the transcriptional control of either the viral *polh* or *p10* gene promoters, the amount of time required to achieve paralysis (i.e., the median response time, or RT_{50}) is about 65–75% of the median survival time (ST_{50}) of larvae infected with the corresponding wild-type virus (Stewart *et al.*, 1991; Maeda *et al.*, 1991; McCutchen *et al.*, 1991). More importantly, the amount of feeding damage caused by larvae infected with these AaIT-expressing baculoviruses is roughly half of that caused by larvae infected with wild-type AcMNPV (Stewart *et al.*, 1991).

Since 1989, AaIT has been one of the cornerstones of a research program at American Cyanamid that is aimed at developing commercially viable baculovirus insecticides. By systematically exploring the use of different viral and cellular promoters and optimizing other aspects of expression vector design, it has been possible to make steady incremental improvements in the speed of action and crop protection performance of the AaIT-expressing viruses. One of the viruses developed under this program uses the AcMNPV DA26 (early) gene promoter to drive expression of a synthetic AaIT gene in an *egt*$^-$ derivative of AcMNPV. The RT_{50} for this virus on *Heliothis virescens* larvae is approximately 40% of the ST_{50} for wild-type AcMNPV. Given that a finite amount of time is required for dose acquisition in these assays and that a systemic infection appears to be required to achieve paralysis, it may very well be that the speed of action of this virus is approaching the limit of what is biologically achievable. The question is, "Is this fast enough?" Although no answer is currently available, the issue is being addressed in an ongoing EPA-approved small-scale field evaluation program that was initiated in late-summer 1995 (American Cyanamid, 1994, 1996).

In addition to excitatory insect toxins, such as AaIT, many Buthinae scorpion venoms contain a second type of insect-selective toxin that causes a slow, progressive flaccid paralysis (Zlotkin *et al.*, 1991, 1993). Like AaIT, these depressant-type insect toxins are small cysteine-rich polypeptides of 60–65 amino acids that interact specifically with insect sodium channels. There is no detectable mammalian or crustacean toxicity (Zlotkin *et al.*, 1991; DuPont, 1996). The close physical proximity of the target sites for the excitatory and depressant toxins is evident from competitive binding studies in which the depressant insect toxin LqhIT2 (isolated from the scorpion *Leiurus quinquestriatus hebreus*) displaced AaIT from its single high-affinity binding site in a variety of insect neuronal preparations, especially those from lepidopteran larvae (Moskowitz *et al.*, 1994). Nevertheless, the pharmacological response to the depressant insect toxins is quite different from that of AaIT: Following a very brief period of excitatory activity (similar to that caused by AaIT), there is a progressive suppression of neuromuscular transmission, resulting in complete relaxation of body musculature (Zlotkin *et al.*, 1991).

In 1996, DuPont Agricultural Products initiated an EPA-approved small-scale field trial program aimed at evaluating the crop protection potential of a recombinant AcMNPV expressing the LqhIT2 depressant insect toxin (DuPont, 1996). The synthetic toxin gene in this case is coupled to the bombyxin signal peptide and is placed under the transcriptional control of an AcMNPV *ie1* (immediate early gene) promoter/*hr5* enhancer complex (DuPont, 1996). Although the native LqhIT2 cDNA sequence encodes three C-terminal amino acids (Gly-Lys-Lys) that are removed from the toxin posttranslationally (Zlotkin *et al.*, 1993), these residues are apparently not required for the formation of active toxin, since they have been successfully omitted from the synthetic LqhIT2 construct. Unfortunately, the crop protection performance data on this viral construct are not yet publicly available.

Another paralytic neurotoxin that holds considerable promise as an active ingredient in viral insecticides is the TxP-I toxin, which is a venom component of the predatory straw itch mite, *Pyemotes tritici* (Tomalski *et al.*, 1988, 1989). Although the mode of action of Txp-I is not known, it is not toxic to mice (Tomalski *et al.*, 1989; J. Hayashi, personal communication) and its potency on lepidopteran larvae is comparable to or better than that of the insect-selective scorpion toxins. TxP-I is encoded as a precursor protein of 291 amino acids by the *tox34* gene (Tomalski and Miller, 1991). Larvae infected with an occlusion-negative derivative of AcMNPV expressing *tox34* under the control of a modified polyhedrin gene promoter became paralyzed and died almost 2 full days before significant mortality was observed in larvae infected with wild-type AcMNPV (Tomalski and Miller, 1991). To further optimize this system, Miller and co-workers examined how secreted toxin production and virus performance were affected by the use of alternate promoters and signal peptides (Tomalski and Miller, 1992; Lu *et al.*, 1996). Although altering the signal peptide clearly impacted secreted toxin production and virus performance, none of the heterologous signal peptides tested worked better than the native signal peptide encoded by the *tox34* gene. On the other hand, significant improvements in virus performance were obtained when the modified *polh* promoter (P_{synXIV}) was replaced by either the AcMNPV p6.9 (basic protein) gene promoter or a *Drosophila hsp70* heat-shock gene promoter. The *p6.9* gene encodes a small highly basic DNA binding protein that is expressed at high levels very early in the late phase of the AcMNPV replicative cycle. Peak Txp-I accumulation was greater and occurred about 24 hr sooner than when P_{synXIV} was replaced by the p6.9 promoter. This translated to a significant reduction in the RT_{50}, which decreased from 55 to 60% of the wild-type ST_{50} for the P_{synXIV} promoter to about 40 to 45% for the p6.9 promoter (Lu *et al.*, 1996). TxP-I production was dramatically lower when *tox34* was driven by the *hsp70* promoter, but it was detectable at much earlier times (12–18 hr postinfection). Despite the relatively low level of toxin formation, the median response time for the *hsp70–tox34* construct on *Spodoptera frugiperda* larvae (42% of wild type) was roughly equivalent to that of the p6.9–*tox34* construct (45% of wild type). The *hsp70–tox34* recombinant did not perform as well on *Trichoplusia ni* larvae, however, and the results obtained with two early viral promoters (DA26 and pETL) were also disappointing when compared to the *polh* promoter (Tomalski and Miller, 1992; Lu *et al.*, 1996). In each case, toxin production was poor. Thus, there appears to be a benefit to producing TxP-I early in the viral life cycle, provided that the level of expression exceeds the threshold required for the establishment of paralysis (Lu *et al.*, 1996).

As can be seen from the foregoing discussion, the assortment of insect-selective neurotoxins that can be used as active ingredients in baculovirus insecticides is extensive and continues to grow. Most recently, Prikhod'ko *et al.* (1996) reported that the speed of action of AcMNPV could be significantly improved by inserting a gene expressing the insect-selective μ-Aga-IV toxin from the funnel web spider, *Agelenopsis aperta*, or either of two small insect-selective toxins from sea anemones. In contrast, other types of toxins have not been able to match the overall performance of the insect-selective neurotoxins. These included two Bt toxin genes (Martens *et al.*, 1990; Merryweather *et al.*, 1990) and a maize mitochondrial gene involved in cytoplasmic male sterility (T-*urf13*) (Korth and Levings, 1993). The failure of the Bt toxins to improve the speed of action of AcMNPV was perhaps not surprising given their mode of action. The speed of action of two AcMNPV recombinants expressing T-*urf13* under the control of the *polh* promoter was significantly enhanced when nonoccluded virus was injected into *Trichoplusia ni* larvae. The problem with this gene, however, is that it appeared to be somewhat toxic to cultured insect cells, and efforts to produce occluded virus by co-occlusion were largely unsuccessful (Korth and Levings, 1993). None of the neurotoxin genes examined to date interfered with cell growth or virus production *in vitro.* It seems quite likely, therefore, that an insect-selective neurotoxin will be the active ingredient in the first commercial recombinant baculovirus insecticides.

It should be pointed out, however, that while insect-selective neurotoxins can dramatically enhance the speed of action of AcMNPV and related baculoviruses, they do not alter the intrinsic infectivity of the virus on permissive and semipermissive host species or alter its natural host range (Possee *et al.,* 1993a; Bishop *et al.,* 1995; American Cyanamid, 1996; DuPont, 1996; Lu *et al.,* 1996; Prikhod'ko *et al.,* 1996). This finding is consistent with the hypothesis that a toxin-expressing recombinant baculovirus must establish a productive systemic infection in order to control its host, just as the wild-type parental virus must. Hence, these genes address problems relating to the speed of action of the acquired dose and have no significant impact on effective dose acquisition.

C. Effective Dose Acquisition and Virus Host Range

A critical factor for the performance of any baculovirus insecticide is how well the host range of the baculovirus matches the spectrum of pest insects that it must be able to control in any given crop treatment. For some viruses, such as the gypsy moth (*Lymantria dispar*) NPV or the beet armyworm (*Spodoptera exigua*) NPV, the host range is confined to a single species or at least to a small group of closely related species (Table II). In the case of the gypsy moth virus, this is not a major concern; however, beet armyworms are frequently found in association with other lepidopteran pests, such as budworms and bollworms, that lie outside the host range of *Spodoptera exigua* NPV. In these situations, a narrow-range baculovirus insecticide is of little use.

At the other end of the spectrum are the members of the *Autographa californica* NPV family, which includes not only all of the common strains of AcMNPV, but also *Anagrapha falcifera* NPV, *Galleria mellonella* NPV, *Rachiplusia ou* NPV, *Heliothis virescens* NPV, *Trichoplusia ni* NPV, and a variety of other NPVs that have been genetically characterized as "Acal-like." These viruses have qualitatively similar host ranges and are able to replicate at some level in a relatively wide range of lepidopteran hosts, especially in the family *Noctuidae.* However, the efficiency of virus replication, as measured by the amount of virus required to cause mortality in 50% of the treated insects (the LD_{50}), varies from host to host and can range over 5 to 6 orders of magnitude. From a crop protection standpoint, therefore, a distinction must be made between the "biological" host range, which encompasses all of the species that can be infected at any dosage in the laboratory, and the "economic" host range, which is restricted to those species

TABLE II. Susceptibililty of Agricultural Pests to Wild-Type Nuclear Polyhedrosis Viruses[a]

	Spodoptera eridania	*Spodoptera exigua*	*Spodoptera frugiperda*	*Heliothis virescens*	*Helicoverpa zea*	*Trichoplusia ni*
Autographa californica NPV	++	++++	++	++++	+	++++
Heliothis virescens NPV	+	+++	++	++++	++	++++
Trichoplusia ni NPV	–	+++	–	++	–	++++
Spodoptera exigua NPV	–	++++	–	–	–	–
Helicoverpa zea NPV	–	–	–	++++	++++	+
Heliothis armigera NPV	–	+++	–	–	++++	++

[a]Based on unpublished bioassay data collected at American Cyanamid. Susceptibility is reflected in a semi-quantitative scale that runs from "–" (nonpermissive to infection) to "++++" (highly permissive to infection).

that can be effectively controlled at cost-effective field application rates. For example, *Heliothis virescens* (tobacco budworm) and *Helicoverpa zea* (cotton bollworm) are both within the biological host range of AcMNPV, but the LD_{50} of AcMNPV for *Helicoverpa zea* is 1,000–10,000 times higher than it is for *Heliothis virescens* (Table II). As a result, AcMNPV is able to provide good control of *Heliothis virescens,* but is not able to protect target crops effectively when the pest complex contains significant pressure from *Helicoverpa zea.* In this case, the economic host range of even a "broad-spectrum" baculovirus such as AcMNPV can look quite narrow.

There are two fundamental ways to address this problem. One is to maximize the usefulness of the intrinsic host range of the virus through product formulation, which is discussed in detail in Section IV. This can include the use of optical brighteners to enhance baculovirus infectivity (i.e., lower the effective LD_{50}) and the addition of other inert components that improve the residual activity of the applied dose. The other way is to modify the host range of the product so that it better accommodates the various pest complexes encountered in key market situations.

The simplest way to modify product host range is to find a naturally occurring baculovirus whose economic host range includes the key members of the pest complex one needs to control. This is possible because of the extraordinary host range diversity that exists among baculoviruses. For example, it is clear from the data in Table II that *Helicoverpa zea* NPV would provide much better control than AcMNPV of a pest complex dominated by *Helicoverpa zea* and *Heliothis virescens,* whereas AcMNPV would likely be the virus of choice to control *Heliothis virescens* either alone or in conjunction with *Trichoplusia ni.* A less attractive, but potentially viable, alternative is to prepare a blend of viruses that accomplishes the same goal. This approach offers greater versatility in dealing with the host range issue, but it is not presently clear if the increased cost of such a product will allow it to compete effectively with broader-spectrum conventional insecticides.

Although the first commercial baculovirus insecticides will likely be constrained by the host range properties of naturally occurring baculoviruses, there is every reason to believe that it will be possible to modify these properties through genetic manipulation to produce baculovirus insecticides that are tailored to the needs of key markets. What is required, of course, is a fairly detailed understanding of virus–host interactions, including the early events in viral pathogenesis (Engelhard *et al.,* 1994; Engelhard and Volkman, 1995; Knebel-Mörsdorf *et al.,* 1996), host responses to virus infection, and how host and viral gene products interact during virus replication. Many of the viral genes involved in the regulation of AcMNPV early gene expression, DNA replication, and late and very late gene expression have been identified (see Chapters 6–8, this volume, for detailed discussion), and there is already evidence that some of these genes can influence virus host specificity. For example, AcMNPV and BmNPV are closely related baculoviruses, but AcMNPV is not able to replicate in *Bombyx mori* cells in culture. AcMNPV can be made to replicate on *Bombyx mori* cells, however, if a small segment of the AcMNPV p143 gene, which is required for DNA replication, is substituted by the cognate segment from BmNPV (Maeda *et al.*, 1993; Croizier *et al.*, 1994). In addition, the AcMNPV *hcf-1* gene has been shown to be required for AcMNPV replication in *Trichoplusia ni* Tn368 cells, but not on *Spodoptera frugiperda* Sf21 cells (Lu and Miller, 1995), and the LdMNPV *hrf-1* gene can confer on AcMNPV the ability to replicate in the normally nonpermissive Ld652Y gypsy moth cell line (Thiem *et al.*, 1996). Poor expression of the AcMNPV *p35* gene has also been implicated as the possible cause of the apoptotic response that limits the ability of AcMNPV to replicate in *Spodoptera littoralis* SL2 cells (Chejanovsky and Gershburg, 1995). Although none of these findings is directly applicable to a problem of commercial significance, they demonstrate the feasibility of manipulating at least the *in vitro* cell line specificity of a baculovirus. More importantly, they lay the groundwork for a directed research effort aimed at solving specific host range problems that are of commercial interest. A more detailed discussion of the genetics of baculovirus host range is presented by Miller and Lu (Chapter 9, this volume).

There is also evidence that the intrinsic infectivity of at least AcMNPV can be improved by gene deletion. Ignoffo *et al.* (1995) demonstrated that the infectivity of AcMNPV on even a highly permissive host (*Trichoplusia ni*) is increased sixfold by deletion of the *pp34* gene. This gene encodes a major constituent of the membrane that normally envelopes the polyhedral inclusion bodies (polyhedra, or PIBs) of AcMNPV. This membrane is missing in the *pp34*-deleted mutants, and it has been postulated that this makes it easier for virions to find their way to the midgut epithelium after the polyhedra dissolve. According to this model, a higher initial hit rate in the insect midgut would effectively reduce the number of polyhedra required to establish a productive systemic infection. Additional research is required to see if this manipulation increases the infectivity of AcMNPV on other species.

III. PRODUCTION

Impressive progress has occurred with regard to the development of baculovirus production methods. *In vivo* production still leads this technology by supplying all of current viral insecticides now in the marketplace, for example Spod-X (*Spodoptera exigua* NPV), GemStar (*Helicoverpa zea* NPV), Gypcheck (*Lymantria dispar* NPV), and *Cydia pomonella* GV. Niche markets have been identified and *in vivo* production is currently filling these needs. On the other hand, cost and field performance relative to chemical treatments currently limit broader application of these wild-type viral insecticides. This situation is changing as newly developed recombinant baculoviruses (rNPVs) expressing insecticidal genes are developed. However, a recombinant baculovirus cannot be economically produced in its insect hosts because it kills the host so rapidly with the consequence of producing relatively few viral progeny. Recent advances in *in vitro* production technology now appear to support the commercialization of rNPVs. Production issues focus entirely on the most efficient means to produce low-cost, efficacious virus. This section on baculovirus production intends to build on two excellent reviews on *in vivo* and *in vitro* production by Shapiro (1986) and by Weiss and Vaughn (1986).

A. *In Vivo*

In vivo-produced virus is the source of private and government-sponsored viral insecticides in commercial use today. Modern *in vivo* production had its start during the 1960s with the development of artificial diets and mass rearing technology pioneered by Vanderzant *et al.* (1962a,b). During the 1970s, one product (Elcar, *Helicoverpa zea* NPV) was developed by private industry and was sold in the retail market. Three other preparations (Gypcheck, TM BioControl-1, and Neocheck-S) were registered by the US Forestry Service and were not for sale to the public. Gypcheck is the only viral insecticide being produced today by the US Forestry Service for insect control with an annual production of around 10,000–20,000 acre treatments. A resurgence of interest in viral insecticides during the 1990s has led to EPA registration and introduction of Spod-X (*Spodoptera exigua* NPV) and GemStar *Helicoverpa zea* NPV by biosys.

In vivo production has several advantages over *in vitro* production particularly with regard to wild-type viruses. Start-up costs are considerably lower than *in vitro* production. Also, *in vivo* production is based on well-known technology that has been producing baculoviruses for years. Conditions and methods for mass insect rearing, insect diets, viral infection, harvesting, and processing are well developed (Shapiro, 1986). The cost effectiveness and feasibility of a virus production campaign can be readily determined. Large-scale production has been additionally improved with the introduction of the Alcatraz system with its moving bed technology developed by Esparo (now Thermo Trilogy). Another production system is the HerD technique licensed to AgriViron, which claims improved diet utilization and reduced manpower costs.

In vivo production is particularly useful for the ability to produce a wide variety of insect viruses. Providing that a host insect can be economically mass reared on artificial diet in the laboratory, it is reasonable to suppose virus production can be feasible. This flexibility provides the opportunity to produce viral insecticides like LdNPV, SeNPV, and CfNPV which cannot be easily produced by large-scale fermentation at this time. In addition, other useful insect viruses like granulosis viruses (*Trichoplusia ni* GV and CpGV), entomopoxviruses, and cytoplasmic polyhedrosis viruses are also amenable to *in vivo* production.

Despite these advantages, *in vivo* production has disadvantages that limit its broad application. Most importantly is the inability for *in vivo* production to produce new recombinant virus strains economically, since the most active of these rNPVs kill their hosts before any substantial viral replication can occur. Another significant disadvantage of *in vivo* production is the inclusion of large amounts of insect parts in the product. For example, "refined" Gypcheck polyhedra contain about 85% insect parts. This gross contamination creates additional problems that limit formulation techniques that can be applied. Homogeneous *in vitro* products are not subject to these limitations. Finally, *in vivo* productivity is not readily increased by experimentation. Hence, yield is fixed.

In vivo production is expected to be less competitive in scale-up than *in vitro* production when supplying virus products to cover large acreages. Scale-up for *in vitro* production is capital-intensive but does not appreciably drive up manpower costs correspondingly. On the other hand, *in vivo* production is labor-intensive. Insects must be handled and harvested by hand. *In vivo* scale-up costs require the addition of staff as well as facility costs for each "unit" increase in production. Large *in vivo* production facilities are also vulnerable to failure due to disease by the introduction of insect pathogens via diet, contamination by workers, or by the activation of a latent virus similar to the *Mamestra brassicae* NPV described by Hughes *et al.* (1993). Larger insect colonies increase the opportunity that an insect pathogen can appear and become established in the production facility. This requires close and continuing surveillance to avoid these problems.

A variety of insects have behavioral and/or physical attributes that create problems during *in vivo* production. For example, the diamondback moth (*Plutella xylostella*) and the spruce budworm (*Choristoneura fumiferana*) are small larvae that make it more difficult to efficiently harvest the biomass needed to sustain commercial production. Insects like the gypsy moth (*Lymantria dispar*) are extremely allergenic and hairy. This poses a risk for insect handlers and the larval hairs contaminate the refined viral product. The preferred host of *Heliothis armigera* NPV and HzNPV are Heliothine Lepidoptera. These Heliothine larvae can be mass reared until their third instar, but afterward must be individually reared since they become cannibalistic. Solutions to these problems can be accommodated, but they often result in reduced production efficiency and increased cost.

Perhaps one of the most difficult problems for *in vivo* production is maintaining viral efficacy. Adventitious viruses can contaminate viral seed stocks and quickly amplify and corrupt an entire production run. Also, as viral pathogenesis progresses, insects become increasingly sensitive to secondary bacterial and fungal infections. These opportunistic secondary infections can reduce production yield and result in a viral product heavily contaminated with microorganisms. For example, Mazzone *et al.* (1971) found the ratio of bacteria to polyhedra was 1:2.5 in crude preparations of the European pine sawfly (*Neodiprion sertifer*) NPV. For these reasons, viral inoculum for *in vivo* production must have a low bacterial load ($\sim 10^4$ aerobes/g) (Shieh and Bohmfalk, 1980) and be free of certain contaminants such as other insect viruses as well as human pathogenic enterobacteria.

In summary, *in vivo* "virus production is the result of complex research in host–pathogen–environment interrelationships resulting in maximization of both virus quantity and quality" (Shapiro, 1986, p. 54). This method is successfully producing wild-type baculovirus products needed for commercial and government use. It affords a flexibility for producing new viruses that is not shared by *in vitro* production but at the same time has little potential to produce rNPVs economically. In addition, *in vivo* production suffers from a lack of process control. Wild-type virus is cheaper by *in vivo* production in small acreage markets, but beyond a certain level of production, *in vitro* production becomes more cost-effective. In areas where the field performance of wild-type viruses is acceptable and where cost is competitive with alternative insect control strategies, one would predict *in vivo* production will continue to be used.

B. *In Vitro*

The development of fast-killing rNPVs has strikingly improved the field performance of viral insecticides, with this faster action translating into improved crop protection. These improvements in coordination with advances in formulation and application technology have created new market possibilities that did not exist prior to these developments. Our experiments on the replication of AaIT-expressing AcNPV in the highly susceptible insect *Heliothis virescens* indicate that larvae infected with this rNPV produce 80–90% less polyhedra when compared to wild-type AcNPV infected larvae (unpublished data). *In vivo* production costs along with this yield reduction in recombinant viruses makes *in vivo* production untenable. A similar polyhedra yield reduction in a rNPV expressing the LqIT gene has been reported (DuPont, 1996). In summary, the better the crop protection provided by a rNPV due to faster killing, the less likely that the rNPV can be economically produced *in vivo*.

While a variety of *in vitro* production methods have been suggested, the production scale for even a modest commercial introduction severely limits the choice of *in vitro* production methods. Even a modest introduction of a viral insecticide into a row crop or forestry market would require a bioreactor vessel exceeding 10,000 liters running at high efficiency. For this reason, it is easy to exclude a list of previously reviewed methods such as spinner culture, roller bottles, microgravity propagation, and shaker culture (Weiss and Vaughn, 1986). None of these methods appear to have any production potential at a scale of 10,000 liters and beyond, especially when one considers the sharp cost constraints imposed on this process. For example, current lepidopteran chemical insecticides cost between \$5 and \$20 per acre treatment, with the bulk of sales in the \$7 to \$10 range. To be commercially successful, viral insecticides must rival these insecticides on a cost and performance basis, or find a niche market that justifies their particular activity and/or higher cost. Conventional stir-tank bioreactors are the only alternatives to meet these production needs. However, it should be noted that even a modest sales program for a major field crop (say, 500,000 acre treatments) would require an extensive production campaign assuming 1 to 10×10^{11} polyhedra/acre.

Selection of viral strains for commercialization must be market-driven and companies are reluctant to divulge specific information during their development. AcNPV and LdNPV are the only viruses at this time that have been *in vitro* produced and extensively field tested. Both of these viruses require cell lines that remain highly productive in fermentation vessels with regard to polyhedra/liter yield during production. While cell–virus systems have been developed for other viruses like CpGV, HzNPV, SeNPV, OpNPV, and CfNPV, it is not clear whether these cell–virus systems can support commercialization at this time. Development of novel cell lines is currently an important objective of academic, federal, and industrial laboratories in anticipation of their eventual use in commercial virus production.

Certified seed stocks must be developed for both the viral strain as well as the producer cell lines to satisfy EPA regulatory requirements. A seed stock bank for both viruses and cells must be created and certified to be genetically homogeneous and free of human pathogens. The viral seed stock must demonstrate genetic stability during a minimum of ten amplification passes that are required to inoculate a production vessel without detectable genetic rearrangements, loss of productivity, or loss of infectivity. Viral genetic stability is strain-specific and must be empirically determined for each viral isolate. For example, Abby624 LdNPV and Hampton LdNPV strains rapidly lose efficacy and genetic stability after only three to four rounds of amplification, while the V8 strain of AcNPV maintains genetic stability and efficacy for 12 rounds of amplification.

Another essential viral trait required for commercialization is the production of high titers of budded virus stocks that retain virulence during storage. AcNPV isolates routinely yield high titers in excess of 10^8 plaque forming units (PFU)/ml that maintain high virulence for over a year in storage at 4°C. In contrast, Abby624:LdNPV produces low budded virus titers of about 10^5 PFU/ml and soon loses infectivity after 3–4 months of storage when stored under identical conditions. This low virulence for LdNPV causes severe production problems. Increased resources (and cost) must be devoted to producing viral inoculum. In addition, large inoculum volumes are required to initiate infection, which also introduces cell lysates and spent media into the production vessel that have an adverse effect on cell viability as well as virus production. The ability of a virus to generate high budded virus titers during an infection is determined by a complex interplay of virus strain, cell line, and media. Hence, each virus strain entering production must be monitored as changes are made with regard to cell line and media optimization.

Like viral stocks, certified cell stocks must be shown to be free of viral, mycoplasmal, and bacterial pathogens. These stocks must be developed, amplified, aliquoted, and stored under liquid nitrogen. Vials are removed as needed at the onset of a production campaign and amplified to the desired level in production bioreactors. Because of the time and expense required to develop and certify production lines, a considerable amount of time and research is expended on identifying and developing suitable cell lines. As many cell lines as is practical need to be screened for their productivity and growth characteristics. For many viruses (like SeNPV, LdNPV, and HzNPV) the choice of available cell lines is limited, while for AcNPV, there is a wide range of candidate cell lines to evaluate.

With few exceptions, insect cell lines have been developed in insect tissue culture medias that rely on fetal bovine serum (FBS) to supply essential growth factors. FBS is unacceptable for *in vitro* production due to its high cost. For this reason, cells must be adapted into various serum-free medias like ExCell401, Sf900, or in various proprietary medias. Adaptation to serum-free medias cannot be assured for a cell line. Successful adaptation typically requires several months to attain the desired growth kinetics and final cell densities required for production. Additional selection must be imposed on these cell lines to adapt to the shear stress and culturing conditions found in commercial bioreactors. Highly productive cell lines with acceptable growth rates can then be certified as production seed stocks.

Successful *in vitro* production of viral insecticides depends on consistent, inexpensive medias that do not rely on FBS. Many of these initial objectives have been met with ExCell401 and Sf900 media. Several media developments in the last 10 years have contributed to the success of *in vitro* production. First is the replacement of FBS with a cod liver oil-based substitute. Without this substitution, it would not be economical to *in vitro* produce viral insecticides. A second major advance was the discovery of pluronic agents (e.g., Pluronic F68) as membrane stabilizing agents to protect cells from shear stress during fermentation (Maiorella *et al.*, 1988). Mixing and mass transfer problems encountered with oxygen delivery to cells in large bioreactors compound shear stress and impose an absolute requirement for the inclusion of these pluronic agents in all production media. The current focus of media development is the creation of consistent quality, low-cost production media formulations. Media inconsistency has important implications for virus production. Nutritionally deficient medias adversely affect viral productivity. Sometimes these effects are cryptic; for example, Carstens *et al.* (1986) and Slavicek *et al.* (1995) have found that deficient medias in unidentified components result in a slightly decreased yield of AcNPV and LdNPV polyhedra per cell with drastically reduced potency. Slavicek showed that LdNPV produced in deficient media had polyhedra with sharply reduced numbers of virions/polyhedra. The high cost of the media as well as large reactor volumes demand highly consistent media. The loss of a single reactor run is a costly affair. Each cell line seems to have its own requirements. For this reason, medium development must parallel the creation of each production cell line.

There are additional scale-up challenges as one moves from pilot bioreactions to full-scale production reactors. Oxygen demand increases sharply after infection. To maintain cells optimally during infection, the shear damage in production must be balanced with oxygen demand. In addition, cells during growth and infection consume larger amounts of glucose and glutamine. The consumption of these nutrients require a replacement strategy to maintain cell growth rates and virus production. Various strategies using batch feeding, continuous feeding, and perfusion have all been used successfully. The method and feeding rates must be determined empirically for these production vessels. Waste metabolites (e.g., lactic acid, ammonia, and asparagine) are generally not deleterious to virus productivity. Removal of waste metabolites when required would favor perfusion technology for both feeding and waste removal.

Amplification of viral seed stock is very sensitive to attenuation. Tramper *et al.* (1990) found that AcNPV infection of Sf9 cells in a bioreactor is very sensitive to multiplicity of infection (MOI). Cells infected with a MOI of 5 rapidly attenuate as defective interfering viral particles become the predominant viral form. A similar observation has been made with LdNPV growing in Ld652Y cells by Slavicek. Infection at a low MOI (0.01) minimized these problems but required close coordination with the cell growth rate. A medium supporting a predictable growth rate is critical for the successful amplification when infecting at low MOI.

In conclusion, both *in vitro* and *in vivo* production show promise for the production of viral insecticides. Many desirable baculoviruses and all granulosis viruses (Gvs) require *in vivo* production for lack of suitable production cell lines at this time. *In vivo* production efficiencies currently support the commercialization of selected wild-type NPVs. The improved crop protection afforded by recombinant baculoviruses that express insect-specific toxins will likely be feasible only by *in vitro* production.

IV. FORMULATION

A. Introduction

An efficacious virus and viable production scheme alone cannot attain commercial success without a formulation that protects the virus from environmental degradation to allow time for the virus to be consumed by insect pests. The requirements for this formulation are typical to pesticides in general. The goal is to preserve biological activity and to deliver the product to the target system using conventional delivery techniques familiar to the end user. Research has been ongoing ever since the introduction of the first baculoviral insecticides in the late 1970s with the introduction of Elcar (HzNPV), Gypcheck (LdNPV), and TM BioControl-1 (OpNPV). Although there has been continuous improvement, much needs to be accomplished to meet the field persistence and storage stability characteristics as we know in chemicals. Viral insecticide formulations in use today are typically flowables and wettable powders that are compatible with conventional spray equipment. Viruses have also been "experimentally" formulated as granules, solid baits, and powders. Much of the formulation research is carried out in the private sector and companies keep this research proprietary, as a means of protecting their research investments.

When pests are feeding, at what stage, and which crop (or forest) is being targeted have important implications for the formulator. For example, forestry formulations targeting hardwoods (e.g., LdNPV) or conifers (e.g., OpNPV) typically use 1–2 gal/acre application rates. Broad coverage at these application volumes require the delivery of the formulated product in micron-sized droplets. Leaf adhesion and spreading properties are largely determined by the mass and falling velocity of these droplets, as well as by the droplet surface tension and viscosity properties. To obtain consistent results, a formulator must consider the potential for a droplet to desiccate. Temperature, humidity, and time of flight can easily modify the physical characteristics, resulting in a failure to adhere (bounce) on the leaf surface as well as inadequate spreading. On the other hand, a sprayable formulation delivered to leafy vegetable crops is typically diluted in water to 20–30 gal/acre. Additional complicating factors are the behavioral and feeding habits of targeted pests. For example, standard fan-jet application gives excellent coverage of the upper leaf surface in cotton, but poorly covers the leaf's undersurface where early instar pests (like *Spodoptera littoralis*) tend to feed. Unfortunately, late-instar larva become much more developmentally resistant to the virus when the larvae grow sufficiently large to eat through the leaf (and acquire the viral insecticides). Similarly, late-instar *Heliothis virescens* and *Helicoverpa zea* larvae become difficult to control after they seek refuge in cotton bolls. Because of the sensitivity of placement and timing, the end user must target the insect pests when they are biologically and behaviorally exposed to the pesticide.

B. Measuring Efficacy

Viruses, like other biological pesticides, pose special challenges that differ from chemical formulations because biological activity of formulated material cannot be determined by straightforward extraction and chemical analysis. The need for industrial and international standardization was recognized early by Dulmage and Bergerjon (1982), who reported that many of the early problems with the introduction of Bt were aggravated because of a lack of standardization. Unfortunately, no solution or common consensus has been adopted at this time for baculoviruses. Baculoviruses share similar standardization problems encountered with Bt but have additional properties that make standardization more difficult. Baculovirus activity is typically determined in the laboratory by adding a defined number of polyhedra to an assay arena. In our experience, we seldom find that two independent laboratories arrive at the same estimate, without extensive cross-training between the two laboratories. Differences in diet, insect colonies, larval instar assayed, cohort variation, and even elapsed time following the last molt prior to virus exposure all profoundly affect the results obtained by these assays. Such differences contribute to the difficulty in creating standardized assays.

Measuring the intrinsic biological activity of a virus in a formulation does not fully define the properties of a particular formulation that contribute to its field performance. For example, specific assays need to be developed to measure rainfastness, UV protection, abrasion resistance, spreading properties, the influence of chemical enhancers, and synergists in addition to standard assays to measure the intrinsic activity of these viruses. All of these properties must be addressed in a formulation to provide a product with consistent activity. Evaluation by a single technique does not reveal the full properties of a formulation. The presence of chemical synergists (like stilbene optical brighteners) complicates these evaluations. Varying with the virus and host insect, stilbene optical brighteners may potentiate viral efficacy 10- to 10,000-fold (Shapiro and Robertson, 1992). Hence, the presence of stilbenes in a formulation complicates any attempt to compare a formulated product to a reference strain. Since the activity of stilbenes is concentration dependent, it is important that the formulation evaluations reflect the presentation of the virus to insects, in a manner that simulates the intended field rates.

The measurement of biological activity is further mandated because although the polyhedra count may remain constant, variable efficacy can result from viral (virion) degradation during the formulation process, storage losses, and also during production by viral attenuation and nutritional defects. A final issue that complicates the development of industrial and international standards is the inability to directly compare different baculovirus strains. For example, insecticidal properties of various Bt strains are compared to an internationally accepted reference strain using a standardized assay protocol. Direct comparisons are impractical for various baculovirus strains (like AcNPV, LdNPV, OpNPV, SeNPV, and CpGV) where host range specificity and efficacy are highly strain specific. These intrinsic characteristics of baculoviruses make it very difficult to create a general international standard that predicts field performance.

One practice employed at American Cyanamid relies on the comparison of a formulated product with an archived reference standard. The reference standard must present an acceptable level of potency/polyhedron within allowable limits of deviation and slope before data are acceptable. This precaution minimizes the effects of assay variability. Formulated product efficacy can then be directly compared to the reference standard. Data show that formulation components that enhance viral potency (like optical brighteners and/or insecticides) inflate the activity of the formulated material, while other formulation components (UV protection, sticking, wetting agents) have the potential to depress the apparent activity. Because viral efficacy is so variable with different insect hosts, these assays are standardized on a single susceptible host species that is included in every bioassay.

C. Methods

Until recently, formulations started with *in vivo*-produced viral polyhedra. This material was typically contaminated with host material. Refined *in vivo*-produced LdNPV is 15–18% polyhedra, with the balance being largely insect debris having a particle range from submicron to around 80 μm. Viral polyhedra are usually stored as lyophilized powders. Lyophilization is expensive to incorporate into a commercial process and lyophilized baculovirus powders may be difficult to resuspend due to their propensity to clump. This clumping problem can be minimized by lyopholizing the virus in a lactose paste without adversely affecting storage stability during lengthy periods of refrigeration (Ignoffo, 1964).

An alternative to lyopholization is a liquid concentrate of *in vivo*-produced Spod-X and GemStar (currently marketed by Thermo Trilogy). In this application, virus is refined from cadavers and mixed with water and various agents to increase viscosity and osmolarity. Liquid formulations such as this are optimally stored under refrigeration and typically require preservatives to prevent microbial growth. This formulation has been most successful in greenhouse applications (i.e., Spod-X) where the impact of UV degradation is minimized. Spod-X (SeNPV) was recently introduced into the vegetable crop market in Thailand, and GemStar (HzNPV) has been evaluated in the Mississippi Delta as part of the USDA area-wide budworm/bollworm suppression program. The product label recommends reapplication every 3 days for Spod-X, which may suggest that the issue of virus inactivation has not been adequately addressed when compared to chemical standards. How both of these liquid formulation products perform under a wide variety of field conditions, as well as their storage stability under normal circumstances, will be critical to the commercial success of the products.

Spray-drying techniques have been successfully applied to a number of baculoviruses. In a process developed by Sandoz, virus was mixed with attapulgite clay and other diluents. This mixture was spray-dried to produce a microencapsulated formulation that readily resuspended in a tank mixture. This formulation of the HzNPV demonstrated increased stability and was more active than previous formulations of the virus (Ignoffo *et al.*, 1976). Sandoz marketed this formulated virus under the trade name Elcar. This spray-drying process was successfully applied to other baculoviruses like AcNPV (Sandoz404) and *T. ni* NPV (Sandoz405), but resulted in a loss of activity and/or loss of stability with *Choristoneura fumiferana* NPV and *Cydia pomonella* GV (Sandoz406) (Bull, 1978; Cunningham *et al.*, 1978).

It is essential that the formulation does not adversely affect the stability of the product. A minimum of 18 months of storage stability is required for a commercial baculoviral insecticide (Couch and Ignoffo, 1981). Storage is required in the stockpiling of the product in anticipation of the growing season, on the distributor's shelf, and in the end user's possession. Distribution and field storage conditions can be quite harsh, with storage temperatures occasionally exceeding 48°C (120°F). These extremes in storage temperature are problematic for any biological insecticide. Baculovirus polyhedra are stable when stored refrigerated or frozen. Baculoviruses are slightly less stable at room temperature, but can remain infective for several years depending on the storage method (Huger, 1963; Lewis and Rollinson, 1978). At temperatures between 38 to 42°C, virus samples lose significant activity over a period of weeks to months (Lewis and Rollinson, 1978; Hunter *et al.*, 1973; David and Gardiner, 1967). Baculoviruses in solution lose viral activity measured in hours to minutes at temperatures exceeding 50°C (Jaques, 1977). In general, increasing moisture decreases the stability of baculoviruses (Couch and Ignoffo, 1981; Jaques, 1977).

Perhaps the most pressing issue facing baculovirus formulation is photodegradation of the virus activity. Baculoviruses are rapidly inactivated by short- and long-wavelength (254–310 nm) UV light. Without formulation, viral insecticides lose most of their field activity within the first 24 hr of exposure. This light sensitivity may be further exacerbated by moisture. Ignoffo and Garcia (1990) report that HzNPV polyhedra in water were three times more sensitive to a simulated sunlight-UV exposure than dried polyhedra. There is a notable exception to this generalization. David (1975) reported that *Pieris brassicae* GV was more stable in wet films when exposed to UV light. This extreme light sensitivity results in very short residual activity in the field.

D. Field Application

The measure of the formulator's success will be largely determined by efficacy coupled with field persistence. Unformulated virus is rapidly lost on plant surfaces by 24–48 hr (Bullock, 1967; Ignoffo and Batzer, 1971; Young and Yearian, 1974). Low persistence is primarily due to degradation by sunlight. Experiments studying HzNPV persistence on cotton, tomato, and soybeans found that protecting plants from sunlight by covering preserved most of the viral activity after 4 days. In contrast, exposing these plants to sunlight resulted in a rapid inactivation of viral activity within 24 hr (Young and Yearian, 1974). As expected, baculoviruses persist significantly longer on shaded plant surfaces. For example, TnNPV and *Pieris brassicae* GV lose most of their activity on the upper leaf surfaces of cabbage within 2 days, while 15% of the virus on the under surfaces persisted for 10 days (Jaques, 1972; David *et al.*, 1968). Similarly, HzNPV persisted significantly longer on cotton leaf under surfaces, square bracts, and square calyces than on upper leaf surfaces.

Viral sensitivity to light has encouraged the experimentation on a variety of UV screening agents to extend field persistence. The US Forestry Service applies LdNPV using either Carrier038 (Abbott Labs) or the Forestry Service's tank mix formulation. Both of these compositions rely on lignosulfates as UV blocking agents. The presence of lignosulfates clearly contributes to LdNPV field persistence. Carbon black has also been widely used as a UV screening agent for a variety of baculovirus formulations, including HzNPV, CfNPV, and LdNPV (Cunningham *et al.*, 1978; Yendol *et al.*, 1978). Relatively high concentrations of carbon black (6% Shade™) are required in tank mixtures to affect field persistence. In addition to the above, other materials with UV protectant properties have been used in tank mixtures, including gelatin, titanium dioxide, and chemical UV adsorbants like Uval (Miles Laboratory, Inc.) (David, 1975). While these materials variably increase field persistence, no combination has been found that protects a viral insecticide in the field for more than 4 days.

Baculoviruses intrinsically stick tightly to plant surfaces. Indeed, it is difficult to remove bound NPV and GV from leaf surfaces by washing. For example, 40% of an *in vivo*-produced TnNPV applied to a cabbage leaf remained after 4 days of continuous washing (David, 1975). Formulation ingredients added for UV protection or to increase spreading properties may interfere with this intrinsic ability for baculoviruses to adhere to plant surfaces. Hence, a formulated viral product may contain many components such as gustatory stimulants, thickening agents, UV screening agents, emulsifiers, wetting agents, emulsifiers, wetting agents, viral synergists, and spreader-sticker agents in addition to the active ingredient (virus). The physical properties of these viruses may now be masked such that the addition of a spreader-sticker agent helps to improve efficacy by improving the coverage on critical plant surfaces where insect pests are feeding.

E. Baculovirus Enhancers and Synergists

In addition to the traditional formulation objectives to prolong persistence and improve the delivery and coverages, Shapiro and Robertson (1992) found that stilbene optical brighteners serve as enhancers of viral infectivity. These optical brighteners must be coadministered with virus to show the biological effect. Such delivery can be accommodated by formulation and/or tank mix additions. Enhancing activity is most easily demonstrated for baculoviruses that show low virulence against their host, like LdNPV against gypsy moth and AcNPV against *H. zea*. Laboratory studies show virus–stilbene combinations required 1000X and 30X less virus to attain a comparable LD_{50} for gypsy moth and *Helicoverpa zea,* respectively (Shapiro and Robertson, 1992; Webb *et al.*, 1996). On the other hand, AcNPV against a very sensitive host, *Heliothis virescens,* only shows a two- to threefold enhanced response. Translation of this elevated activity to the field has not been so dramatic. Webb found LdNPV formulated in a 1% solution of stilbene optical brightener, Blankophor BBH, in the spray tank only proved to be about tenfold more effective than virus alone.

The advent of recombinant NPVs has created new opportunities where insects succumb to expressed insect-specific toxins rather than to viral pathogenesis. This faster killing directly translates into improved crop protection. Building on this idea, McCutchen *et al.* (1997) reported benefits of combining recombinant AcNPV expressing the AaIT insect toxin (AcNPV–AaIT) and low rates (LC_{10}–LC_{20}) of a pyrethroid insecticide in reducing the time needed to kill *Heliothis virescens.* AcNPV–AaIT in combination with cypermethrin resulted in a LT_{50} of 30 hr compared to 71 hr for AcNPV–AaIT alone, a 58% reduction in kill time over virus alone. The combination of a recombinant NPV with a low rate of pyrethroid insecticide may give growers the cost-effective insect control they have come to expect without needing to use high rates of synthetic insecticides. It is possible that combinations of recombinant viruses expressing other neurotoxins (like LqHIT3) with low rates of chemical insecticides will also produce similar increases in performance.

In conclusion, there are a variety of formulations currently in the marketplace; however, field and storage stability of the compositions still fall short of the goals needed for wide application. The extent that baculovirus formulations can protect viral activity from UV degradation and increase persistence will greatly influence the eventual market size that viral insecticides will occupy.

V. RECOMBINANT NPV FIELD TESTING

A. History

Field testing is critical to determining the commercial potential of an insecticide, whether it be chemical or biological. Field testing of rNPVs began in 1986 by the Natural Environment Research Council (NERC), Institute of Virology, Oxford, England. The baculovirus construct tested was an AcMNPV that had a "junk" piece of DNA inserted downstream of the *polh* gene coding region. Government approval was obtained. The test site was highly contained using fencing, netting, and so forth to prevent the entry and release of any vertebrates or arthropods to or from the treated area. At the end of the test, the site was decontaminated by spraying with formalin (Bishop *et al.*, 1988).

In 1987, NERC conducted a second release utilizing an AcMNPV virus that had the *polh* gene deleted and then replaced with a β-galactosidase gene. Studies were conducted to determine the persistence of this nonoccluded virus in soil. The results clearly showed that the virus disappeared over time, therefore, proving the hypothesis that the lack of polyhedrin results in a virus that cannot persist in the field (Bishop *et al.*, 1988).

In 1989, the Boyce Thompson Institute for Plant Research at Cornell University conducted the first US field application of a genetically altered baculovirus (Wood *et al.*, 1994). The purpose of this release was to study the environmental persistence of an AcMNPV isolate that had the *polh* gene open reading frame removed and had then been co-occluded with a wild-type AcMNPV (Hamblin *et al.*, 1990). This was accomplished by producing both viruses in a common tissue culture preparation.

B. Regulatory Issues

In order to conduct small-scale (< 10 acres) field trials in the United States with a gene deleted/inserted recombinant baculovirus, a notification of intent to field test must be filed with the US EPA before any testing can take place. The notification document must include information regarding: (1) identity of the microorganisms, (2) description of the natural habitat of the parental strain, (3) information on the host range with an assessment of infectivity and pathogenicity to nontarget organisms, (4) information on the survival and the ability of the microbial pesticide to replicate in the environment, (5) identity of possible transmission vectors, (6) data on the relative environmental competitiveness compared to the parental strain, (7) description of genetic modification methods, (8) data on the potential for genetic exchange and on genetic stability of inserted sequences, (9) a description of the proposed field program, and (10) a statement of composition for the formulation to be tested. Thus, obtaining approval from the US EPA may involve the submission of data from numerous studies on nontarget insects, environmental fitness, soil persistence, and host range. This equates to numerous hours of research and development of data before even the first field study can take place. Large-scale field trials (10–5000 acres) require to experimental use permit. This would be the next step after a notification, and requires more extensive toxicology and ecotoxicology information.

Numerous concerns were voiced about the notification, which had to be answered prior to receiving any approval to conduct the field tests. The US EPA questioned effect on nontarget species, specificity of the AaIT toxin, competitiveness of the AaIT containing baculovirus versus the wild type, and environmental persistence of the AaIT form and genetic exchange with other viruses. Numerous meetings were held with the Biopesticides and Pollution Prevention Division (BPPD) of the US EPA's Office of Pesticide Programs to conclude what information would be needed to address these concerns.

The effect on nontargets was addressed both in the literature (Possee *et al.*, 1993b, McNitt *et al.*, 1995; McCutchen *et al.*, 1996) and via specific studies conducted at Cyanamid (Tracy *et al.*, 1997). These studies clearly showed that the AaIT recombinant baculovirus had no impact on nonpermissive species, which included predators and parasitoids.

The insect specificity of the AaIT toxin was well covered in the literature (Zlotkin *et al.*, 1985, 1991, 1993). These data showed that the toxin was clearly insect specific, which gave the construct two levels of safety: toxin specificity and baculovirus specificity.

Demonstration of the noncompetitiveness of the AaIT baculovirus was accomplished by designing and carrying out a laboratory experiment (Dierks *et al.*, manuscript in preparation) where permissive insects were exposed to a mixture of the AaIT and wild-type baculovirus on the food source. All dead and infected larvae from a cohort were harvested and fed to a succeeding cohort of permissive insects. After doing this for six generations, the AaIT gene was undetectable in the population. The recombinant produces 75–95% fewer polyhedra than wild type, which (due to its superior multiplicative advantage) outcompeted the recombinant. This experiment certainly represents the "worst case" basis.

Baculoviruses are sensitive to degradation by UV light and persist only a few hours when exposed to direct sunlight if left unprotected. Polyhedra can persist for some time in the soil, which serves as a reservoir. Studies conducted at American Cyanamid (1994, 1996) showed the persistence to be high shortly after a soil application, but that the persistence decreases over time to an undetectable levels using the method of Wood *et al.*, (1994).

Genetic exchange certainly represented the biggest challenge to answer. It is a fact that this does happen in nature, but that barriers do arise to maintain genetic integrity of a species. One such barrier is the degree of homology of the donor to the recipient. One would expect that any recombination would be limited to closely related baculoviruses. Studies have been reported (Bishop *et al.*, 1995) that support this premise. Even if a highly unlikely successful genetic exchange does result in viable progeny, the new microorganism would itself be ecologically disadvantaged due to the inclusion of the AaIT gene and the subsequent decrease in polyhedron production. Very clearly, the AaIT gene confers a competitive disadvantage; therefore, its presence in another closely related baculovirus would doom it to extinction.

The US EPA evaluated the data generated by American Cyanamid, in addition to the literature available in the public domain, and approved the small-scale field testing.

C. American Cyanamid Field Trials

American Cyanamid met the EPA's requirements for conducting a field trial with an *egt* gene-deleted form and conducted field trials in 1993. Additional trials were conducted in 1994 and 1995.In 1995 American Cyanamid was successful in obtaining approval for testing a gene-inserted construct and conducted the first field trials of a recombinant baculovirus engineered to express an insect-specific toxin. Additional field studies to determine efficacy of the gene-inserted construct were conducted in 1996. Extensive field research must occur to determine the effectiveness of a genetically modified baculovirus and its fit in various commercial insect control scenarios.

The 1993 field trial program consisted of five test sites, one each in Arizona, California, Florida, New Jersey, and Texas. Subsequent *egt*$^-$ studies were conducted in 1994 and 1995. Results showed that the deletion of the *egt* gene produced a construct that would kill target insect species about 15–20% faster than wild type. This equated to a reduction in the overall damage as compared to wild type.

In 1995, American Cyanamid received permission from the US EPA to conduct two field trials, one in Georgia and one in Texas, with a new AcMNPV construct that still had the *egt* gene deleted and had a gene coding for an insect-specific toxin (AaIT) inserted. This gene was derived from the scorpion, *Androctonus australis,* a common mid-Eastern scorpion. This was the first US field trial of toxin gene-expressing baculovirus designed to evaluate the commercial potential of this technology as a means to control insect pests. The data derived from these studies showed an even faster kill of the target insect species, which resulted in an increased level of control over the wild type and the *egt*$^-$ forms. It was concluded that this rNPV had possible commercial potential, and that the field test regimen needed to be expanded to include other locations and crops.

In order to better characterize the performance of this rNPV for insect control, American Cyanamid applied for an expanded field trial regimen in 1996. The request was approved by the US EPA to conduct up to 20 trials on 7.4 total acres in 12 states. The studies are in progress at the time of this writing.

VI. SAFETY CONCERNS

A. Safety to Vertebrates

The long and close relationship between baculoviruses and arthropods has led to specific evolutionary adaptations by this family of viruses. These adaptations, which allow efficient exploitation of the arthropods by baculoviruses, have also resulted in a restriction of their host range. The best illustration of this coevolution is the existence of the occlusion body, a proteinaceous crystal that provides environmental stability to the fragile virions but is extremely sensitive to the high pH found in the insect midgut. This allows the baculovirus occlusion body to survive on a leaf surface long enough to infect a host insect yet readily release the encased infectious virions during its short transit through the insect digestive tact. In the neutral or acidic environment of vertebrate digestive tracts, the occlusion bodies pass through the digestive tract and are either excreted intact or become inactivated during digestion.

During the past 40 years, extensive testing of the safety of baculoviruses to vertebrate species has been carried out. A compilation of data indicates that 26 different insect baculoviruses have been tested for pathogenicity. These tests were performed on 10 different mammalian species including rats, mice, dogs, guinea pigs, monkeys, and humans. The baculoviruses in these tests were administered by a variety of routes including orally, intravenous injection, intracerebral injection, intramuscular injection, and by topical application. In all cases, there were no indications of toxicity, allergic response, or evidence of pathogenicity due to the baculovirus (Ignoffo and Heimpel, 1965). When adjusted to the weight of a 70-kg (160 lb) man, the doses in many of the tests were 10–100 times the per-acre field rate. In no case did exposure to the baculovirus result in deleterious effects to test animals. The results of these tests have been summarized in a number of reviews (Doller, 1985; Burges *et al.*, 1980; Ignoffo, 1973, 1975).

In one study, 230 white mice and guinea pigs were exposed to *Helicoverpa zea* NPV polyhedra, occlusion-derived virus, or polyhedrin protein. Routes of administration included inhalation, feeding, and intradermal, intraperitoneal, or intracerebral injection. A single death occurred due to acute pneumonia. All other animals remained healthy (Ignoffo and Heimpel, 1965). Long-term studies on *Helicoverpa zea* baculovirus administered orally and parenterally to rats were also conducted (Barnes *et al.*, 1970). These tests included a 2-year feeding study with Sprague-Dawley rats. No baculovirus-related deaths were observed, and the incidence of neoplasia between the baculovirus-fed and control groups was not significantly different. Rats that received the highest dose of baculovirus received an amount (on a weight basis) that would equate to feeding a man sufficient virus to treat 100 acres.

The safety of baculoviruses in humans has been demonstrated both directly and by indirect evidence. *Helicoverpa zea* NPV polyhedra ingested by ten men and women at a dose of almost 6 billion polyhedra over a 5-day period (Heimpel and Buchanan, 1967) caused no ill effects. Six persons exposed to the *Helicoverpa zea* NPV during 26 months of production also showed no adverse effects (Huang *et al.*, 1977). Analysis and bioassay of blood samples did not detect infectious baculovirus, baculoviral antigens, or baculoviral antibodies.

It is important to note that the indigenous baculovirus load in the environment is quite high. Heimpel *et al.* (1973) have shown that polyhedron counts made on cabbage taken from store shelves or collected from the field vary between 2×10^6 polyhedra/in^2 on nonepizootic plots to 7×10^7 polyhedra/in^2 on epizootic plots. Using these numbers, it has been estimated that a typical cole slaw serving (16 in^2 of cabbage) contains an average 1.12×10^8 polyhedra. Therefore, not only are large amounts of baculoviruses present in the environment, but people who eat raw cabbage in any form consume very large numbers of baculovirus particles. This ubiquitous and harmless association between baculoviruses and humans underscores their safety.

It is clear from the extensive testing summarized above that wild-type baculoviruses pose no safety hazard to man, fish, birds, or other vertebrates. The safety of baculoviruses to vertebrates should not be changed by the addition of a gene for an insect-selective toxin, enzyme, or hormone. However, prior to registration of a recombinant baculovirus as a biopesticide, the US EPA requires data to confirm vertebrate safety. Tests designed to supply these data are specified in a set of guidelines published by the US EPA (United States Environmental Protection Agency, 1996).

Some testing on an AaIT-expressing AcMNPV has already been reported (Possee *et al.*, 1993b). These tests included a subcutaneous injection of 1×10^6 polyhedra into rats, oral feeding of 1×10^6 polyhedra also to rats, and acute dermal exposure of the recombinant virus to guinea pigs. No adverse effects were recorded for any of the test animals.

B. Safety to Nontarget Invertebrates: Nonlepidopteran Arthropods

With the safety of baculoviruses to vertebrates firmly established, attention has focused on the safety of baculoviruses to nontarget invertebrates. Most of the wild-type baculoviruses being considered for use as biopesticides and all of the genetically engineered baculoviruses have host ranges limited to the order Lepidoptera. This specificity makes baculoviruses an excellent tool for use in integrated pest management programs. By controlling specific insect pests in the field and leaving the complement of beneficial arthropod predators and parasites unharmed, baculoviruses augment natural control of pests rather than supplant it as do many chemical pesticides.

The addition of a toxin, enzyme, or hormone gene to the genome of a baculovirus raises the question, "Would beneficial insects such as predatory spiders or parasitic wasps be harmed by the ingestion or parasitism of an insect larvae infected with the recombinant baculovirus?" For example, could the finite amount of toxin expressed in a target insect by such a recombinant virus result in death of a predatory arthropod or larvae of a parasitic wasp? To address this issue, a number of studies were conducted with various recombinant AcMNPV viruses that express the insect-specific toxin AaIT. In one study, *Spodoptera frugiperda* larvae infected with an AaIT-expressing recombinant AcMNPV were fed for approximately 3 months to colonies of the social predatory wasp *Polistes metricus* (McNitt *et al.*, 1995). No adverse effects on fecundity, development, or behavior of the wasps were observed. In a second study, the effect of two AcMNPV recombinants (one expressing AaIT and a second expressing a modified juvenile hormone esterase gene) on the development of the parasitic wasp, *Microplitis croceipes* were evaluated (McCutchen *et al.*, 1996). It was found that parasitism of larvae infected with either recombinant baculovirus resulted in a shortened developmental time for the wasp. The resulting adult wasps that developed inside the recombinant virus-infected larvae were significantly smaller than wasps that had more time to develop in uninfected larvae or larvae infected by wild-type AcMNPV. This decrease in size is most probably due to a decreased time for development provided by the recombinant virus infected larvae. Although smaller, there was no significant increase in mortality for wasps developing inside these infected larvae, and the emerging adult wasps were capable of successfully mating and parasitizing a new round of host larvae.

Several studies have been conducted that examine the effect of feeding larvae infected with a recombinant baculovirus to several generalized insect predators. The predatory arthropods include a carabid beetle (*Pterostichus madidus*) (Possee *et al.*, 1993b), the green lacewing (a predatory Neuroptera; *Chrysopa carnae*), a predatory Hemiptera (*Orius insidiosus*) (Heinz *et al.*, 1995), a funnel web spider (*Ixeuticus* spp.), and the Chinese mantid (*Tenodera aridifolia sinensis*) (American Cyanamid, 1994; Treacy *et al.*, 1997). No adverse effects were observed. Furthermore, honeybees (*Apis mellifera*) injected with the budded form of AaIT-expressing recombinant AcMNPV also showed no ill effects (Heinz *et al.*, 1995).

Indirect data on the safety of a recombinant baculovirus to nontarget arthropods also comes from surveys of their population densities during two small field trials of an AaIT-expressing AcMNPV conducted by American Cyanamid during 1995. Results showed that population levels of nontarget arthropods (18 different nonlepidopteran insect families in addition to various spiders) were not adversely affected by weekly applications of doses of up to 2×10^{12} polyhedra per hectare (Treacy and All, 1996; Treacy *et al.*, 1997).

C. Nontarget Lepidopterans

Most baculoviruses have host ranges limited to a few closely related species within the same family of lepidopterous insects. The host range of AcMNPV appears to be much broader, infecting several dozen lepidopteran species in several different families. This host range for AcMNPV has been determined by laboratory bioassays where insects are exposed to high doses of virus under controlled conditions. The sensitivity of these various insect species to the virus varies greatly. Insect species that are efficiently infected only at high doses are often included in the host range of the virus as semipermissive species. Thus, laboratory testing can overestimate the number of species that are infected under natural conditions. The effective host range of AcMNPV in nature is probably much more restrictive, with only the rare semipermissive individual acquiring a high enough dose to become infected under field conditions.

It has been postulated that the expression of a highly active toxin by a recombinant baculovirus could increase the effective host range of the virus. In a practical sense, this would be a benefit, since rarely does only one insect pest species invade a farmers field. However, any change in host range would need to be carefully evaluated. A comparison of the LD_{50} of recombinant AcMNPVs expressing AaIT to the wild-type parental virus has been made for a large number of lepidopteran species. In total, 48 species that exhibit varying degrees of sensitivity to AcMNPV have been tested for their relative sensitivity to an AaIT-expressing AcMNPV compared to wild-type AcMNPV. These species belong to nine separate families of the order Lepidoptera. The data from these studies do not support the hypothesis that expression of the toxin increases the inherent host range in recombinant baculoviruses. In fact, the addition of the AaIT gene to AcMNPV did not alter the relative infectivity of the baculovirus (Possee *et al.*, 1993b; American Cyanamid, 1994, 1996).

The data comparing the LC_{50} for the recombinant AaIT-expressing virus versus the wild-type virus indicate a four- to tenfold decrease in LC_{50} for the recombinant virus on the semipermissive insect, *Helicoverpa zea*. Interestingly, a similar result was obtained for a second recombinant AcMNPV that expresses a different scorpion toxin gene, designated LqhIT2 (DuPont, 1996). In each case, determination of the LC_{50} was carried out at 12 and 6 days postinfection, respectively, timepoints that give an accurate estimate of field performance. However, detailed observation of the course of AcMNPV infection in *Helicoverpa zea* larvae indicated that these early timepoints may not be predictive of the final mortality for the wild-type virus on this and perhaps other semipermissive species. A test was carried out on *Helicoverpa zea* where relative LC_{50}'s for the AaIT-expressing recombinant and the wild-type virus were calculated at various timepoints up through emergence of the treated larvae as adult moths. Results show that the true LC_{50} was established early in the test for the recombinant due to its faster speed of kill. Larvae infected with the wild-type virus continued to die throughout the test. At 33 days, the timepoint where all surviving adults had emerged, the LC_{50} for the wild-type virus reached the equivalent to that of the AaIT-expressing recombinant (American Cyanamid, 1996). These data indicate that, in some species, the addition of an insect-specific toxin gene may result in an incorrect LC_{50} calculation if care is not taken to determine mortality at a point where the wild-type infection process has run its full course in all of the test larvae.

D. Establishment of the Virus in the Environment

It is well known that the occlusion bodies of baculoviruses can persist for many years in the soil of areas that have experienced an epizootic outbreak. Several wild-type baculoviruses have been registered and used as biopesticides around the world, and their use has not led to deleterious effects on the environment. Since the addition of an insect-specific toxin does not affect the physical characteristics of the virus that are responsible for its persistence in the environment, it can be assumed that use of recombinant baculoviruses will result in some persistence of the recombinant viruses in the soil of treated fields. Just as with the wild-type baculoviruses, simple persistence would not be expected to have a negative environmental impact. A more important issue is whether the recombinant virus could become established in the environment, increasing in number and perhaps initiating epizootics unrelated to its deliberate application as a biopesticide. A related question is whether the recombinant could supplant its parental wild-type virus in the environment. Results of research combined with biological principles as described below argue against either of these possibilities. Data to back up these arguments are just now becoming available.

The main goal of the genetic engineering of a baculovirus as a biopesticide is to increase the speed of kill of target pests. If this result is achieved, and in many cases the reduction in LT_{50} has been significant, then many fewer cycles of baculovirus replication will occur with a dramatic reduction in the number of progeny polyhedra produced. This puts the recombinant baculovirus at a severe replicative disadvantage compared to wild-type baculoviruses. For example, the deletion of the *egt* gene from the AcMNPV genome reduces the LT_{50} by 1–2 days. The consequence of shortening the LT_{50} is that the baculovirus is unable to undergo as many rounds of replication within its host, and, thus, produces 30% fewer polyhedra when compared to the parental strain (O'Reilly and Miller, 1991). Insertion of the AaIT gene into the *egt* gene-deleted baculovirus results in a further decrease in the LT_{50}. This AaIT-expressing baculovirus (vEGTDEL–AaIT) will kill the target species 60% faster than the wild-type AcMNPV. Experiments on the replication of vEGTDEL–AaIT in the permissive insect, *Heliothis virescens,* indicate that larvae infected with this recombinant baculovirus produce 75–93% less polyhedra compared to wild-type AcNPV-infected larvae (American Cyanamid, 1994). This decrease is not additive to the decrease in polyhedron production seen in the *egt*$^{-}$ baculovirus but is solely due to the expression of the AaIT gene. The reduction of progeny polyhedra to 7–25% of wild-type levels would make it difficult for this recombinant to compete with indigenous AcMNPV in the environment. A similar reduction in the number of progeny polyhedra produced by an AcMNPV recombinant expressing the LqIT gene has also been reported (DuPont, 1996).

A simple experiment designed to test the hypothesis that a recombinant virus that kills insect larvae significantly faster and produces a significantly reduced number of polyhedra is at a competitive disadvantage to its wild-type parent was designed and carried out at American Cyanamid. Under laboratory conditions, a "competition" study was conducted. A 1:10 mixture of wild type to AaIT-expressing recombinant AcMNPV was fed to larvae of the permissive insect *Heliothis virescens* at either an LC_{50} dose or a LC_{95} dose. After 6 days, all the larvae for each dose were harvested and processed for polyhedra. Each mixture of polyhedra was then fed to a second generation of larvae at the appropriate dose, either LC_{50} or LC_{95}. This was continued for several generations. Two methods were used to track the amount of AaIT-expressing virus; the contractile phenotype exhibited by larvae infected with an AaIT-expressing baculovirus, and restriction endonuclease analysis of viral genomes followed by Southern blotting specific for the region of the genome containing the AaIT gene. The results demonstrated a rapid decrease in the AaIT-expressing virus as a percentage of the total virus population. This supports the hypothesis that a recombinant virus that kills its host too quickly and produces fewer offspring will be at a competitive disadvantage to its wild-type parent. This recombinant should produce less of an environmental impact than its wild-type parent and should be unable to displace the parental strain from the environment. Compensating factors such as an increased host range, greater infectivity for sensitive species, or increased environmental stability could compensate for the decreased number of progeny polyhedra produced. However, as discussed elsewhere in this chapter, these conditions are not present in any of the recombinant viruses currently under development. If in the future these viral traits are altered, a reevaluation of the fitness of the recombinant would be necessary.

A second line of evidence that addresses possible establishment of the virus in the environment comes from soil sampling for the presence of the recombinant virus in the 1994 field tests (an *egt*$^-$AcMNPV) and 1995 field tests (vEGTDEL–AaIT). In these tests, soil samples were collected from the site of the field trial before, immediately after, and at various timepoints after completion of the test. Polyhedra were extracted from the samples according to the method of Evans *et al.* (1980). The extracted material was fed to the highly sensitive insect, *Heliothis virescens.* Larvae that died were examined microscopically for the presence of polyhedra. All dead larvae were processed to recover polyhedra and polymerase chain reaction (PCR) was used to amplify the viral DNA to determine if the recombinant baculovirus was present. Data from these trials indicate that immediately following spray application, the recombinant virus was present in the soil. In the months following the completion of the test, a small amount of recombinant virus persisted in the soil as expected, but gradually decreased over time (American Cyanamid, 1994, 1996).

There are additional factors that affect the fate of a recombinant virus in the environment. In field trials conducted by the NERC Institute of Virology in England (Cory *et al.*, 1994), secondary transmission of an AaIT-expressing AcMNPV was significantly reduced from that of wild-type AcMNPV. This was attributed to the decreased number of viral offspring produced by the recombinant, as well as to the contractile paralysis phenotype of this virus. This contractile phenotype results in infected larvae dropping off plant surfaces and being lost as an immediate source of inoculum to spread infection to subsequent generations of larvae on the plants. This loss of infected larvae from plant surfaces has also been observed by Hoover *et al.* (1995). In contrast, a larvae infected by a wild-type AcMNPV liquefies and deposits a large amount of virus on the plant surface that serves as inoculum for a second generation of insects.

E. Genetic Recombination

Two issues must be considered when evaluating the potential risk of movement of an inserted gene from a recombinant baculovirus into another organism. The first issue involves the likelihood of a genetic exchange between the recombinant baculovirus and another organism. The second deals with the consequences of such an event. It is a fact that genetic exchange between organisms can and does occur in nature. However, the genetic integrity of a species is protected by significant barriers to gene movement, which limit the likelihood and types of genetic exchange. One important barrier to genetic exchange is the requirement for a shared location of replication for both the donor and recipient. A genetic exchange involving a baculovirus must occur within cells of an infected insect. Therefore, the baculovirus and the potential recipient organism must share at least one common insect host and the recipient must replicate within the insect cell. These requirements generally limit the possibilities for exchange to another virus. Exchange is also limited by the cellular compartment in which the viruses replicate. Since baculoviruses replicate within the nucleus of the insect cell, the most likely partner for the exchange of genetic material is another virus that also replicates within the nucleus of a lepidopteran insect cell.

Another important barrier to genetic exchange is determined by the nature of the genetic material of the donor and the prospective recipient. Even if two viruses infect the same host and replicate in the same cellular compartment, differences in the mode of replication and the composition of the viral genome can limit the exchange of genetic information. In addition, the degree of homology between the donor and recipient will directly affect the probability of genetic exchange. Therefore, the most realistic situation would be an exchange of genetic information between two highly related baculoviruses.

The second and more important issue is whether there would be any negative consequences of a possible genetic exchange. New genetic traits become fixed in populations if they confer a selective advantage to the species. As discussed earlier in this chapter, the addition of an insect-specific toxin gene such as AaIT, which increases the speed of kill of the recombinant baculovirus, decreases the fitness of the baculovirus that contains it. It follows that any baculovirus acquiring the AaIT gene would also have decreased environmental fitness and would be at a strong selective disadvantage. Finally, if the product of the inserted gene is selective for insects, as is the AaIT protein, an added level of safety is present.

F. Summary

In assessing the risk associated with the proposed use of a specific recombinant baculovirus, a number of key questions must be answered. As each recombinant is developed, experiments must be designed and data collected to satisfactorily address these issues. However, even at this time, some general principles are emerging.

At present, all available data demonstrate that the addition of an insect-selective toxin does not alter the effective host range of a recombinant baculovirus. Although expansion of the host range of a recombinant baculovirus to include the complete complex of lepidopteran pests that are often present in a particular crop would increase the utility of a baculoviral biopesticide, these designer biopesticides are still in the future.

A second general principle emerging is that the engineering of a virus for faster kill will result in infected larvae that produce less crop damage, are smaller at the time of death, and produce fewer progeny polyhedra than larvae infected with the parental, wild-type virus. This engineered trait puts the recombinant virus at a selective disadvantage to its wild-type counterpart and limits the establishment of the recombinant in the environment.

VII. PUBLIC RELATIONS ISSUES

A. Perception of Biotechnology

Biotechnology has been projected to have significant economic and societal benefits in agriculture. We are beginning to reap some of these benefits through the herbicide-tolerant and the Bt-producing crops. Biotechnology has been developing within a larger context of consumer concerns about the impact of this technology on their health and environment. Biotechnology has come under public scrutiny and certain environmental activist groups have attacked biotechnology from scientific, ethical, and personal viewpoints.

Although scientists have long been accepted as credible sources, commercial applications of nuclear physics and chemistry, as related to negative experiences, have caused the loss of some credibility. Congressman George Brown (1987) developed principles for dealing with the public sector that have great applicability to communicating aspects of biotechnology. These are: scientists dealing with the public should avoid arrogantly dismissing lay hypotheses, ignoring nonscientific points about biotechnology, and/or underestimating the intelligence or power of the public. Other research conducted at this time indicated that the US population was moderately aware that genetic engineering was being applied to agriculture. As many as four of ten Americans had heard something about gene splicing and recombinant DNA. This 40% broke down to 29% among those with less than a high school degree to a high of 62% among those with a college degree [Office of Technology Assessment (OTA), 1987]. Certainly, a higher percentage of the US population has heard something about biotechnology today. In this same OTA report, the release of a genetically modified microorganism was viewed as the most important potential danger. This finding indicated to American Cyanamid scientists that a significant effort in the development of recombinant baculovirus insecticides had to involve a plan and an information system to educate the general public.

Obviously, field testing was a major concern, as it would involve the release of a genetically engineered microorganism into the environment. Although the OTA report identified certain concerns about biotechnology, the OTA report also indicated that eight of ten Americans thought that small-scale field testing of a genetically altered organism to increase agricultural activity should be permitted. However, when considering moving biotechnology from small-scale experimental field trials to large-scale field trials, a majority of the people (53%) indicated that commercial firms should not be permitted large-scale applications, even if risk to the environment was considered extremely small. There was no clear reason (in view) for this difference. Public perception of biotechnology, as revealed by company experience and various surveys (OTA, 1987; Lacey *et al.*, 1991; Hoban and Kendall, 1992; Weber *et al.*, 1995) indicated that the issue was complex and included social, economic, moral, environmental, and health issues.

As we have been discussing in earlier sections of this chapter, American Cyanamid has been committed to the vision that there is a future for genetically manipulated baculoviruses as potent weapons in the insect control arsenal. The effort to develop an effective baculoviral insecticide was initiated by American Cyanamid in 1989. However, involvement in the biotechnology field dated back much earlier, to bovine somatotropin. During this time period, much controversy was raised about the ethical, moral, and economic issues of this type of product. Having been involved, American Cyanamid scientists and management decided that they should be extremely cautious and critically evaluate each step in the introduction of the rNPV insecticides.

It became very clear that a dialogue needed to be developed with those individuals and organizations that had the power to positively or negatively impact upon the project, and that American Cyanamid had to be proactive in communicating the technology. Significant effort was expended in evaluation and understanding aspects of risk communication to determine what would be the best way to introduce the subject to the public "at large."

B. Public Communication

The first step in the process was to identify the key groups to whom the technology should be presented. After some serious, careful consideration, the following groups were identified:

- Regulatory
- General public
- Academia
- Media
- Environmentalists

The next step was to identify what information needed to be developed and how to present it in an understandable format to each respective identified group. To facilitate the message development process and to put it into an understandable format for each identified group, focus groups were developed within the Cyanamid Agricultural Research Center (CARC). CARC had employees consisting of a cross-section of individuals (clerical, chemists, entomologists, molecular biologists, etc.) and it was felt that focus groups could be formed that were representative of the above groups. The technology was presented to the internal groups and feedback was solicited as to what questions they would need answered to understand and feel comfortable with the technology. A list of key questions were developed and answers were assembled to address the questions as best as possible. Questions for which answers were not available in literature or in our studies were deemed as research issues that needed to be addressed. After spending several months addressing the questions coming from the focus groups, a campaign was initiated to educate and inform the public groups that could impact the project.

It was decided that the best way to initiate the communication process was to develop media pieces that would be usable with all the above identified groups. These media pieces consisted of a videotape and a brochure containing detailed information about the baculovirus, the AaIT toxin gene, and the engineering process, as well as numerous questions and answers that came out of the focus groups regarding the safety of the technology and potential benefits to agriculture and society. The video and brochure proved to be extremely useful in the education process.

Communication lines were opened among American Cyanamid, the US EPA, key scientists working with baculoviruses (worldwide), environmental groups, and the public sector. Information about the project was shared openly with all groups. This educational and dialogue process established strong credibility with each of the different communities and paved the way for the first successful field testing in the United States of an rNPV that had been engineered with a gene deletion and/or gene insertion.

Even after the approval of field testing was secured from the US EPA, it was felt that informing the local public in the area where a field test was scheduled was an important step. The importance of this consideration was borne out when field studies with an rNPV in the United Kingdom (Cory *et al.*, 1994) were met with strong local opposition, even though all legal obligations had been satisfied. In the UK study, the public was informed only via the obligation to publish the notice in the local press and no attempt was made to inform the public directly. American Cyanamid (from the outset) planned to inform the local public of our intent. This was and still is being accomplished via the formation of local citizen's advisory panels. Before any rNPV field test is initiated, American Cyanamid personnel invite key individuals from the local area to meet and hear a presentation on the technology. At these meetings a clear and complete presentation is made, followed by an open "question and answer" session. The baculovirus videotape may be shown and the brochure is made available to those desiring more information. This format has received very positive reviews.

VIII. CONCLUSIONS

At the onset of the 1990s, baculoviral insecticides were found in only a few minor niche markets. Even today, the contribution of viral insecticides constitutes a minor component of the world's global insecticide usage. But, there is much excitement and the expectations are high for recombinant baculovirus insecticides. An "explosion" of research to further commercial aspects of all critical components of this technology has accelerated during the last 6 years. Sufficient progress has been made to increase the chances that a new class of viral insecticides, rNPVs, will be introduced to the marketplace. These products will possess new attributes for a crop protection product that will redefine the role of baculoviruses in agriculture and provide a desirable nonchemical alternative for insect control.

Despite impressive advances, the field continues to grow. While improvement has been made with regard to the speed of action of baculoviruses using insect-selective neurotoxins, these improvements do not alter the intrinsic infectivity of these viruses on permissive and semipermissive hosts. Today, the performance of rNPVs is limited by the spectrum of pest insects that can be effectively controlled in the field. Improving the viral host targeting for a particular pest–crop complex is currently driven by improved virus strain selection. Over time, one might expect that research concerning the genetic modification of the "economic" host range of baculoviruses will contribute to a further increase in the value of recombinant viral insecticides for managing pest populations in agriculture and forestry. These improvements need to be paralleled by the development of suitable cell lines for viral production. Further refinement in *in vitro* production is needed to reduce the manufacturing costs.

While the preceding comments address technical issues pertinent to the commercialization of baculoviral insecticides, other issues such as public education, acceptance, safety concerns pertaining to both current and future rNPVs, and regulatory issues need to be satisfied. The safer pesticide policies embraced around the world mandate lower chemical inputs to control insect pests in the environment. Recombinant NPVs offer an attractive alternative to chemicals. For this reason, regulatory agencies are generally supportive of this technology. On the other hand, this technology is new and governmental agencies may be sensitive to risk—real, perceived, and/or political. Hence, reluctance or delay in making critical decisions and product approvals result in increased costs to companies investing in product development. It is clearly to the benefit of all parties (public, government, and private industry) to collect all needed data to address concerns and benefits of this technology to agriculture and society on a global basis. In this way, we will reap the full benefits offered by baculoviruses that have eluded us for so long.

ACKNOWLEDGMENT. The authors would like to acknowledge the assistance of Tina Holston in the typing and proofing of this manuscript. Her considerable efforts have contributed significantly to the accuracy and consistency of this multiauthored work.

REFERENCES

American Cyanamid, 1994, Notification to Conduct Small-Scale Field Testing of a Genetically Altered Baculovirus, EPA No. 241-NMP-2, EPA-OPP Public Docket No. 50799.

American Cyanamid, 1996, Notification to Conduct Small-Scale Field Testing of a Genetically Altered Baculovirus, EPA No. 241-NMP-3, EPA-OPP Public Docket No. 50816.

Barnes, R. W., Mienecke, C. F., McLane, W. C., and Rehnborg, C. S., 1970, Long-term feeding and other toxicity–pathogenicity studies on rats using a commercial preparation of the nuclear-polyhedrosis virus of *Heliothis zea*, *J. Invertebr. Pathol.* **16:**112–115.

Bishop, D. H. L., Entwistle, P. F., Cameron, I. R., Allen, C. J., and Possee, R. D., 1988, Field trial of genetically engineered baculovirus insecticides, in: *The Release of Genetically Engineered Organisms* (M. Sussman, C. H. Collins, F. A. Skinner, and D. E. Stewart-Tull, eds.), pp. 143–179, Academic Press, New York.

Bishop, D. H. L., Hirst, M. L., Possee, R. D., and Cory, J. S., 1995, Genetic engineering of microbes: Virus insecticides—A case study, in: *Fifty Years of Antimicrobials: Past Perspectives and Future Trends* (P. A. Hunter, G. K. Darby, and N. J. Russell, eds.), pp. 249–277, Cambridge University Press, Cambridge, England.

Bonning, B. C., and Hammock, B. D., 1992, Development and potential of genetically engineered viral insecticides, *Biotechnol. Genet. Eng. Rev.* **10:**455–489.

Bonning, B. C., and Hammock, B. D., 1996, Development of recombinant baculoviruses for insect control, *Annu. Rev. Entomol.* **41:**191–210.

Bonning, B. C., Hoover, K., Booth, T. F., Duffey, S., and Hammock, B. D., 1995, Development of a recombinant baculovirus expressing a modified juvenile hormone esterase with potential for insect control, *Arch. Insect Biochem. Physiol.* **30:**177–194.

Bougis, P. E., Rochat, H., and Smith, L. A., 1989, Precursors of *Androctonus australis* scorpion neurotoxins, *J. Biol. Chem.* **264:**19259–19265.

Brown, Jr., G. E., 1987, Social and political aspects of risk assessment, in: *Public Perceptions of Biotechnology* (L. A. Batra and W. Klassen, eds., pp. 73–80, Agricultural Research Institute, Bethesda, MD.

Bull, D. L., 1978, Formulations of microbial insecticides: Microencapsulation and adjuvants, *Misc. Publ. Entomol. Soc. Am.* **10:**11.

Bullock, H. R., 1967, Persistence of *Heliothis* nuclear polyhedrosis virus on cotton foliage, *J. Invertebr. Pathol.* **9:**434–439.

Burges, H. D., Croizier, G., and Huger, J., 1980, A review of safety tests on baculoviruses, *Entomophaga* **25:**329–340.

Carbonell, L. F., Hodge, M. R., Tomalski, M. D., and Miller, L. K., 1988, Synthesis of a gene coding for an insect-specific scorpion neurotoxin and attempts to express it using baculovirus vectors, *Gene* **73:**409–418.

Carstens, E. B., Krebs, A., and Gallerneault, C. E., 1986, Identification of an amino acid essential to the normal assembly of *Autographa californica* nuclear polyhedrosis virus polyhedra, *J. Virol.* **58:**684–688.

Chejanovsky, N., and Gershburg, E., 1995, The wild-type *Autographa californica* nuclear polyhedrosis virus induces apoptosis of *Spodoptera littoralis* cells, *Virology* **209:**519–525.

Cory, J. S., Hirst, M. L., Williams, T., Hails, R. S., Goulson, D., Green, B. M., Carty, T. M., Possee, R. D., Cayley, P. J., and Bishop, D. H. L., 1994, Field trial of a genetically improved baculovirus insecticide, *Nature* **370:**138–140.

Couch, T. L., and Ignoffo, C. M., 1981, Formulation of insect pathogens, in: *Microbial Control of Pests and Plant Diseases, 1970–1980,* (H. D. Burges, ed.), pp. 621–643, Academic Press, New York.

Croizier, G., Croizier, L., Argaud, O., and Poudevigne, D., 1994, Extension of *Autographa californica* nuclear polyhedrosis virus host range by interspecific replacement of a short DNA sequence in p143 helicase gene, *Proc. Natl. Acad. Sci. USA* **91:**48–52.

Cunningham, J. C., Kaupp, W. J., House, G. M., McPhee, J. R., and deGroot, P., 1978, Aerial application of spruce budworm baculovirus: Tests of virus strains, dosages and formulations in 1977, *Can. For. Serv. Info. Rpt.* No. FPM-X-3.

Darbon, H., Zlotkin, E., Kopeyan, C., Van Rietschoten, J., and Rochat, H., 1982, Covalent structure of the insect toxin of the North African scorpion *Androctonus australis* Hector, *Int. J. Peptide Protein Res.* **20:**320–330.

David, W. A. L., 1975, The granulosis virus of *Pieris brassicae L.* and relationship with its host, in: *Advances in Virus Research* (M. A. Lauffer, F. R. Bang, K. Maramososch, and K. M. Smith, eds.), pp. 111–161, Academic Press, New York.

David, W. A. L., and Gardiner, B. O. C., 1967, The effect of heat, cold, and prolonged storage on a granulosis virus of *Pieris rapae, J. Invertebr. Pathol.* **9:**555–563.

David, W. A. L., Gardiner, B. O. C., and Woolner, M., 1968, The effects of sunlight on a purified granulosis virus of *Pieris brassicae* applied to cabbage leaves, *J. Invertebr. Pathol.* **11:**496–506.

Doller, G., 1985, The safety of insect viruses as biological control agents, in: *Viral Insecticides for Biological Control* (K. Maramorosch and K. E. Sherman, eds.) pp. 399–439, Academic Press, New York.

Dulmage, H., and Bergerjon, A., 1982, Industrial and international standardization of microbial pesticides, II. Insect viruses, *Entomophaga* **22:**131–144.

DuPont, 1996, Notification to Conduct Small-Scale Field Testing of a Genetically Altered Baculovirus, EPA No. 352-NMP-4.

Eldridge, R., O'Reilly, D. R., Hammock, B. D., and Miller, L. K., 1992a, Insecticidal properties of genetically engineered baculoviruses expressing an insect juvenile hormone esterase gene, *Appl. Environ. Microbiol.* **58:**1583–1591.

Eldridge, R., O'Reilly, D. R., and Miller, L. K., 1992b, Efficacy of a baculovirus insecticide expressing an eclosion hormone gene, *Biol. Control* **2:**104–110.

Engelhard, E. K., and Volkman, L. E., 1995, Developmental resistance in fourth instar *Trichoplusia ni* orally inoculated with *Autographa californica* M nuclear polyhedrosis virus, *Virology* **209:**384–389.

Engelhard, E. K., Kam-Morgan, L. N. W., Washburn, J. O., and Volkman, L. E., 1994, The insect tracheal system: A conduit for the systemic spread of *Autographa californica* M nuclear polyhedrosis virus, *Proc. Natl. Acad. Sci. USA* **91:**3224–3227.

Evans, H. F., Bishop, J. M., and Page, E. A., 1980, Methods for the quantitative assessment of nuclear polyhedrosis virus in the soil, *J. Invertebr. Pathol.* **35:**1–8.

Flipsen, J. T. M., Mans, R. M. W., Kleefsman, A. W. F., Knebel-Mörsdorf, D., and Vlak, J. M., 1995, Deletion of the baculovirus UDP–glucosyltransferase gene induces early degeneration of Malpighian tubules in infected insects, *J. Virol.* **69:**4592–4532.

Gopalakrishnan, K., Muthukrishnan, S., and Kramer, K. J., 1995, Baculovirus-mediated expression of a *Manduca sexta* chitinase gene: Properties of the recombinant protein, *Insect Biochem. Mol. Biol.* **25:**255–265.

Hamblin, M., van Beek, N. A. M., Hughes, P. R., and Wood, H. A., 1990, Co-occlusion and persistence of a baculovirus mutant lacking the polyhedrin gene, *Appl. Environ. Microbiol.* **56:**3057–3062.

Hammock, B. D., Bonning, B. C., Possee, R. D., Hanzlik, T. N., and Maeda, S., 1990, Expression and effects of juvenile hormone esterase in a baculovirus vector. *Nature* **344:**458–461.

Hammock, B. D., McCutchen, B. F., Beetham, J., Choudary, P. V., Fowler, E., Ichinose, R., Ward, V. K., Vickers, J. M., Bonning, B. C., Harshman, L. G., Grant, D., Uematsu, T., and Maeda, S., 1993, Development of recombinant viral insecticides by expression of an insect-specific toxin and insect-specific enzyme in nuclear polyhedrosis viruses, *Arch. Insect Biochem. Physiol.* **22:**315–344.

Heimpel, A. M., and Buchanan, L. K., 1967, Human feeding tests using a nuclear-polyhedrosis virus of *Heliothis zea, J. Invert. Pathol.* **9:**55–57.

Heimpel, A. M., Thomas, E. D., Adams, J. R., and Smith, L. J., 1973, The presence of nuclear polyhedrosis virus of *Trichoplusia ni* on cabbage from the market shelf, *Environ. Entomol.* **2:**72–75.

Heinz, K. M., McCutchen, B. F., Herrmann, R., Parella, M. P., and Hammock, B. D., 1995, Direct effects of recombinant nuclear polyhedrosis viruses on selected non-target organisms, *J. Econ. Entomol.* **88:**259–264.

Hoban, T. J., and Kendall, P. A., 1992, *Consumer Attitudes About the Use of Biotechnology in Agriculture and Food Production,* North Carolina State University, Raleigh.

Hoover, K., Schultz, C. M., Lane, S. S., Bonning, B. C., Duffey, S. S., McCutchen, B. F., and Hammock, B. D., 1995, Reduction in damage to cotton plants by a recombinant baculovirus that knocks moribund larvae of *Heliothis virescens* off the plant, *Biol. Control* **5:**419–426.

Huang, H., Ignoffo, C. M., and Shapiro, M., 1977, Physical and clinical examinations of personnel involved in production of the insect virus, *Baculovirus Heliothis, J. Kansas Entomol. Soc.* **50:**200–202.

Huger, A., 1963, Granulosis of insects, in: *Insect Pathology: An Advanced Treatise* (E. A. Steinhaus, ed.), Vol. I, pp. 531–575, Academic Press, New York.

Hughes, D. S., Possee, R. D., and King, L. A., 1993, Activation and detection of a latent baculovirus resembling *Mamestra brassicae* nuclear polyhedrosis virus in *Mamestra brassicae* insects, *Virology* **194**(2):608–615.

Hunter, D. K., Hoffman, D. F., and Collier, S. J., 1983, Pathogenicity of a nuclear polyhedrosis virus of the almond moth, *Cadra cautella, J. Invertebr. Pathol.* **21**:282–294.

Ignoffo, C. M., 1964, Production and virulence of a nuclear polyhedrosis virus from larvae of *Trichoplusia ni* (Hubner) reared on a semisynthetic diet, *J. Insect Pathol.* **6**:318–329.

Ignoffo, C. M., 1973, Effects of entomopathogens on vertebrates, *Ann. NY Acad. Sci.* **217**:141–164.

Ignoffo, C. M., 1975, Entomopathogens as insecticides, *Environ. Lett.* **8**:23–40.

Ignoffo, C. M., and Batzer, O. F., 1971, Microencapsulation and ultraviolet protectants to increase sunlight stability of an insect virus, *J. Econ. Entomol.* **64**:850–853.

Ignoffo, C. M., and Garcia, C., 1990, Influence of physical factors (water, pH, temperature) on simulated sunlight-UV inactivation of occluded HzSNPV, *Abstracts International:* 189.

Ignoffo, C. M., and Heimpel, A. M., 1965, The nuclear-polyhedrosis virus of *Heliothis zea* (Boddie) and *Heliothis virescens* (Fabricius). V. Toxicity–pathogenicity of virus to white mice and guinea pigs, *J. Invert. Pathol.* **7**:329–340.

Ignoffo, C. M., Hostetter, D. L., and Smith, D. B., 1976, Gustatory stimulant, sunlight protectant, evaporation retardant: Three characteristics of a microbial insecticidal adjuvant, *J. Econ. Entomol.* **69**:207–210.

Ignoffo, C. M., Garcia, C., Zuidema, D., and Vlak, J. M., 1995, Relative *in vivo* activity and simulated sunlight-UV stability of inclusion bodies of a wild-type and an engineered polyhedral envelope-negative isolate of the nucleopolyhedrosis virus of *Autographa californica, J. Invertebr. Pathol.* **66**:212–213.

Jaques, R. P., 1972, The inactivation of foliar deposits of viruses of *Trichoplusia ni* and *Pieris rapae* and tests on protectant activities, *Can. Entomol.* **104**:1985–1993.

Jaques, R. P., 1977, Stability of entomopathogenic viruses, *Entomol. Soc. Am.* **10**:99–105.

Knebel-Mörsdorf, D., Flipsen, J. T. M., Roncarati, R., Jahnel, F., Kleefsman, A. W. F., and Vlak, J. M., 1996, Baculovirus infection of *Spodoptera exigua* larvae: *lacZ* expression driven by promoters of early genes pe38 and me53 in larval tissue, *J. Gen. Virol.* **77**:815–824.

Korth, K. L., and Levings, III, C. S., 1993, Baculovirus expression of the maize mitochondrial protein URF13 confers insecticidal activity in cell cultures and larvae, *Proc. Natl. Acad. Sci. USA* **90**:3388–3392.

Lacey, W. B., Busch, L., and Lacey, L. R., 1991, Public perceptions of agricultural biotechnology, in: *Agricultural Biotechnology: Issues and Choices* (B. R. Baumgardt and M. A. Martin, eds.), pp. 139–161, Purdue University, Lafayette, Indiana.

Lewis, F. B., and Rollinson, W. D., 1978, Effect of storage on the virulence of gypsy moth nucleopolyhedrosis inclusion bodies, *J. Econ. Entomol.* **71**:719–722.

Lu, A., and Miller, L. K., 1995, Differential requirements for baculovirus late expression factor genes in two cell lines, *J. Virol.* **69**:6265–6272.

Lu, A., Seshagiri, S., and Miller, L. K., 1996, Signal sequence and promoter effects on the efficacy of toxin expressing baculoviruses as biopesticides, *Biol. Control* **7**:320–332.

Maeda, S., 1989, Increased insecticidal effect by a recombinant baculovirus carrying a synthetic diuretic hormone gene, *Biochem. Biophys. Res. Commun.* **165**:1177–1183.

Maeda, S., Kamita, S. G., and Kondo, A., 1993, Host range expression of *Autographa californica* nuclear polyhedrosis virus (NPV) following recombination of a 0.6-kilobase-pair DNA fragment originating from *Bombyx mori* NPV, *J. Virol.* **67**:6234–6238.

Maeda, S., Volrath, S. L., Hanzlik, T. N., Harper, S. A., Majima, K., Maddox, D. W., Hammock, B. D., and Fowler, E., 1991, Insecticidal effects of an insect-specific neurotoxin expressed by a recombinant baculovirus, *Virology* **184**:777–780.

Maiorella, B., Inlow, D., Shauger, A., and Harano, D., 1988, Large-scale insect cell-culture for recombinant protein production, *Bio/Technology* **6**:1406–1410.

Martens, J. W. M., Honée, G., Zuidema, D., van Lent, J. W. M., Visser, B., and Vlak, J. M., 1990,

Insecticidal activity of a bacterial crystal protein expressed by a recombinant baculovirus in insect cells, *Appl. Environ. Microbiol.* **56:**2764–2770.

McCutchen, B. F., Choudary, P. V., Crenshaw, R., Maddox, D., Kamita, S. G., Pelekar, N., Volrath, S., Fowler, E., Hammock, B. D., and Maeda, S., 1991, Development of a recombinant baculovirus expressing an insect-selective neurotoxin: Potential for pest control, *Bio/Technology* **9:**848–852.

McCutchen, B. F., Herrmann, R., Heinz, K. M., Parrella, M. P., and Hammock, B. D., 1996, Effects of recombinant baculoviruses on a nontarget endoparasitoid of *Heliothis virescens*, *Biol. Control* **6:**45–50.

McCutchen, B. F., Hoover, K., Preisler, H. K., Betana, M. D., Herrmann, R., Robertson, J. L., and Hammock, B. D., Interactions of recombinant and wild-type baculoviruses with classical insecticides and pyrethroid-resistant tobacco budworm (Lepidoptera: Noctuidae), *J. Econ. Entomol.*, in press.

McNitt, L., Espielie, K. E., and Miller, L. K., 1995, Assessing the safety of toxin-producing baculovirus biopesticides to a non-target predator, the social wasp *Polistes metricus* Say, *Biol. Control* **5:**267–278.

Merryweather, A. T., Weyer, U., Harris, M. P. G., Hirst, M., Booth, T., and Possee, R. D., 1990, Construction of genetically engineered baculovirus insecticides containing the *Bacillus thuringiensis* spp. *Kurstaki* HD-73 delta endotoxin, *J. Gen Virol.* **71:**1535–1544.

Miller, L. K., 1995, Genetically engineered insect virus pesticides: Present and future, *J. Invertebr. Pathol.* **65:**211–216.

Moskowitz, H., Hermann, R., Zlotkin, E., and Gordon, D., 1994, Variability among insect sodium channels revealed by selective neurotoxins, *Insect Biochem. Mol. Biol.* **24:**13–19.

Office of Technology Assessment (OTA), US Congress, 1987, *New Developments in Biotechnology: Public Perceptions of Biotechnology*, OTA-BP-BA, US Government Printing Office, Washington, DC.

O'Reilly, D. R., 1995, Baculovirus-encoded ecdysteroid UDP-glucosyltransferases, *Insect Biochem. Mol. Biol.* **25:**541–550.

O'Reilly, D. R., and Miller, L. K., 1989, A baculovirus blocks molting by producing ecdysteroid UDP–glucosyltransferase, *Science* **245:**1110–1112.

O'Reilly, D. R., and Miller, L. K., 1990, Regulation of expression of a baculovirus ecdysteroid UDP–glucosyltransferase gene, *J. Virol.* **64:**1321–1328.

O'Reilly, D. R., and Miller, L. K., 1991, Improvement of a baculovirus pesticide by deletion of the *egt* gene, *Biotechnology* **9:**1086–1089.

O'Reilly, D. R., Howarth, O. W., Rees, H. W., and Miller, L. K., 1991, Structure of the ecdysone glucoside formed by a baculovirus ecdysteroid UDP–glucosyltransferase, *Insect Biochem.* **21:**795–801.

O'Reilly, D. R., Brown, M. R., and Miller, L. K., 1992, Alteration of ecdysteroid metabolism due to baculovirus infection of the fall armyworm *Spodoptera frugiperda:* Host ecdysteroids are conjugated with galactose, *Insect Biochem. Mol. Biol.* **22:**313–320.

O'Reilly, D. R., Kelly, T. K., Masler, E. P., Thyafaraja, B. S., Robson, R. M., Shaw, R. C., and Miller, L. K., 1995, Overexpression of *Bombyx mori* prothoracicotropic hormone using baculovirus vectors, *Insect Biochem. Mol. Biol.* **25:**475–485.

Possee, R. D., Cayley, P. J., Cory, J. S., and Bishop, D. H. L., 1993a, Genetically engineered viral insecticides: New insecticides with improved phenotypes, *Pestic. Sci.* **39:**109–115.

Possee, R. D., Hirst, M., Jones, J. D., Bishop, D. H. L., and Cayley, P. J., 1993b, Field tests of genetically engineered baculoviruses, *Br. Crop Protect. Council* **55:**23–36.

Prikhod'ko, G. G., Robson, M., Warmke, J. W., Cohen, C. J., Smith, M. M., Wang, P., Warren, V., Kaczorowski, G., Van der Ploeg, L. H. T., and Miller, L. K., 1996, Properties of three baculovirus-expressing genes that encode insect-selective toxins: μ-Aga-IVB, As II, and Sh I, *Biol. Control* **7:**236–244.

Shapiro, M., 1986, *In vivo* production of baculoviruses, in: *The Biology of Baculoviruses* (R. R. Granados and B. A. Federici, eds.), Vol. II, pp. 31–61, CRC Press, Boca Raton, FL.

Shapiro, M., and Robertson, J. L., 1992, Enhancement of gypsy moth (Lepidoptera: Lymantriidae) baculovirus activity by optical brighteners, *J. Econ. Entomol.* **85:**1120–1124.
Shieh, T. T., and Bohmfalk, G. T., 1980, Production and efficacy of baculoviruses, *Biotechnol. Bioeng.* **22:**1357–1364.
Slavicek, J. M., Hayes-Plazolles, N., and Kelly, M. E., 1995, Rapid formation of few polyhedra mutants of *Lymantria dispar* multinucleocapsid nuclear polyhedrosis virus during serial passage in cell culture, *Biol. Control* **5:**1–11.
Stewart, L. M. D., Hirst, M., Ferber, M. L., Merryweather, A. T., Cayley, P. J., and Possee, R. D., 1991, Construction of an improved baculovirus insecticide containing an insect-specific toxin gene, *Nature* **352:**85–88.
Teitelbaum, Z., Lazarovici, P., and Zlotkin, E., 1979, Selective binding of the scorpion venom insect toxin to insect nervous tissue, *Insect Biochem.* **9:**343–346.
Thiem, S. M., Du, X., Quentin, M. E., and Berner, M. M., 1996, Identification of a baculovirus gene that promotes *Autographa californica* nuclear polyhedrosis virus replication in a non-permissive insect cell line, *J. Virol.* **70:**2221–2229.
Tomalski, M., and Miller, L. K., 1991, Insect paralysis by baculovirus-mediated expression of a mite neurotoxin gene, *Nature* **352:**82–85.
Tomalski, M., and Miller, L. K., 1992, Expression of a paralytic neurotoxin gene to improve insect baculoviruses as biopesticide, *Bio/Technology* **10:**545–549.
Tomalski, M. D., Bruce, W. A., Travis, J., and Blum, M. S., 1988, Preliminary characterization of toxins from the straw itch mite, *Pyemotes tritici,* which induce paralysis in the larvae of a moth, *Toxicon* **26:**127–132.
Tomalski, M. D., Kutney, R., Bruce, W. A., Brown, M. R., Blum, M. S., and Travis, J., 1989, Purification and characterization of insect toxins derived from the mite, *Pyemotes tritici, Toxicon* **27:**1151–1167.
Tramper, J., van den End, E. J., de Gooijer, C. D., Kompier, R., van Lier, F. L. J., and Vlak, J. M., 1990, Production of baculovirus in a continuous insect-cell culture, bioreactor design, operation, and modeling, *Ann. NY Acad. Sci.* **5829:**423–430.
Treacy, M. F., and All, J. N., 1996, Impact of insect-specific AaHIT gene insertion on inherent bioactivity of baculovirus against tobacco budworm, *Heliothis virescens,* and cabbage looper, *Trichoplusia ni,* in: *Proceedings of the Beltwide Cotton Production and Research Conference* (P. Dugger & D. Richter, eds.), pp. 911–917, National Cotton Council, Memphis, TN.
Treacy, M. F., All, J. N., and Ghidiu, G. M., Impact of ecdysteroid UDP–glucosyltransferase gene deletion on efficacy of a baculovirus against *Heliothis virescens* and *Trichoplusia ni* (Lepidoptera: Noctuidae), *J. Econ. Entomol,* accepted.
Treacy, M. F., All, J. N., and Kukel, C. F., 1997, Invertebrate selectivity of a recombinant baculovirus: Case study on AaHIT gene-inserted *Autographa californica* nuclear polyhedrosis virus, in: *New Developments in Entomology* (K. Bondari, ed.), pp. 57–68, Research Signpost, London.
United States Environmental Protection Agency, 1996, Series 885—*Microbial Pesticide Test Guidelines,* EPA 712-C-96, US Government Printing Office, Washington, DC.
Vanderzant, E. S., Pool, M. C., and Richardson, C. D., 1962a, The role of ascorbic acid in the nutrition of three cotton insects, *J. Insect Physiol.* **8:**287–297.
Vanderzant, E. S., Pool, M. C., and Richardson, C. D., 1962b, Rearing of the bollworm on artificial diet, *J. Econ. Entomol.* **55:**140–161.
Walther, C., Zlotkin, E., and Rathmayer, W., 1976, Action of different toxins from the scorpion *Androctonus australis* on a locust nerve-muscle preparation. *J. Insect Physiol.* **22:**1187–1194.
Webb, R. E., Dill, N. H., McLaughlin, J. M., Kershaw, L. S., Podgwaite, J. D., Cook, S. P., Thorpe, K. W., Farrar, R. R., Ridgway, R. L., Fuester, R. W., Shapiro, M., Argaver, R. J., Venables, L., and White, G. B., 1996, Blankophor BBH as an enhancer of nuclear polyhedrosis virus in arborist treatments against the gypsy moth, *Forest Entomol.* **89:**957–962.
Weber, G. M., Hoban, T. J., Kendall, P. A., and Bull, L. S., 1995, Consumer concerns about

modern technologies in agriculture: Considerations for undergraduate and graduate teaching, *J. Animal Sci.* **73**(9):2727–2732.

Weiss, S. A., and Vaughn, J. L., 1986, Cell culture methods for large-scale propagation of baculoviruses, in: *The Biology of Baculoviruses* (R. R. Granados and B. A. Federici, eds.), Vol. II, pp. 63–87, CRC Press, Boca Raton, FL.

Wood, H. A., Hughes, P. R., and Shelton, A., 1994, Field studies of the co-occlusion strategy with a genetically altered isolate of *Autographa californica* nuclear polyhedrosis virus, *Environ. Entomol.* **23**:211–219.

Yendol, W. G., Heldung, R. C., and Lewis, F. B., 1978, Field investigation of a baculovirus of the gypsy moth, *J. Econ. Entomol.* **70**:598–602.

Young, S. Y., and Yearian, W. C., 1974, Persistence of *Heliothis* NPV of foliage of cotton, soybean, and tomato, *Environ. Entomol.* **3**:253–260.

Zilberberg, N., Zlotkin, E., and Gurevitz, M., 1991, The cDNA sequence of a depressant insect selective neurotoxin from the scorpion *Buthotus judaicus, Toxicon* **29**:1155–1158.

Zlotkin, E., 1983, Insect selective toxins derived from scorpion venoms: An approach to insect neuropharmacology, *Insect Biochem.* **13**:219–236.

Zlotkin, E., 1986, The interaction of insect selective neurotoxins from scorpion venoms with insect neuronal membranes, in: *Neuropharmacology and Pesticide Action* (Neurotox '85) (M. G. Ford, G. G. Lunt R. C. Reay, and P. N. R. Usherwood, eds.), pp. 352–383, Ellis Horwood, Chichester, England.

Zlotkin, E., Rochat, H., Kopeyan, C., Miranda, F., and Lissitzky, S., 1971, Purification and properties of the insect toxin from the venom of the scorpion *Androctonus australis* Hector, *Biochimie* **53**:1073–1078.

Zlotkin, E., Kadouri, D., Gordon, E., Pelhate, M., Martin, M. F., and Rochat, H., 1985, An excitatory and depressant insect toxin from scorpion venom—Both affect sodium conductance and possess a common binding site, *Arch. Biochem. Biophys.* **240**:877–887.

Zlotkin, E., Eitan, M., Bindokas, M. E., Adams, M. E., Moyer, M., Burkhart, W., and Fowler, E., 1991, Functional duality and structural uniqueness of depressant insect-selective neurotoxins, *Biochemistry* **30**:4814–4821.

Zlotkin, E., Gurevitz, M., Fowler, E., and Adams, M. E., 1993, Depressant insect selective neurotoxins from scorpion venom: Chemistry, action, and gene cloning, *Arch. Insect Biochem. Physiol.* **22**:55–73.

CHAPTER 14

Baculovirus Expression Vectors

Donald L. Jarvis

I. OVERVIEW

A baculovirus expression vector (BEV) is a recombinant baculovirus with a double-stranded circular DNA genome that has been genetically modified to include a foreign gene of interest. BEVs are viable and can infect susceptible hosts, usually cultured lepidopteran insect cells or larvae, in a helper-independent fashion. Therefore, BEVs can efficiently transfer foreign genes into these eukaryotic host cells. The foreign gene is usually a chimeric construct with the sequence encoding a protein of interest placed under the transcriptional control of a viral promoter. This arrangement enables viral functions to transcribe the gene during infection. The resulting mRNA is translated and the newly synthesized protein modified by host-encoded biosynthetic machinery. In essence, then, BEVs and their insect cell hosts are two separate components of a binary eukaryotic expression system, which will be called simply *the BEV system* throughout the remainder of this chapter.

The BEV system is among the best tools currently available for the expression of recombinant genes in a eukaryotic host. The BEV system has contributed immensely to basic research, since it has been used to produce hundreds of different recombinant proteins for further studies. This system also holds great promise for the industrial production of proteins with direct applications as vaccines, therapeutic agents, and/or diagnostic reagents. Fi-

DONALD L. JARVIS • Department of Entomology and Center for Advanced Invertebrate Molecular Sciences, Texas A&M University, College Station, Texas 77843.

The Baculoviruses, edited by Lois K. Miller. Plenum Press, New York, 1997.

nally, BEVs are being developed as improved biological pest control agents, as detailed in Chapter 13 (this volume).

The most significant advantage of the BEV system over other expression systems is that it can be used to produce exceptionally large amounts of functional foreign proteins. The production levels provided by the BEV system are often comparable to those provided by prokaryotic systems, and at late times after infection, the recombinant protein usually constitutes a significant proportion of the total protein in the host cell. Unlike prokaryotic expression systems, however, the BEV system has eukaryotic protein processing capabilities, which enables it to produce more authentic foreign proteins. Thus, it is the potential of this system to provide prokaryotic levels of foreign gene expression in a eukaryotic background that makes it so powerful and attractive. Another advantage of this system is that BEVs are noninfectious for vertebrates, so they are relatively safe for laboratory manipulation and industrial production of pharmaceutically important proteins. Finally, the actual process of isolating BEVs has become increasingly fast, simple, and efficient as more sophisticated molecular tools have been developed.

While it must be appreciated that the BEV system has been used with routine success to produce many different foreign gene proteins for over a decade, it also must be recognized that this system has some limitations that need to be addressed through further research and development. One limitation is that there is significant protein-to-protein variation in the production levels provided by this system. Generally, some classes of recombinant proteins, such as membrane-bound and secreted glycoproteins, are produced at much lower levels than others. Another limitation is that insect cell protein processing pathways are not necessarily identical to those of higher eukaryotes, and as a result, covalent chemical modifications of recombinant proteins produced in this system may differ from those on the native protein. Finally, the BEV system provides only transient expression, because the foreign gene is delivered into the host cell by a virus that ultimately kills it. A related problem is that certain protein processing pathways are compromised by baculovirus infection, and this can decrease the efficiency and kinetics of recombinant protein processing well before the cell dies.

The purpose of this chapter is to describe the BEV system and its applications, capabilities, and limitations in further detail. I will provide minimal background on baculoviruses, as further details are readily available in the preceding chapters of this volume. Also, space constraints preclude comprehensive citation of the literature and I apologize to those whose work I fail to cite. For additional background on baculoviruses, the BEV system, and more comprehensive reference lists, the reader can refer to several other sources (Montreuil *et al.*, 1995; Richardson, 1995; King and Possee, 1992; O'Reilly *et al.*, 1992; Adams and McClintock, 1991; Blissard and Rohrmann, 1990; Luckow and Summers, 1988; Miller, 1988; Summers and Smith, 1987), as well as the preceding chapters of this volume.

II. INTRODUCTION AND BACKGROUND

A. Baculoviruses

1. Classification and General Features

The family *Baculoviridae* consists of a large group of double-stranded DNA-containing viruses that infect arthropods (Volkman *et al.,* 1995). The majority of these viruses infect insects and the type species is *Autographa californica* nuclear polyhedrosis virus (AcMNPV). There are two phenotypically distinct forms of AcMNPV, occluded virus (OV) and budded virus (BV). OV consists of rod-shaped nucleocapsids enclosed by an envelope and embedded within a polyhedral-shaped crystalline matrix, or polyhedron. The "M" in AcMNPV indicates that one enveloped OV particle can contain multiple nucleocapsids. BV also consists of rod-shaped nucleocapsids enclosed by an envelope, but BV particles contain only one nucleocapsid and are released as free, nonoccluded virions by budding from the surface of the infected cell. OV and BV have different relative infectivities for insect larvae or cultured insect cells, which reflects their distinct roles in baculovirus infections, as discussed in Section II.A.2.

AcMNPV was originally isolated (Vail *et al.,* 1971) as a mixture of genotypic variants with different restriction patterns (Lee and Miller, 1978), and different variants were used by investigators who went on to study this baculovirus and develop the first BEVs. Other nucleopolyhedroviruses also have been used to develop BEVs. Among these, BEVs derived from *Bombyx mori* nuclear polyhedrosis virus (BmNPV) are notable for their frequent use for foreign gene expression in insect larvae (Maeda, 1989). The remainder of this chapter will focus on generic "baculoviruses" and "BEVs," except where it is important to be more specific.

2. The Baculovirus Life Cycle

Natural baculovirus infections begin when a susceptible insect ingests OV in the form of polyhedra-contaminated food. The crystalline polyhedral matrix dissociates in the insect midgut and the liberated virus particles infect columnar epithelial and regenerative cells (Keddie *et al.,* 1989). The infected midgut cells produce BV progeny that invade the insect circulatory and respiratory systems (Engelhard *et al.,* 1994; Keddie *et al.,* 1989), where they initiate secondary infections and produce both BV and OV. BV progeny are produced when newly assembled viral nucleocapsids migrate from the nucleus and bud from the infected cell surface. These virions acquire a lipid envelope and plasma membrane-bound glycoproteins during the process of budding. OV progeny are produced when nucleocapsids are enveloped within the nucleus (Stoltz *et al.,* 1973) and the resulting virions are surrounded by the polyhedral matrix. Polyhedra remain in the nuclei of infected cells until being released when the cell dies. Baculovirus-infected insect larvae are ultimately liquefied by virus-encoded digestive enzymes (Hawtin *et al.*, 1995; Ohkawa *et al.*, 1994), and when the cuticle ruptures, BV and polyhedra are released into the environment. Occluded virions are protected by the crystalline polyhedral matrix, and as a result are relatively resistant to inactivation by adverse environmental conditions. Moreover, this form of the virus is significantly more infectious than BV for orally inoculated insect larvae (Volkman and Summers, 1977). Accordingly, OV is responsible for horizontal transmission of baculovirus infection in nature and is used to experimentally infect larvae via the oral route. Conversely, BV is significantly more infectious than OV for cultured insect cells (Volkman and Summers, 1977; Volkman *et al.*, 1976) and is used to experimentally infect these cells in the laboratory.

3. Baculovirus Genes and Gene Expression

The complete nucleotide sequences of AcMNPV (C6 isolate; Ayres *et al.*, 1994; Acc. No. L22858) and BmNPV (S. Maeda, personal communication; Acc. No. L33180) have been determined. The AcMNPV genome is a double-stranded circular DNA molecule consisting of 133,894 nucleotides that probably encode about 150 proteins (see Chapter 5, this volume). Viral genes are distributed evenly throughout the genome on both strands of the DNA and are expressed in a temporally regulated fashion after infection. The immediately early genes, such as *ie1* (Guarino and Summers, 1987), are expressed immediately after infection in the absence of other viral functions. At least some of the immediately early genes encode transcription factors that function to stimulate the expression of other early genes like *39K* (Guarino and Smith, 1990). The early phase of baculovirus infection is followed by viral DNA replication and the onset of late viral gene expression. Many baculovirus late genes encode proteins needed for the assembly of progeny virions. For example, *vp39* is a late gene that encodes the major nucleocapsid structural protein (Thiem and Miller, 1989), and *p6.9* is a late gene that encodes a basic, protaminelike protein found in association with viral DNA (Wilson *et al.*, 1987). Late gene expression is followed by expression of the very late genes, including *polh* (Hooft van Iddekinge *et al.*, 1983) and *p10* (Kuzio *et al.*, 1984), which are needed for the assembly and envelopment of polyhedra, respectively.

There have been significant advances in our understanding of baculovirus gene expression during the past decade (see Chapters 6 and 8, this volume). It is appropriate to simply state here that immediately early genes are transcribed by host cell factors and transcription of later classes of viral genes is increasingly influenced by virus-encoded factors. In particular, studies have identified nearly 20 virus-encoded proteins that are required for late and very late gene expression (Lu and Miller, 1995; Todd *et al.*, 1995; McLachlin and Miller, 1994), and at least some of these must assemble into a virus-modified or -encoded transcriptional complex (Beniya *et al.*, 1996;

Passarelli *et al.*, 1994; Fuchs *et al.*, 1983). This is an important concept, because it is the nature of the transcriptional complex that is assembled during the very late phase of baculovirus infection, together with the structure of the promoters in baculovirus very late genes (Qin *et al.*, 1989; Weyer and Possee, 1989; Rankin *et al.*, 1988; Possee and Howard, 1987), that determines one of the most attractive properties of BEVs—their ability to express foreign genes at extremely high levels.

4. Polyhedrin

By the end of the baculovirus replication cycle, the nucleus occupies most of the volume of the infected host cell and it is literally stuffed with polyhedra. The major component of polyhedra is a protein called *polyhedrin* (Rohrmann, 1986), which forms the crystalline matrix. Thus, baculoviruses must be able to produce copious amounts of polyhedrin protein. In fact, polyhedrin comprises at least 25% of the total protein in baculovirus-infected insect cells during the very late phase of infection (Smith *et al.*, 1983c). The ability to produce such large amounts of polyhedrin reflects the availability of a huge pool of polyhedrin mRNA, which represents about 25% of the total polyadenylated RNA in the cell during the very late phase of infection (Adang and Miller, 1982). The *polh* promoter contains a critical core sequence, TAAG, which also serves as the transcriptional initiation site and is conserved in baculovirus very late and late promoters (Rohrmann, 1986). However, very late promoters are transcribed later in infection and produce more mRNA than late promoters (Thiem and Miller, 1990). Therefore, the very late promoters and/or the very late transcriptional complex must have additional features, in addition to the TAAG sequence, that account for these differences. Indeed, studies have identified a region in the *polh* promoter (Ooi *et al.*, 1989) and a virus-encoded factor (McLachlin and Miller, 1994) that specifically enhance very late transcription, and transcriptional complexes have been isolated that exhibit preferential activity toward late or very late promoters *in vitro* (Xu *et al.*, 1995). Furthermore, various differences in the behavior of the *polh* and *p10* promoters have been reported (Tomita *et al.*, 1995; McLachlin and Miller, 1994; Roelvink *et al.*, 1992; Min and Bishop, 1991), indicating that even individual very late promoters are not functionally homologous.

B. The BEV Concept

The ability to produce large amounts of polyhedrin during infection was the fundamental property of baculoviruses that led to their development as expression vectors. The AcMNPV *polh* gene was mapped, cloned, and sequenced (Hooft van Iddekinge *et al.*, 1983; Adang and Miller, 1982; Vlak *et al.*, 1981), and together with the establishment of marker rescue in the baculovirus system (Miller, 1981), these studies provided some of the molecular tools that were needed to develop BEVs. In addition, the *polh* gene was shown to be nonessential for virus replication in cultured insect cells (Smith *et al.*, 1983a). This indicated that one could create a BEV simply by replacing the *polh* open reading frame in the wild-type viral genome with a nucleotide sequence encoding a foreign protein of interest. The resulting BEV could be phenotypically distinguished from wild type by its inability to produce polyhedra. Moreover, the BEV would be viable and could be used to infect cultured insect cells, in which the foreign coding sequences would be expressed under the transcriptional control of the *polh* promoter and large quantities of the foreign protein would be produced. The feasibility of this concept was demonstrated when it was shown that BEVs could be isolated and used to express human β-interferon (Smith *et al.*, 1983b) or *Escherichia coli* β-galactosidase (Pennock *et al.*, 1984) in cultured insect cells.

III. PRODUCING BEVs

A. Introduction

The first step in the original procedure used to produce a BEV is to clone the sequences encoding a protein of interest into a suitable "transfer plasmid." A classic transfer plasmid contains the *polh* promoter and long upstream and downstream flanking sequences, but lacks some or all of the *polh* open reading frame, which is usually replaced by a convenient cloning site. Once the DNA sequence encoding a protein of interest has been inserted into this site, the recombinant transfer plasmid is purified and mixed with genomic DNA from wild-type baculovirus, and the mixture is cotransfected into cultured insect cells. Upon entering the cell, the viral DNA will be replicated and wild-type progeny will be produced. In addition, homologous recombination can occur between the *polh* flanking sequences in the transfer plasmid and the same sequences in the viral DNA. This process, called *allelic replacement*, produces recombinant viral DNAs in which the *polh* open reading frame in the parental virus has been replaced by sequences from the recombinant transfer plasmid encoding the protein of interest. These viral DNAs are replicated and packaged to produce recombinant viral progeny. Allelic replacement of *polh* occurs at a maximum frequency of only about 1% (Smith *et al.*, 1983a), but this is no problem because it is relatively easy to distinguish wild-type and recombinant progeny by their plaque phenotypes. The mixture of viruses is simply harvested from the cotransfected cells and resolved in a plaque assay. The wild-type progeny, which retain the polyhedrin gene, produce polyhedron-positive plaques, whereas the recombinant progeny, which lack the polyhedrin gene, produce polyhedron-negative plaques. Once identified, recombinant viral clones can be further plaque purified, amplified, and characterized, and then large BV stocks can be produced and used to infect cultured insect cells for foreign gene expression and foreign protein production.

B. Transfer Plasmids

There are many different transfer plasmids that can be used to construct a chimeric gene and insert it into the baculovirus genome by allelic replacement. The following sections will describe the general features of some of these plasmids. Maps and more detailed descriptions are available in several other places, including the primary literature (cited below), technical manuals (Richardson, 1995; King and Possee, 1992; O'Reilly *et al.*, 1992; Summers and Smith, 1987), and commercial literature and catalogs from various scientific supply houses, including Clontech (Palo Alto), InVitroGen (San Diego), Novagen (Madison), Pharmingen (San Diego), and Stratagene (La Jolla). Many different transfer plasmids can be purchased from these companies.

1. Very Late Promoters

The first transfer plasmids were designed to produce BEVs in which the *polh* gene had been replaced by a new, chimeric gene consisting of the very late *polh* promoter and the sequence encoding the protein of interest positioned downstream, as described above (Pennock *et al.*, 1984; Smith *et al.*, 1983b). These transfer plasmids were constructed before it was clearly determined precisely which sequences from the 5′ untranslated region of the *polh* gene were needed for optimal levels of *polh* promoter-mediated transcription (Rankin *et al.*, 1988; Matsuura *et al.*, 1987; Possee and Howard, 1987). Once this information became available, it was easier to ensure that the transfer plasmid included all of the critical promoter sequences and to decide exactly where to place multiple cloning sites to facilitate subcloning of the foreign coding sequence. The *polh* promoter and multiple cloning site in the transfer plasmid must be flanked by sequences that normally flank the *polh* gene in the viral genome, since these sequences are needed for homologous recombination between the transfer plasmid and the viral DNA. This process mediates allelic replacement of the *polh* gene, as described above. Transfer plasmids containing the *polh* promoter are still the most widely used tools for the production of BEVs. However, the increased popularity of the BEV system has led to the design and construction of many variations on this basic theme.

For example, there are transfer plasmids that can be used to produce BEVs that will express a foreign gene under the control of the very late *p10* promoter. Some of these transfer plasmids contain both the *p10* promoter and *p10* flanking sequences, and after the sequence encoding the protein of interest has been inserted, the resulting plasmid is used to replace the nonessential baculovirus *p10* gene (Vlak *et al.*, 1990). This approach requires special screening procedures to identify recombinant viruses, since loss of the viral *p10* gene is not accompanied by an easily distinguishable change in plaque phenotype. Usually, this type of transfer plasmid includes a marker gene, such as *E. coli lacZ,* which is incorporated together with the chimeric gene of interest into the recombinant virus genome during allelic replacement. Another approach is to use a hybrid transfer plasmid, in which the *p10* promoter is embedded within *polh* flanking sequences, to construct the chimeric gene and insert it into the *polh* region of the baculovirus genome (Weyer *et al.,* 1990). This latter approach takes advantage of the simple visual screen that can be used to distinguish between parental (occlusion-positive plaques) and recombinant (occlusion-negative plaques) viruses whenever *polh* is used as the target for allelic replacement. The *p10* promoter has been used to produce BEVs for foreign protein production (Bozon *et al.,* 1995; Tomita *et al.,* 1995; van Lier *et al.,* 1994; Roelvink *et al.,* 1992; Vlak *et al.,* 1990), but it provides lower levels of foreign gene expression and is used much less frequently than the *polh* promoter for this purpose. The *p10* promoter is more commonly used to construct BEVs for biopesticide applications, because allelic replacement of *p10* is one way to produce recombinants that can express a foreign gene without deleting the *polh* gene (McCutchen *et al.,* 1991; Stewart *et al.,* 1991; Merryweather *et al.,* 1990). This is important because BEVs intended for use as pesticides must be able to produce polyhedra to infect insect larvae naturally in the field.

2. Alternative Promoters

Other transfer plasmids can be used to construct BEVs that will express foreign genes under the transcriptional control of alternative promoters, including both viral and cellular promoters. These plasmids typically contain the promoter of choice embedded within *polh* flanking sequences and are used for allelic replacement of the *polh* gene as described above. Baculoviral promoters that have been used most commonly for this purpose include the late *p6.9* (Hill-Perkins and Possee, 1990) and *vp39* (Thiem and Miller, 1980) promoters and the early *etl* (Morris and Miller, 1992) and *iel* (Jarvis *et al.,* 1996b) promoters. Cellular promoters include *Drosophila hsp70* (Vlak *et al.,* 1990) and *B. mori actin* (Johnson *et al.,* 1992). Transfer plasmids with a hybrid *vp39–polh* promoter (Thiem and Miller, 1990) or tandem *polh–etl* promoters (Xia *et al.,* 1993) also have been described. Except for the last two, none of these promoters are as strong as the *polh* promoter (Jarvis *et al.,* 1996b; Morris and Miller, 1992; Thiem and Miller, 1990).

Considering that high-level expression is one of the most attractive features of the BEV system, it might seem foolish to use any promoter besides *polh* for baculovirus-mediated foreign gene expression. But, there are some good reasons to do this. BEVs cannot produce all classes of foreign proteins in equal abundance or quality under *polh* control. Generally, secretory pathway proteins are produced at much lower levels than other types of proteins and some are biologically inactive and/or insoluble when expressed under *polh* control (Pajot-Augy *et al.*, 1995; Rankl *et al.*, 1994; Arp *et al.*, 1993; Xie *et al.*, 1992; Tsao *et al.*, 1990). This might reflect the adverse effects of baculovirus infection on host cell secretory pathway functions, which have already begun to decay by the start of the very late phase of infection (Murphy *et al.*, 1990; Jarvis and Summers, 1989). Alternatively, it might reflect saturation of the protein folding and secretory capacity of the host cell due to high-level foreign gene expression, as has been documented in a yeast system (Parekh *et al.*, 1995). Either way, it has been shown that the *p6.9* (Chazenbalk and Rapoport, 1995; Lawrie *et al.*, 1995; Bonning *et al.*, 1994; Rankl *et al.*, 1994; Sridhar *et al.*, 1993) and *ie1* (Jarvis *et al.*, 1990, 1996b) promoters, which are expressed earlier and produce less mRNA, and even the *p10* promoter (Bozon *et al.*, 1995), which is expressed only slightly earlier and produces only slightly less mRNA, can sometimes be used to produce larger amounts of biologically active and/or soluble foreign protein than the *polh* promoter. The use of these alternative promoters to produce BEVs that can express foreign genes earlier in infection also represents a good approach for biopesticide applications, since the main idea is to produce a virus with a new gene that will help it to kill insects or stop their feeding more quickly (Jarvis *et al.*, 1996a,b).

3. Multiple Promoters

Sometimes, it is important to be able to express more than one foreign protein simultaneously to study protein–protein interactions, assemble functional protein complexes, or reconstruct biochemical pathways. The BEV system is especially useful for this purpose. One approach is to use a mixture of two or more BEVs, each containing one foreign gene of interest, to coinfect host cells (St. Angelo *et al.*, 1987). Theoretically, this approach can be used to vary the ratios of the different proteins being produced. However, an obvious problem with this approach is that it is difficult to obtain a reasonably uniform population of cells coinfected with each virus. Another approach is to use transfer plasmids containing multiple promoters to produce a single BEV that can express multiple foreign genes. The first transfer plasmids of this type contained two copies of the *polh* promoter and were used to produce "dual" BEVs that could express two different foreign genes in a single cell (Emery and Bishop, 1987). Later, transfer plasmids were constructed that contained various combinations of different viral promoters, including the *polh*, *p10*, and *vp39* promoters, and these could be used to produce BEVs capable of expressing up to five different foreign proteins in the same infected cell (Belyaev *et al.*, 1995; Wang *et al.*, 1991; Weyer and Possee, 1991).

4. Specialized Functions

a. Identification of BEVs

One of the first specialized functions to be built into transfer plasmids was a marker gene that could be used to identify BEVs. The marker was designed to be incorporated along with the gene of interest during allelic replacement. The resulting BEV would express this marker and produce a protein that could be detected with a chromogenic substrate and identify recombinant viral plaques. This was mandatory for the identification of BEVs with allelic replacements in *p10* (Vlak *et al.*, 1990), but it also facilitated the identification of BEVs with allelic replacements in *polh* (Vialard *et al.*, 1990), because many investigators had trouble seeing the occlusion-negative plaques produced by those recombinants. The first marker used for this purpose was *E. coli lacZ,* and its expression was controlled variously by baculovirus *p10, ie1, et1,* and *Drosophila hsp70* promoters. Other markers that have been used for this purpose include luciferase (Oker-Blom *et al.*, 1993) and β-glucuronidase (Bishop *et al.*, 1995).

b. Simplified Subcloning

The sheer diversity of transfer plasmids that have become available, with their diverse array of multiple cloning sites, has greatly simplified the process of subcloning a foreign coding sequence for insertion into the baculovirus genome. Some transfer plasmids include translational initiation signals and can be used to construct chimeric genes encoding fusion proteins. Others lack these signals and are used to construct genes encoding nonfused proteins. Still others can be used either way, depending on where the foreign coding sequence is inserted. Some transfer plasmids also have been streamlined by shortening the flanking sequences to the minimum lengths needed for efficient homologous recombination. This approach has been used to reduce the size of a transfer plasmid by nearly half, which facilitates the subcloning process by leaving more room for the insertion of larger foreign protein coding sequences (Pharmingen and InVitrogen catalogs).

Some of the newest transfer plasmids are designed for ligation-independent cloning of a foreign protein coding sequence to be incorporated into a BEV (Bishop *et al.*, 1995; Pharmingen catalog). This approach circumvents the need to subclone the sequence of interest into the transfer plasmid and amplify it in *E. coli* prior to cotransfection with viral DNA. Transfer plasmids designed for ligation-independent cloning have long single-stranded overhangs that can anneal to complementary single-stranded overhangs on a polymerase chain reaction (PCR) amplification product encoding the foreign protein of interest. The annealed products are mixed with viral DNA and the mixture is used to cotransfect insect cells for production of BEVs by allelic replacement. This ligation-independent cloning approach is also called "direct" cloning in the commercial literature. However, a transfer plasmid serves as an intermediate, which differentiates this approach from truly direct cloning by ligation of a DNA fragment encoding a protein of interest with restriction enzyme-digested viral DNA (Lu and Miller, 1996; Ernst *et al.*, 1994; see also Section III.G.4).

c. Generic Protein Purification

One of the biggest challenges faced by investigators using any expression system is purification of the overexpressed heterologous protein. This has led to the development of various generic protein purification methods (Ford *et al.*, 1991). The usual approach is to design the expression vector to include a short nucleotide sequence that encodes a short amino acid sequence that can be used as an affinity "tag" to purify any protein. The sequence encoding the protein of interest is inserted into the vector in such a way that it will be expressed as a fusion protein with the affinity tag on its N- or C-terminus. Then, the fusion protein can be solubilized and affinity purified with a reagent that specifically recognizes and binds to the tag. An obvious problem with this approach is that the affinity tag sometimes interferes with the biological activities of the protein of interest. Hence, some vectors also include sequences that encode a short amino acid sequence between the affinity tag and the protein of interest that can be specifically cleaved by treating the purified fusion protein with a protease.

A variety of different transfer plasmids have been constructed and used to isolate BEVs that produce affinity-tagged proteins. Many different affinity tags have been used in the BEV system, and it is beyond the scope of this chapter to discuss the specific results obtained with each different tag. However, it is fair to state that various affinity tags can be used effectively in this system to purify recombinant proteins to varying degrees, ranging from about 80% to homogeneity. Also, antibodies against some affinity tags can be used to monitor the expression and localization of foreign proteins in BEV-infected insect cells by immunocytochemical techniques. This idea has been taken one step further with the development of transfer plasmids that can be used to produce BEVs that encode fusion proteins containing a naturally fluorescent protein, such as green fluorescent protein (Pharmingen catalog). Fluorescent protein tags permit direct visualization of the fusion protein in unfixed BEV-infected cells with a fluorescence microscope.

d. Protein Secretion

Some transfer plasmids include sequences that encode signal peptides that can mediate secretion of a foreign protein from BEV-infected insect cells. The coding sequence of interest is inserted downstream and in-frame and the resulting construct encodes the protein of interest with a cleavable N-terminal signal peptide. The signal peptides used to direct secretion are sometimes derived from insect glycoproteins, including honeybee prepromellitin (Tessier *et al.*, 1991) and baculovirus *gp64* and *egt* (Murphy *et al.*, 1993), and it has been found that insect-derived signal peptides can increase the efficiency of secretion of some recombinant proteins from BEV-infected insect cells. However, mammalian signal peptides also can be used, sometimes with better success, to direct secretion of recombinant proteins in this system (Mroczkowski *et al.*, 1994; Jarvis *et al.*, 1993; Andersons *et al.*, 1991). Some transfer plasmids provide both an N-terminal signal peptide and a C-terminal affinity tag to facilitate purification of recombinant proteins from the growth medium (Kuhn and Zipfel, 1995).

e. Virus Display

Finally, some transfer plasmids can be used to produce BEVs that will express a foreign protein that can be incorporated into the envelope of recombinant BV particles (Boublik *et al.*, 1995). These BEVs are eukaryotic versions of bacteriophage "display" vectors, which have been used to select rare recombinants from mixed virus populations by using ligands that bind to the protein of interest (Winter *et al.*, 1994). The transfer vectors contain the sequence encoding *gp64*, the major BV envelope glycoprotein embedded within *polh* flanking sequences. The sequence encoding the protein of interest is inserted between the sequences encoding the signal peptide and mature portions of *gp64*. The resulting plasmid is used for allelic replacement of *polh* to produce a BEV that will express the protein of interest as a fusion protein that can be incorporated into the BV envelope and "displayed" for interactions with specific ligands or antibodies.

C. Sequences Encoding Foreign Proteins

1. cDNA or Genomic DNA?

The promoter in the transfer plasmid is only one component of the chimeric gene that must be assembled and inserted into the baculovirus genome in order to produce a BEV by allelic replacement. Another obvious requirement is the sequence encoding the protein of interest, which could be either a cDNA or genomic DNA sequence. The vast majority of BEVs contain cDNA inserts, but there are a few reports of BEVs that contain and can express foreign DNA sequences containing introns (Davrinche *et al.*, 1993; Iatrou *et al.*, 1989; Jeang *et al.*, 1987) and at least one baculovirus gene has introns that are removed by splicing in infected insect cells (Kovacs *et al.*, 1991; Chisholm and Henner, 1988). However, there is evidence that RNA splicing is very inefficient at late times of baculovirus infection (Kovacs *et al.*, 1991), and it has been reported that a human papillomavirus gene failed to be spliced when expressed in the BEV system (Park *et al.*, 1993). Thus, it is probably prudent to use cDNAs to construct BEVs.

2. Translational Initiation Sites and Codon Usage

Publication of the entire AcMNPV nucleotide sequence was accompanied by the identification of 337 open reading frames of at least 150 basepairs (bp) in length (Ayres *et al.*, 1994). Among these, 154 were included in a selected set deemed most likely to be expressed during baculovirus infection. Analysis of the predicted translational initiation sites in this selected set of open reading frames revealed that 41% did not conform to Kozak's rules (Kozak, 1986). This might indicate that biosynthesis of some viral proteins is down-regulated by the absence of an optimal translational initiation site. Or, it might indicate that Kozak's rules do not accurately define the signals required for optimal translational initiation in baculovirus-infected insect cells. The latter interpretation is supported by experimental evidence that showed that Kozak's rules could be broken with no reduction in the levels of foreign protein produced by a BEV (Hills and Crane-Robinson, 1995). Furthermore, three different proteins were expressed at higher levels by BEVs when fused to a bacterial leader sequence that had a pyrimidine instead of a purine at position −3 relative to the initiation codon (Peakman *et al.*, 1992a). A putative baculovirus translational initiation consensus sequence has been elucidated by computer analysis of 23 viral genes and this sequence includes a purine at position −3 (Ranjan and Hasnain, 1995). However, considering the experimental observations cited above, it is difficult to know how accurately this consensus sequence defines a translational initiation site in the BEV system. Overall, it seems that the rules governing translational initiation in this system remain unclear. Similarly, analysis of codon usage in the selected set of AcMNPV open reading frames revealed some bias, but it is not clear whether this influences the levels of foreign protein production in the BEV system (Ayres *et al.*, 1994).

One clear finding is that AUU can sometimes serve as an inefficient translational initiation codon in baculovirus-infected insect cells (Beames *et al.*, 1991). This was discovered when a BEV was used to express hepatitis B virus core protein and about one fifth of the end product had a *polh* amino acid sequence fused to its N-terminus. The BEV had been produced with a transfer plasmid in which the polyhedrin ATG was changed to ATT, and the sequence encoding the core protein, which had its own ATG, was inserted downstream and in-frame. It was discovered that the fusion protein was produced as a result of translational initiation at the upstream ATT (AUU in the transcript). This problem might be related to the fact that the AUU was located in precisely the same position formerly occupied by the *polh* AUG; in any case, it can be avoided simply by cloning coding sequences out-of-frame with respect to this upstream ATT or by using a transfer plasmid that does not have the ATT in this position.

3. Transcriptional Termination and RNA Processing Signals

The chimeric gene in a BEV also must have appropriate signals for transcriptional termination and RNA processing. These signals might be included in the 3′ untranslated region of the foreign sequence inserted into the transfer plasmid. Or, they can be provided by the transfer plasmid itself, which includes the 3′ untranslated region of the *polh* gene. mRNA processing signals have not been extensively analyzed in the baculovirus system, but several baculovirus genes, including *polh,* contain the 3′ motif (AAUAAA) that serves as a polyadenylation signal in higher eukaryotes (Westwood *et al.*, 1993). As in higher eukaryotes, this motif alone is insufficient for mRNA processing in baculovirus-infected insect cells. Also, heterologous mRNA processing signals from the SV40 early region or a rabbit β-globin gene have been included in some transfer plasmids and are functional in the BEV system (Westwood *et al.*, 1993). These findings suggest that baculovirus-infected insect cells and higher eukaryotes have the same or similar mechanisms for transcriptional termination and mRNA processing.

D. Viral DNAs for Allelic Replacement

Besides the transfer plasmid, the other critical component in an allelic replacement reaction is baculoviral DNA. Baculoviral DNA is usually isolated from BV particles partially purified from the extracellular medium of wild-type baculovirus-infected insect cells, as described elsewhere (Richardson, 1995; O'Reilly *et al.*, 1992; Summers and Smith, 1987). Historically, a major problem with using wild-type viral DNA to isolate BEVs was that the frequency of allelic replacement is low and many investigators had trouble finding recombinant virus plaques among the high background of parental virus plaques. This problem was addressed by incorporating marker genes into transfer plasmids, as described above. However, this approach did not reduce parental virus background. This problem was solved by the development of linearizable viral DNAs that could be used as the targets for allelic replacement. The first linearizable viral DNA was created by constructing an occlusion-negative recombinant virus with a unique Bsu36I site in the *polh* region (Kitts *et al.*, 1990). The circular genomic DNA from this recombinant could be linearized by digestion with Bsu36I, which significantly reduced its infectivity. Thus, when linearized viral DNA is mixed with a transfer plasmid and used to cotransfect insect cells, the recircularized recombinant viral DNAs produced by allelic replacement have a strong replicative advantage over the linear parental DNA molecules. The net result is an increase in the maximum efficiency of BEV production from about 1% to about 30%. An occlusion-positive linearizable viral DNA has been described that provides the same high efficiency of BEV production while preserving the ability to identify recombinants by using the classic visual screen (Hartig *et al.*, 1992). There also is a viral DNA that can be linearized at a unique Bsu36I site in the *p10* region and used for high-efficiency production of BEVs with allelic replacements in that region (Martens *et al.*, 1995).

The development of linearizable viral DNAs was followed by the development of a viral DNA that can provide even higher efficiencies of BEV production (Kitts and Possee, 1993). This viral DNA has an *E. coli lacZ* insert in the *polh* region and Bsu36I sites in the two flanking genes on each side. Therefore, Bsu36I digestion actually deletes a fragment of the viral DNA, including part of the ORF 1629 gene located downstream of *polh,* which encodes an essential nucleocapsid-associated phosphoprotein (Vialard and Richardson, 1993). This effectively inactivates the viral DNA, but it can be rescued by homologous recombination with the transfer plasmid, which simultaneously introduces the coding sequence of interest, restores ORF 1629, and recircularizes the viral genome. The efficiency of BEV production with these Bsu36I-gapped viral DNAs is routinely over 90%.

Another type of baculovirus DNA that is currently being developed as a target for allelic replacement is one in which various "auxiliary" genes have been deleted (Bishop *et al.,* 1995). These genes encode proteins that are needed for baculoviruses to infect insect larvae but are nonessential for the replication of these viruses in cultured insect cells. The idea behind the development of these viral DNAs is that expression of the auxiliary genes might interfere with the production of a protein of interest by a BEV. For example, it would be a good idea to delete the baculoviral cathepsinlike protease gene, since this protease could degrade some BEV-expressed recombinant proteins. Similarly, deletion of the viral protein phosphatase gene (Kim and Weaver, 1993) might improve the quality of phosphoproteins produced in the BEV system.

E. Producing BEVs by Allelic Replacement

After being subcloned into an appropriate transfer plasmid or annealed to a ligation-independent transfer plasmid, a sequence encoding a protein of interest is incorporated into the baculovirus genome by mixing it with viral DNA and transferring the mixture into cultured insect cells, where allelic replacement can occur by homologous recombination, as previously discussed.

The production of BEVs by the allelic replacement method relies upon the homology between sequences flanking the *polh* (or *p10*) genes in the viral DNA and the same sequences flanking the gene of interest in the transfer plasmid. The frequency with which BEVs are produced by this method is determined by the choice of target viral DNA, as described above, and selectable markers are not usually used. However, some selectable markers have been used to increase the frequency of BEV production, including negative selection of parental viruses containing the herpes simplex virus thymidine kinase gene (Godeau *et al.,* 1992) and positive selection of recombinant viruses containing neomycin resistance or *p35* genes (Lerch and Friesen, 1993).

F. Isolating, Identifying, and Characterizing BEVs

The most common way to resolve the mixture of parental and recombinant baculovirus progeny obtained from cotransfected cells is by using a conventional viral plaque assay (Summers and Smith, 1987). However, limiting dilution (Reed and Muench, 1938) and fluorescence-activated cell sorting (Peng *et al.*, 1993) also have been used for this purpose. The classic approach used to identify BEVs in viral plaque assays is direct visualization of occlusion-negative plaques, as previously described. Obviously, this screen can be used only if the parental virus had an intact *polh* gene and the transfer plasmid was designed to replace that gene. Other ways to identify BEVs in plaque assays are to do plaque lifts with nucleic acid or antibody probes (Capone, 1989; Summers and Smith, 1987) or to rely on visualization of a marker gene that was introduced during allelic replacement, as described above. Conversely, BEVs produced using Bsu36I-gapped viral DNA can be tentatively identified by the loss of the *lacZ* marker in the parental viral DNA (Kitts and Possee, 1993). When limiting dilution is used as the isolation method, BEVs are usually identified by using nucleic acid or antibody probes in dot–blot formats (Manns and Grosse, 1991; Pen *et al.*, 1989; Summers and Smith, 1987). Finally, BEVs can be identified by direct sequencing (Slightom and Sieu, 1992; Wang and Fraser, 1991) or PCR analysis (Sisk *et al.*, 1992; Malitschek and Schartl, 1991; Webb *et al.*, 1991) of viral DNAs.

After a BEV has been isolated and identified by one of the above methods, it can be amplified in insect cells, titered, and used to produce the foreign protein of interest by infecting a fresh cell culture. However, it is important to carefully characterize BEVs before proceeding with foreign gene expression, as there are some traps in the isolation and screening process. The most serious trap is that recombinants are produced far more frequently by single crossovers than by allelic replacement, which requires a double crossover between the viral DNA and transfer vector (see O'Reilly *et al.*, 1992, for an excellent discussion of this topic). Single-crossover recombinants contain the entire transfer plasmid at a random site in the viral genome and are genetically unstable. Depending on which screening method was used, single-crossover recombinants can be mistakenly identified as BEVs in which the target gene in the parental viral DNA has been properly replaced by the gene of interest. True allelic replacement must be confirmed by verifying the location of the inserted gene in the BEV genome and/or the absence of the gene it was intended to replace. This can be done by using any of several different methods to analyze the viral DNA, including restriction mapping, Southern blotting, or PCR. Using linearized or gapped viral DNAs for BEV production minimizes the problem of single crossover recombination because double crossover recombination at the appropriate site is necessary to regenerate a circular viral DNA molecule that can efficiently replicate. However, single-crossover recombinants can be obtained if digestion of the viral DNA is incomplete.

Standard methods of protein analysis can be used to assess foreign protein production by BEV-infected insect cells. If expression levels are high, as expected, foreign proteins can be identified in electrophoretic profiles of total lysates prepared at late times after infection. In the best cases, this can be done simply by staining protein gels with Coomassie Brilliant Blue. More sensitive methods can be used to detect proteins produced at lower levels, including electrophoretic analysis of total protein lysates from radiolabeled cells, Western blotting, radiolabeling and immunoprecipitation, or specific activity assays.

G. Alternative Ways to Produce BEVs

The development of new transfer plasmids and linearizable and gappable viral DNAs for the production of BEVs by allelic replacement significantly increased the rate and efficiency and simplified the process of BEV production. Alternatives to the allelic replacement approach also have been developed to facilitate BEV production.

1. Homologous Recombination and Selection in Yeast

The first alternative method described for producing BEVs involved homologous recombination and selection of recombinant viral DNAs in yeast (Patel *et al.*, 1992). A recombinant baculovirus was constructed with a yeast autonomous replication sequence, centromere, and selectable marker in the *polh* region. This viral DNA was introduced into yeast, where it replicates as a stable, low copy number episome. A second marker, SUP4-o, which can be selected either for or against, was then added to produce the viral DNA molecule that serves as the target for homologous recombination. Recombination occurs when yeast cells containing this viral DNA are transfected with a transfer plasmid designed to replace the SUP4-o marker with a sequence encoding the foreign protein of interest. Yeast transformants are counterselected for the absence of SUP4-o and used as a source of the recombinant viral DNA, which can be isolated and transfected onto insect cells to produce BEVs.

2. Enzymatic Recombination *in Vitro*

Another alternative method that can be used to produce BEVs is enzymatic recombination with the Cre-lox system (Peakman *et al.*, 1992b). These investigators constructed a recombinant baculovirus and transfer plasmid with LoxP sites that can mediate site-specific *in vitro* recombination by purified Cre recombinase. This method produces recombinant viral DNAs at a frequency of up to 50%, and BEVs can be isolated by cotransfecting insect cells and resolving the recombinant and parental progeny by plaque assay. BEVs can be identified by the presence of a *lacZ* marker donated by the transfer plasmid. This approach provided a way to produce

BEVs with high frequency before gapped viral DNAs became available. It also circumvents the single-crossover problem associated with conventional allelic replacements *in vivo*.

3. Site-Specific Transposition in *E. coli*

Site-specific transposition of a foreign coding sequence into the baculovirus genome is another method that can be used for highly efficient and rapid production of BEVs (Luckow *et al.*, 1993). This approach requires the use of a "bacmid," which is a recombinant baculoviral DNA containing a mini-F replicon, selectable marker, and Tn7 transposition site. Thus, a bacmid can replicate autonomously in *E. coli* and strains harboring the bacmid and a helper plasmid that encodes the Tn7 transposase functions can be used to produce BEVs. This is done by introducing a donor plasmid containing the desired cDNA sequence and a second selectable marker positioned between the left and right arms of Tn7. The transposition functions provided by the helper plasmid will move the cDNA and selectable marker from the donor plasmid to the bacmid. This produces a recombinant bacmid that can be selected, isolated, and transfected onto cultured insect cells to produce BEVs. The bacmid system is available commercially from Life Technologies (Gaithersburg). In addition, a modified bacmid system that uses an *E. coli* host strain with an occupied Tn7 attachment site and a temperature-sensitive selection step has been described (Leusch *et al.*, 1995).

4. Cloning Directly into the Baculovirus Genome

Finally, baculovirus DNAs designed for the direct insertion of a foreign gene into unique cloning sites have been described (Lu and Miller, 1996; Ernst *et al.*, 1994). Due to their large size (~130 kb), it is difficult to construct viral DNAs with unique restriction sites and to efficiently ligate linearized viral DNA with a foreign DNA fragment *in vitro*. However, due to the replicative advantage enjoyed by the recircularized viral DNA, direct cloning can be used successfully for highly efficient production of BEVs. Direct cloning of cDNAs into baculovirus vectors containing an appropriate promoter upstream of the insertion site has been proposed as a way to produce baculovirus-based cDNA expression libraries (Lu and Miller, 1996).

IV. USING BEVs

A. Expression Levels

The ability to produce foreign proteins at exceptionally high levels is one of the hallmark features of the BEV system. These high production levels mainly reflect the ability of the transcriptional complex and *polh* promoter to produce large pools of mRNA during the very late phase of infection. Thus, BEVs that express a foreign coding sequence under the control of a different promoter will usually provide lower production levels. The production levels provided by *polh*-based BEVs are usually measured in the hundreds of milligrams of recombinant protein per liter of infected cells (about 1×10^9 cells). However, this crude generalization must be immediately qualified with a reminder that production levels vary widely from protein to protein. Secretory pathway proteins are produced at lower levels, often only 1 to 5 mg/liter of infected cells. The block to high-level production of secretory pathway proteins is probably posttranscriptional, as BEVs encoding these proteins can produce large amounts of the foreign mRNA (Jarvis *et al.*, 1993). However, the nature of this block is unknown. Possibilities include saturation of secretory pathway functions, malfolding of newly synthesized proteins, and adverse effects of baculovirus infection (Jarvis *et al.*, 1993; Jarvis and Summers, 1989). It is unlikely that high-level production of all secretory pathway proteins is blocked at one key step, however, because this system can produce some secretory pathway proteins at high levels. Finally, even though they produce less mRNA than the *polh* promoter, alternative promoters can sometimes produce larger amounts of biologically active secretory pathway proteins, as discussed above.

B. Hosts

Besides the protein-to-protein variation, production levels in the BEV system depend on the host being used (Hink *et al.*, 1991). Historically, the most widely used hosts were the established insect cell lines IPLB-Sf21-AE ("Sf21"), originally derived from *Spodoptera frugiperda* ovaries (Vaughn *et al.*, 1977), and its clonal derivative, Sf9 (Summers and Smith, 1987). However, in 1992 it was reported that BTI-TN-5B1-4, an insect cell line derived from *Trichoplusia ni* eggs, provided higher levels of foreign protein production (Wickham *et al.*, 1992). Subsequent studies on a larger sample of recombinant proteins generally supported this claim. As a result, BTI-TN-5B1-4 cells, more commonly known as High Five cells (a trade name of InVitroGen), have become another widely used host for BEVs. Recent data suggest that subclones of yet another established insect cell line, BTI-EaA, derived from *Estigmene acrea* (Granados and Naughton, 1975), can provide more extensive N-glycosylation of foreign glycoproteins expressed by BEVs (Ogonah *et al.*, 1996). Thus, in addition to providing different levels of foreign protein production, different cell lines also can provide different levels of processing in the BEV system.

Insect cells must be perfectly healthy to provide optimal levels of BEV-mediated foreign protein production. This requires high-quality growth media, routine subculturing, and careful monitoring of cell doubling times and viabilities. Insect cell growth media are available from several different companies and serum-free media have been developed (Maiorella *et al.*,

1988). Insect cell lines may be grown as monolayer cultures in T-flasks, as suspension cultures in shake flasks or spinner flasks, or as large-scale cultures in stirred tank or airlift bioreactors (Shuler *et al.*, 1995; Weiss *et al.*, 1995a,b). The development of scaleup methods was challenging because insect cells have an unusually high oxygen demand and large cultures must be aerated, but aeration can damage the cells because they are extremely sensitive to shear stress. This problem was solved when it was found that a nonionic surfactant, Pluronic F68, could protect insect cells from shear stress (Maiorella *et al.*, 1988; Murhammer and Goochee, 1988). The physical and nutritive conditions needed for optimal foreign protein production by large-scale insect cell cultures have been studied extensively (Shuler *et al.*, 1995; Taticek *et al.*, 1994; Tramper *et al.*, 1993). Among many other interesting results, it has been found that perfusion techniques can be used to obtain extremely high-density cultures ($>5 \times 10^7$ cells/ml) that can produce larger amounts of recombinant protein (Deutschmann and Jager, 1994) and that oxygen demand rises even higher after baculovirus infection (Wong *et al.*, 1994).

Two disadvantages associated with the use of established insect cell lines as hosts for foreign protein production by BEVs are that animal cell culture media are expensive and individual insect cell lines might not have all the protein processing capabilities found in higher eukaryotes (see Section IV.C). Sometimes, these problems can be circumvented by using insect larvae as an alternative host. BmNPV has been extensively developed for this purpose and is commonly used as a BEV to express foreign proteins in silkworm larvae (Maeda, 1989; Maeda *et al.*, 1985). Methods for rearing and infecting larvae have been described (Choudary *et al.*, 1995; O'Reilly *et al.*, 1992). However, most investigators use established cell lines because this method is more familiar and it is easier to purify recombinant proteins from cultured cells than from insect larvae. Moreover, although larvae are usually cheaper to cultivate than established insect cell lines, their use does not always solve protein processing problems (Pajot-Augy, 1995).

C. Protein Processing

The other hallmark feature of the BEV system is its ability to process proteins. Biosynthesis of many eukaryotic proteins includes co- and/or post-translational processing, which can be critical for protein solubility and function, and insect cells have most of the protein processing pathways associated with higher eukaryotes. In fact, development of the BEV system has contributed immensely to our knowledge of protein processing in insect cells. Most reports of foreign protein production in the BEV system include structural analyses of the end product and its covalent chemical modifications. These are often simple qualitative analyses designed to determine if a specific modification took place and whether there are gross differences in the structures of the recombinant and native proteins (e.g., differences in electrophoretic mobility). However, some studies include more detailed structural analyses, which have provided extremely valuable information on protein processing pathways in insect cells. This information comes with a caveat, though, which is that it must be applied specifically to the protein processing pathway of a specific baculovirus-infected host. This is important because baculovirus infection probably alters cellular protein processing pathways in various ways (Velardo *et al.*, 1993; Davidson *et al.*, 1991; Murphy *et al.*, 1990; Jarvis and Summers, 1989) and the specific protein processing capabilities of different hosts, including different insect cell lines, can be quite different (Ogonah *et al.*, 1996; Kuroda *et al.*, 1989). The chemical modifications found on many different recombinant proteins have been cataloged elsewhere (Luckow, 1991; Luckow and Summers, 1988). A few selected results that have provided the most definitive information are summarized below.

1. Proteolytic Cleavages

Many secretory pathway proteins have short, N-terminal signal peptides that are proteolytically cleaved during biosynthesis. Signal peptide cleavage has been carefully evaluated by directly sequencing the N-termini of many different foreign proteins produced in the BEV system. The results indicate that insect cells can accurately remove native signal peptides of plant or animal secretory pathway proteins. Insect cells also can accurately cleave heterologous signal peptides that are encoded by some transfer plasmids and used to direct secretion (discussed above). Thus, insect cells clearly have secretory signal peptide recognition and cleavage machinery. However, the relationship between this machinery and that of other eukaryotes remains unclear, because some foreign signal peptides are nonfunctional in insect cells (Pajot-Augy *et al.*, 1995) and insect cell-derived signal peptides sometimes but not always provide better secretion (discussed above).

Many eukaryotic proteins also have prosequences, which are usually short amino acids sequences that need to be removed to convert a protein to its biologically active form. Prosequences may or may not be accurately and efficiently cleaved from foreign proteins produced in the BEV system. For example, the N-terminal prosequence of human tissue plasminogen activator was efficiently cleaved (Furlong *et al.*, 1988), but the N-terminal prosequences of other proteins, including a frog α-amidating enzyme (Suzuki *et al.*, 1990), were not. The C-terminal prosequence of the gastrin-releasing peptide precursor was cleaved at the proper site, but also at several other sites (Lebacq-Verheyden *et al.*, 1988). These problems probably reflect limiting levels of the appropriate proteases in baculovirus-infected insect cells, as well as differences in their substrate specificities.

Proteolytic cleavages at internal dibasic amino acid sites occur inefficiently in cultured insect cells. For example, influenza virus hemagglutinin

(Kuroda *et al.*, 1986) and HIV gp160 (Hu *et al.*, 1987) were cleaved slowly and/or inefficiently in the BEV system. This suggests that insect cell lines have limiting amounts of the kex-2 family of processing proteases needed for these internal cleavages. This conclusion is supported by the findings that hemagglutinin was cleaved more efficiently in insect larvae (Kuroda *et al.*, 1989) and gp160 was cleaved more efficiently when furin is coexpressed in this system (Yamshchikov *et al.*, 1995).

2. Glycosylation

Many eukaryotic proteins are modified by the covalent addition of carbohydrate side chains. There are three well-defined protein glycosylation pathways in eukaryotic cells: *N*-glycosylation, *O*-glycosylation, and addition of *O*-linked *N*-acetylglucosamine (*O*-GlcNAc) (Montreuil *et al.*, 1995; Hart, 1992). Insect cells have all of these pathways, but they are not necessarily the same as those found in higher eukaryotes (Marz *et al.*, 1995).

a. N-Glycosylation

N-glycosylation begins with cotranslational addition of a preassembled oligosaccharide precursor, $Glc_3Man_9GlcNAc_2$, to a nascent polypeptide chain (Kornfeld and Kornfeld, 1985). The glucose residues are removed by glucosidases I and II to produce a "high-mannose" side chain, which can be the finished end product. In mammalian cells, however, high-mannose side chains can be converted to "complex" structures. This requires class I α-mannosidases, which remove the α-1, 2-linked mannose residues to produce $Man_5GlcNAc_2$, GlcNAc transferase I, which adds a GlcNAc residue to produce $GlcNAcMan_5GlcNAc_2$, and α-mannosidase II, which removes two mannose residues to produce $GlcNAcMan_3GlcNAc_2$. This key intermediate is then elongated by glycosyltransferases that add GlcNAc, galactose, fucose, and sialic acid residues to complete the complex side chain.

The overall conclusion from most structural data on *N*-linked oligosaccharides from glycoproteins produced in insect cells, together with biochemical data on the processing activities of these cells, is that the insect cell *N*-glycosylation pathway is a truncated version of the mammalian pathway (Jarvis and Finn, 1995; Marz *et al.*, 1995; and references therein). Insect cells clearly can add *N*-linked precursors to newly synthesized proteins and convert them to trimmed and fucosylated structures. But, these cells usually do not elongate the side chains further to produce complex structures. Interestingly, insect cells have GlcNAc transferase I and II activities (Altmann *et al.*, 1993; Velardo *et al.*, 1993), indicating that they should be able to add GlcNAc residues to the trimmed structures. This idea is supported by the finding that some glycoproteins produced in insect cells have *N*-linked side chains consisting of $GlcNAcMan_3GlcNAc_2$. It has been proposed that $GlcNAcMan_3GlcNAc_2$ is a transient intermediate needed for fucosylation and that the terminal GlcNAc is subsequently removed to produce the fucosylated $Man_3GlcNAc_2$ structure seen on many insect cell-derived glycoproteins. This possibility is supported by the discovery that some insect cells contain a membrane-bound *N*-acetylglucosaminidase activity (Altmann *et al.*, 1995). Alternatively, if they have the requisite glycosyltransferases, insect cells could convert $GlcNAcMan_3GlcNA_2$ to complex structures with penultimate galactose and terminal sialic acids, like those produced by mammalian cells. This possibility is supported by structural data on two foreign glycoproteins produced in BEV-infected insect cells that showed that they had *N*-linked side chains with terminal galactose (Ogonah *et al.*, 1996) or sialic acid (Davidson *et al.*, 1990).

It has been proposed that the synthesis of complex *N*-linked oligosaccharide side chains by insect cells requires induction of cellular processing enzymes resulting from baculovirus infection (Velardo *et al.*, 1993; Davidson *et al.*, 1991). However, the total glycopeptide profiles of uninfected and infected Sf9 cells are similar (Kretzschmar *et al.*, 1994), and it is generally thought that baculoviruses turn cellular gene expression off (Ooi and Miller, 1988). Another proposal is that only a select few recombinant glycoproteins can acquire complex *N*-linked side chains because of special structural properties that make them excellent substrates for extremely low levels of glycosyltransferase activities or poor substrates for the processing *N*-acetylglucosaminidase in baculovirus-infected insect cells (Jarvis and Finn, 1995,1996).

Although they might have different *N*-linked oligosaccharide side chains, most foreign glycoproteins produced in the BEV system are biologically active and antigenically authentic (Luckow and Summers, 1988). However, it has been shown that a recombinant human glycoprotein with insect cell-derived *N*-linked glycans is cleared more rapidly from the mammalian circulatory system (Sareneva *et al.*, 1993). Another problem with *N*-glycosylation in the BEV system is that it can be inefficient, particularly at later times of infection, and the nonprocessed subpopulation of the protein of interest is often insoluble and/or inactive (Jarvis *et al.*, 1996a; Murphy *et al.*, 1990).

b. O-Glycosylation

O-glycosylation occurs in the Golgi apparatus and results in the posttranslational addition of relatively small glycans *O*-linked through *N*-acetylgalactosamine (GalNAc) to serine or threonine (Montreuil *et al.*, 1995). Structural analyses of recombinant *O*-glycosylated proteins produced in the BEV system reveal that the major *O*-linked side chain consists of just the monosaccharide, GalNAc (Marz *et al.*, 1995). A subpopulation of these proteins have *O*-linked Gal-GalNAc, but the sialylated Gal-GalNAc trisaccharides produced by mammalian cells have not been detected. Further analysis showed that various insect cell lines have comparable levels of UDP-GalNAc–polypeptide GalNAc transferase activity, but different levels of UDP-Gal–GalNAc β-1,3 galactosyltransferase activity (Thompsen *et al.*,

1990). Sf9 cells have the lowest levels of the latter activity, which probably explains why GalNac is the major *O*-linked side chain produced by these cells. Overall, the *O*-glycosylation pathway in baculovirus-infected insect cells appears to be incomplete, due to the absence of sialyltransferases, and inefficient, due to limiting levels of transferase activities.

c. *O-GlcNAc*

Many proteins are posttranslationally glycosylated by a cytoplasmic enzyme that adds a single GlcNAc via *O*-linkage to serine or threonine residues (Hart, 1992). This process is reversible and probably regulates the functions of many nuclear and cytosolic proteins. The BEV system clearly can add *O*-GlcNAc to foreign proteins, as demonstrated for mammalian keratins (Ku and Omary, 1994), nucleoporins (Bailer *et al.*, 1995), and *c-myc* (Chou *et al.*, 1995), among others. However, the addition of *O*-GlcNAc to one keratin was less efficient to this system than in mammalian cells.

3. Phosphorylation

Phosphorylation is another reversible covalent chemical modification that can regulate protein function. Phosphorylation of many different foreign proteins has been documented in the BEV system. The more detailed studies have led to the general conclusion that foreign phosphoproteins can be accurately phosphorylated in this system, but phosphorylation of specific sites is sometimes inefficient. For example, the same sites were phosphorylated on SV40 large T antigen produced in SV40-infected monkey or BEV-infected insect cells, but certain serine residues recognized by nuclear kinases were relatively underphosphorylated in insect cells (Hoss *et al.*, 1990). Considering that T antigen is localized in the nucleus of BEV-infected insect cells, it appeared that the levels of these kinases might be too low to process all of the T antigen being produced by these cells. However, another factor to consider is that baculoviruses encode a protein phosphatase and kinase (Ayres *et al.*, 1994), which could alter the structures of recombinant phosphoproteins in unexpected ways. In fact, one recent study concluded that underphosphorylation of recombinant proteins in the BEV system may result from unusually high phosphatase activities rather than saturation of kinases with recombinant substrate (Fuchs *et al.*, 1995). Another study showed that recombinant kinases involved in cellular signaling were activated by site-specific phosphorylation in the absence of kinases that lie upstream in the signaling pathway, probably by baculovirus-encoded or -induced kinase activity (Kozma *et al.*, 1993).

Finally, it is worth mentioning that various growth factor receptors expressed in the BEV system are accurately tyrosine phosphorylated, since this has led people to use this system to express multiple proteins that functionally reconstitute cellular signaling pathways (Agarwal *et al.*, 1995).

4. Other Covalent Chemical Modifications

a. Acylation

Many eukaryotic proteins are modified by the addition of lipid side chains, including myristate, palmitate, isoprenoids, and glycosylphosphatidylinositols (GPIs). Studies of various foreign proteins have shown that each of these lipid modifications can occur in BEV-infected insect cells. Myristylation is dependent on protein synthesis, occurs at the expected sites, and can produce side chains with the expected structures, as shown by site-directed mutagenesis, hydroxylamine resistance, and direct structural analyses (Risinger *et al.*, 1992; Delchambre *et al.*, 1989). Palmitylation is independent of protein synthesis and can produce structurally authentic, hydroxylamine-sensitive side chains, but it can be quite inefficient in BEV-infected insect cells (Veit *et al.*, 1994; Page *et al.*, 1989). One study showed that palmitylation is sensitive to pharmacological regulation, suggesting that the insect and mammalian cell pathways are similar (Mouillac *et al.*, 1992). This conclusion is supported by the finding that palmitylation of some recombinant proteins produced in the BEV system is dependent upon prior isoprenylation, as in higher eukaryotes (Lowe *et al.*, 1992). Isoprenylation is inefficient in BEV-infected insect cells, but occurs at the same recognition sites and produces the same side chains, suggesting that prenyltransferase functions are conserved in insect cells and higher eukaryotes (Kalman *et al.*, 1995; Buss *et al.*, 1991). Finally, studies on several recombinant proteins, including human CD59 antigen (Davies and Morgan, 1993), have shown that BEV-infected insect cells can produce GPI anchors. Like many other types of protein processing provided by this system, GPI addition was inefficient, and large subpopulations of these products were unanchored and secreted into the extracellular growth medium.

b. N-Terminal Acetylation

N-Terminal protein modifications are often a nuisance in protein sequencing projects, but *N*-acetylation is sometimes required for protein function. The BEV system can produce *N*-acetylated proteins, and this system was used to show that *N*-acetylation is required for the function of α-tropomyosin (Urbancikova and Hitchcock-DeGregori, 1994). Another N-terminal modification that can occur in the BEV system is removal of an N-terminal methionine, followed by acetylation of the formerly penultimate alanine (Han *et al.*, 1995) or serine (Becker *et al.*, 1994).

c. C-Terminal Methylation

About half of the Kirsten-ras p21 protein produced in the BEV system was modified by C-terminal methylation, as well as isoprenylation (Lowe *et al.*, 1991).

d. α*-Amidation*

α-Amidation is one of the few covalent chemical modifications that does not occur in BEV-infected Sf9 (Lebacq-Verheyden *et al.*, 1988) or other insect cell lines, including High Five (Vakharia *et al.*, 1995). This covalent chemical modification results from a complex, multistep pathway that begins with proteolytic cleavage of a C-terminal prosequence, followed by the removal of additional amino acids to produce a C-terminal glycine. The glycine residue is then hydroxylated and amidated by two distinct activities. The C-terminal prosequence can be removed in BEV-infected insect cells, but α-amidation does not occur, suggesting that these cells lack a subsequent step(s) in this processing pathway. By contrast, α-amidation can occur in BEV-infected insect larvae (Hellers *et al.*, 1991).

5. Supramolecular Assembly

Supramolecular protein assembly is one of the best-documented capabilities of the BEV system. Individual proteins can assemble into higher-order structures via disulfide bond formation, which is virtually complete in some cases (Giese *et al.*, 1989), but not in others (Domingo and Trowbridge, 1988). This indicates that BEV-infected insect cells have protein disulfide isomerase activity, but perhaps too little to completely process highly expressed foreign proteins. Proteins also can assemble by noncovalent interactions in BEV-infected insect cells at rates similar to (Lanford, 1988) or lower than (Kuroda *et al.*, 1991) native rates.

Hetero-oligomeric protein complexes can be assembled by infecting insect cells with multiple BEVs (St. Angelo *et al.*, 1987) or with a single BEV encoding multiple foreign genes (Emery and Bishop, 1987). Both of these approaches have been used to produce functional IgG heterodimers with normal heavy and light chain composition in the BEV system (Hasemann and Capra, 1990). The ability of this system to produce larger, even more complex oligomeric assemblies, including viruslike particles, is well documented. Subviral particles consisting of hepatitis B virus, bluetongue virus, or rotavirus proteins were among the first to be described (Urakawa and Roy, 1988; Estes *et al.*, 1987; Kang *et al.*, 1987). This was followed by a remarkable example of protein processing in which a BEV was used to express the poliovirus genome in insect cells, and the resulting polyprotein was properly cleaved and assembled into stable, noninfectious polioviruslike particles (Urakawa *et al.*, 1989).

These early studies led to widespread use of the BEV system to express multiple viral proteins and produce many different types of viruslike particles that hold great promise as noninfectious vaccines and diagnostic reagents (Pearson and Roy, 1993). Most recently, potential multivalent vaccines have been produced by incorporating heterologous epitopes into the viruslike particles or other supramolecular protein complexes that can be assembled in this system (Garnier *et al.*, 1995b; Belyaev and Roy, 1992).

Viruslike particles containing exogenous DNA also have been produced and used to efficiently transfer DNA into eukaryotic cells, indicating that these particles could be good tools for gene therapy (Forstova *et al.*, 1995).

6. Protein Targeting

Protein targeting signals had not been widely investigated in insect systems when the BEV system was first developed, so it was not known whether insect cells would be able to recognize protein targeting signals in foreign proteins. Today, we know that insect cells can recognize heterologous targeting signals, which suggests that these cells have similar protein trafficking machinery. Foreign proteins destined for secretion or the cell surface can enter the insect cell secretory pathway and native or heterologous signal peptides can be cleaved accurately. Polarized cell surface expression of foreign proteins can occur in midgut epithelial cells of insect larvae (Kuroda *et al.*, 1989). Mitochondrial proteins localize to the mitochondria (Takagi *et al.*, 1992) and nuclear proteins localize to the nuclei of BEV-infected insect cells, and two baculovirus-encoded nuclear proteins contain nuclear targeting signals similar to those found in higher eukaryotes (Broussard *et al.*, 1996; Jarvis *et al.*, 1991). However, lysosomal enzymes produced in the BEV system lack the lysosomal targeting signal, mannose-6-phosphate, and are secreted by default (Boose *et al.*, 1990; Martin *et al.*, 1988). Biochemical assays indicate that Sf9 cells lack the phosphotransferase activity responsible for this modification (Aeed and Elhammer, 1994). A curious feature of protein trafficking in insect cells is that they secrete the intracellular domain of rabbit prolactin receptor, which lacks a typical signal peptide (Garnier *et al.*, 1995a). Furthermore, this domain can mediate secretion and ubiquitination of heterologous proteins, which suggests that lepidopteran insect cells have an unusual secretory pathway that can respond to this novel targeting signal.

7. Variation and the Art of Molecular Genetics

Structural analyses of recombinant proteins and biochemical analyses of cellular protein processing activities have produced a wealth of information on protein processing pathways in the BEV system. However, it is important to recognize the limitations of these biochemical approaches. They are indirect and provide only a retrospective view of protein processing pathways, which must be inferred from the structures of the end products or the presence or absence of processing activities. These conclusions can be confused by degradative pathways, which might alter the product of the biosynthetic pathway and lead to misinterpretations. Also, the inability to detect a processing activity clearly does not prove the absence of that activity. Conclusions drawn from structural data on any one recombinant protein need to be applied only to that protein and not to the pathway in general. These conclusions also should be applied only to the specific host that was used to produce the recombinant protein under analysis. Finally, the possible effects of baculovirus infection on the host need to be considered. Like most investigators, I broke these rules in this chapter and overgeneralized my discussion of "insect cell" protein processing capabilities!

An alternative approach that circumvents many of these problems is to use molecular genetics to isolate genes encoding insect cell processing enzymes. This makes it possible to study these genes, their expression, the conditions that influence their expression, and, ultimately, the properties of the enzymes they encode. My laboratory has begun to utilize this approach with these goals in mind. We recently reported the isolation and characterization of class I and II α-mannosidase cDNAs from Sf9 cells (Jarvis *et al.*, 1997; Kawar *et al.*, 1997) and will describe class II α-mannosidase cDNAs from two other lepidopteran insect cell lines in the near future. We hope that the combination of biochemical and molecular genetic data will provide a clearer picture of protein processing capabilities in "insect cells" and in the BEV system.

8. Summary

Overall, the biochemical evidence suggests that foreign proteins can be appropriately processed in the BEV system. However, it is important to recognize that there are some exceptions and caveats to this generalization. BEV-infected cell lines clearly lack certain protein processing capabilities, like α-amidation and lysosomal targeting. There also are clear differences in the structures of the *N*-linked glycans found on most recombinant glycoproteins produced in this system. Inefficient processing of recombinant proteins is a common problem in the BEV system. Nonetheless, this problem does not preclude the use of this expression system, because if adequate yields of a perfectly processed recombinant protein can be obtained, a high background of unprocessed material can be irrelevant. The "inefficiency" of protein processing in this system might reflect adverse effects of baculovirus infection or the inability of the cellular processing machinery to cope with the high levels of foreign gene transcription provided by BEVs. There also might be subtle differences in the protein processing machinery of insect cells and higher eukaryotes, which reduce the functional efficiency of foreign protein processing in this expression system. Theoretically, any of these protein processing problems can be addressed by metabolic engineering, which could be used to improve the BEV system, as discussed further below.

D. Selected Applications of BEV-Expressed Proteins

1. Vaccines

One of the best-recognized applications of the BEV system is vaccine production. There are far too many examples of this application to discuss here and a comprehensive list with references is available (Luckow, 1991). However, it is appropriate to summarize some general conclusions and new discoveries that have arisen from immunologic studies on recombinant proteins produced in this system. Recombinant proteins from BEV-infected insect cells are invariably recognized by antibodies produced against the corresponding native proteins, indicating that they are antigenically authentic. Moreover, recombinant proteins from the BEV system usually induce protective immunity in laboratory animals. Human immunodeficiency virus (HIV) gp160 produced in this system has been used extensively for human clinical trials and the results indicate that this product is safe and immunogenic. Unfortunately, although gp160 can induce both humoral and cell-mediated immune responses, these are usually weak, transient, and nonneutralizing. On the positive side, clinical trials on BEV-expressed recombinant gp160 showed that it has a therapeutic effect in previously infected individuals, and the new concept of "vaccine therapy" emerged from these results (Redfield *et al.*, 1991). It also was found that stronger humoral and cell-mediated immune responses could be obtained by using a combined vaccine regimen involving priming with a live recombinant vaccinia virus encoding gp160 and boosting with BEV-expressed recombinant gp160 (Cooney *et al.*, 1993; Graham *et al.*, 1993).

2. Diagnostic Tests

Another exciting application of recombinant proteins produced in the BEV system is diagnostic testing. Recombinant proteins produced in the BEV system have been used to develop diagnostic tests for many different infectious agents, including viruses, protozoa, rickettsia, and bacteria, as well as tests for human autoantibodies and cancer markers. Diagnostic testing with recombinant proteins from the BEV system has progressed beyond the developmental state, as these proteins have been used for several large epidemiological studies (e.g., Numata *et al.*, 1994).

3. Three-Dimensional Structural Analyses

The BEV system is being used with increasing frequency to produce recombinant proteins for three-dimensional structural analyses. Rat acid phosphatase was one of the first recombinant proteins to be crystallized (Vihko *et al.*, 1993), and subsequently many recombinant proteins have been crystallized and analyzed by X-ray diffraction. In addition, the three-dimensional structures of many of the viruslike particles produced in this system have been examined by X-ray crystallography (Roy, 1996; Agbandje *et al.*, 1991) and electron cryomicroscopy (Prasad *et al.*, 1994).

V. DIRECTIONS FOR THE FUTURE

A. Preventing Adverse Effects of Baculovirus Infection

BEVs ultimately kill the host that is producing the foreign protein of interest. Thus, recombinant protein production with the BEV system is a "batch" process, and fresh cells and virus must be used to produce each batch of recombinant protein. BEVs also have adverse effects on host protein processing pathways long before they kill the cells and encode at least one protease and phosphatase that might degrade or dephosphorylate the foreign protein being produced.

One way to address these problems is to use viral promoters to develop insect cell expression systems that avoid using a BEV altogether. This can be done by producing stably transformed insect cells that express a foreign gene constitutively under the control of baculovirus early promoters (Jarvis *et al.*, 1990). The production levels obtained with transformed insect cells are usually lower than those obtained by infecting insect cells with *polh*-based BEVs. However, where *polh*-based BEVs fail to produce high levels of a foreign protein, as in the case of many secretory pathway proteins, transformed insect cells can provide similar levels of recombinant product. Furthermore, stably transformed cells can produce these proteins continuously over a long time period and process them faster and more efficiently than infected cells. Thus, stably transformed insect cells hold some promise for foreign protein production, but this approach would be significantly more attractive if it could provide higher production levels. In the future, this could be accomplished by developing gene amplification methods or by using stronger promoters.

Another way to circumvent the adverse effects of baculovirus infection might be to develop BEVs that can provide high-level foreign gene expression without the cytopathic effects. One can imagine a "miniviral replicon": a baculovirus-based DNA molecule encoding only the *cis*- and *trans*-acting functions needed for autonomous replication and transcription of strong viral promoters in insect cells. This replicon would lack all other viral functions. It would replicate as an episome and produce a foreign protein(s) of interest at high levels, but it would have no adverse side effects on the host cell and would produce no progeny virus. Efforts to identify the functions required for baculovirus DNA replication and late/very late transcription are proceeding at a rapid pace (as described in preceding chapters, this volume). In the future, these efforts could lead to the development of a miniviral replicon expression vector.

B. Metabolic Engineering

Another possible reason for the relatively lower levels of secretory pathway protein production in baculovirus-infected insect cells is that host protein processing activities might be saturated. If specific cellular processing activities are limiting, it should be possible to improve the system by "metabolic engineering" to increase available levels of these activities. One approach would be to use insect cells that have been stably transformed to overexpress processing activities as modified hosts for conventional BEVs, as proposed previously (Jarvis *et al.*, 1990). However, this approach is complicated by the unexpected finding that baculovirus infection shuts down expression of integrated genes, even when their expression is controlled by a viral promoter that is normally active throughout infection (Jarvis, 1993). An alternative approach is to use modified BEVs that include and can overexpress genes encoding cellular protein processing enzymes.

This latter approach was used to try to improve the assembly of steroid receptors (Alnemri and Litwack, 1993) and immunoglobulins (Hsu *et al.*, 1994), both of which were inefficiently processed and formed insoluble protein aggregates when overexpressed in BEV-infected insect cells. Coexpression of the steroid receptors and hsp90 or hsp70 failed to enhance assembly and prevent aggregation. Coexpression of immunoglobulins with immunoglobulin binding protein produced higher levels of soluble intracellular immunogloboulin but failed to increase secretion. These results indicate that metabolic engineering of protein processing pathways in the BEV system might be less straightforward than anticipated. A better understanding of protein processing mechanisms in insect cells will minimize this problem, and the use of insect cell genes encoding endogenous protein processing activities might facilitate future metabolic engineering efforts.

Metabolic engineering of protein glycosylation pathways is another potential way to improve the BEV system. Insect cell glycosylation pathways could be modified by increasing the levels of existing processing activities or by adding new processing activities thought to be missing in these cells. Modified host cells or modified viruses could be used for either purpose. Two recent reports demonstrate the feasibility of these goals. The effect of overexpressing an existing processing activity was demonstrated when *N*-linked oligosaccharides on influenza hemagglutinin were elongated by using a BEV to coexpress mammalian GlcNAc transferase I under *polh* control (Wagner *et al.*, 1996). The effect of adding a new processing activity was demonstrated when *N*-linked oligosaccharides on the major BV envelope glycoprotein were differentially galactosylated in insect cells infected with a BEV that expressed mammalian β-1,4-galactosyltransferase earlier in infection, under *ie1* control (Jarvis *et al.*, 1996b). Immediate early BEVs can be used to modify host cell protein processing pathways before the protein of interest is expressed, which should provide more efficient processing than temporal coexpression.

ACKNOWLEDGMENTS. The author gratefully acknowledges the support of the National Institutes of Health (GM49734), the USDA NRICGP (95-37302-1921), the Texas Advanced Research and Technology Programs (999902-039 and 999902-027), and DuPont Agricultural Products.

REFERENCES

Adams, J. R., and McClintock, J. T., 1991, Nuclear polyhedrosis viruses of insects, in: *Atlas of Invertebrate Viruses* (J. R. Adams and J. R. Bonami, eds.), pp. 87–204, CRC Press, Boca Raton, FL.

Adang, M. J., and Miller, L. K., 1982, Molecular cloning of DNA complementary to mRNA of the baculovirus *Autographa californica* nuclear polyhedrosis virus: Location and gene products of RNA transcripts found late in infection, *J. Virol.* **44:**782.

Aeed, P. A., and Elhammer, A. P., 1994, Glycosylation of recombinant prorenin in insect cells: The insect cell line Sf9 does not express the mannose 6-phosphate recognition signal, *Biochemistry* **33:**8793.

Agarwal, S., Corbley, M. J., and Roberts, T. M., 1995, Reconstitution of signal transduction from the membrane to the nucleus in a baculovirus expression system: Activation of Raf-1 leads to hypermodification of *c-jun* and *c-fos* via multiple pathways, *Oncogene* **11:**427.

Agbandje, M., McKenna, R., Rossmann, M. G., Kajigaya, S., and Young, N. S., 1991, Preliminary X-ray crystallographic investigation of human parvovirus B19, *Virology* **184:**170.

Alnemri, E. S., and Litwack, G., 1993, The steroid binding domain influences intracellular solubility of the baculovirus overexpressed glucocorticoid and mineralocorticoid receptors, *Biochemistry* **32:**5387.

Altmann, F., Kornfeld, G., Dalik, T., Staudacher, E., and Glossl, J., 1993, Processing of asparagine-linked oligosaccharides in insect cells. *N*-acetylglucosaminyltransferase I and II activities in cultured lepidopteran cells, *Glycobiology* **3:**619.

Altmann, F., Schwihla, H., Staudacher, E., Glossl, J., and Marz, L., 1995, Insect cells contain an unusual, membrane-bound β-*N*-acetylglucosaminidase probably involved in the processing of protein *N*-glycans, *J. Biol. Chem.* **270:**17344.

Andersons, D., Engstrom, A., Josephson, S., Hansson, L., and Steiner, H., 1991, Biologically active and amidated cecropin produced in a baculovirus expression system from a fusion construct containing the antibody-binding part of protein A, *Biochem. J.* **280:**219.

Arp, J., Ford, C. M., Palker, T. J., King, E. E., and Dekaban, G. A., 1993, Expression and immunogenicity of the entire human T cell leukaemia virus type I envelope protein produced in a baculovirus system, *J. Gen. Virol.* **74:**211.

Ayres, M. D., Howard, S. C., Kuzio, J., Lopez-Ferber, M., and Possee, R. D., 1994, The complete DNA sequence of *Autographa californica* nuclear polyhedrosis virus, *Virology* **202:**586.

Bailer, S. M., Berlin, W. K., Starr, C. M., and Hanover, J. A., 1995, Characterization of nuclear pore protein p62 produced using baculovirus, *Protein Expr. Purif.* **6:**546.

Beames, B., Braunagel, S., Summers, M. D., and Lanford, R. E., 1991, Polyhedrin initiator codon altered to AUU yields unexpected fusion protein from a baculovirus vector, *Biotechniques* **11:**378.

Becker, G. W., Miller, J. R., Kovacevic, S., Ellis, R. M., Louis, A. I., Small, J. S., Stark, D. H., Roberts, E. F., Wyrick, T. K., Hoskins, J., Chiou, X. G., Sharp, J. D., McClure, D. B., Riggin, R. M., and Kramer, R. M., 1994, Characterization by electrospray mass spectrometry of human Ca^{2+}-sensitive cytosolic phospholipase A2 produced in baculovirus-infected insect cells, *Bio/Technology* **12**:69.

Belyaev, A. S., and Roy, P., 1992, Presentation of hepatitis B virus preS2 epitope on bluetongue core-like particles, *Virology* **190:**840.

Belyaev, A. S., Hails, R. S., and Roy, P., 1995, High-level expression of five foreign genes by a single recombinant baculovirus, *Gene* **156:**229.

Beniya, H., Funk, C. J., Rohrmann, G. F., and Weaver, R. F., 1996, Purification of a virus-induced RNA polymerase from *Autographa californica* nuclear polyhedrosis virus-infected *Spodoptera frugiperda* cells that accurately initiates late and very late transcription in vitro, *Virology* **216:**12.

Bishop, D. H. L., Novy, R., and Mierendorf, R., 1995, The BacVector system: Simplified cloning and protein expression using novel baculovirus vectors, *inNovations* **4:**1.

Blissard, G. W., and Rohrmann, G. F., 1990, Baculovirus diversity and molecular biology, *Annu. Rev. Entomol.* **35:**127.

Bonning, B. C., Roelvink, P. W., Vlak, J. M., Possee, R. D., and Hammock, B. D., 1994, Superior expression of juvenile hormone esterase and beta-galactosidase from the basic protein promoter of *Autographa californica* nuclear polyhedrosis virus compared to the p10 protein and polyhedrin promoters, *J. Gen. Virol.* **75:**1551.

Boose, J. A., Tifft, C. J., Proia, R. L., and Myerowitz, R., 1990, Synthesis of a human lysosomal enzyme, beta-hexosaminidase B, using the baculovirus expression system, *Protein Expr. Purif.* **1:**111.

Boublik, Y., Di Bonito, P., and Jones, I. M., 1995, Eukaryotic virus display: Engineering the major surface glycoprotein of the *Autographa californica* nuclear polyhedrosis virus (AcNPV) for the presentation of foreign proteins on the virus surface, *Bio/Technology* **13:**1079.

Bozon, V., Remy, J. J., Pajot-Augy, E., Couture, L., Biache, G., Severini, M., and Salesse, R., 1995, Influence of promoter and signal peptide on the expression and secretion of recombinant porcine LH extracellular domain in baculovirus/lepidopteran cells or the caterpillar system, *J. Mol. Endocrinol.* **14:**277.

Broussard, D. R., Guarino, L. A., and Jarvis, D. L., 1996, Mapping functional domains in AcMNPV pp31, *Virology* **222:**318.

Buss, J. E., Quilliam, L. A., Kato, K., Casey, P. J., Solski, P. A., Wong, G., Clark, R., McCormick, F., Bokoch, G. M., and Der, C. J., 1991, The COOH-terminal domain of the Rap1A (Krev-1) protein is isoprenylated and supports transformation by an H-Ras:Rap1A chimeric protein, *Mol. Cell. Biol.* **11:**1523.

Capone, J., 1989, Screening recombinant baculovirus plaques *in situ* with antibody probes, *Genet. Anal. Tech.* **6:**62.

Chazenbalk, G. D., and Rapoport, B., 1995, Expression of the extracellular domain of the thyrotropin receptor in the baculovirus system using a promoter active earlier than the polyhedrin promoter. Implications for the expression of functional highly glycosylated proteins, *J. Biol. Chem.* **270:**1543.

Chisholm, G. E., and Henner, D. J., 1988, Multiple early transcripts and splicing of the *Autographa californica* nuclear polyhedrosis virus IE-1 gene, *J. Virol.* **62:**3193.

Chou, T. Y., Dang, C. V., and Hart, G. W., 1995, Glycosylation of the c-Myc transactivation domain, *Proc. Natl. Acad. Sci. USA* **92:**4417.

Choudary, P. V., Kamita, S. G., and Maeda, S., 1995, Expression of foreign genes in *Bombyx mori* larvae using baculovirus vectors, in: *Baculovirus Expression Protocols* (C. D. Richardson ed.), pp. 243–264, Humana Press, Clifton, NJ.

Cooney, E. L., McElrath, M. J., Corey, L., Hu, S. L., Collier, A. C., Arditti, D., Hoffman, M., Coombs, R. W., Smith, G. E., and Greenberg, P. D., 1993, Enhanced immunity to human immunodeficiency virus (HIV) envelope elicited by a combined vaccine regimen consisting of priming with a vaccinia recombinant expressing HIV envelope and boosting with gp160 protein, *Proc. Natl. Acad. Sci. USA* **90:**1882.

Davidson, D. J., Fraser, M. J., and Castellino, F. J., 1990, Oligosaccharide processing in the expression of human plasminogen cDNA by lepidopteran insect (*Spodoptera frugiperda*) cells, *Biochemistry* **29:**5584.

Davidson, D. J., Bretthauer, R. K., and Castellino, F. J., 1991, alpha-Mannosidase-catalyzed trimming of high-mannose glycans in noninfected and baculovirus-infected *Spodoptera frugiperda* cells (IPLB-SF-21AE). A possible contributing regulatory mechanism for assembly of complex-type oligosaccharides in infected cells, *Biochemistry* **30:**9811.

Davies, A., and Morgan, B. P., 1993, Expression of the glycosylphosphatidylinositol-linked complement-inhibiting protein CD59 antigen in insect cells using a baculovirus vector, *Biochem. J.* **295:**889.

Davrinche, C., Pasquier, D., Cerutti, M., Serradell, L., Clement, D., Devauchelle, G., Michelson, S., and Davignon, J. L., 1993, Expression of human cytomegalovirus immediate early protein IE1 in insect cells: Splicing of RNA and recognition by CD4+ T-cell clones, *Biochem. Biophys. Res. Commun.* **195:**469.

Delchambre, M., Gheysen, D., Thines, D., Thiriart, C., Jacobs, E., Verdin, E., Horth, M., Burny, A., and Bex, F., 1989, The GAG precursor of simian immunodeficiency virus assembles into virus-like particles, *EMBO J.* **8:**2653.

Deutschmann, S. M., and Jager, V., 1994, Optimization of the growth conditions of Sf21 insect cells for high-density perfusion culture in stirred-tank bioreactors, *Enzyme Microb. Tech.* **16:**506.

Domingo, D. L., and Trowbridge, I. S., 1988, Characterization of the human transferrin receptor produced in a baculovirus expression system, *J. Biol. Chem.* **263:**13386.

Emery, V. C., and Bishop, D. H. L., 1987, The development of multiple expression vectors for high level synthesis of eukaryotic proteins: Expression of LCMV-N and AcNPV polyhedrin protein by a recombinant baculovirus, *Protein Eng.* **1:**359.

Engelhard, E. K., Kam-Morgan, L. N., Washburn, J. O., and Volkman, L. E., 1994, The insect tracheal system: A conduit for the systemic spread of *Autographa californica* M nuclear polyhedrosis virus, *Proc. Natl. Acad. Sci. USA* **91:**3224.

Ernst, W. J., Grabherr, R. M., and Katinger, H. W., 1994, Direct cloning into the *Autographa californica* nuclear polyhedrosis virus for generation of recombinant baculoviruses, *Nucleic Acids Res.* **22:**2855.

Estes, M. K., Crawford, S. E., Penaranda, M. E., Petrie, B. L., Burns, J. W., Chan, W. K., Ericson, B., Smith, G. E., and Summers, M. D., 1987, Synthesis and immunogenicity of the rotavirus major capsid antigen using a baculovirus expression system, *J. Virol.* **61:**1488.

Ford, C. F., Suominen, I., and Glatz, C. E., 1991, Fusion tails for the recovery and purification of recombinant proteins, *Protein Expr. Purif.* **2:**95.

Forstova, J., Krauzewicz, N., Sandig, V., Elliott, J., Palkova, Z., Strauss, M., and Griffin, B. E., 1995, Polyoma virus pseudocapsids as efficient carriers of heterologous DNA into mammalian cells, *Hum. Gene Ther.* **6:**297.

Fuchs, B., Hecker, D., and Scheidtmann, K. H., 1995, Phosphorylation studies on rat p53 using the baculovirus expression system. Manipulation of the phosphorylation state with okadaic acid and influence on DNA binding, *Eur. J. Biochem.* **228:**625.

Fuchs, L. Y., Woods, M. S., and Weaver, R. F., 1983, Viral transcription during *Autographa californica* nuclear polyhedrosis virus infection: A novel RNA polymerase induced in infected *Spodoptera frugiperda* cells, *J. Virol.* **48:**641.

Furlong, A. M., Thomsen, D. R., Marotti, K. R., Post, L. E., and Sharma, S. K., 1988, Active human tissue plasminogen activator secreted from insect cells using a baculovirus vector, *Biotech. Appl. Biochem.* **10:**454.

Garnier, L., Cahoreau, C., Devauchelle, G., and Cerutti, M., 1995a, The intracellular domain of the rabbit prolactin receptor is able to promote the secretion of a passenger protein via an unusual secretory pathway in lepidopteran cells, *Bio/Technology* **13:**1101.

Garnier, L., Ravallec, M., Blanchard, P., Chaabihi, H., Bossy, J. P., Devauchelle, G., Jestin, A., and Cerutti, M., 1995b, Incorporation of pseudorabies virus gD into human immunodeficiency virus type 1 Gag particles produced in baculovirus-infected cells, *J. Virol.* **69:**4060.

Giese, N., May-Siroff, M., LaRochelle, W. J., van Wyke Coelingh, K., and Aaronson, S. A., 1989, Expression and purification of biologically active *v-sis*/platelet-derived growth factor B protein by using a baculovirus vector system, *J. Virol.* **63:**3080.

Godeau, F., Saucier, C., and Kourilsky, P., 1992, Replication inhibition by nucleotide analogues of a recombinant *Autographa californica* multicapsid nuclear polyhedrosis virus harboring the herpes thymidine kinase gene driven by the IE-1(0) promoter: A new way to select recombinant baculoviruses, *Nucleic Acids Res.* **20:**6239.

Graham, B. S., Matthews, T. J., Belshe, R. B., Clements, M. L., Dolin, R., Wright, P. F., Gorse, G. J., Schwartz, D. H., Keefer, M. C., Bolognesi, D. P., *et al.*, 1993, Augmentation of human immunodeficiency virus type 1 neutralizing antibody by priming with gp160 recombinant vaccinia and boosting with rgp160 in vaccinia-naive adults. The NIAID AIDS Vaccine Clinical Trials Network, *J. Infect. Dis.* **167**:533.

Granados, R. R., and Naughton, M., 1975, Development of *Amsacta moorei* entomopoxvirus in ovarian and haemocyte cultures from *Estigmene acrea* larvae, *Intervirology* **5**:62.

Guarino, L. A., and Smith, M. W., 1990, Nucleotide sequence of the 39K gene region of *Autographa californica* nuclear polyhedrosis virus, *Virology* **179**:1.

Guarino, L. A., and Summers, M. D., 1987, Nucleotide sequence and temporal expression of a baculovirus regulatory gene, *J. Virol.* **61**:2091.

Han, B. D., Livingstone, L. R., Pasek, D. A., Yablonski, M. J., and Jones, M. E., 1995, Human uridine monophosphate synthase: Baculovirus expression, immunoaffinity column purification and characterization of the acetylated amino terminus, *Biochemistry* **34**:10835.

Hart, G. W., 1992, Glycosylation, *Curr. Opin. Cell Biol.* **4**:1017.

Hartig, P. C., Cardon, M. C., and Kawanishi, C. Y., 1992, Effect on baculovirus on selected vertebrate cells, *Dev. Biol. Stand.* **76**:313.

Hasemann, C. A., and Capra, J. D., 1990, High-level production of a functional immunoglobulin heterodimer in a baculovirus expression system, *Proc. Natl. Acad. Sci. USA* **87**:3942.

Hawtin, R. E., Arnold, K., Ayres, M. D., Zanotto, P. M., Howard, S. C., Gooday, G. W., Chappell, L. H., Kitts, P. A., King, L. A., and Possee, R. D., 1995, Identification and preliminary characterization of a chitinase gene in the *Autographa californica* nuclear polyhedrosis virus genome, *Virology* **212**:673.

Hellers, M., Gunne, H., and Steiner, H., 1991, Expression of posttranslational processing of preprocecropin A using a baculovirus vector, *Eur. J. Biochem.* **199**:435.

Hill-Perkins, M. S., and Possee, R. D., 1990, A baculovirus expression vector derived from the basic protein promoter of *Autographa californica* nuclear polyhedrosis virus, *J. Gen. Virol.* **71**:971.

Hills, D., and Crane-Robinson, C., 1995, Baculovirus expression of human basic fibroblast growth factor from a synthetic gene: Role of the Kozak consensus and comparison with bacterial expression, *Biochim. Biophys. Acta* **1260**:14.

Hink, W. F., Thomsen, D. R., Davidson, D. J., Meyer, A. L., and Castellino, F. J., 1991, Expression of three recombinant proteins using baculovirus vectors in 23 insect cell lines, *Biotech. Progr.* **7**:9.

Hooft van Iddekinge, B. J. L., Smith, G. E., and Summers, M. D., 1983, Nucleotide sequence of the polyhedrin gene of *Autographa californica* nuclear polyhedrosis virus, *Virology* **131**:561.

Hoss, A., Moarefi, I., Scheidtmann, K. H., Cisek, L. J., Corden, J. L., Dornreiter, I., Arthur, A. K., and Fanning, E., 1990, Altered phosphorylation pattern of simian virus 40 T antigen expressed in insect cells by using a baculovirus vector, *J. Virol.* **64**:4799.

Hsu, T. A., Eiden, J. J., Bourgarel, P., Meo, T., and Betenbaugh, M. J., 1994, Effects of co-expressing chaperone BiP on functional antibody production in the baculovirus system, *Protein Expr. Purif.* **5**:595.

Hu, S.-L., Kosowski, S. G., and Schaaf, K. F., 1987, Expression of envelope glycoproteins of human immunodeficiency virus by an insect virus vector, *J. Virol.* **61**:3617.

Iatrou, K., Meidinger, R. G., and Goldsmith, M. R., 1989, Recombinant baculoviruses as vectors for identifying proteins encoded by intron-containing members of complex multigene families, *Proc. Natl. Acad. Sci. USA* **86**:9129.

Jarvis, D. L., 1993, Effects of baculovirus infection on IE1-mediated foreign gene expression is stably transformed insect cells, *J. Virol.* **67**:2583.

Jarvis, D. L., and Finn, E. E., 1995, Biochemical analysis of the *N*-glycosylation pathway in baculovirus-infected lepidopteran insect cells, *Virology* **212**:500.

Jarvis, D. L., and Finn, E. E., 1996, Modifying the insect cell *N*-glycosylation pathway with immediate early baculovirus expression vectors, *Nature-Biotechnology* **14**:1288.

Jarvis, D. L., and Summers, M. D., 1989, Glycosylation and secretion of human tissue plasminogen activator in recombinant baculovirus-infected insect cells, *Mol. Cell. Biol.* **9**:214.

Jarvis, D. L., Fleming, J. A., Kovacs, G. R., Summers, M. D., and Guarino, L. A., 1990, Use of early baculovirus promoters for continuous expression and efficient processing of foreign gene products in stably transformed lepidopteran cells, *Bio/Technology* **8**:950.

Jarvis, D. L., Bohlmeyer, D. A., and Garcia, J. A., 1991, Requirements for nuclear localization and supramolecular assembly of a baculovirus polyhedrin protein, *Virology* **185**:795.

Jarvis, D. L., Summers, M. D., Garcia, A., Jr., and Bohlmeyer, D. A., 1993, Influence of different signal peptides and prosequences on expression and secretion of human tissue plasminogen activator in the baculovirus system, *J. Biol. Chem.* **268**:16754.

Jarvis, D. L., Reilly, L. M., Hoover, K., Hammock, B., and Guarino, L. A., 1996a, Construction and characterization of immediate-early baculovirus insecticides, *Biol. Control* **7**:228.

Jarvis, D. L., Weinkauf, C., and Guarino, L. A., 1996b, Immediate early baculovirus vectors for foreign gene expression in transformed or infected insect cells, *Protein Expr. Purif.* **8**:191.

Jarvis, D. L., Bohlmeyer, D. A., Liao, Y.-F., Lomax, K., Merkle, R., Weinkauf, C., and Moremen, K. W., 1997, Isolation and characterization of a class II alpha-mannosidase cDNA from lepidopteran insect cells, *Glycobiology* **7**:113.

Jeang, K. T., Holmgren-Koning, M., and Khoury, G., 1987, A baculovirus vector can express intron-containing genes, *J. Virol.* **61**:1761.

Johnson, R., Meidinger, R. G., and Iatrou, K., 1992, A cellular promoter-based expression cassette for generating recombinant baculoviruses directing rapid expression of passenger genes in infected insects, *Virology* **190**:815.

Kalman, V. K., Erdman, R. A., Maltese, W. A., and Robishaw, J. D., 1995, Regions outside of the CAAX motif influence the specificity of prenylation of G protein gamma subunits, *J. Biol. Chem.* **270**:14835.

Kang, C. Y., Bishop, D. H. L., Seo, J. S., Matsuura, Y., and Choe, M., 1987, Secretion of particles of hepatitis B surface antigen from insect cells using a baculovirus vector, *J. Gen. Virol.* **68**:2607.

Kawar, Z., Herscovics, A., and Jarvis, D. L., 1997, Isolation and characterization of an a-1,2-mannosidase from the lepidopteran insect cell line Sf9, *Glycobiology,* **7**:433.

Keddie, B. A., Aponte, G. W., and Volkman, L. E., 1989, The pathway of infection of *Autographa californica* nuclear polyhedrosis virus in an insect host, *Science* **243**:1728.

Kim, D., and Weaver, R. F., 1993, Transcription mapping and functional analysis of the protein tyrosine/serine phosphatase (PTPase) gene of the *Autographa californica* nuclear polyhedrosis virus, *Virology* **195**:587.

King, L. A., and Possee, R. D., 1992, *The Baculovirus System: A Laboratory Guide,* Chapman and Hall, London.

Kitts, P. A., and Possee, R. D., 1993, A method for producing recombinant baculovirus expression vectors at high frequency, *Biotechniques* **14**:810.

Kitts, P. A., Ayres, M. D., and Possee, R. D., 1990, Linearization of baculovirus DNA enhances the recovery of recombinant virus expression vectors, *Nucleic Acids Res.* **18**:5667.

Kornfeld, R., and Kornfeld, S., 1985, Assembly of asparagine-linked oligosaccharides, *Annu. Rev. Biochem.* **54**:631.

Kovacs, G. R., Guarino, L. A., Graham, B. L., and Summers, M. D., 1991, Identification of spliced baculovirus RNAs expressed late in infection, *Virology* **185**:633.

Kozak, M., 1986, Point mutations define a sequence flanking the AUG initiator codon that modules translation by eucaryotic ribosomes, *Cell* **44**:283.

Kozma, S. C., McGlynn, E., Siegmann, M., Reinhard, C., Ferrari, S., and Thomas, G., 1993, Active baculovirus recombinant p70s6k and p85s6k produced as a function of the infectious response, *J. Biol. Chem.* **268**:7134.

Kretzschmar, E., Geyer, R., and Klenk, H. D., 1994, Baculovirus infection does not alter *N*-glycosylation in *Spodoptera frugiperda* cells, *Biol. Chem. Hoppe-Seyler* **375**:23.

Ku, N. O., and Omary, M. B., 1994, Expression, glycosylation, and phosphorylation of human keratins 8 and 18 in insect cells, *Exp. Cell Res.* **211:**24.

Kuhn, S., and Zipfel, P. F., 1995, The baculovirus expression vector pBSV-8His directs secretion of histidine-tagged proteins, *Gene* **162:**225.

Kuroda, K., Hauser, C., Rott, R., Klenk, H. D., and Doerfler, W., 1986, Expression of the influenza virus haemagglutinin in insect cells by a baculovirus vector, *EMBO J.* **5:**1359.

Kuroda, K., Groner, A., Frese, K., Drenckhahn, D., Hauser, C., Rott, R., Doerfler, W., and Klenk, H. D., 1989, Synthesis of biologically active influenza virus hemagglutinin in insect larvae, *J. Virol.* **63:**1677.

Kuroda, K., Veit, M., and Klenk, H. D., 1991, Retarded processing of influenza virus hemagglutinin in insect cells, *Virology* **180:**159.

Kuzio, J., Rohel, D. Z., Curry, C. J., Krebs, A., Carstens, E. B., and Faulkner, P., 1984, Nucleotide sequence of the p10 polypeptide gene of *Autographa californica* nuclear polyhedrosis virus, *Virology* **139:**414.

Lanford, R. E., 1988, Expression of simian virus 40 T antigen in insect cells using a baculovirus expression vector, *Virology* **167:**72.

Lawrie, A. M., King, L. A., and Ogden, J. E., 1995, High level synthesis and secretion of human urokinase using a late gene promoter of the *Autographa californica* nuclear polyhedrosis virus, *J. Biotech.* **39:**1.

Lebacq-Verheyden, A. M., Kasprzyk, P. G., Raum, M. G., Van Wyke Coelingh, K., Lebacq, J. A., and Battey, J. F., 1988, Posttranslational processing of endogenous and of baculovirus-expressed human gastrin-releasing peptide precursor, *Mol. Cell. Biol.* **8:**3129.

Lee, H. H., and Miller, L. K., 1978, Isolation of genotypic variants of *Autographa californica* nuclear polyhedrosis virus, *J. Virol.* **27:**754.

Lerch, R. A., and Friesen, P. D., 1993, The 35-kilodalton protein gene (p35) of *Autographa californica* nuclear polyhedrosis virus and the neomycin resistance gene provide dominant selection of recombinant baculoviruses, *Nucleic Acids Res.* **21:**1753.

Leusch, M. S., Lee, S. C., and Olins, P. O., 1995, A novel host–vector system for direct selection of recombinant baculoviruses (bacmids) in *Escherichia coli, Gene* **160:**191.

Lowe, P. N., Page, M. J., Bradley, S., Rhodes, S., Sydenham, M., Paterson, H., and Skinner, R. H., 1991, Characterization of recombinant human Kirsten-ras (4B) p21 produced at high levels in *Escherichia coli* and insect baculovirus expression systems, *J. Biol. Chem.* **266:**1672.

Lowe, P. N., Skinner, R. H., Cooper, D. J., Bradley, S., Sydenham, M., and Page, M. J., 1992, Expression of polyisoprenylated Ras proteins in the insect/baculovirus system, *Biochem. Soc. Trans.* **20:**484.

Lu, A., and Miller, L. K., 1995, The role of eighteen baculovirus late expression factor genes in transcription and DNA replication, *J. Virol.* **69:**975.

Lu, A., and Miller, L. K., 1996, Generation of recombinant baculoviruses by direct cloning, *Biotechniques* **21:**63.

Luckow, V. A., Lee, S. C., Barry, G. F., and Olins, P. O., 1993, Efficient generation of infectious recombinant baculoviruses by site-specific transposon-mediated insertion of foreign genes into a baculovirus genome propagated in *Escherichia coli, J. Virol.* **67:**4566.

Luckow, V. L., 1991, Cloning and expression of heterologous genes in insect cells with baculovirus vectors, in: *Recombinant DNA Technology and Application* (A. Prokop, R. K., Bajpai, and C. S. Ho, eds.), pp. 97–152, McGraw-Hill, New York.

Luckow, V. L., and Summers, M. D., 1988, Trends in the development of baculovirus expression vectors, *Bio/Technology* **6:**47.

Maeda, S., 1989, Expression of foreign genes in insects using baculovirus vectors, *Annu. Rev. Entomol.* **34:**351.

Maeda, S., Kawai, T., Obinata, M., Fujiwara, H., Horiuchi, T., Saeki, Y., Sato, Y., and Furusawa, M., 1985, Production of human alpha-interferon in silkworm using a baculovirus vector, *Nature* **315:**592.

Maiorella, B., Inlow, D., Shauger, A., and Harano, D., 1988, Large-scale insect cell culture for recombinant protein production, *Bio/Technology* **6:**1406.

Malitschek, B., and Schartl, M., 1991, Rapid identification of recombinant baculoviruses using PCR, *Biotechniques* **11:**177.

Manns, A., and Grosse, F., 1991, Fast isolation of recombinant baculovirus by antibody screening, *Biotechniques* **10:**154.

Martens, J. W., van Oers, M. M., van de Bilt, B. D., Oudshoorn, P., and Vlak, J. M., 1995, Development of a baculovirus vector that facilitates the generation of p10-based recombinants, *J. Virol. Meth.* **52:**15.

Martin, B. M., Tsuji, S., LaMarca, M. E., Maysak, K., Eliason, W., and Ginns, E. I., 1988, Glycosylation and processing of high levels of active human glucocerebrosidase in invertebrate cells using a baculovirus expression vector, *DNA* **7:**99.

Marz, L., Altmann, F., Staudacher, E., and Kubelka, V., 1995, Protein glycosylation in insects, in: *Glycoproteins* (J. Montreuil, J. F. G. Vliegenthart, and H. Schachter, eds.) Vol. 29a, pp. 543–563, Elsevier, Amsterdam.

Matsuura, Y., Possee, R. D., Overton, H. A., and Bishop, D. H. L., 1987, Baculovirus expression vectors: The requirements for high level expression of proteins, including glycoproteins, *J. Gen. Virol.* **68:**1233.

McCutchen, B. F., Choudary, P. V., Crenshaw, R., Maddox, D., Kamita, S. G., Palekar, N., Volrath, S., Fowler, E., Hammock, B. D., and Maeda, S., 1991, Development of a recombinant baculovirus expressing an insect-selective neurotoxin: Potential for pest control, *Bio/Technology* **9:**848.

McLachlin, J. R., and Miller, L. K., 1994, Identification and characterization of *vlf-1*, a baculovirus gene involved in very late gene expression, *J. Virol.* **68:**7746.

Merryweather, A. T., Weyer, U., Harris, M. P., Hirst, M., Booth, T., and Possee, R. D., 1990, Construction of genetically engineered baculovirus insecticides containing the *Bacillus thuringiensis* subsp. kurstaki HD-73 delta encotoxin, *J. Gen. Virol.* **71:**1535.

Miller, L. K., 1981, Construction of a genetic map of the baculovirus *Autographa californica* nuclear polyhedrosis virus by marker rescue of temperature-sensitive mutants, *J. Virol.* **39:**973.

Miller, L. K., 1988, Baculoviruses as gene expression vectors, *Annu. Rev. Microbiol.* **42:**177.

Min, M. K., and Bishop, D. H. L., 1991, Transcriptional analyses of baculovirus polyhedrin and foreign gene expression relative to baculovirus p10 mRNA levels, *J. Gen Virol.* **72:**2551.

Montreuil, J., Vliegenthart, J. F. G., and Schachter, H. (eds.), 1995, *Glycoproteins,* Vol 29A, *New Comprehensive Biochemistry* (A. Neuberger and L. L. M. Van Deenen, series eds.), Elsevier, Amsterdam.

Morris, T. D., and Miller, L. K., 1992, Promoter influence on baculovirus-mediated gene expression in permissive and nonpermissive insect cell lines, *J. Virol.* **66:**7397.

Mouillac, B., Caron, M., Bonin, H., Dennis, M., and Bouvier, M., 1992, Agonist-modulated palmitoylation of beta 2-adrenergic receptor in Sf9 cells, *J. Biol. Chem.* **267:**21733.

Mroczkowski, B. S., Huvar, A., Lernhardt, W., Misono, K., Nielson, K., and Scott, B., 1994, Secretion of thermostable DNA polymerase using a novel baculovirus vector, *J. Biol. Chem.* **269:**13522.

Murhammer, D. W., and Goochee, C. F., 1988, Scaleup of insect cell cultures: Protective efforts of pluronic F-68, *Bio/Technology* **6:**1411.

Murphy, C. I., Lennick, M., Lehar, S. M., Beltz, G. A., and Young, E., 1990, Temporal expression of HIV-1 envelope proteins in baculovirus-infected insect cells: Implications for glycosylation and CD4 binding, *Genet. Anal. Tech. Appl.* **7:**160.

Murphy, C. I., McIntire, J. R., Davis, D. R., Hodgdon, H., Seals, J. R., and Young, E., 1993, Enhanced expression, secretion, and large-scale purification of recombinant HIV-1 gp120 in insect cell using the baculovirus *egt* and *gp67* signal peptides, *Protein Expr. Purif.* **4:**349.

Numata, K., Nakata, S., Jiang, X., Estes, M. K., and Chiba, S., 1994, Epidemiological study of Norwalk virus infections in Japan and Southeast Asia by enzyme-linked immunosorbent assays with Norwalk virus capsid protein produced by the baculovirus expression system, *J. Clin. Microbiol.* **32:**121.

Ogonah, O. W., Freedman, R. B., Jenkins, N., Patel, K., and Rooney, B., 1996, Isolation and characterization of an insect cell line able to perform complex N-linked glycosylation on recombinant proteins, *Bio/Technology* **14:**197.

Ohkawa, T., Majima, K., and Maeda, S., 1994, A cysteine protease encoded by the baculovirus *Bombyx mori* nuclear polyhedrosis virus, *J. Virol.* **68:**6619.

Oker-Blom, C., Jansson, C., Karp, M., Lindqvist, C., Savola, J. M., Vlak, J., and Akerman, K., 1993, Functional analysis of the human alpha 2C-C4 adrenergic receptor in insect cells expressed by a luciferase-based baculovirus vector, *Biochim. Biophys. Acta* **1176:**269.

Ooi, B. G., and Miller, L. K., 1988, Regulation of host RNA levels during baculovirus infection, *Virology* **166:**515.

Ooi, B. G., Rankin, C., and Miller, L. K., 1989, Downstream sequences augment transcription from the essential initiation site of a baculovirus polyhedrin gene, *J. Mol. Biol.* **210:**721.

O'Reilly, D. R., Miller, L. K., and Luckow, V. A., 1992, *Baculovirus Expression Vectors,* W. H. Freeman, New York.

Page, M. J., Hall, A., Rhodes, S., Skinner, R. H., Murphy, V., Sydenham, M., and Lowe, P. N., 1989, Expression and characterization of the Ha-ras p21 protein produced at high levels in the insect/baculovirus system, *J. Biol. Chem.* **264:**19147.

Pajot-Augy, E., Couture, L., Bozon, V., Remy, J. J., Biache, G., Severini, M., Huet, J. C., Pernollet, J. C., and Salesse, R., 1995, High-level expression of recombinant porcine LH receptor in baculovirus-infected insect cells or caterpillars, *J. Mol. Endocrinol.* **14:**51.

Parekh, R., Forrester, K., and Wittrup, D., 1995, Multicopy overexpression of bovine pancreatic trypsin inhibitor saturates the protein folding and secretory capacity of *Saccharomyces cerevisiae, Protein Expr. Purif.* **6:**537.

Park, D. S., Selvey, L. A., Kelsall, S. R., and Frazer, I. H., 1993, Human papillomavirus type 16 E6, E7 and L1 and type 18 E7 proteins produced by recombinant baculoviruses, *J. Virol. Meth.* **45:**303.

Passarelli, A. L., Todd, J. W., and Miller, L. K., 1994, A baculovirus gene involved in late gene expression predicts a large polypeptide with a conserved motif of RNA polymerases, *J. Virol.* **68:**4673.

Patel, G., Nasmyth, K., and Jones, N., 1992, A new method for the isolation of recombinant baculovirus, *Nucleic Acids Res.* **20:**97.

Peakman, T. C., Charles, I. G., Sydenham, M. A., Gewert, D. R., Page, M. J., and Makoff, A. J., 1992a, Enhanced expression of recombinant proteins in insect cells using a baculovirus vector containing a bacterial leader sequence, *Nucleic Acids Res.* **20:**6111.

Peakman, T. C., Harris, R. A., and Gewert, D. R., 1992b, Highly efficient generation of recombinant baculoviruses by enzymatically medicated site-specific in vitro recombination, *Nucleic Acids Res.* **20:**495.

Pearson, L. D., and Roy, P., 1993, Genetically engineered multicomponent virus-like particles as veterinary vaccines, *Immunol. Cell Biol.* **71:**381.

Pen, J., Welling, G. W., and Welling-Wester, S., 1989, An efficient procedure for the isolation of recombinant baculovirus, *Nucleic Acids Res.* **17:**451.

Peng, S., Sommerfelt, M. A., Berta, G., Berry, A. K., Kirk, K. L., Hunter, E., and Sorscher, E. J., 1993, Rapid purification of recombinant baculovirus using fluorescence-activated cell sorting, *Biotechniques* **14:**274.

Pennock, G. D., Shoemaker, C., and Miller, L. K., 1984, Strong and regulated expression of *Escherichia coli* beta-galactosidase in insect cells with a baculovirus vector, *Mol. Cell. Biol.* **4:**399.

Possee, R. D., and Howard, S. C., 1987, Analysis of the polyhedrin gene promoter of the *Autographa californica* nuclear polyhedrosis virus, *Nucleic Acids Res.* **15:**10233.

Prasad, B. V., Rothnagel, R., Jiang, X., and Estes, M. K., 1994, Three-dimensional structure of baculovirus-expressed Norwalk virus capsids, *J. Virol.* **68:**5117.

Qin, J., Liu, A., and Weaver, R. F., 1989, Studies on the control region of the p10 gene of the *Autographa californica* nuclear polyhedrosis virus, *J. Gen. Virol.* **70:**1273.

Ranjan, A., and Hasnain, S. E., 1995, Influence of codon usage and translational initiation codon context in the AcNPV-based expression system: Computer analysis using homologous and heterologous genes, *Virus Genes* **9:**149.

Rankin, C., Ooi, B. G., and Miller, L. K., 1988, Eight base pairs encompassing the transcriptional start point are the major determinant for baculovirus polyhedrin gene expression, *Gene* **70:**39.

Rankl, N. B., Rice, J. W., Gurganus, T. M., Barbee, J. L,. and Burns, D. J., 1994, The production of an active protein kinase C-delta in insect cells is greatly enhanced by the use of the basic protein promoter, *Protein Expr. Purif.* **5:**346.

Redfield, R. R., Birx, D. L., Ketter, N., Tramont, E., Polonis, V., Davis, C., Brundage, J. F., Smith, G., Johnson, S., Fowler, A., and *et al.*, 1991, A phase I evaluation of the safety and immunogenicity of vaccination with recombinant gp160 in patients with early human immunodeficiency virus infection. Military Medical Consortium for Applied Retroviral Research, *N. Engl. J. Med.* **324:**1677.

Reed, L., and Muench, H., 1938, A simple method for estimating fifty percent endpoints, *Am. J. Hyg.* **27:**493.

Richardson, C. D. (ed.), 1995, *Baculovirus Expression Protocols*, Vol. 39, *Methods in Molecular Biology* (J. M. Walker, series ed.), Humana Press, Totowa, NJ.

Risinger, M. A., Dotimas, E. M., and Cohen, C. M., 1992, Human erythrocyte protein 4.2, a high copy number membrane protein, is *N*-myristylated, *J. Biol. Chem.* **267:**5680.

Roelvink, P. W., van Meer, M. M., de Kort, C. A., Possee, R. D., Hammock, B. D., and Vlak, J. M., 1992, Dissimilar expression of *Autographa californica* multiple nucleocapsid nuclear polyhedrosis virus polyhedrin and p10 genes, *J. Gen. Virol.* **73:**1481.

Rohrmann, G. F., 1986, Polyhedrin structures, *J. Gen. Virol.* **67:**1499.

Roy, P., 1996, Orbivirus structure and assembly, *Virology* **216:**1.

Sareneva, T., Cantell, K., Pyhala, L., Pirhonen, J., and Julkunen, I., 1993, Effect of carbohydrates on the pharmacokinetics of human interferon-gamma, *J. Interfon. Res.* **13:**267.

Shuler, M. L., Wood, H. A., Granados, R. R., and Hammer, D. A., 1995, *Baculovirus Expression Systems and Biopesticides*, Wiley-Liss, New York.

Sisk, W. P., Bradley, J. D., Seivert, L. L., Vargas, R. A., and Horlick, R. A., 1992, An improved method for rapid screening of baculovirus recombinant plaques by PCR amplification, *Biotechniques* **13:**186.

Slightom, J. L., and Sieu, L. C., 1992, Direct sequencing of baculovirus genomic DNA: Sequence determination of the engineered respiratory syncytial virus chimeric FG gene, *Biotechniques* **13:**94.

Smith, G. E., Fraser, M. J., and Summers, M. D., 1983a, Molecular engineering of the *Autographa californica* nuclear polyhedrosis virus genome: Deletion mutations within the polyhedrin gene, *J. Virol.* **46:**584.

Smith, G. E., Summers, M. D., and Fraser, M. J., 1983b, Production of human beta interferon in insect cells infected with a baculovirus expression vector, *Mol. Cell. Biol.* **3:**2156.

Smith, G. E., Vlak, J. M., and Summers, M. D., 1983c, Physical analysis of *Autographa californica* nuclear polyhedrosis virus transcripts for polyhedrin and a 10,000-molecular-weight protein, *J. Virol.* **45:**215.

Sridhar, P., Panda, A. K., Pal, R., Talwar, G. P., and Hasnain, S. E., 1993, Temporal nature of the promoter and not relative strength determines the expression of an extensively processed protein in a baculovirus system, *FEBS Lett.* **315:**282.

St. Angelo, C., Smith, G. E., Summers, M. D., and Krug, R. M., 1987, Two of the three influenza viral polymerase proteins expressed by using baculovirus vectors form a complex in insect cells, *J. Virol.* **61:**361.

Stewart, L. M., Hirst, M., Lopez Ferber, M., Merryweather, A. T., Cayley, P. J., and Possee, R. D.,

1991, Construction of an improved baculovirus insecticide containing an insect-specific toxin gene, *Nature* **352:**85.

Stoltz, D. B., Pavan, C., and DaCuhna, A. B., 1973, Nuclear polyhedrosis virus: A possible example of *de novo* intranuclear membrane morphogenesis, *J. Gen. Virol.* **19:**145.

Summers, M. D., and Smith, G. E., 1987, *A Manual of Methods for Baculovirus Vectors and Insect Cell Culture Procedures,* Texas Agricultural Experiment Station Bulletin No. 1555.

Suzuki, K., Shimoi, H., Iwasaki, Y., Kawahara, T., Matsuura, Y., and Nishikawa, Y., 1990, Elucidation of amidating reaction mechanism by frog amidating enzyme, peptidylglycine alpha-hydroxylating monooxygenase, expressed in insect cell culture, *EMBO J.* **9:**4259.

Takagi, Y., Ogawa, H., Harada, N., Shimada, H., Ishimura, Y., and Takagi, Y., 1992, Expression and transport into mitochondria of bovine cytochrome P-450(SCC) in insect cells using the baculovirus expression system, *Biochem. Biophys. Res. Commun.* **184:**471.

Taticek, R. A., Lee, C. W., and Shuler, M. L., 1994, Large-scale insect and plant cell culture, *Curr. Opin. Biotech.* **5:**165.

Tessier, D. C., Thomas, D. Y., Khouri, H. E., Laliberte, F., and Vernet, T., 1991, Enhanced secretion from insect cells of a foreign protein fused to the honeybee melittin signal peptide, *Gene* **98:**177.

Thiem, S. M., and Miller, L. K., 1989, Identification, sequence, and transcriptional mapping of the major capsid protein gene of the baculovirus *Autographa californica* nuclear polyhedrosis virus, *J. Virol.* **63:**2008.

Thiem, S. M., and Miller, L. K., 1990, Differential gene expression mediated by late, very late and hybrid baculovirus promoters, *Gene* **91:**87.

Thompsen, D. R., Post, L. E., and Elhammer, A. P., 1990, Structure of *O*-glycosidically linked oligosaccharides synthesized by the insect cell line Sf9, *J. Cell. Biochem.* **43:**67.

Todd, J. W., Passarelli, A. L., and Miller, L. K., 1995, Eighteen baculovirus genes, including lef-11, p35, 39K, and p47, support late gene expression, *J. Virol.* **69:**968.

Tomita, S., Kanaya, T., Kobayashi, J., and Imanishi, S., 1995, Isolation of p10 gene from *Bombyx mori* nuclear polyhedrosis virus and study of its promoter activity in recombinant baculovirus vector system, *Cytotechnology* **17:**65.

Tramper, J., de Gooijer, K. D., and Vlak, J. M., 1993, Scale-up considerations and bioreactor development for animal cell cultivation, *Bioproc. Tech.* **17:**139.

Tsao, T., Hsieh, J. C., Durkin, M. E., Wu, C. Y., Chakravarti, S., Dong, L. J., Lewis, M., and Chung, A. E., 1990, Characterization of the basement membrane glycoprotein entactin synthesized in a baculovirus expression system, *J. Biol. Chem.* **265:**5188.

Urakawa, T., and Roy, P., 1988, Bluetongue virus tubules made in insect cells by recombinant baculoviruses: Expression of the NS1 gene of bluetongue virus serotype 10, *J. Virol.* **62:**3919.

Urakawa, T., Ferguson, M., Minor, P. D., Cooper, J., Sullivan, M., Almond, J. W., and Bishop, D. H. L., 1989, Synthesis of immunogenic, but non-infectious, poliovirus particles in insect cells by a baculovirus expression vector, *J. Gen. Virol.* **70:**1453.

Urbancikova, M., and Hitchcock-DeGregori, S. E., 1994, Requirement of amino-terminal modification for striated muscle alpha-tropomyosin function, *J. Biol. Chem.* **269:**24310.

Vail, P., Sutter, G., Jay, D., and Gough, D., 1971, Reciprocal infectivity of nuclear polyhedrosis viruses of the cabbage looper and alfalfa looper, *J. Invertebr. Pathol.* **17:**383.

Vakharia, V. N., Raina, A. K., Kingan, T. G., and Kempe, T. G., 1995, Synthetic pheromone biosynthesis activating neuropeptide gene expressed in a baculovirus expression system, *Insect Biochem. Mol. Biol.* **25:**583.

van Lier, F. L., van Duijnhoven, G. C., de Vaan, M. M., Vlak, J. M., and Tramper, J., 1994, Continuous beta-galactosidase production in insect cells with a p10 gene based baculovirus vector in a two-stage bioreactor system, *Biotech. Progr.* **10:**60.

Vaughn, J. L., Goodwin, R. H., Thompkins, G. J., and McCawley, P., 1977, The establishment of two insect cell lines from the insect *Spodoptera frugiperda* (Lepidoptera: Noctuidae), *In Vitro* **13:**213.

Veit, M., Nurnberg, B., Spicher, K., Harteneck, C., Ponimaskin, E., Schultz, G., and Schmidt, M. F., 1994, The alpha-subunits of G-proteins G12 and G13 are palmitoylated, but not amidically myristoylated, *FEBS Lett.* **339:**160.

Velardo, M. A., Bretthauer, R. K., Boutaud, A., Reinhold, B., Reinhold, V. N., and Castellino, F. J., 1993, The presence of UDP-*N*-acetylglucosamine:alpha-3-D-mannoside beta, 1,2,-*N*-acetylglucosaminyltransferase I activity in *Spodoptera frugiperda* cells (IPLB-SF-21AE) and its enhancement as a result of baculovirus infection, *J. Biol. Chem.* **268:**17902.

Vialard, J., Lalumiere, M., Vernet, T., Briedis, D., Alkhatib, G., Henning, D., Levin, D., and Richardson, C., 1990, Synthesis of the membrane fusion and hemagglutinin proteins of measles virus, using a novel baculovirus vector containing the beta-galactosidase gene, *J. Virol.* **64:**37.

Vialard, J. E., and Richardson, C. D., 1993, The 1,629-nucleotide open reading frame located downstream of the *Autographa californica* nuclear polyhedrosis virus polyhedrin gene encodes a nucleocapsid-associated phosphoprotein, *J. Virol.* **67:**5859.

Vihko, P., Kurkela, R., Porvari, K., Herrala, A., Lindfors, A., Lindqvist, Y., and Schneider, G., 1993, Rat acid phosphatase: Overexpression of active, secreted enzyme by recombinant baculovirus-infected insect cells, molecular properties, and crystallization, *Proc. Natl. Acad. Sci. USA* **90:**799.

Vlak, J., Smith, G. E., and Summers, M. D., 1981, Hybridization selection and *in vitro* translation of *Autographa californica* nuclear polyhedrosis virus mRNA, *J. Virol.* **40:**762.

Vlak, J. M., Schouten, A., Usmany, M., Belsham, G. J., Klinge-Roode, E. C., Maule, A. J., Van Lent, J. W., and Zuidema, D., 1990, Expression of cauliflower mosaic virus gene I using a baculovirus vector based upon the p10 gene and a novel selection method, *Virology* **179:**312.

Volkman, L. E., and Summers, M. D., 1977, *Autographa californica* nuclear polyhedrosis virus: Comparative infectivity of the occluded, alkali-liberated, and nonoccluded forms, *J. Invertebr. Pathol.* **30:**102.

Volkman, L. E., Summers, M. D., and Hsieh, C. H., 1976, Occluded and nonoccluded nuclear polyhedrosis virus grown in *Trichooplusia ni:* Comparative neutralization comparative infectivity, and *in vitro* growth studies, *J. Virol.* **19:**820.

Volkman, L. E., Blissard, G. W., Friesen, P., Keddie, B. A., Possee, R., and Theilmann, D. A., 1995, Virus taxonomy, in: *Sixth Report of the International Committe on Taxonomy of Viruses* (F. A. Murphy, C. M. Fauquet, D. H. L. Bishop, S. A. Ghabrial, A. W. Jarvis, G. P. Martelli, M. A. Mayo, and M. D. Summers, eds.), p. 104, Springer-Verlag, Wien.

Wagner, R., Liedtke, S., Kretzschmar, E., Geyer, H., Geyer, R., and Klenk, H. D., 1996, Elongation of the *N*-glycans of fowl plague virus hemagglutinin expressed in *Spodoptera frugiperda* (Sf9) cells by coexpression of human β1,2-*N*-acetylglucosaminyltransferase I, *Glycobiology* **6:**165.

Wang, H. H., and Fraser, M. J., 1991, Direct double-stranded DNA sequencing with baculovirus genomes, *J. Virol. Methods* **31:**113.

Wang, X. Z., Ooi, B. G., and Miller, L. K., 1991, Baculovirus vectors for multiple gene expression and for occluded virus production, *Gene* **100:**131.

Webb, A. C., Bradley, M. K., Phelan, S. A., Wu, J. Q., and Gehrke, L., 1991, Use of the polymerase chain reaction for screening and evaluation of recombinant baculovirus clones, *Biotechniques* **11:**512.

Weiss, S. A., Godwin, G. P., Gorfien, S. F., and Whitford, W. G., 1995a, Insect cell culture in serum-free media, in: *Baculovirus Expression Protocols* (C. D. Richardson ed.), pp. 79–95, Humana Press, Clifton, NJ.

Weiss, S. A., Whitford, W. G., Gorfien, S. F., and Godwin, G. P., 1995b, Insect cell-culture techniques in serum-containing medium, in: *Baculovirus Expression Protocols* (C. D. Richardson ed.), pp. 65–78, Humana Press, Clifton, NJ.

Westwood, J. A., Jones, I. M., and Bishop, D. H. L., 1993, Analyses of alternative poly(A) signals for use in baculovirus expression vectors, *Virology* **195**:90.

Weyer, U., and Possee, R. D., 1989, Analysis of the promoter of the *Autographa californica* nuclear polyhedrosis virus p10 gene, *J. Gen. Virol.* **70**:203.

Weyer, U., and Possee, R. D., 1991, A baculovirus dual expression vector derived from the *Autographa californica* nuclear polyhedrosis virus polyhedrin and p10 promoters: Co-expression of two influenza virus genes in insect cells, *J. Gen. Virol.* **72**:2967.

Weyer, U., Knight, S., and Possee, R. D., 1990, Analysis of very late gene expression by *Autographa californica* nuclear polyhedrosis virus and the further development of multiple expression vectors, *J. Gen. Virol.* **71**:1525.

Wickham, T. J., Davis, T., Granados, R. R., Shuler, M. L., and Wood, H. A., 1992, Screening of insect cell lines for the production of recombinant proteins and infectious virus in the baculovirus expression system, *Biotech. Progr.* **8**:391.

Wilson, M. E., Mainprize, T. H., Friesen, P. D., and Miller, L. K., 1987, Location, transcription, and sequence of a baculovirus gene encoding a small arginine-rich polypeptide, *J. Virol.* **61**:661.

Winter, G., Griffiths, A. D., Hawkins, R. E., and Hoogenboom, H. R., 1994, Making antibodies by phage display technology, *Annu. Rev. Immunol.* **12**:433.

Wong, T. K., Nielsen, L. K., Greenfield, P. F., and Reid, S., 1994, Relationship between oxygen uptake rate and time of infection of Sf9 insect cells infected with a recombinant baculovirus, *Cytotechnology* **15**:157.

Xia, Y., Van Etten, J. L., Dobos, P., Ling, Y. Y., and Krell, P. J., 1993, Adenine DNA methyltransferase M. CviRI expression accelerates apoptosis in baculovirus-infected insect cells, *Virology* **196**:817.

Xie, Y. B., Sui, Y. P., Shan, L. X., Palvimo, J. J., Phillips, D. M., and Janne, O. A., 1992, Expression of androgen receptor in insect cells. Purification of the receptor and renaturation of its steroid-and DNA-binding functions, *J. Biol. Chem.* **267**:4939.

Xu, B., Yoo, S., and Guarino, L. A., 1995, Differential transcription of baculovirus late and very late promoters: Fractionation of nuclear extracts by phosphocellulose chromatography, *J. Virol.* **69**:2912.

Yamshchikov, G. V., Ritter, G. D., Vey, M., and Compans, R. W., 1995, Assembly of SIV virus-like particles containing envelope proteins using a baculovirus expression system, *Virology* **214**:50.

Index

Entries found in tables are indicated by *t* following the page number.

THE VIRUSES

COMPLETE LISTING OF VOLUMES

Series Editors
HEINZ FRAENKEL-CONRAT, *University of California*
Berkeley, California

ROBERT R. WAGNER, *University of Virginia School of Medicine*
Charlottesville, Virginia

THE ADENOVIRUSES
Edited by Harold S. Ginsberg

THE ARENAVIRIDAE
Edited by Maria S. Salvato

THE BACTERIOPHAGES
Volumes 1 and 2 • Edited by Richard Calendar

THE BACULOVIRUSES
Edited by Lois K. Miller

THE BUNYAVIRIDAE
Edited by Richard M. Elliott

THE CORONAVIRIDAE
Edited by Stuart G. Siddell

THE HERPESVIRUSES
Volumes 1–3 • Edited by Bernard Roizman
Volume 4 • Edited by Bernard Roizman and Carlos Lopez

THE INFLUENZA VIRUSES
Edited by Robert M. Krug

THE PAPOVAVIRIDAE
Volume 1 • Edited by Norman P. Salzman
Volume 2 • Edited by Norman P. Salzman and Peter M. Howley

THE PARAMYXOVIRUSES
Edited by David W. Kingsbury

THE PARVOVIRUSES
Edited by Kenneth I. Berns

THE PLANT VIRUSES
Volume 1 • Edited by R. I. B. Francki
Volume 2 • Edited by M. H. V. Van Regenmortel and Heinz Fraenkel-Conrat
Volume 3 • Edited by Renate Koenig
Volume 4 • Edited by R. G. Milne
Volume 5 • Edited by B. D. Harrison and A. F. Murant

THE REOVIRIDAE
Edited by Wolfgang K. Joklik

THE RETROVIRIDAE
Volumes 1–4 • Edited by Jay A. Levy

THE VIRUSES
COMPLETE LISTING OF VOLUMES

THE RHABDOVIRUSES
Edited by Robert R. Wagner

THE TOGAVIRIDAE AND FLAVIVIRIDAE
Edited by Sondra Schlesinger and Milton J. Schlesinger

THE VIROIDS
Edited by T. O. Diener

THE VIRUSES: Catalogue, Characterization, and Classification
Heinz Fraenkel-Conrat